Recommendations of the Committee for Waterfront Structures Harbours and Waterways
EAU 1996

A Wiley Company

Recommendations of the Committee for Waterfront Structures Harbours and Waterways EAU 1996

7th English Edition
English Translation of the 9th German Edition

Issued by the Committee for Waterfront Structures of the Society for Harbour Engineering and the German Society for Soil Mechanics and Foundation Engineering

The original German edition was published under the title
*Empfehlungen des Arbeitsausschusses „Ufereinfassungen"
Häfen und Wasserstraßen EAU 1996*
by Ernst & Sohn, Berlin

With 243 figures and 45 tables

Die Deutsche Bibliothek – CIP Cataloguing-in-Publication-Data
A catalogue record for this publication is available from Die Deutsche Bibliothek

ISBN 3-433-01790-5

© 2000 Ernst & Sohn, Verlag für Architektur und technische Wissenschaften GmbH, Berlin

All rights reserved (including those of translation into other languages). No part of this book may be reproduced in any form – by photoprint, microfilm, or any other means – nor transmitted or translated into a machine language without written permission from the publisher.

The quotation of trade descriptions, trade names or other symbols in this book shall not justify the assumption that they may be freely used by anyone. On the contrary, these may be registered symbols even if they are not expressly marked as such.

Typesetting: Manuela Treindl, Regensburg
Printing: betz-druck, Darmstadt
Binding: Wilh. Osswald + Co., Neustadt

Printed in Germany

Members of the Committee for Waterfront Structures

At present the working committee "Waterfront Structures" has the following members:

(since 1995)   Professor Dr.-Ing. *V. Rizkallah*, Hanover,
               Chairmain of the Committee since 1997
(since 1994)   Civil Eng., Chief Design-Eng. *E. Clausen,* Esbjerg/Denmark
(since 1990)   Erster Baudirektor Dr.-Ing. *H. P. Dücker*, Hamburg
(since 1992)   Project manager Ir. *J. G. de Gijt*, Rotterdam/Netherlands
(since 1997)   Baudirektor Dr.-Ing. *M. Heibaum*, Karlsruhe
(since 1993)   Head of Department Dipl.-Ing. *W. Hering*, Rostock
(since 1995)   Professor Dr.-Ing. *B. Mazurkiewicz*, Danzig/Poland
(since 1998)   Dr.-Ing. K. *Morgen*, Hamburg
(since 1991)   Managing director Dr.-Ing. *F. W. Oeser*, Hamburg
(since 1998)   Managing director Dr.-Ing. *H. Otten*, Dortmund
(since 1998)   Director Dipl.-Ing. *U. Reinke*, Hanover
(since 1998)   Professor Dr.-Ing. *W. Richwien*, Essen
(since 1985)   Managing director Dr.-Ing. *H. Salzmann*, Hamburg
(since 1999)   Managing director Dr.-Ing. *R. Schlim*, Esch-sur-Alzette/
               Luxembourg
(since 1986)   Professor Dr.-Ing. *H. Schulz*, Munich
(since 1998)   Technical Manager Dr.-Ing. *M. Stocker*, Schrobenhausen
(since 1993)   Ministerialrat Dipl.-Ing. *H.-P. Tzschucke,* Bonn
(since 1994)   Leitender Baudirektor Dr.-Ing. *H.-W. Vollstedt*, Bremerhaven

The following committee members have participated in establishing the EAU 1996 until they left the committee:

(1990–1997)   Director Dipl.-Ing. *P. Arz*, Mannheim/Hamburg
(1980–1994)   Baudirektor Dipl.-Ing. *G. Gerdes*, Bremen
(1993–1996)   Managing director Dipl.-Ing. *G. Hachenberg*, Dortmund
(1987–1993)   Managing director Dipl.-Ing. *S. Roth*, Dortmund

Moreover the following members still supports the committee in puplishing the English Edition of the EAU 1996:

(1990–1997) Retired managing director Dipl.-Ing. *M. Borchert*, Bremen
(1976–1997) Ministerialrat a.D. Professor Dr.-Ing. *M. Hager*, Bonn,
 Chairman of the Committee until the end of 1996
(1983–1997) Professor Dr.-Ing. *A. Horn*, Munich
(1977–1998) Port director Professor Dr.-Ing. *J. Müller*, Duisburg
(1989–1993) President Professor Dipl.-Ing. *D. Schröder*, Hanover

Preface to the 7th Revised English Edition

This 7th English Edition of the Recommendations of the Committee for Waterfront Structures (EAU 1996 – Harbours and Waterways) incorporates essentially the entire integration of the safety concept, which was already announced in the previous edition (EAU 1990), into the calculation rules of the EAU, with partial safety factors based on the applicable Eurocodes and the corresponding adjusted national standards, prestandards and draft standards, taking account of the National Application Document (NAD) for geotechnical calculations.
In the choice of partial safety values and calculation methods, sufficient empirical values to satisfy the safety requirements have been introduced in the EAU 1996, making use of the scope allowed in these rules.
The main target of the Committee "Waterfront Structures" is the harmonisation and simplification of the design and calculation of waterfront structures in harbours and waterways. From the very beginning, the committee's work has concentrated primarily on such areas where there is a need for standardisation, or where there are no or only inadequate regulations.
Recommendations or changes respectively new versions are initially announced in the form of Annual Technical Reports for general discussion, and finally concluded after the integration of objections or suggestions. Recommendations are summarised or abridged to the extent that the items being regulated therein have meanwhile been covered by national or international standards.
As part of implementation of the Single European Market, the Eurocodes (EC), to which the national standards have to be adjusted, have been released initially only for trial purposes, as harmonised directives for general safety requirements for buildings and structures. At present it is not possible to predict when the final European standards and adjusted national standards will be available.
By taking account of the new safety concept, the EAU 1996 makes a considerable contribution to European harmonisation of the regulations. The Committee presumes that during the trial phase of the European prestandards, civil engineers will make use of the possibility of implementing the new recommendations and thus provide verification of their practical suitability.
In order to prevent repetitions, as regards further details we refer to the general presentations in section 0.1 and to the following preface to the 9th Revised German Edition (EAU 1990 – Harbours and Waterways).
In the interests of easier handling, this 7th Revised English Edition has retained the previous system in sections and individual recommendations. Compared to the previous edition, the number of recommendations which have been altered and supplemented to bring them in line with state-of-the-art developments, has been reduced from 205 to 185 recommendations by streamlining and summarising. Consideration of work results and suggestions from other specialist Committees

and from standardisation bodies in other countries also means that the recommendations continue to justify their claim to international acceptance and use in the construction of harbour and waterway structures according to uniform rules.

The chairman wishes to express his gratitude to the financial support for translation given by the German civil engineering industry (Philipp Holzmann AG, Hauptniederlassung Nord, Hanover; Aug. Prien Bauunternehmung GmbH & Co, Hamburg) and the sheet pile industry (HSP Hoesch Spundwand und Profil, Dortmund; Stahlwerke Peine Salzgitter AG, Salzgitter; Profil Arbed, Esch-sur-Alzette/Luxembourg) and the HTG and DGGT associations. This support is acknowledged with appreciation.

Hanover, January 2000   Professor Dr.-Ing. *Victor Rizkallah*

Excerpt from the Preface to the 9th Revised German Edition

Deviating from the previously practised method of on-going continuation of the recommendations of the EAU, this 9th edition contains a complete revision of the collected publications of the recommendations as part of European harmonisation of the regulations. The foreground here is taken up by the complete integration of the concept of partial safety values in the calculation rules of the EAU according to the Eurocodes, European standards and correspondingly adjusted national standards. The EAU 1996 now takes consideration of all pertinent standards or prestandards available up to the end of 1996. Further details for application of the new rules are indicated in section 0 and the corresponding recommendations for static calculation of waterfront structures.

At the same time, various recommendations have been streamlined and summarised as far as possible to improve handling. But new rulings have also been included in line with state-of-the-art developments; aspects concerning environmental compatibility and ecology have been given increasing consideration along side demands for structural and operational safety and economy.

Streamlining has resulted in the omission of 28 recommendations; new rulings are contained in 4 new recommendations. For clearer orientation, the previous recommendation designations and numbers and the structure of the EAU have been essentially retained.

The 9th edition with integration of the European standardisation concept fulfils future requirements and thus complies with the announcements of the 8th edition, EAU 1990, which were notified with the EU commission under no. 1991/117/D. The 9th edition, EAU 1996, has also been notified with the EU commission under notification number 1997/552/D.

Object of the notification is the equivalence clause which has to act as a basis for contracts in which the EAU or individual provisions of the EAU are an integral part. The wording is as follows:

"Products from other member states of the European Communities and original goods from member states of the European Economic Region, which do not conform with these Technical Specifications, are treated as equivalents including the tests and monitoring procedures carried out in the manufacturing state, if these achieve the required safety level – safety, health and serviceability – in an equivalent permanent manner".

In addition, numerous contributions which have been presented by the professional world, as well as recommendations from other committees and international technical-scientific associations have been taken into consideration.

With these contributions and results of the revision work, the EAU 1996 now conforms with today's international standard. It provides the construction profession with an adapted, updated version brought into line with European standards which will continue to act in future as a valuable help for design, calling for tenders, placing or orders, technical processing, economic and ecological construction, quality control and settlement of contracts, so that harbour and waterway constructions can be made according to the latest state of the art and according to uniform conditions.

The committee thanks all those whose contributions and suggestions have helped to bring the version to its present state, and wishes the EAU 1996 the same success as its earlier editions.

Special thanks are owed to the publishers Ernst & Sohn for the meticulous care in preparation of all drawings, tables and formulae, providing once again an excellent printing quality and layout of the EAU 1996.

Bonn, March 1997   Professor Dr.-Ing. *Martin Hager*

Contents

Members of the Committee for Waterfront Structures V
Preface to the 7th Revised English Edition VII
Excerpt from the Preface to the 9th Revised German Edition IX
List of Recommendations in the 7th English Edition XIX

Recommendations 1

0   **Statical Calculations** 1

0.1   General 1
0.2   Safety Concept 2
0.3   Calculations of Waterfront Structures 9

1   **Subsoil** 11

1.1   Mean Characteristic Soil Properties (R 9) 11
1.2   Layout and Depth of Borings and Penetrometer Tests (R 1) 11
1.3   Preparation of Reports and Expert Opinions on Subsoil Examinations
       Under Difficult Conditions (R 150) 15
1.4   Determination of Undrained Shear Strength c_u (R 88) 16
1.5   Shear Parameters of the Drained Soil in the State of Failure or Sliding
       (R 131) 17
1.6   Investigation of the Degree of Density of Non-cohesive Backfill for
       Waterfront Structures (R 71) 19
1.7   Degree of Density of Hydraulically Filled, Non-cohesive Soils (R 175) 20
1.8   Degree of Density of Dumped, Non-cohesive Soils (R 178) 22
1.9   Assessment of the Subsoil for the Driving of Sheet Piles and Steel Piles
       (R 154) 23

2   **Active and Passive Earth Pressures** 28

2.0   General 28
2.1   Assumed Cohesion in Cohesive Soils (R 2) 29
2.2   Assumed Apparent Cohesion in Sand (R 3) 29
2.3   Assumed Angle of Wall Friction and Adhesion (R 4) 29
2.4   Determination of the Active Earth Pressure Using the CULMANN Method
       (R 171) 29
2.5   Determination of Active Earth Pressure in a Steep, Paved Embankment
       of a Partially Sloping Bank Construction (R 198) 34

2.6	Determination of Active Earth Pressure in Saturated, Non- or Partially Consolidated, Soft Cohesive Soils (R 130)	36
2.7	Effect of Artesian Water Under the Harbour Bottom or River Bed, on Active and Passive Earth Pressures (R 52)	40
2.8	Action of Active Earth Pressure and Water Pressure Difference and Construction Hints for Waterfront Structures with Soil Replacement and Fouled or Disturbed Dredge Pit Bottom (R 110)	42
2.9	Effect of Percolating Groundwater on Water Pressure Difference, Active and Passive Earth Pressures (R 114)	46
2.10	Determination of the Amount of Displacement Required for the Mobilisation of Partial Passive Earth Pressures in Non-cohesive Soils (R 174)	52
2.11	Measures for Increasing the Passive Earth Pressure in Front of Waterfront Structures (R 164)	54
2.12	Passive Earth Pressure in Front of Sheet Piles in Soft Cohesive Soils, with Rapid Loading on the Land Side (R 190)	56
2.13	Effects of Earthquakes on the Design and Dimensioning of Waterfront Structures (R 124)	59

3	**Overall Stability, Foundation Failure and Sliding**	66
3.1	Relevant Standards	66
3.2	Safety Against Failure by Heave (R 115)	66
3.3	Foundation Failure Due to Erosion; its Occurrence and its Prevention (R 116)	69
3.4	Verification of Overall Stability of Structures on Elevated Pile-founded Structures (R 170)	71

4	**Water Levels, Water Pressure, Drainage**	73
4.1	Mean Groundwater Level (R 58)	73
4.2	Water Pressure Difference in the Water-side Direction (R 19)	73
4.3	Water Pressure Difference on Sheet Piling in Front of Built-over-embankments in Tidal Areas (R 65)	76
4.4	Design of Filter Weepholes for Sheet Piling Structures (R 51)	77
4.5	Design of Flap Valves for Waterfront Structures in Tidal Areas (R 32)	78
4.6	Relieving Artesian Pressure Under Harbour Bottoms (R 53)	80
4.7	Assessment of Groundwater Flow (R 113)	83
4.8	Temporary Stabilisation of Waterfront Structures by Groundwater Lowering (R 166)	91
4.9	Flood Protection Walls in Seaports (R 165)	94

5	**Ship Dimensions and Loading of Waterfront Structures**	102
5.1	Ship Dimensions (R 39)	102
5.2	Assumed Berthing Pressure of Vessels at Quays (R 38)	108

5.3	Berthing Velocities of Vessels Transverse to Berth (R 40)	108
5.4	Load Cases (R 18)	109
5.5	Vertical Live Loads (R 5)	110
5.6	Determining the "Design Wave" for Maritime and Port Structures (R 136) .	113
5.7	Wave Pressure on Vertical Waterfront Structures in Coastal Areas (R 135) .	122
5.8	Loads Arising from Surging and Receding Waves Resulting from the Introduction or Extraction of Water (R 185)	128
5.9	Effects of Waves from Movements of Ships (R 186)	129
5.10	Wave Pressure on Pile Structures (R 159)	132
5.11	Wind Loads on Moored Ships and Their Influences on the Dimensioning of Mooring and Fendering Facilities (R 153)	140
5.12	Layout and Loading of Bollards for Seagoing Vessels (R 12)	143
5.13	Layout, Design and Loading of Bollards in Inland Harbours (R 102)	144
5.14	Quay Loads from Cranes and Other Transhipment Equipment (R 84)	147
5.15	Impact and Pressure of Ice on Waterfront Structures, Fenders and Dolphins in Coastal Areas (R 177)	151
5.16	Impact and Pressure of Ice on Waterfront Structures, Piers and Dolphins in Inland Areas (R 205)	156
6	**Configuration of Cross-section and Equipment of Waterfront Structures**	159
6.1	Standard Dimensions of Cross-section of Waterfront Structures in Seaports (R 6)	159
6.2	Top Elevation of Waterfront Structures in Seaports (R 122)	160
6.3	Standard Cross-sections of Waterfront Structures in Inland Harbours (R 74)	162
6.4	Sheet Piling Waterfront on Canals for Onland Vessels (R 106)	165
6.5	Partially Sloped Waterfront Construction in Inland Harbours with Extreme Water Level Fluctuations (R 119)	168
6.6	Design of Waterfront Areas in Inland Ports According to Operational Aspects (R 158)	171
6.7	Nominal Depth and Design Depth of Harbour Bottom (R 36)	172
6.8	Reinforcement of Waterfront Structures to Deepen Harbour Bottoms in Seaports (R 200)	174
6.9	Re-design of Waterfront Structures in Inland Harbours (R 201)	179
6.10	Equipping of Berths for Large Vessels with Quick Release Hooks (R 70) . .	182
6.11	Layout, Design and Loading of Access Ladders (R 14)	184
6.12	Layout and Design of Stairs in Seaports (R 24)	186
6.13	Equipment of Waterfront Structures in Seaports with Supply and Disposal Facilities (R 173)	187
6.14	Fenders for Berths for Large Vessels at Quays (R 60)	191
6.15	Fenders in Inland Harbours (R 47)	204
6.16	Foundation of Craneways at Waterfront Structures (R 120)	205

XIII

6.17	Installation of Crane Rails on Concrete (R 85)	207
6.18	Connection of Expansion Joint Seal in a Reinforced Concrete Bottom to Bearing External Steel Sheet Piling (R 191)	215
6.19	Connection of Steel Sheet Piling to a Concrete Structure (R 196)	216
6.20	Floating Wharves in Seaports (R 206)	218

7 Earthwork and Dredging ... 221

7.1	Dredging in Front of Quay Walls in Seaports (R 80)	221
7.2	Dredging and Hydraulic Fill Tolerances (R 139)	223
7.3	Hydraulic Filling of Port Areas for Planned Waterfront Structures (R 81)	226
7.4	Backfilling of Waterfront Structures (R 73)	230
7.5	Dredging of Underwater Slopes (R 138)	232
7.6	Scour and Scour Protection at Waterfront Structures (R 83)	236
7.7	Vertical Drains to Accelerate the Consolidation of Soft Cohesive Soils (R 93)	243
7.8	Subsidence of Non-cohesive Soils (R 168)	247
7.9	Soil Replacement Procedure for Waterfront Structures (R 109)	248
7.10	Calculation and Design of Rubble Mound Moles and Breakwaters (R 137)	254
7.11	Light Backfilling for Sheet Piling Structures (R 187)	264
7.12	Soil Compaction Using Heavy Drop Weights (R 188)	264
7.13	Consolidation of Soft, Cohesive Soils by Preloading (R 179)	265
7.14	Installation of Mineral Bottom Seals Under Water and Their Connection to Waterfront Structures (R 204)	270

8 Sheet Piling Structures ... 274

8.1	Material and Construction	274
8.1.1	Design and Driving of Timber Sheeting (R 22)	274
8.1.2	Design and Driving of Reinforced Concrete Sheet Piling (R 21)	277
8.1.3	Steel Sheet Piling (R 34)	280
8.1.4	Combined Steel Sheet Piling (R 7)	280
8.1.5	Shear-resistant Interlock Joining in Steel Sheet Piling (R 103)	283
8.1.6	Quality Requirements for Steels and Interlock Dimension Tolerances for Steel Sheet Piles (R 67)	287
8.1.7	Acceptance Conditions for Steel Sheet Piles and Steel Piles at the Site (R 98)	292
8.1.8	Corrosion of Steel Sheet Piling and Counter-measures (R 35)	292
8.1.9	Danger of Sand Abrasion on Sheet Piling (R 23)	298
8.1.10	Driving Assistance for Steel Sheet Piling by Means of Loosening Blasting (R 183)	299
8.1.11	Driving Corrugated Steel Sheet Piles (R 118)	301
8.1.12	Driving of Combined Steel Sheet Piling (R 104)	306
8.1.13	Observations During the Installation of Steel Sheet Piles, Tolerances (R 105)	310

8.1.14	Noise Protection, Low-noise Driving (R 149)	312
8.1.15	Driving of Steel Sheet Piles and Steel Piles at Low Temperatures (R 90)	317
8.1.16	Repair of Interlock Damage to Driven Steel Sheet Piling (R 167)	317
8.1.17	Design of Pile Driving Trestles (R 140)	321
8.1.18	Design of Welded Joints in Steel Sheet Piles and Driven Steel Piles (R 99)	324
8.1.19	Burning Off the Tops of Driven Steel Sections for Load-bearing Welded Connections (R 91)	328
8.1.20	Watertightness of Steel Sheet Piling (R 117)	329
8.1.21	Waterfront Structures in Regions Subject to Mining Subsidence (R 121)	331
8.1.22	Vibration of U- and Z-shaped Steel Sheet Piles (R 202)	335
8.1.23	Jetting when Driving Steel Sheet Piles (R 203)	339
8.2	Calculation and Design of Sheet Piling	341
8.2.0	General	341
8.2.1	Sheet Piling Structures, Fully Fixed in the Ground Without Anchorage (R 161)	342
8.2.2	Calculation of Single-anchored Sheet Piling Structures (R 77)	344
8.2.3	Calculation of Double-anchored Sheet Piling (R 134)	346
8.2.4	Assumed Angle of Wall Friction and of Adhesion (R 4)	348
8.2.5	Assumed Angle of Wall Friction of Passive Earth Pressure for an Inclined Embankment of Non-cohesive Soil in Front of Sheet Piling (R 199)	348
8.2.6	Bearing Stability Verification of Sheet Piling Structures (R 20)	348
8.2.7	Consideration of Axial Loads in Sheet Piling (R 44)	350
8.2.8	Selection of the Embedment Depth of Sheet Piling (R 55)	351
8.2.9	Determination of Embedment Depth with Partial or Full Fixity of the Point of the Sheet Piling (R 56)	351
8.2.10	Staggered Embedment Depth of Steel Sheet Piling (R 41)	353
8.2.11	Vertical Load Bearing Capacity of Sheet Piling (R 33)	355
8.2.12	Capacity of Steel Sheet Piling to Absorb Horizontal Longitudinal Forces Acting Parallel to the Shore (R 132)	356
8.2.13	Calculation of an Anchor Wall Fixed in the Earth (R 152)	359
8.2.14	Staggered Design of Anchor Walls (R 42)	360
8.2.15	Steel Sheet Piling Driven into Bedrock or Rock-like Soils (R 57)	361
8.2.16	Waterfront Sheet Piling in Unconsolidated, Soft Cohesive Soils, Especially in Connection with Undisplaceable Structures (R 43)	362
8.2.17	Effects of Earthquakes on the Design and Dimensioning of Waterfront Structures (R 124)	363
8.2.18	Design and Dimensioning of Single-anchored Sheet Piling Structures in Earthquake Areas (R 125)	363
8.3	Calculation and Design of Cofferdams	364
8.3.1	Cellular Cofferdams as Excavation Enclosures and as Permanent Waterfront Structures (R 100)	364
8.3.2	Double-wall Cofferdams as Excavation Enclosures and as Permanent Waterfront Structures (R 101)	370
8.3.3	Narrow Partition Moles in Sheet Piling Construction (R 162)	373

8.4	Anchorings, Stiffeners	374
8.4.1	Design of Steel Wales for Sheet Piling (R 29)	374
8.4.2	Verification of Bearing Capacity of Steel Wales (R 30)	376
8.4.3	Wales of Reinforced Concrete for Sheet Piling with Driven Steel Anchor Piles (R 59)	377
8.4.4	Steel Capping Beams for Waterfront Structures (R 95)	383
8.4.5	Reinforced Concrete Capping Beams for Waterfront Structures (R 129)	386
8.4.6	Top Steel Nosing for Reinforced Concrete Walls and Capping Beams at Waterfront Structures (R 94)	392
8.4.7	Auxiliary Anchoring at the Top of Steel Sheet Piling Structures (R133)	394
8.4.8	Threads of Sheet Piling Anchors (R 184)	396
8.4.9	Verification of Stability of Anchoring at Lower Failure Plane (R 10)	398
8.4.10	Sheet Piling Anchorings in Unconsolidated, Soft Cohesive Soils (R 50)	403
8.4.11	Design and Calculation of Protruding Corner Structures with Round Steel Anchoring (R 31)	406
8.4.12	Design and Calculation of Protruding Quay Wall Corners with Batter Pile Anchoring (R 146)	409
8.4.13	High Prestressing of Anchors of High-strength Steels for Waterfront Structures (R 151)	411
8.4.14	Hinged Support of Quay Wall Superstructures on Steel Sheet Piling (R 64)	413
8.4.15	Hinged Connection of Driven Steel Anchor Piles to Steel Sheet Piling Structures (R 145)	417
8.4.16	Armoured Steel Sheet Piling (R 176)	426

9	**Anchor Piles**	**433**
9.1	General	433
9.2	Anchoring Elements, General Explanations	433
9.3	Safety Factors for Anchoring (R 26)	435
9.4	Limit Tension Load of Anchor Piles (R 27)	436
9.5	Design and Embedment of Driven Steel Piles (R 16)	437
9.6	Design and Loading of Driven Piles With Grouted Skin (VM Piles) (R 66)	439
9.7	Design and Loading of Tubular Grouted Piles (R 207)	443
9.8	Anchoring with Piles of Small Diameter (R 208)	444
9.9	Connections of Anchor Piles to Reinforced Concrete and Steel Structures	444
9.10	Transmission of Horizontal Loads via Pile Bents, Slotted Wall Plates, Frames and Large Bored Piles (R 209)	445

10	**Waterfront Structures, Quays and Superstructures of Concrete and Reinforced Concrete**	**448**
10.1	Design of Quays and Superstructures, Structures on Pile Foundations (R 17)	448
10.2	Construction of Reinforced Concrete Waterfront Structures (R 72)	450

10.3	Formwork in Marine Environment (R 169)	453
10.4	Design of Reinforced Concrete Roadway Slabs on Piers (R 76)	454
10.5	Box Caissons as Waterfront Structures in Seaports (R 79)	455
10.6	Pneumatic Caissons as Waterfront Structures in Seaports (R 87)	458
10.7	Design and Dimensioning of Quay Walls in Blockwork Construction (R 123)	461
10.8	Design and Dimensioning of Quay Walls in Blockwork Construction in Earthquake Areas (R 126)	467
10.9	Construction and Design of Quay Walls Using the Open Caisson Method (R 147)	468
10.10	Construction and Design of Quay Walls with Open Caissons in Earthquake Regions (R 148)	471
10.11	Application and Design of Bored Pile Walls (R 86)	471
10.12	Application and Design of Diaphragm Walls (R 144)	475
10.13	Application and Construction of Impermeable Diaphragm Walls and Impermeable Thin Walls (R 156)	480
10.14	Inventory Before Repairing Concrete Components in Hydraulic Engineering (R 194)	488
10.15	Repair of Concrete Components in Hydraulic Engineering (R 195)	491

11 Pile-founded Structures — 501

11.1	General	501
11.2	Determining the Active Earth Pressure Shielding on a Wall Below a Relieving Platform Under Average Ground Surcharges (R 172)	501
11.3	Active Earth Pressure on Sheet Piling in Front of Pile-founded Structures (R 45)	503
11.4	Calculation of Plane Pile-founded Structures (R 78)	510
11.5	Design and Calculation of General Pile-founded Structures (R 157)	512
11.6	Wave Pressure on Pile Structures (R 159)	518
11.7	Verification of Overall Stability of Structures on Elevated Pile-founded Structures (R 170)	518
11.8	Design and Dimensioning of Pile-founded Structures in Earthquake Areas (R 127)	518
11.9	Bracing of the Tops of Steel Pipe Driving Piles (R 192)	519

12 Embankments — 521

12.1	Embankments in Seaports and Inland Harbours with Tide (R 107)	521
12.2	Embankments Under Quay Wall Superstructures Behind Tight Sheet Piling (R 68)	528
12.3	Partially Sloped Bank Construction in Inland Harbours with Extreme Water Level Fluctuations (R 119)	529
12.4	Use of Geotextile Filters in Embankment and Bottom Protection (R 189)	529

13	**Dolphins**	534
13.1	Design of Resilient Multi-pile and Single-pile Dolphins (R 69)	534
13.2	Spring Constant for the Calculation and Dimensioning of Heavy Fendering and Heavy Berthing Dolphins (R 111)	537
13.3	Impact Forces and Required Energy Absorption Capacity of Fenders and Dolphins in Seaports (R 128)	540
13.4	Use of Weldable Fine-grained Structural Steels for Elastic Berthing and Mooring Dolphins in Marine Construction (R 112)	544

14	**Experience with Waterfront Structures**	547
14.1	Average Service Life of Waterfront Structures (R 46)	547
14.2	Operational Damage to Steel Sheet Piling (R 155)	547
14.3	Steel Sheet Piling Waterfront Structures Under Fire Loads (R 181)	549

15	**Supervision of Structures**	553
15.1	Observation and Inspection of Waterfront Structures in Seaports (R 193)	553

Annex I Bibliography 557

I.1	Annual Technical Reports	557
I.1	Books and Papers	557
I.3	Technical Provisions	567
I.3.1	Standards	567
I.3.2	Regulations of the German Federal Railways (DS)	573
I.3.3	Regulations of the German Committee for Steel Constructions (DASt-Ri)	573
I.3.4	Steel-Iron Material Table of the "Verein Deutscher Eisenhüttenleute" (SEW)	574

Annex II List of Conventional Symbols 575

II.1	Symbols	575
II.2	Abbreviations	580
II.3	Symbols for Water Levels	580

Annex III List of Key Words 581

List of Recommendations in the 7th English Edition

		Section	Page
R 1	Layout and depth of borings and penetrometer tests	1.2	11
R 2	Assumed cohesion in cohesive soils	2.1	29
R 3	Assumed apparent cohesion in sand	2.2	29
R 4	Assumed angle of wall friction and of adhesion	8.2.4	348
R 5	Vertical live loads	5.5	110
R 6	Standard dimensions of cross-section of waterfront structures in seaports	6.1	159
R 7	Combined steel sheet piling	8.1.4	280
R 9	Mean characteristic soil properties	1.1	11
R 10	Verification of stability of anchoring at lower failure plane	8.4.9	398
R 12	Layout and loading of bollards for seagoing vessels	5.12	143
R 14	Layout, design and loading of access ladders	6.11	184
R 16	Design and embedment of driven steel piles	9.5	437
R 17	Design of quays and superstructures, structures on pile foundations	10.1	448
R 18	Load classes	5.4	109
R 19	Water pressure difference in the water-side direction	4.2	73
R 20	Bearing stability verification of sheet piling structures	8.2.6	348
R 21	Design and driving of reinforced concrete sheet piling	8.1.2	277
R 22	Design and driving of timber sheeting	8.1.1	274
R 23	Danger of sand abrasion on sheet piling	8.1.9	298
R 24	Layout and design of stairs in seaports	6.12	186
R 26	Safety factors for anchoring	9.3	435
R 27	Limit tension load of anchor piles	9.4	436
R 29	Design of steel wales for sheet piling	8.4.1	374
R 30	Verification of bearing capacity for steel wales	8.4.2	376
R 31	Design and calculation of protruding corner structures with round steel anchoring	8.4.11	406
R 32	Design of flap valves for waterfront structures in tidal areas	4.5	78
R 33	Vertical load bearing capacity of sheet piling	8.2.11	355
R 34	Steel sheet piling	8.1.3	280
R 35	Corrosion of steel sheet piling and counter-measures	8.1.8	292
R 36	Nominal depth and design depth of harbour bottom	6.7	172
R 38	Assumed berthing pressure of vessels at quays	5.2	108
R 39	Ship dimensions	5.1	102
R 40	Berthing velocities of vessels transverse to berth	5.3	108
R 41	Staggered embedment depth of steel sheet piling	8.2.10	353
R 42	Staggered design of anchor walls	8.2.14	360

		Section	Page
R 43	Waterfront sheet piling in unconsolidated, soft cohesive soils, especially in connection with undisplaceable structures	8.2.16	362
R 44	Consideration of axial loads in sheet piling	8.2.7	350
R 45	Active earth pressure on sheet piling in front of pile-founded structures	11.3	503
R 46	Average service life of waterfront structures	14.1	547
R 47	Fenders in inland harbours	6.15	204
R 50	Sheet piling anchorings in unconsolidated, soft cohesive soils	8.4.10	403
R 51	Design of filter weepholes for sheet piling structures	4.4	77
R 52	Effect of artesian water under the harbour bottom or river bed, on active and passive earth pressures	2.7	40
R 53	Relieving artesian pressure under harbour bottoms	4.6	80
R 55	Selection of the embedment depth of sheet piling	8.2.8	351
R 56	Determination of embedment depth with partial or full fixity of the point of the sheet piling	8.2.9	351
R 57	Steel sheet piling driven into bedrock or rock-like soils	8.2.15	361
R 58	Mean groundwater level	4.1	73
R 59	Wales of reinforced concrete for sheet piling with driven steel anchor piles	8.4.3	377
R 60	Fenders for berths for large vessels at quays	6.14	191
R 64	Hinged support of quay wall superstructures on steel sheet piling	8.4.14	413
R 65	Water pressure difference on sheet piling in front of built-over embankments in tidal areas	4.3	76
R 66	Design and loading of driven piles with grouted skin (VM piles)	9.6	439
R 67	Quality requirements for steels and interlock dimension tolerances for steel sheet piles	8.1.6	287
R 68	Embankments under quay wall superstructures behind tight sheet piling	12.2	528
R 69	Design of resilient multi-pile and single-pile dolphins	13.1	534
R 70	Equipping of berths for large vessels with quick release hooks	6.10	182
R 71	Investigation of the degree of density of non-cohesive backfill for waterfront structures	1.6	19
R 72	Construction of reinforced concrete waterfront structures	10.2	450
R 73	Backfilling of waterfront structures	7.4	230
R 74	Standard cross-sections of waterfront structures in inland harbours	6.3	162
R 76	Design of reinforced concrete roadway slabs on piers	10.4	454
R 77	Calculation of single-anchored sheet piling structures	8.2.2	344
R 78	Calculation of plane pile-founded structures	11.4	510
R 79	Box caissons as waterfront structures in seaports	10.5	455

		Section	Page
R 80	Dredging in front of quay walls in seaports	7.1	221
R 81	Hydraulic filling of port areas for planned waterfront structures	7.3	226
R 83	Scour and scour protection at waterfront structures	7.6	236
R 84	Quay loads from cranes and other transhipment equipment	5.14	147
R 85	Installation of crane rails on concrete	6.17	207
R 86	Application and design of bored pile walls	10.11	471
R 87	Pneumatic caissons as waterfront structures in seaports	10.6	458
R 88	Determination of undrained shear strength c_u	1.4	16
R 90	Driving of steel sheet piles and steel piles at low temperatures	8.1.15	317
R 91	Burning off the tops of driven steel sections for load-bearing welded connections	8.1.19	328
R 93	Vertical drains to accelerate the consolidation of soft cohesive soils	7.7	243
R 94	Top steel nosing for reinforced concrete walls and capping beams at waterfront structures	8.4.6	392
R 95	Steel capping beams for waterfront structures	8.4.4	383
R 98	Acceptance conditions for steel sheet piles and steel piles at the site	8.1.7	292
R 99	Design of welded joints in steel sheet piles and driven steel piles	8.1.18	324
R 100	Cellular cofferdams as excavation enclosures and as permanent waterfront structures	8.3.1	364
R 101	Double-wall cofferdams as excavation enclosures and as permanent waterfront structures	8.3.2	370
R 102	Layout, design and loading of bollards in inland harbours	5.13	144
R 103	Shear-resistant interlock joining in steel sheet piling	8.1.5	283
R 104	Driving of combined steel sheet piling	8.1.12	306
R 105	Observations during the installation of steel sheet piles, tolerances	8.1.13	310
R 106	Sheet piling waterfront on canals for inland vessels	6.4	165
R 107	Embankments in seaports and inland harbours with tide	12.1	521
R 109	Soil replacement procedure for waterfront structures	7.9	248
R 110	Action of active earth pressure and water pressure difference and construction hints for waterfront structures with soil replacement and fouled or disturbed dredge pit bottom	2.8	42
R 111	Spring constant for the calculation and dimensioning of heavy fendering and heavy berthing dolphins	13.2	537
R 112	Use of weldable fine-grained structural steels for elastic berthing and mooring dolphins in marine construction	13.4	544
R 113	Assessment of groundwater flow	4.7	83
R 114	Effect of percolating groundwater on water pressure difference, active and passive earth pressure	2.9	46

		Section	Page
R 115	Safety against failure by heave	3.2	66
R 116	Foundation failure due to erosion; its occurrence and its prevention	3.3	69
R 117	Watertightness of steel sheet piling	8.1.20	329
R 118	Driving corrugated steel sheet piles	8.1.11	301
R 119	Partially sloped waterfront construction at inland harbours with extreme water level fluctuations	6.5	168
R 120	Foundation of craneways at waterfront structures	6.16	205
R 121	Waterfront structures in regions subject to mining subsidence	8.1.21	331
R 122	Top elevation of waterfront structures in seaports	6.2	160
R 123	Design and dimensioning of quay walls in blockwork construction	10.7	461
R 124	Effects of earthquakes on the design and dimensioning of waterfront structures	2.13	59
R 125	Design and dimensioning of single-anchored sheet piling structures in earthquake areas	8.2.18	363
R 126	Design and dimensioning of quay walls in blockwork construction in earthquake areas	10.8	467
R 127	Design and dimensioning of pile-founded structures in earthquake areas	11.8	518
R 128	Impact forces and required energy absorption capacity of fenders and dolphins in seaports	13.3	540
R 129	Reinforced concrete capping beams for waterfront structures	8.4.5	386
R 130	Determination of active earth pressure in saturated, non- or partially consolidated, soft cohesive soils	2.6	36
R 131	Shear parameters of the drained soil in the state of failure or sliding	1.5	17
R 132	Capacity of steel sheet piling to absorb horizontal longitudinal forces acting parallel to the shore	8.2.12	356
R 133	Auxiliary anchoring at the top of steel sheet piling structures	8.4.7	394
R 134	Calculation of double-anchored sheet piling	8.2.3	346
R 135	Wave pressure on vertical waterfront structures in coastal areas	5.7	122
R 136	Determining the "design wave" for maritime and port structures	5.6	113
R 137	Calculation and design of rubble mound moles and breakwaters	7.10	254
R 138	Dredging of underwater slopes	7.5	232
R 139	Dredging and hydraulic fill tolerances	7.2	223
R 140	Design of pile driving trestles	8.1.17	321
R 144	Application and design of diaphragm walls	10.12	475
R 145	Hinged connection of driven steel anchor piles to steel sheet piling structures	8.4.15	417

		Section	Page
R 146	Design and calculation of protruding quay wall corners with batter pile anchoring	8.4.12	409
R 147	Construction and design of quay walls using the open caisson method	10.9	465
R 148	Construction and design of quay walls with open caissons in earthquake regions.	10.10	471
R 149	Noise protection, low-noise driving	8.1.14	312
R 150	Preparation of reports and expert opinions on subsoil examinations under difficult conditions	1.3	15
R 151	High prestressing of anchors of high-strength steels for waterfront structures	8.4.13	411
R 152	Calculation of an anchor wall fixed in the earth	8.2.13	359
R 153	Wind loads on moored ships and their influences on the dimensioning of mooring and fendering facilities	5.11	140
R 154	Assessment of the subsoil for the driving of sheet piles and steel piles	1.9	23
R 155	Operational damage to steel sheet piling.	14.2	547
R 156	Application and construction of impermeable diaphragm walls and impermeable thin walls	10.13	480
R 157	Design and calculation of general pile-founded structures	11.5	512
R 158	Design of waterfront areas in inland ports according to operational aspects.	6.6	171
R 159	Wave pressure on pile structures	5.10	132
R 161	Sheet piling structures, fully fixed in the ground without anchorage.	8.2.1	342
R 162	Narrow partition moles in sheet piling construction	8.3.3	373
R 164	Measures for increasing the passive earth pressure in front of waterfront structures	2.11	54
R 165	Flood protection walls in seaports.	4.9	94
R 166	Temporary stabilisation of waterfront structures by groundwater lowering	4.8	91
R 167	Repair of interlock damage to driven steel sheet piling	8.1.16	317
R 168	Subsidence of non-cohesive soils	7.8	247
R 169	Formwork in marine environment.	10.3	453
R 170	Verification of overall stability of structures on elevated pile-founded structures	3.4	71
R 171	Determination of the active earth pressure using the Culmann method	2.4	29
R 172	Determining the active earth pressure shielding on a wall below a relieving platform under average ground surcharges	11.2	501
R 173	Equipment of waterfront structures in seaports with supply and disposal facilities.	6.13	187

		Section	Page
R 174	Determination of the amount of displacement required for the mobilisation of partial passive earth pressures in non-cohesive soils	2.10	52
R 175	Degree of density of hydraulically filled, non-cohesive soils	1.7	20
R 176	Armoured steel sheet piling	8.4.16	426
R 177	Impact and pressure of ice on waterfront structures, fenders and dolphins in coastal areas	5.15	151
R 178	Degree of density of dumped, non-cohesive soils	1.8	22
R 179	Consolidation of soft, cohesive soils by preloading	7.13	265
R 181	Steel sheet piling waterfront structures under fire loads	14.3	549
R 183	Driving assistance for steel sheet piling by means of loosening blasting	8.1.10	299
R 184	Threads of sheet piling anchors	8.4.8	396
R 185	Loads arising from surging and receding waves resulting from the introduction or extraction of water	5.8	128
R 186	Effects of waves from movements of ships	5.9	129
R 187	Light backfilling for sheet piling structures	7.11	264
R 188	Soil compaction using heavy drop weights	7.12	264
R 189	Use of geotextile filters in embankment and bottom protection	12.4	529
R 190	Passive earth pressure in front of sheet piles in soft cohesive soils, with rapid loading on the land side	2.12	56
R 191	Connection of expansion joint seal in a reinforced concrete bottom to bearing external steel sheet piling	6.18	215
R 192	Bracing of the tops of steel pipe driving piles	11.9	519
R 193	Observation and inspection of waterfront structures in seaports	15.1	553
R 194	Inventory before repairing concrete components in hydraulic engineering	10.14	488
R 195	Repair of concrete components in hydraulic engineering	10.15	491
R 196	Connection of steel sheet piling to a concrete structure	6.19	216
R 198	Determination of active earth pressure in a steep, paved embankment of a partially sloping bank construction	2.5	34
R 199	Assumed angle of wall friction of passive earth pressure for an inclined embankment of non-cohesive soil in front of sheet piling	8.2.5	348
R 200	Reinforcement of waterfront structures to deepen harbour bottoms in seaports	6.8	174
R 201	Re-design of waterfront structures in inland harbours	6.9	179
R 202	Vibration of U- and Z-shaped steel sheet piles	8.1.22	335
R 203	Jetting when driving steel sheet piles	8.1.23	339
R 204	Installation of mineral bottom seals under water and their connection to waterfront structures	7.14	270
R 205	Impact and pressure of ice on waterfront structures, piers and dolphins in inland areas	5.16	156

		Section	Page
R 206	Floating wharves in seaports	6.20	218
R 207	Design and loading of tubular grouted piles	9.7	443
R 208	Anchoring with piles of small diameter	9.8	444
R 209	Transmission of horizontal loads via pile bents, slotted wall plates, frames and large bored piles	9.10	445

Recommendations

0 Statical Calculations

0.1 General

Up until the 8th German edition (EAU 1990), respectively the 6th English edition, the earth-static calculations in the recommendations of the "Committee for Waterfront Structures", which have been developed in over 40 years of work by the Committee, were based on reduced values of soil properties, known as "calculation values", with the prefix "cal". Results of calculations using these calculation values must then fulfil the global safety criteria, in accordance with R 96, section 1.13.2a) of EAU 1990. This safety concept used up to now distinguished three different load cases (R 18, section 5.4), and has proved its worth over the years.

If the previous safety concept, which remains valid until the end of the trial period of the ENV, is still used for calculations, the corresponding recommendations of the EAU 1990 apply.

As part of the completion of the Single European Market, the "Eurocodes" (EC) have been introduced for trials as general safety requirements for structural works, and national standards will then have to be brought into line. The national application document (NAD) stipulates the context between EC 7 and the German standards. EC 7, DIN V ENV 1997-1[1] is of significance for the EAU. EAU 1996 will now take account of the quantitative statements contained in this Eurocode regarding calculation procedures and partial safety factors.

Stability and serviceability are prerequisites for safe structural work. In the European Union EU, verifications are to be provided according to the European standards:

DIN V ENV 1991-1, EC 1:	Basis of design and actions on structures, Part 1: Basis of design,
DIN V ENV 1992-1, EC 2:	Design of concrete structures, Part 1: General rules,
DIN V ENV 1993-1, EC 3:	Design of steel structures, Part 1: General rules,

---

[1] Note: "V" stands for "pre-standard in the trial phase"

DIN V ENV 1994-1, EC 4:		Design of composite steel and concrete structures, Part 1: General rules,
DIN V ENV 1995-1, EC 5:		Design of timber structures, Part 1: General rules,
DIN V ENV 1996-1, EC 6:		Design of masonry structures, Part 1: General rules,
DIN V ENV 1997-1, EC 7:		Geotechnical design, Part 1: General rules,
DIN V ENV 1998-1, EC 8:		Earthquake resistant design of structures, Part 1: General rules,
DIN V ENV 1999-1, EC 9:		Design of aluminium structures, Part 1: General rules.

EC2, EC 3, EC 4 and in particular EC 7 are of significance for statical calculations to EAU 1996. In combination with the national application document (NAD) to EC 7, the following German pre-standards can be used for geotechnical calculations:

DIN V 1054-100:	Soil –	Verification of the safety of earthworks and foundations,
DIN V 4017-100:	Soil –	Calculation of design bearing capacity of soil beneath shallow foundations,
DIN V 4019-100:	Soil –	Analysis of settlements,
DIN V 4084-100:	Soil –	Calculation of slope and embankment failure and overall stability of retaining structures,
DIN V 4085-100:	Soil –	Calculation of earth pressure[2].

DIN 18 800 and DIN 18 807 are to be taken into account for verification of load bearing safety of steel structures.

0.2 Safety Concept

0.2.1 General

The failure of a structure can occur as a result of exceeding the limit state of the bearing capacity (LS 1) (failure in the soil or structure) or the limit state of the serviceability (LS 2) (excessive deformation).

The safety limits are stipulated in DIN V ENV 1991-1 for actions and in DIN V ENV 1992-1 to DIN V ENV 1999-1 for actions and resistances, or alternatively according to the NAD in DIN V 1054-100, for both actions and resistances.

---

[2] Part 100: Analysis in accordance with the partial safety factor concept

The relationship between DIN V ENV 1997-1 and DIN V 1054-100 with the calculation standards DIN V 4017-100, DIN V 4019-100, DIN V 4084-100 and DIN V 4085-100 is illustrated by the national application document (NAD).

EAU 1996 fundamentally takes account of the European pre-standards, particularly DIN V ENV 1997-1. However, in view of the fact that these still contain considerable discretionary scope so that DIN V ENV 1991-1 prescribes national application documents (NAD), EAU 1996 reverts back to the calculation standards according to the German NAD for the verification of safety of earthworks and foundations.

For the verification of safety DIN V ENV 1991-1 differentiates between 3 ultimate limit states of the bearing capacity in limit state 1 (case A, case B, case C). These are illustrated in table R 0-1.

DIN V 1054-100 basically adopts these limit states as LS 1A, 1B and 1C (table R 0-3 and R 0-4). However, according to the NAD there are differences between DIN V ENV 1997-1 and DIN V 1054-100 in treating the limit state of bearing capacity and thus in ascertaining the outer dimensions of the foundation structure. Usually, DIN V ENV 1997-1 treats this case as limit state 1C, DIN V 1054-100 as limit state 1B.

There are also differences between DIN V ENV 1997-1 and DIN V 1054-100 regarding the verification of stability.

In calculating the bearing capacity of soil and verification of slide stability, ENV always verifies the safety for limit state 1C. According to DIN V 1054-100, the safety of individual foundations is ascertained according to limit state 1B from the characteristic values of the shearing parameters and subsequent reduction of the calculated soil failure load with partial safety factor γ_s. The same procedure is used for verification of slide stability. The soil bearing capacity for limit state 1C is only ascertained with the design values of the shearing parameter for shallow foundations, grill foundations and for individual and strip foundations which are joined to form foundation groups by a rigid superstructure and act as an uniform foundation structure for the whole ground surface of the structure.

The difference is to be found, for example, in the fact that in LS 1C, the partial safety factors are to be applied to the shearing parameters φ_k and c_k, in LS 1B to the actions and resistances calculated with φ_k and c_k. The characteristic values c_{uk}, c'_k and φ'_k are usually stipulated by the test institution and designated as such.

There are also essential differences between DIN V ENV 1997-1 and DIN V 1054-100 in the fact that according to German experience, the "load cases" 1, 2 and 3 (table R 0-3 and R 0-4) remain introduced for different safety requirements. This is a formal difference to DIN V EN V 1991-1 and 1997-1, where only partial safety factors are indicated for one load case (corresponding to load case 1, see table R 0-2).

Limit state 1 (LS) [1]	Action		Symbol	Situations	
				P/T	A
Case 1A Loss of static equilibrium: strength of structural material or ground insignificant (see ENV 1991-1: 1994, 9.4.1)	Permanent actions: self weight of structural and non structural components, permanent actions caused by ground, groundwater and free water				
		– unfavourable – favourable	$\gamma_{Gsup}^{2,4}$ $\gamma_{Ginf}^{2,4}$	[1.10] [0.90]	[1.00] [1.00]
	Variable actions – unfavourable		γ_Q	[1.50]	[1.00]
	Accidental actions		γ_A		[1.00]
Case 1B [5] Failure of structure or structural elements, including those of the footing, piles, basement walls etc., governed by strength of structural material (see ENV 1991-1:1994, 9.4.1)	Permanent actions [6] (see above)				
		– unfavourable – favourable	$\gamma_{Gsup}^{3,4}$ $\gamma_{Ginf}^{3,4}$	[1.35] [1.00]	[1.00] [1.00]
	Variable actions – unfavourable		γ_Q	[1.50]	[1.00]
	Accidental actions		γ_A		[1.00]
Case 1C [5] Failure in the ground	Permanent actions (see above)				
		– unfavourable – favourable	γ_{Gsup}^{4} γ_{Ginf}^{4}	[1.00] [1.00]	[1.00] [1.00]
	Variable actions – unfavourable		γ_Q	[1.30]	[1.00]
	Accidental actions		γ_A		[1.00]

P: Persistent situation, T: Transient situation, A: Accidental situation
1) The design should be verified for each case 1A, 1B and 1C separately as relevant.
2) In this verification the characteristic value of the unfavourable part of the permanent action is multiplied by the factor [1.1] and the favourable part by the factor [0.9]. More refined rules are given in ENV 1993 and ENV 1994.
3) In the verification, the characteristic values of all permanent actions from one source are multiplied by [1.35] if the total resulting action effect is unfavourable and by [1.0] if the total resulting action effect is favourable.
4) In cases when the limit state is very sensitive to variations of permanent actions, the upper and lower characteristic values of these actions should be taken according to 4.2 (3)P of ENV 1991-1.
5) For cases B and C the design ground properties may be different, see ENV 1997-1-1.
6) Instead of using γ_G [1.35] and γ_Q [1.50] for lateral earth pressure actions, the design ground properties may be introduced in accordance with ENV 1997 and a model factor γ_{Sd} is applied.
[ ] Recommended values for the trial phase, partial safety factors in [ ] are values recommended in ENV.

Table R 0-1. Partial safety factors for buildings in the ultimate limit states of bearing capacity (corresponds to table 9.2 from EC 1 (DIN V ENV 1991-1))

LS	Actions			Resistances (ground properties)			
	Permanent		Variable	$\tan \varphi'$	c'	c_u	q_u [1)]
	Un-favour-able	Favour-able	Un-favour-able				
1A	[1.00]	[0.95]	[1.50]	[1.10]	[1.30]	[1.20]	[1.20]
1B	[1.35]	[1.00]	[1.50]	[1.00]	[1.00]	[1.00]	[1.00]
1C	[1.00]	[1.00]	[1.30]	[1.25]	[1.60]	[1.40]	[1.40]

[1)] Cylinder compressive strength of soil and rock.
[ ] Recommended values for the trial phase, partial safety factors in [ ] are values recommended in ENV.

Table R 0-2. Partial safety factors for ultimate limit states of the bearing capacity in persistent and transient situations (corresponds to table 2.1 of EC 7 (DIN V ENV 1997-1)

LS	Actions	Symbol	Load case (LC)		
			1	2	3
1A	permanent actions, unfavourable	γ_{Gsup}	1.00	1.00	1.00
	permanent actions, favourable	γ_{Ginf}	0.90	0.95	1.00
	hydrostatic pressure	γ_F	1.00	1.00	1.00
	variable actions, unfavourable	γ_{Qsup}	1.05	1.00	1.00
1B	permanent actions, unfavourable	γ_{Gsup}	1.35	1.20	1.00
	permanent actions, favourable	γ_{Ginf}	1.00	1.00	1.00
	hydrostatic pressure	γ_F	1.35	1.20	1.00
	variable actions, unfavourable	γ_{Qsup}	1.50	1.30	1.00
	lateral pressure, permanent	γ_H	1.35	1.20	1.00
	skin friction, permanent	γ_M	1.35	1.20	1.00
	earth pressure, permanent	γ_{Eg}	1.35	1.20	1.10
	earth pressure, variable, unfavourable	γ_{Eq}	1.50	1.30	1.10
1C	permanent actions	γ_G	1.00	1.00	1.00
	hydrostatic pressure	γ_F	1.00	1.00	1.00
	variable actions, unfavourable	γ_{Qsup}	1.30	1.20	1.10
	lateral pressure, permanent	γ_H	1.00	1.00	1.00
	skin friction, permanent	γ_M	1.00	1.00	1.00
	earth pressure, permanent earth pressure, variable, unfavourable	using shearing parameters with partial safety factors according to table R 0-2			
2	1.00 for permanent actions, favourable or unfavourable 1.00 for variable actions, unfavourable				

Table R 0-3. Partial safety factors for actions (corresponds to table 2 from DIN V 1054-100)

PEUTE BAUSTOFF GMBH

HYDRAULIC CONSTRUCTION STONE

Made of iron silicate stone in accordance with German Technical Delivery Specifications for Hydraulic Construction Stone (TLW; DIN 4301-MHS-1)

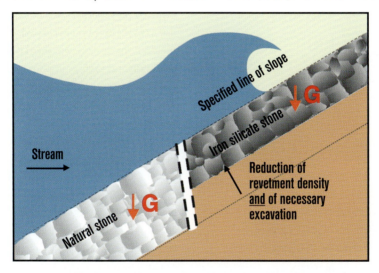

High specific weight (3.7 metric ton/m^3)
High area weight and low installation thickness
High bulk weight (2.0 ... 2.2 metric ton/m^3) – Reduction in excavation
Absolute resistance to frost – High positional stability
Very good grain form – Extremely long service life
Allows rapid growth of grass through the stone

Please don't hesitate to get in touch with us
if you have any questions on our product
or its application.

Peutestraße 79 · 20539 Hamburg
Tel.: (+49 40) 78 91 60-0 · Fax: (+49 40) 78 91 60-19

Structural Dynamics

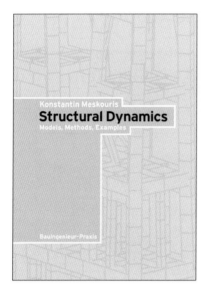

Konstantin Meskouris
Structural Dynamics
Models, Methods, Examples
2000. Approx. 250 pages.
Format: 17 x 24 cm.
Hb. approx. DM 150,-
öS 1095,-/sFr 133,-
ISBN 3-433-01327-6
Publication date: March 2000

Vibration problems are not uncommon in many civil engineering structures and an insight into the principles involved plus some degree of expertise in modern, computer-based methods for dealing with them is becoming increasingly important for the engineering profession.
Written for engineering students as well as for practising engineers, this book provides a comprehensive introduction to the theory of structural dynamics with special emphasis on practical issues, illustrating applications by a plethora of worked examples. As an innovative feature, it includes a large number of computer programs as ready-to-use executables on a CD-ROM, complete with detailed input/output descriptions and auxiliary software. In the spirit of „learning by doing", it encourages readers to immediately apply these tools to their own problems, becoming familiar with the broad field of structural dynamic response in the process.

Ernst & Sohn
Verlag für Architektur
und technische Wissenschaften GmbH
Bühringstraße 10, D-13086 Berlin
Tel. (030) 470 31-284
Fax (030) 470 31-240
E-mail: mktg@ernst-und-sohn.de
www.ernst-und-sohn.de

INTENTIONS TO REALITY
50 YEARS ENGINEERING SERVICES WITH DESIGN FOR SOLUTIONS

- Port and Harbour Installations
- Docks, Jetties, Waterways
- Hydraulic and Civil Engineering
- Traffic Facilities, Roads, Bridges
- Infrastructural Planning, Services
- Urban Development, Buildings
- Consulting Engineers and Architects

INROS
Rosa - Luxemburg - Strasse 16 - 18 • 18055 Rostock / Germany
Phone: +49 - 381 - 45 67 80 • Fax: +49 - 381 - 45 67 919
eMail: mail@inros.de • Internet: http://www.inros.de

planning · construction · execution

40 years experience in special civil engineering and geotechnique.
permanent and temporary anchors working load up to 12500 kN, micropiles, foundation piles, sheetpile walls, construction pits, rock and soil nails, shotcrete, chemical and cement grouting, jet grouting "Stump-Jetting", masonry and concrete improvement, core drilling, Electro-Osmose.

Headquarter Langenfeld Max-Planck-Ring 1 · 40764 Langenfeld Tel. 02173/7902-0 · Fax 02173/7902-20	**NL Berlin** Stralauer Allee 2–16 · 10245 Berlin Tel. 030/754904-0 · Fax 030/75490420	**GS Niedersachswerfen** Northeimer Str. · 99762 Niedersachswerfen Tel. 03 63 31/9 42-0 · Fax 03 63 31/9 42-20
NL Langenfeld Max-Planck-Ring 1 · 40764 Langenfeld Tel. 02173/7902-99 · Fax 02173/7902-90	**ZN Hannover** Fränkische Straße 11 · 30455 Hannover Tel. 0511/94999-0 · Fax 0511/499498	**e-mail:** InfoStump@Stump.de
NL München Am Lenzenfleck 1–3 · 85737 Ismaning Tel. 089/960701-0 · Fax 089/963151	**ZN Chemnitz** Blankenauer Straße 99 · 09113 Chemnitz Tel. 0371/4900-266 · Fax 0371/4900-268	**Internet:** http://www.Stump.de

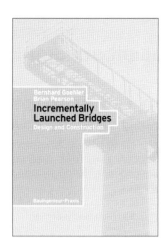

Bernhard Göhler, Brian Pearson
Incremental Launched Bridges
2000. Approx. 200 pages with 200 illustrations.
Format: 17 x 24 cm.
Pb approx. Br. 128,-/öS 934,-/sFr 113,-
ISBN 3-433-01793-X
Publication date: April 2000

With the incremental launching method the design is connected to the practise of construction more than with any other method of bridge construction. The book therefore should help to clarify the principles and suppositions for both design and construction. Appropriate detailing of super structure and substructure is illustrated and the calculation of forces on the substructure components during launching is explained by examples.

Ernst & Sohn Verlag für Architektur und technische Wissenschaften GmbH
Bühringstraße 10, D-13086 Berlin, Tel. (030) 470 31-284, Fax (030) 470 31-240
E-mail: mktg@ernst-und-sohn.de   www.ernst-und-sohn.de

LS	Resistance	Symbol	Load case (LC) 1	2	3
1B	Earth resistance	γ_{Ep}	1.40	1.30	1.20
	Bottom pressure resistance (soil failure)	γ_s	1.40	1.30	1.20
	Bottom shear resistance (sliding)	γ_{St}	1.50	1.35	1.20
	Individual piles (pressure and tension, axial)	γ_P	1.40	1.20	1.10
	Pressure-grouted anchor	γ_A	1.10	1.05	1.00
	Ground bolts	γ_N	1.20	1.10	1.05
	Flexible armouring elements	γ_B	1.40	1.30	1.20
1C	Friction value (tan φ)	γ_φ	1.25	1.15	1.10
	Cohesion, drained soil	γ_c	1.60	1.50	1.40
	Shearing strength, undrained soil	γ_{cu}	1.40	1.30	1.20
	Individual piles (pressure and tension, axial)	γ_P	1.40	1.20	1.10
	Pressure-grouted anchor	γ_A	1.30	1.20	1.10
	Ground bolts	γ_N	1.30	1.20	1.10
	Flexible armouring elements	γ_B	1.40	1.30	1.20

Table R 0-4. Partial safety factors for ground resistances (corresponds to table 3 from DIN V 1054-100)

In the ENVs, reductions in the actions can be taken into account with the combination factors. This procedure is impracticable for waterfront structures.

According to the NAD, the trial phase of the ENV should not result in essentially more uneconomic dimensions of the structures than in the previously usual procedure (here: EAU 1990), neither should there be any risk of clearly insufficient dimensions.

During the trial phase this means in practice, that the partial safety factors in the EAU 1996 according to current knowledge and experience can deviate from the partial safety factors stated in the ENV and in DIN V 1054-100.

0.2.2 Ultimate Limit State LS 1: Bearing Capacity

The computed verification of adequate stability is always provided using the method of limit states with the aid of partial safety factors, which are to be used for the characteristic actions and resistances:

- characteristic actions are usually multiplied by the partial safety factors,
- characteristic resistances are divided by the partial safety factors,

in order to obtain the design values. Design values can also be stipulated directly (nominal values).

The safety verification is carried out according to the following fundamental equation:

$$R_d - S_d \geq 0$$

R_d is the design value of resistances, produced as function of the characteristic soil properties and the characteristic resistance of structural elements, divided by the corresponding partial safety factors according to the corresponding calculation method (e.g. soil failure to DIN V 4017-100),

S_d is the design value of actions, produced from the characteristic values of the actions multiplied by the corresponding partial safety factors (e.g. foundation load).

The partial safety factors to be used are to be taken from tables R 0-1 to R 0-4 and from the corresponding construction material and structure standards, in so far as EAU 1996 does not recommend any other partial safety factors.

The characteristic value of the parameters defining resistance is the value of a physical variable used or introduced in the calculations, primarily the friction angle φ' and the cohesion c' respectively c_u. It is the anticipated mean value on the safe side and is represented with the index k as the characteristic value, e.g. c_k. It is stipulated according to DIN V ENV 1997-1 respectively DIN 4020.

The design value (index d) is the value of a scattering physical variable resulting from the characteristic value and the partial safety factor. It is used for the limit state equation, e.g.: characteristic value for cohesion c_k divided by the partial safety factor for cohesion γ_c:

$$c_d = \frac{c_k}{\gamma_c},$$

for the friction factor

$$\tan \varphi_d = \frac{\tan \varphi_k}{\gamma_\varphi}.$$

0.2.3 Limit State LS 2: "Serviceability" (Deformation Safety)

Deformation verification is to be provided for all structures whose function can be impaired or rendered useless by deformation.

These deformations are calculated with the characteristic values of actions and resistances and must be less than the deformations permissible for perfect functioning of the partial structure or whole building. Where applicable, the calculations should include the upper and lower limit values of the characteristic values.

In the case of deformation verification in particular, the progression of influences in time must be taken into account in order to cover critical deformation states during various operating and construction stages.

0.2.4 Geotechnical Categories

The minimum requirements in terms of scope and quality of geotechnical investigations, calculations and supervisory measures are graded in three geotechnical categories in accordance with EC 7, these designating a low (category 1), normal (category 2) and high (category 3) geotechnical risk.

Waterfront structures are to be allocated to category 2, or category 3 in the case of difficult subsoil conditions. A skilled subsoil expert should always be consulted.

0.3 Calculations of Waterfront Structures

The statical calculation is an essential component of the design of waterfront structures. It must provide verification that no failure will occur for any of the failure and deformation mechanisms using the design values of actions and resistances.

In order to provide this calculated verification as simply and directly as possible, the design of the waterfront structure must be uncomplicated and clearly understandable, while fulfilling requirements for economy and simplicity in construction. The less uniform the soil, the greater the need for statically determinate designs to avoid as far as possible any additional stresses from unequal deformations, which cannot be properly taken into account.

The stability verification of a waterfront structure must contain in particular the following:

– use of the structure,
– drawing of the structure with all essential planned dimensions,
– design value of the bottom depth,
– characteristic values of all actions,
– soil layers and corresponding soil characteristic values,

- important free water levels, referred to SKN or MSL or a local gauge zero, together with corresponding groundwater levels (out-of-reach of high water, out-of reach of flooding),
- combinations of actions or load cases,
- required or introduced partial safety factors,
- brief description of the structure, particularly with all information not clearly visible in the drawings,
- intended construction materials and their strength respectively resistance values,
- all data about building schedules and building execution with principle structural states,
- description and justification of the intended verification procedures,
- information about used literature and other calculation aids.

The actual statical calculation with subsequent dimensioning must take into account that in foundation and hydraulic engineering, appropriate soil investigations, shearing parameters, load assumptions, the ascertaining of hydrodynamic influences and unconsolidated states, a favourable bearing system and a realistic computing model are more important than an exaggeratedly accurate numerical calculation.

Furthermore, reference is made to the Additional Technical Contract Conditions ZTV-K 96 for man-made structures [80], ZTV-W (LB 215) for hydraulic structures of concrete and reinforced concrete [118] and ZTV-W (LB 202) for technical processing [164].

1 Subsoil

1.1 Mean Characteristic Soil Properties (R 9)

1.1.1 The values designated with the index k are characteristic values in accordance with DIN 4020. This index is always to be added when using the "characteristic values" as per table R 9-1. These are mean empirical values on the safe side for a larger area of soil, whereby the values determined later on after the assessment of the various soil investigations for the relevant structure can lie both above and below.
The designs are always to be based on these locally measured soil characteristics (see DIN 4020 and R 88, section 1.4).
Loose degree of density must be presumed for natural sands without verification.
Medium-dense degree of density is only to be expected after densification, apart from older geological settlements. The effective shear parameters φ' and c' of a cohesive soil (shear parameters of the drained soil) are ascertained on undisturbed soil samples in the triaxial test according to DIN 18 137, part 1 and 2, and where applicable in the direct shear test.
As per [142], it can be presumed that the friction angle φ' for non-cohesive soils in plane strain state amounts to 9/8 of the friction angle in the triaxial shear test. Accordingly, the triaxial friction angle φ' for dense soils can be increased by up to 10 % for calculation of long waterfront structures, with consent from the test laboratory.
The characteristic values for the friction angle φ'_k and the characteristic values for cohesion c'_k in cohesive soils are selected mean values on the safe side for the area concerned by the building activities, or for part thereof. They apply to calculation of the final stability (consolidated condition = final strength).
The characteristic values of the shear parameters of the undrained soil, φ_u and c_u, consist of the shear parameters which a cohesive soil has, given increase or decrease of the loads, when no consolidation can occur or consolidation is prevented (initial state). φ_u equals 0 for water-saturated soils.

1.2 Layout and Depth of Borings and Penetrometer Tests (R 1)

1.2.1 General

Subsoil investigations are performed in accordance with DIN 4020: Geotechnical investigations for civil engineering purposes. DIN 4014, DIN 4026 and DIN 4128 must also be taken into account for pile structures.
When the subsoil conditions are generally known, subsoil exploratory work frequently begins with an orientation exploration by means of static

No.	1	2	3	4	5		6		7	8	9	10
	Soil type	Soil group as per DIN 18 196[1)]	Peak penetrometer resistance	Degree of density resp. consistency in initial state	Weight density and submerged weight density		Compressibility[2)] (initial load)[3)] $E_S = v_e \sigma_{at}(\sigma/\sigma_{at})^{w_e}$		Shear parameters of the drained soil		Shear parameters of the undrained soil	Permeability factor from ... to k
			q_c		γ	γ'	v_e	w_e	φ'	c'	c_u	
			MN/m²		kN/m³	kN/m³			degrees	kN/m²	kN/m²	m/s
2	Gravel, uniform grains	GE: $U^{4)} = 2$ $U^{4)} = 5$	–	–	16.0 19.0	9.5 10.5	900	0.6 0.4				$2 \cdot 10^{-1}$ to $1 \cdot 10^{-2}$
3	Sandy gravel	GW GI	–	–	21.0 23.0	11.5 13.5	1100	0.7 0.5	35–40°	0		$1 \cdot 10^{-2}$ $1 \cdot 10^{-6}$
4	Sandy gravel with $d < 0.06$ mm < 15 %	GU, GT	–	–	21.0 22.0 23.0	11.5 13.0 13.5	1200	0.7 0.6 0.5	35–40°	0		$1 \cdot 10^{-5}$ to $1 \cdot 10^{-6}$
5	Gravel-sand-fine grain mixture, $d < 0.06$ mm > 15 %	$\overline{GU}, \overline{GT}$	–	–	20.0 21.0 22.5	10.5 12.0 13.0	400	0.9 0.8 0.7	30–35°	0		$1 \cdot 10^{-7}$ to $1 \cdot 10^{-11}$
6	Sand, uniform grains, coarse sand	SE	< 6 6–11 > 11	loose medium-dense dense	17.0 18.0	9.0 10.0	700	0.7 0.6	30–35° 35–37.5°	0		$5 \cdot 10^{-3}$ to $1 \cdot 10^{-4}$
7	Sand, uniform grains, fine sand	SE	as line 7	as line 7	19.0	11.0	700	0.55	37.5–40°			
8			as line 7	as line 7	as line 7		300	0.75 0.65 0.6	as line 7			$1 \cdot 10^{-4}$ to $2 \cdot 10^{-5}$
9a	Sand, gravely, well graduated	SW/SI: $U^{4)} = 6$	as line 7	as line 7	as line 7		600	0.7 0.6 0.55	as line 7			$5 \cdot 10^{-4}$ to $2 \cdot 10^{-5}$
9b	Sand, gravely, well graduated	SW/SI: $U^{4)} = 15$	as line 7	as line 7	18.0 19.5 21.0	10.0 11.5 12.0	600	0.7 0.6 0.55	as line 7			$1 \cdot 10^{-4}$ to $1 \cdot 10^{-5}$
10	Sand $d < 0.006$ mm < 15 %	SU ST	as line 7	as line 7	as line 7		500	0.8 0.7 0.65	as line 7			$2 \cdot 10^{-5}$ to $5 \cdot 10^{-7}$
11	Sand $d < 0.006$ mm > 15 %	$\overline{SU}, \overline{ST}$	–	–	18.0 21.5	9.0 11.0	250	0.9 0.75	30–35°	0		$2 \cdot 10^{-6}$ $1 \cdot 10^{-9}$

12	Coarse silt, difficult to knead	UL	–	soft stiff semi-firm/firm	17.0 18.5 20.0	9.0 10.5 12.0			30–35°	0 2.5 5	5–20 20–50 >50	$1 \cdot 10^{-5}$ to $1 \cdot 10^{-7}$
13	Coarse silt, medium to knead, e.g. harbour basin silt	UM, UA	–	soft stiff semi-firm/firm	17.0 18.5 20.0	9.0 10.0 11.0	110	0.8 0.6	27.5° to 32.5°	5 7.5 10	10–40 40–150 <150	$2 \cdot 10^{-6}$ to $1 \cdot 10^{-7}$
14	Silt or clay, organic, e.g. loam	OU, OT	–	very soft soft stiff	14.0 15.5 17.0	4.0 5.5 7.0	70	0.9 0.7	30–35°	0 2.5 5	<20 20–60 >60	$1 \cdot 10^{-9}$ to $1 \cdot 10^{-11}$
15	Clay, difficult to knead, e.g. boulder clay, tillite[5]	TL	–	soft stiff semi-firm	20.0 21.0 22.0	10.0 11.0 12.0	20 50	1.0 0.85	30–35°	0 5 10	5–20 20–50 >50	$1 \cdot 10^{-7}$ to $2 \cdot 10^{-9}$
16	Clay, medium to knead, e.g. harbour basin clay	TM	–	soft stiff semi-firm	19.0 20.0 21.0	9.0 10.0 11.0	10 30	1.0 0.9	25–30°	10 15 20	10–40 40–150 >150	$5 \cdot 10^{-8}$ to $1 \cdot 10^{-10}$
17	Clay, extremely easy to knead, e.g. Lauenburg clay, Septaria clay	TA	–	soft stiff semi-firm	18.0 19.0 20.0	8.0 9.0 10.0	20	1.0 0.95	22.5 to 27.5°	15 20 25	15–60 60–200 >200	$1 \cdot 10^{-9}$ to $1 \cdot 10^{-11}$
18	Peat	HN, HZ	–	very soft soft stiff semi-firm	10.5 11.0 12.0 13.0	0.5 1.0 2.0 3.0	3 8	1.0 1.0 1.0	–		–	$1 \cdot 10^{-5}$ to $1 \cdot 10^{-8}$
19	Mud, faulschlamm	F	–	very soft soft	12.5 16.0	2.5 6.0	15	1.0 0.9	35°	0	<6 6–60	$1 \cdot 10^{-7}$ $1 \cdot 10^{-9}$

Explanations:
[1] Code letters for the main and secondary components:
G gravel   U silt   O organic components   F mud
S sand   T clay   H peat (humus)   K lime

Codes for characteristic physical soil properties:
Grain size distribution   *Plastic properties*
W well-graded grain size distribution   L low plasticity
E uniform grain size distribution   M medium plasticity
I intermittent graded grain size distribution   A high plasticity

Degree of decomposition of peat:
N not decomposed or scarcely decomposed peat
Z decomposed peat

[2] v_e: stiffness factor, empirical parameter
w_e: empirical parameter
σ_e: load in kN/m^2

[3] σ_{at}: atmospheric pressure (= 100 kN/m^2)
v_e-values for repeated load until 10-times higher,

[4] w_e goes towards 1
U uniformity coefficient

[5] The strengths in line 16 apply to boulder clay

Table R 9-1. Characteristic soil properties (empirical values)

or dynamic penetrometer tests. They enable an initial, rough evaluation of the types of soil. The principal boring program, together with a further penetrometer programme, if necessary, can be stipulated on the basis of the results of such penetrometer tests. It should be borne in mind that borings can also provide information about any possible hindrances in the subsoil. In many cases, static penetrometer tests can frequently be carried out economically and quickly so that it is frequently justified to use these for pre-drafts. The penetrometer resistance of sand thus allows for allocation as per table R 9-1, section 1.1. Reference is also made to DIN 4020, DIN 4094 and [1] and [2] for evaluation of the shear strengths and for the coefficient of compressibility.

1.2.2 Principal Borings

Principal borings, which lie in the shore line, are, in the case of non-anchored walls, sunk to double the height of the difference in ground surface elevation or until encountering a known geological stratum. Regular borehole spacing = 50 m. For expediency, in heavily stratified soils, above all in varved soils, hose core bores according to DIN 4020 are sunk, also for taking extensively undisturbed soil samples of quality grade 2 (DIN 4021), or quality grade 1 in favourable conditions. In some principle borings, piezometers and pore water pressure gauges can be subsequently installed.

1.2.3 Intermediate Borings

Depending on the results of the principal borings or the preferentially applied penetrometer tests, intermediate borings will also be sunk to the depth of the principal borings, or to a depth at which a known, uniform soil stratum is encountered. Typical borehole spacing is again 50 m.

1.2.4 Penetrometer Tests

Penetrometer tests are generally executed according to the plan in fig. R 1-1.

Further penetrometer tests are generally made to the same depth but at least adequately deep into a known, load-bearing geological stratum. Reference is made to DIN 4094 with regard to the instruments and execution of the penetrometer tests, as well as their use. In the case of static sounding devices, where possible local skin friction and the respective pore water pressure should also be measured.

In the case of soft, cohesive soils, vane testing in accordance with DIN 4096 is to be undertaken to determine the undrained shear strength (τ_{FS} resp. c_u).

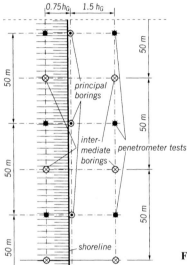

Fig. R 1-1. Example for layout of borings and penetrometer tests for waterfront structures

1.3 Preparation of Reports and Expert Opinions on Subsoil Examinations Under Difficult Conditions (R 150)

Rules for compiling expert opinions on subsoil and foundations are contained in DIN V ENV 1997-1 (EC 7), DIN V 1054-100 and the "Grundbau-Taschenbuch", part 1 [7].

In addition to stability questions, the subsoil expert shall describe the subsoils in question also with respect to their behaviour when excavated, loaded, backfilled and compacted, and draw attention to special properties which can affect construction work (impediments for installation procedures (e.g. R 154, section 1.9), swelling potential, particular weather sensitivity, sensitivity, etc.).

Subsoil examinations are to be carried out to such an extent that the scope for discretion can be narrowed down to a degree which allows for the construction of a safe, economical structure.

1.4 Determination of Undrained Shear Strength c_u (R 88)

1.4.1 Laboratory Tests

The undrained shear strength c_u of water-saturated, cohesive soils is usually determined in the laboratory using axial compression tests (DIN 18 136), unconsolidated and undrained triaxial tests (UU) (DIN 18 137), or laboratory vane shear tests (DIN 4096). For these soils, φ_u equals 0 when they are normally consolidated. Normally consolidated soils which are not saturated usually show a friction angle of the undrained soil of $\varphi_u > 0$.

1.4.2 Field Tests

a) Vane shear test

The vane shear test in accordance with DIN 4096 is suitable for stone-free, soft cohesive soils. It can either be pressed directly into the soil or applied at the borehole bottom. Experience has proven that the vane shear test gives reliable results for normal consolidated as well as for slightly overconsolidated soils.

The measured τ_{FS} value must be reduced as a function of the plasticity coefficient I_p:

$$c_u = \mu \cdot \tau_{FS}.$$

According to [141] respectively DIN 4014, the following μ values are to be used:

I_p	0	30	60	90	120
μ	1.0	0.8	0.65	0.58	0.50

Table R 88-1. μ values

b) Cone penetration test

The following relationships exist for example between the vane shearing strength τ_{FS} and the cone resistance q_c:

in clay: $\tau_{FS} \approx \dfrac{1}{14} \cdot q_c$,

in overconsolidated clay: $\tau_{FS} \approx \dfrac{1}{20} \cdot q_c$,

in soft clay: $\tau_{FS} \approx \dfrac{1}{12} \cdot q_c$,

c) Plate load bearing tests

The c_u values for soil strata close to the surface can be determined in the field by rapidly made load bearing tests using plates of at least 30 cm diameter.

Several tests should be made using plates of various sizes as a control. The respective c_u value for saturated soils is derived from the following equation.

$c_u = 1/6\, p_{failure}$

$p_{failure}$ = mean normal bottom stress at ground failure

If a pronounced failure point does not appear on the load-settlement curve of the test, a settlement = 1/10 of the plate diameter will be assumed as having caused failure.

However, the plate load bearing tests give correct values only when they are made on the ground surface, or on the bottom of a test pit or excavation, provided the area has a diameter of at least three times that of the plate.

d) Pressuremeter tests

The c_u values can also be determined in the field by means of pressuremeter tests. However, laboratories with appropriate experience must be consulted for final determination.

1.5 Shear Parameters of the Drained Soil in the State of Failure or Sliding (R 131)

1.5.1 General

The application of a shearing force under steady normal load on either cohesive or non-cohesive soil will cause shear stresses which depend on the displacement. The shear value in dense, non-cohesive soils, at least firm and generally in overconsolidated cohesive soils will initially rise to a maximum at failure before decreasing to a constant critical limit state (DIN 18 137, part 1) (fig. R 131-1). According to DIN 18 137, part 1,

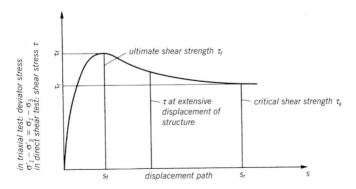

Fig. R 131-1. Stress-displacement diagram for dense, non-cohesive soils, and for soils of minimum firmness or in general for overconsolidated soils (not true-to-scale)

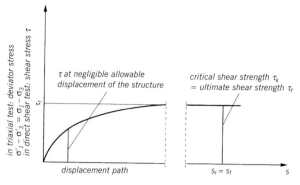

Fig. R 131-2. Stress-displacement diagram for loose, non-cohesive or soft cohesive soils (not true to scale)

the maximum shear stress is called the ultimate shear strength τ_f and the shear stress in the critical limit state is called the critical shear stress τ_k (fig. R 131-1).
In the triaxial test, the same is similarly applicable for the deviator stress $\sigma'_1 - \sigma'_3 = \sigma_1 - \sigma_3$ (fig. R 131-1).
Stress displacement diagrams as shown in fig. R 131-2 result for loose non-cohesive or soft cohesive soils.

1.5.2 Shear Parameters in State of Failure and of Sliding

The friction angle at failure can be determined by constructing the shear line by the use of the ultimate shear strength values; the friction angle at sliding by constructing the shear line by the use of the critical shear strength values (fig. R 131-1 and R 131-2).

1.5.3 Suggestions for Using the Shear Parameters

In calculations according to the EAU, φ'_f and c'_f are to be used for the state of failure and φ'_r and c'_r for the state of sliding.
The shear parameters mentioned here: φ'_f and c'_f, and φ'_r and c'_r, are characteristic values corresponding to section 0.2.1.
For verification of the limit state LS 1, the shear parameters for the state of failure may be used. The shear deformation properties of the soil are included in verification of LS 2. This is to be taken into account for soft cohesive and loose non-cohesive soils in the context of using the shear parameters as per table R 9-1, section 1.1.

1.6 Investigation of the Degree of Density of Non-cohesive Backfill for Waterfront Structures (R 71)

1.6.1 Investigation Methods

A given Proctor density can signify completely different degrees of density, depending on the granulometric composition of the soil.

It is therefore necessary to check the compaction achieved in a backfill of non-cohesive soil by determining its degree of density D in accordance with DIN 18 126 – and not the degree of compaction (DIN 18 127) according to Proctor D_{pr} – by soil samples.

$$D = \frac{\max n - n}{\max n - \min n},$$

$\max n$ = porosity of soil in loosest state, in the dry
$\min n$ = porosity of soil in densest state
n = porosity of soil being tested

The following relationship exists between the degree of density D and the degree of compaction D_{pr} according to Proctor:

$$D = A + B \cdot D_{pr}$$

with

$$A = \frac{\max n - 1}{\max n - \min n},$$

$$B = \frac{1 - n_{pr}}{\max n - \min n},$$

$$D_{pr} = \frac{\rho_d}{\rho_{pr}} = \frac{1 - n}{1 - n_{pr}},$$

ρ_d = dry mass density
ρ_{pr} = mass density with optimum water content according to Proctor test
n_{pr} = porosity at optimum water content in Proctor test

These relations may be used for checking the degree of density achieved in the structure.

The degree of density can also be roughly determined at test depths of > 1 m both above and under water by the use of cone penetration and dynamic penetration tests, if these are calibrated for the conditions under consideration. This also applies to electric and radiometric sounding devices, but these can only be used under water.

1.6.2 Checking

It is necessary and important to check backfill compaction particularly when subject to special requirements, resulting for example from the design of the quay, crane runways with shallow foundations, foundations of other structures, installation of bollards, traffic areas, etc. The same applies when anchors, other structural members or facilities can be endangered by excessive settling or subsidence. Compaction of the backfill layers is to be checked regularly (see section 1.7.4) especially in the immediate vicinity of structures where compaction is generally difficult to attain. The extent of these investigations must be determined as to confirm continuously uniform, adequate compaction. Reference is made to R 175, section 1.7 for hydraulic sand backfill.

Subsidence of non-cohesive soils takes place chiefly when the apparent cohesion in soil placed in a moist state disappears due to rising ground water or flooding of structures. In such cases, highly compacted sand can still subside by approx. 1 % of the layer thickness. The degree of settlement of very loosely deposited, uniformly shaped fine sand can be up to approx. 8 % of the layer thickness (see R 168, section 7.8). Where dynamic stresses or earthquakes occur, the degree of subsidence is still higher.

1.7 Degree of Density of Hydraulically Filled, Non-cohesive Soils (R 175)

1.7.1 General

The usefulness of port areas is determined extensively by the degree of density of the uppermost 1.5 to 2.0 m of the hydraulically filled ground. The latter is contingent above all on the following factors:

- granulometric composition, especially silt content of the fill material,
- type of extraction and further processing of the fill material,
- shaping and installation of the depositing site,
- location and type of the hydraulic flush water discharge.

During hydraulic filling above water, a greater degree of density is generally achieved without additional measures than below water. It is essential thereby to limit the silt content of the fill sand. The content of fine particles with a grain diameter < 0.06 mm should be 10 % at the most. It can be decreased through definite measures while proceeding with the hydraulic filling work, for example through:

- correct scow loading (R 81, section 7.3),
- correct shaping and installation of the hydraulic fill field,
- correct type of the hydraulic filling process.

1.7.2 Influence of the Hydraulic Fill Material

The choice of the hydraulic fill material is generally held within limits by economic and technical demands. Since the granulometric composition and silt content of the material do not remain the same during execution of the work, the degree of density of the hydraulic filling varies as well.

Therefore the sounding resistances indicated in section 1.7.4, table R 175-2, depending on the type of sand, and the corresponding number of blows of the penetration soundings, are only to be regarded as rough guide values.

During hydraulic filling below water, the following approximate degrees of density D are achieved:

Fine sand with different uniformity coefficients and a mean grain size of $d_{50} < 0.15$ mm:

$D = 0.35$ to 0.55

Medium sand with different uniformity coefficients and a mean grain size $d_{50} = 0.25$ to 0.50 m:

$D = 0.15$ to 0.35

1.7.3 Required Degree of Density

The required degree of density D depends on utilisation of the respective port area and should achieve the following values approximately:

Type of utilisation	D	
	Fine sand $d_{50} < 0.15$ mm	Medium sand $d_{50} = 0.25$ to 0.50 mm
Storage areas	0.35–0.45	0.20–0.35
Traffic areas	0.45–0.55	0.25–0.45
Structure areas	0.55–0.75	0.45–0.65

Table R 175-1. Degrees of density D

With these values, stability investigations, bearing capacity calculations and liquefaction questions should also take account of the fact that the angle of internal friction of the fine sand can be smaller than of the medium sand, even when the degree of density D of the fine sand is higher than that of the medium sand.

1.7.4 Checking the Degree of Density

The degree of density achieved in the hydraulic filling can be determined on the ground surface by the customary tests for density determi-

Type of utilisation		Storage areas	Traffic areas	Structure areas
Degree of density D	fine sand	0.35–0.45	0.45–0.55	0.55–0.75
	medium sand	0.15–0.35	0.25–0.45	0.45–0.65
Cone penetration test q_c MN/m²	fine sand	2–5	5–10	10–15
	medium sand	3–6	6–10	> 15
Heavy dynamic penetration test SRS 15, n_{10}	fine sand	2–5	5–10	10–15
	medium sand	3–6	6–15	> 15
Light dynamic penetration test LRS 10, n_{10}	fine sand	6–15	15–30	30–45
	medium sand	9–18	18–45	> 45
Light dynamic penetration test LRS 5, n_{10}	fine sand	4–10	10–20	20–30
	medium sand	6–12	12–30	> 30

Table R 175-2. Correlation between the degree of density D and the cone resistance q_c of the cone penetration test in hydraulically filled sands and dynamic penetration resistances, where the values for fine sand apply above all to non-uniform soils, and the values for medium sand apply above all to uniform soils.

nation, as a rule using equivalent methods, as well as using plate load-bearing tests or radiometric penetration sounding devices. At greater depths, it can be determined through cone penetration or dynamic soundings as per DIN 4094, or with a radiometric depth sounder. The cone penetration sounding device is ideal for the usual grain sizes of hydraulically filled sand. Moreover, in general, the light dynamic penetration device is used for exploration depths of just a few meters, and the heavy dynamic penetration sounder for greater depths and a high degree of density. The values according to table R 175-2 only apply from the critical depth, i.e. from about 1.0 m below the starting point of the sounding device.

1.8 Degree of Density of Dumped, Non-cohesive Soils (R 178)

1.8.1 General

This recommendation is essentially an addition to recommendations R 81, section 7.3 and R 73, section 7.4, as well as R 175, section 1.7. The dumping of non-cohesive soils generally leads to a more or less pronounced segregation of the material.

Dumped, non-cohesive soils can be disturbed by failures in the foundation, slope or embankment. Embankments with a slope of 1 : 5 or less are fairly stable.

1.8.2 Effects on the Achievable Degree of Density

The degree of density of dumped, non-cohesive soils depends primarily on the following factors:

a) Granular composition and silt content of the dumped material. Generally, an non-uniform granular structure produces a higher degree of density than a uniform one. The silt content should not exceed 10 %.

b) Water depth.
The greater the depth of water, the greater the segregation, especially in non-cohesive soils with a uniformity coefficient $U > 5$, which leads to an alteration in grain distribution.

c) Flow velocity in the dumped area.
The greater the flow velocity in the dumped area, the greater the segregation and the more irregular the settlement of the soil.
Reference is made particularly to R 109, section 7.9 with regard to dumping in silty flowing water.

d) Method of dumping.
A somewhat higher degree of density is achieved with barges with slotted flaps than with split hopper barges.

1.8.3 Achievable Degree of Density

The factors listed above explain why the density of dumped, non-cohesive soils can vary greatly. Even with increasing overburden height, there is generally scarcely any change to the degree of density. For small overburden height, the density will generally be only low.

1.9 Assessment of the Subsoil for the Driving of Sheet Piles and Steel Piles (R 154)

1.9.1 General

Construction materials, shape, size, length and driving batter of the sheet piles and steel piles initially play a major role in the assessment of the subsoil for driving sheet piles and piles.

Essential references may be found in:

R 16, section 9.5
R 21, section 8.1.2
R 22, section 8.1.1
R 34, section 8.1.3

R 104, section 8.1.12
R 105, section 8.1.13
R 118, section 8.1.11

In view of the great significance, particular reference is made to the need for the selection of the driving element (material, cross section etc.) to take account not only of statical requirements and economic aspects but also the stresses involved in driving in the respective subsoil.

1.9.2 Report and Expert Opinion on the Subsoil

Soil exploration together with field and laboratory investigations are to furnish information about the following aspects, to enable proper embedment of sheet piles and steel piles:

- Stratification of the subsoil,
- Grain shape,
- Existing inclusions, such as rocks > 63 mm, blocks, old backfill, tree trunks or other obstacles and their depth,
- Shear parameters,
- Porosity and void ratio,
- Weight density above water and submerged weight density,
- Degree of density,
- Compactability of the soil at the embedding of driving elements,
- Cementation of non-cohesive soils, incrustation,
- Preloading and swelling characteristics of cohesive soils,
- Level of the ground water at embedment,
- Artesian ground water in certain strata,
- Water permeability of the soil,
- Degree of saturation in cohesive soils, especially in silt,
- Dynamic and static penetrometer test results and results of standard penetration tests,
- Penetrometer test results with overheavy dynamic penetrometer SR 200.

The shear parameters indicated in the report on the subsoil have only restricted relevance on the behaviour of the subsoil during the driving of sheet piles and steel piles. For example, a rocky lime marl can posses comparatively low shear parameters due to its open seams, but must be viewed as difficult driving soil.

Increasingly more difficult driving must be expected when the number of blows exceeds 30 for 10 cm penetration with the heavy penetrometer (DPH) (DIN 4094) or 50 for 30 cm with the standard penetrometer (SPT). In general it can be presumed that driving elements can be embedded under difficult driving to depths with approx. 80 to 100 blows/10 cm penetration depth. Deeper driving is possible in individual cases.

For more detailed data see [5] and [6]. In the case of non-homogeneous soils and when the penetrometer encounters larger inclusions, the results can deviate strongly and lead to erroneous conclusions.
Static penetration tests in uniform fine and medium sands at a depth of at least 5 m furnish good information on the degree of density.
For further data, see DIN 4094.

Cone resistance MN/m^2	Degree of density	
below 6	loose	($D < 0.3$)
6 to 11	medium dense	($D = 0.3$ to 0.7)
above 11	dense	($D > 0.7$)

Table R 154-1. Degree of density in relation to cone resistance in cone penetration test

Experience shows that cone penetration tests also provide information about:

– the driving capability of the soil,
– the required or possible length of the driving element, and
– the bearing capacity of piles.

1.9.3 Evaluation of Types of Soil as Regards the Embedment Method

1.9.3.1 Driving

Easy driving may be expected in soft, very soft soils, such as bog, peat, slit, clay containing sea-silt, etc. Furthermore, easy driving may generally also be anticipated in loosely deposited medium and coarse sands, as well as in gravels without rock inclusions, unless cemented layers are interspersed.

Medium-difficult driving occurs in medium densely deposited medium and coarse sands, as well as in stiff clay and loam.

Difficult to most difficult driving may be expected in most cases in densely deposited medium and coarse gravels, densely deposited fine-sandy and silty soils, interposed cemented strata, semi-firm and firm clays, rubble and moraine layers, boulder clay, disintegrated and soft to medium-hard rock. Soil moisture or dry soils evoke higher penetration resistance than those under uplift. The same applies to not fully saturated cohesive soils, primarily silts.

1.9.3.2 Vibration

Under successful vibration, the subsoil is placed in a state in which the point resistance is strongly reduced. A higher rate of penetration is achieved during vibration by means of a vibratory pile driver placed at the head of the element. Here it is important for the vibrations to arrive at the point of the driving element for rapid penetration [135].

Gravels and sands of round grain shape, as well as very soft, soft soil strata with low plasticity, are ideal for vibration.

Gravels and sands of angular grain shape or strongly cohesive soils are much less suited. Dry soils are especially critical.

Soils which tend to spring and oscillate reduce the result of the vibration. These include, for example, dry fine sands and stiff marly and clayey soils.

If non-cohesive soil is compacted during vibration, which is possible particularly with close spacing of the driving elements, mainly in the case of the piles of sheet pile walls, the penetration resistance increases so sharply that the rate of penetration may fall to zero. In such cases, the vibration work should be stopped unless auxiliary means as described in section 1.9.3.4 are available.

In non-cohesive soils in particular, the driving of elements by vibration can as a rule cause local settlements, the size and lateral range of which depends on the output of the vibrator, vibration depth and duration and the degree of density of the soil. When the construction work approaches existing building structures, these circumstances should be given particular consideration, and where applicable another embedment method should be used (driving, jacking).

The shape of the embedment element also exerts a remarkable influence. Solid elements or hollow sections with a plug at the pile point are not suitable for vibration, as the peak resistance is too high, unless this is decreased by drilling or similar in the hollow section. Wider sections tend strongly to shimmying, especially when the element is only connected to one clamping shoe.

1.9.3.3 Jacking

In cohesive soils, slim sections can generally be jacked in. In case of non-cohesive soil, jacking may also be employed on the sections if the soil is loosely deposited or properly loosened.

A prerequisite for jacking is that there are no obstacles in the soil, or these are removed before embedding starts.

1.9.3.4 Auxiliary Means

In fine-sandy soils in particular, jetting can be carried out to decrease the amount of energy required or make embedment possible at all. This applies to both driving and vibration.

Further auxiliary means can consist of loosening borings or local soil replacement with advancing large-diameter boreholes or the like.

In rocky soils, planned blasting can make the soil so driveable that the intended depth can be achieved with an appropriate choice of section.

1.9.4 Embedment Gear, Embedment Elements, Embedment Methods

Embedment gear, embedment elements and embedment methods are to be adapted to the respective subsoils.

Slow-stroke free drop hammers or diesel hammers are suited for cohesive and non-cohesive soils. The rapid-stroke hammer and vibration hammer apply gentle stress to the driving element, but are generally only particularly effective in non-cohesive soils with round-shaped grains.

When driving in rocky soil, even after previous disintegration blasting, rapid stroke hammers or heavy pile hammers with small drop heights are to be given preference.

Interruptions in the embedment of the driving element, for example between predriving and re-driving, can facilitate or aggravate continued driving, depending on the type of soil, water saturation and the duration of interruption. As a rule, preliminary tests can make the respective tendency apparent.

The evaluation of the subsoil for the embedment of sheet piles and steel piles presumes that special knowledge and relevant experience is available in the field of embedment. Information on construction sites with comparable conditions, especially regarding the subsoil, can prove to be very useful.

1.9.5 Testing the Embedment Method and the Bearing Capacity Under Difficult Conditions

If there are any doubts in large-scale projects with sheet piles that the statically required depth cannot be achieved producing the duly proper condition of the sheeting, or with steel piles that the rated pile length for absorbing the working load is not adequate, then the intended embedment methods should be tested and test loadings carried out. At least two tests should be carried out per embedment method in order to furnish meaningful information.

Such testing can also be necessary to forecast settlement of the soil and the spread and extent of vibrations caused by the embedment method.

2 Active and Passive Earth Pressures

2.0 General

Verification of the stability of waterfront structures according to these recommendations is usually provided according to DIN V ENV 1997-1. The corresponding earth pressures are calculated according to DIN V 4085-100. Verification of slope failure safety is provided according to DIN V 4084-100, unless simplifications are stated in these recommendations.

The verification of stability of a waterfront structure must indicate that failure does not occur either for limit state 1B (LS 1B) nor for limit state 1C (LS 1C). According to DIN V ENV 1997-1, section 2.4.2, LS 1B is of decisive importance for constructional design. LS 1C is of decisive importance when the strength of a structural element is not at risk but when failure can be caused by failure of the soil in which the structure has its foundations or embedment. LS 1C is as a rule of decisive importance for determining the geometric dimensions of foundations or supportive structures respectively structural parts. But if verification as per LS 1C results in larger cross sections than LS 1B, then LS 1C prevails. According to DIN V ENV 1997-1, LS 1B is of decisive importance in determining the sectional variables and for design the cross sections. This limit state ascertains active and passive earth pressures with characteristic soil properties. Here it is possible to decide whether only part of the characteristic passive earth pressure should be used, for reasons of compatibility of deformations, e.g. according to R 174, section 2.10. The resulting earth pressure figure is then formed and applied to the partial safety factors for influences, table R 0-2. The partial safety factor for the permanent earth pressure is to be taken as 1.35 when the resulting earth pressure is unfavourable, or 1.0 when it is favourable. This results in an increased load in proportion to the increase in foot support force, so that the increase in permanent active earth pressure with the partial safety factor only affects the sectional forces of the support wall and the anchor force, but not the integration depth.

The variable unfavourable earth pressures are multiplied with a partial safety factor of 1.50.

When applying DIN V ENV 1997-1, the definition "characteristic earth pressure" is to include the characteristic water pressure influencing the support structure, together with the stresses involved in the characteristic soil properties and surcharges.

In the case of support structures whose stability depends on the earth resistance in front of the structure, a lowering of the ground surface must be taken into account amounting to 10 % of the wall height on the passive side, respectively 10 % of the height underneath the lowest support in the case of supported walls, but 0.5 m at the most. This does not apply when the draft depth of the port bottom is stipulated according to R 36, section 6.7.

The procedures and principles for determining active and passive earth pressure stated in the following sections apply to determining the earth pressures with both characteristic values and design values of the shear parameters. The shear parameters are therefore featured in diagrams and formulae without any index, unless the characteristic value or design value is referred to explicitly (index k for the characteristic value, index d for the design value).

A recommendation for an appropriate procedure when calculating waterfront walls is featured in section 8.2

2.1 Assumed Cohesion in Cohesive Soils (R 2)

Cohesion in cohesive soils may be considered in active and passive earth pressure calculations when the following conditions are fulfilled:

2.1.1 The soil must be undisturbed. In backfills with cohesive material, the soil must be compacted free of hollow spaces.

2.1.2 The soil must be permanently protected against drying-out and freezing.

2.1.3 The soil must not become pulpy when kneaded.

If the requirements given in sections 2.1.1 and 2.1.2 cannot be met or only fulfilled in part, cohesion should be considered only if justified on the basis of special investigations.

2.2 Assumed Apparent Cohesion in Sand (R 3)

Apparent cohesion c_c (capillary cohesion according to DIN 18 137, part 1) in sand is caused by the surface tension of the pore water. It is lost when the soil is completely wet or dry. As a rule it is not included in the calculation, but referred to as an reserve factor for stability.

2.3 Assumed Angle of Wall Friction and Adhesion (R 4)

Dealt with in section 8.2.4

2.4 Determination of the Active Earth Pressure Using the CULMANN Method (R 171)

2.4.1 Solution with Uniform Soil without Cohesion (fig. R 171-1)

In the CULMANN method, the COULOMB vector polygon (fig. R 171-1) is turned through an angle of $90° - \varphi'$ against the perpendicular, whereby the inherent load G is applied in the slope line. If a parallel to the "active earth pressure line" is applied at the start of the inherent load G, its point of intersection with the relevant sliding line is a point of CULMANN active earth pressure line (fig. R 171-1). When using the characteristic value of φ, one obtains the characteristic value of the earth pressure for verification in the context of LS 1B; when using the design value of φ, one obtains the design value of the earth pressure for verification in the context of LS 1C.

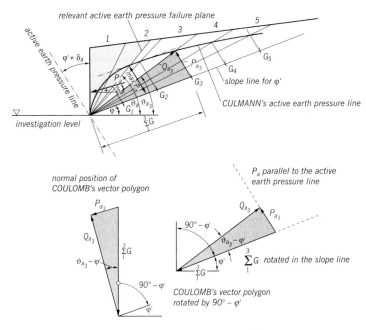

Fig. R 171-1. System sketch for determining the active earth pressure according to CULMANN in uniform soil without cohesion

The distance of this point of the intersection from the slope line measured in direction of the active earth pressure line is the respective active earth pressure for the investigated sliding wedge at the selected angle of wall friction δ_a. This is now repeated for various failure planes. The maximum of CULMANN's active earth pressure line represents the relevant earth pressure sought for. With uniform soil, this can be determined for every optional form of the ground surface and the surcharges prevalent there. Any groundwater table present in any case is also allowed for with corresponding application of the sliding wedge loads with γ or γ'. The same also applies to any other changes of the density, when only φ' and δ_a remain uniform. The earth pressure distribution for a wall is then determined by sections, starting at the top, and preferably plotted by steps.

2.4.2 Solution with Uniform Soil with Cohesion (fig. R 171-2)

In the case with cohesion, besides the soil reaction force Q, the cohesion force $C' = c'_k \cdot l$ also acts in the failure plane with length l. In the COULOMB vector polygon of forces, C' is applied before the inherent load G. In the CULMANN method, C' rotated through an angle of $90° - \varphi'$ is also placed on the slope line of the inherent load G. The parallel to the active earth pressure line is conducted through the initial point of C' and brought to intersection with the appertaining sliding line, through which the now appertaining point of CULMANN's earth pressure line has been found. After investigation of several failure planes, the relevant active earth pressure results as maximum distance of CULMANN's earth pressure line from the connecting line of the initial point of C', measured in direction of the active earth pressure line (fig. R 171-2).

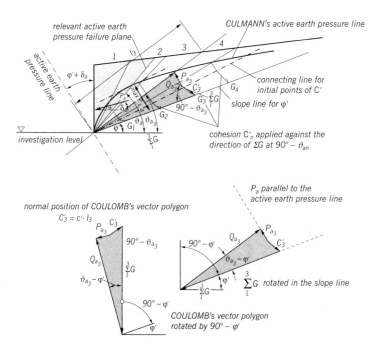

Fig. R 171-2. System sketch for determining the active earth pressure according to CULMANN in uniform soil with cohesion

For greater cohesion, particularly with sloping ground, it must be verified whether straight failure lines are allowed. In the case of great cohesion, curved or bent failure lines frequently cause greater earth pressure loads. See also R 198, section 2.5.

31

2.4.3 Solution with Stratified Soil According to DIN V 4084-100 (fig. R 171-3)

In stratified soil, the active earth pressure is generally determined with a continuous straight failure line, with exceptions in the top strata with high cohesion. Here broken failure lines are to be examined, where the overall failure line must tend to be concave towards the top, as long as curved failure lines are not applicable (see R 198).

For irregular surface, groundwater tables falling away toward the waterfront wall or when additional loads are to be included in calculation of the active earth pressures, the earth pressure may be calculated according to one of the methods stated in DIN V 4084-100.

Fig. R 171-3a shows an example for determining the results of the active earth pressure on a wall with 3 layers and a straight failure line. In this example, the internal earth pressure forces are set horizontally at the lamella limits (see fig. R 171-3b).

The relevant failure line combination is that for which the earth pressure load P_a is the greatest (not examined in fig. R 171-3). A weighted wall friction angle respectively weighted adhesion is to be used for the gradient of the resulting earth pressures at the supporting wall. The weighting can be obtained at the easiest from a layer-by-layer determination of the resulting earth pressures. This produces starting points for the corresponding earth pressure distribution, also from the movement form of the wall and in combination with the data on earth pressure distributions in DIN V 4084-100.

The analytical solution for straight failure lines to fig. R 171-3 is:

$$P_a = \left[\sum_{i=1}^{n}\left(V_i \frac{\sin(\vartheta_a - \varphi_i)}{\cos \varphi_i} - \frac{c_i \cdot b_i}{\cos \vartheta_a}\right)\right] \cdot \frac{\cos \overline{\varphi}}{\cos(\vartheta_a - \overline{\varphi} - \overline{\delta}_a + \alpha)}$$

with

i = serial number of the lamellas
n = quantity of lamellas
V_i = mass forces under uplift including surcharges of the lamellas
ϑ_a = inclination of the failure line to the horizontal
φ_i = friction angle in the failure line in the lamellas i
c_i = cohesion in the lamellas i
b_i = width of the lamellas i
α = inclination of the waterfront wall, defined according to DIN V 4085-100
$\overline{\varphi}$ = mean of the friction angle along the failure line:

$$\overline{\varphi} = \arctan \frac{\sum_{i=1}^{n} V_i \cos \vartheta_a \cdot \tan \varphi_i}{\sum_{i=1}^{n} V_i \cos \vartheta_a}$$

$\bar{\delta}_a$ = mean value of the wall friction angle over wall height. δ_a from $\bar{\delta}_a = \frac{2}{3}\bar{\varphi}$ may be approximated for horizontal strata and comparatively low surcharges. For more precise examinations, the mean must be formed using the earth pressure calculated layer by layer.

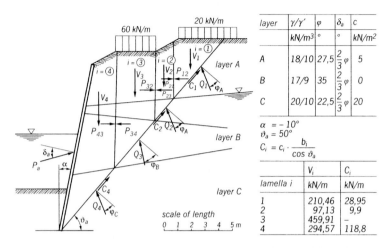

layer	γ/γ' kN/m³	φ °	δ_a °	c kN/m²
A	18/10	27,5	$\frac{2}{3}\varphi$	5
B	17/9	35	$\frac{2}{3}\varphi$	0
C	20/10	22,5	$\frac{2}{3}\varphi$	20

$\alpha = -10°$
$\vartheta_a = 50°$
$C_i = c_i \cdot \dfrac{b_i}{\cos \vartheta_a}$

lamella i	V_i kN/m	C_i kN/m
1	210,46	28,95
2	97,13	9,9
3	459,91	–
4	294,57	118,8

Fig. R 171-3a. Example for determining active earth pressures in stratified soil using the lamella method. Geometry and projection of the forces

Fig. R 171-3b. Vector polygon of the graphical determination of the earth pressure force divided into lamella

Fig. R 171-3. Determining active earth pressure for stratified soil

2.5 Determination of Active Earth Pressure in a Steep, Paved Embankment of a Partially Sloping Bank Construction (R 198)

A case with steep embankment exists if the inclination of the slope β is greater than the characteristic value of friction φ'_k of the exposed soil. The stability of the embankment is only guaranteed if the soil has permanently sufficient cohesion c'_k, and is protected from surface erosion, for example by means of dense turf. A theoretical corresponding case exists when the stability verification is provided with the design values of the shear parameter and φ_d is smaller than the angle β. Calculation of safety from embankment failure can then be made in accordance with DIN V 4084-100, for example.

If cohesion is inadequate to verify the relevant limit states of stability (1B or 1C) for the embankment, the embankment requires fortification, for example with paving, which must be able to support downslope forces to the waterfront wall. The safety of the embankment fortification must be permanently verified. This fortification must be designed in such a way that the resultant force of the active loads lies in the core of the embankment fortification in all cross sections. The active earth pressure for the embankment area down to the top elevation of the backshore for the embankment (active earth pressure reference line fig. R 198-1) can be calculated for non-preponderant cohesion

$$\frac{c'_d}{\gamma_d \cdot h} \leq 0.1$$

according to R 171, section 2.4, whereby the dead weight of the embankment fortification is not taken into consideration.

In doing so, any water pressure difference must be allowed for in addition to the design value P_{ad} of the active earth pressure, with its design value corresponding to the considered limit state. Fig. R 198-1a) illustrates this for impermeable paving. In the case of permeable paving, it is somewhat lower. The load formulae for an embankment fortification are shown in fig. R 198-1b). The reaction force R_d between the embankment fortification and the waterfront wall is produced by the vector polygon in fig. R 198-1c).

The resultant force R_d must be taken into account fully in the calculation of the waterfront wall and its anchoring. The active earth pressure P_{adu} can be determined by analogy in accordance with fig. R 171-3 from the active pressure reference line (imagined stratum limit) downwards in the case:

$$\frac{c'_d}{\gamma_d \cdot h} \leq 0.1$$

In doing so, it should be noted that the design value of the active earth pressure force P_{ado} and the dead weight of the embankment fortification are already contained in the reaction force R_d and are directly subtracted from the waterfront wall including anchoring. Further calculation proceeds according to R 171, section 2.4. By way of approximation, the active earth pressure P_{adu} below the active earth pressure reference line of fig. R 198-1 can also be determined with a wall projecting above the active earth pressure reference line by the fictitious height:

$$\Delta h = \frac{1}{2} \cdot h_B \cdot \left(1 - \frac{\tan \varphi'_d}{\tan \beta}\right).$$

with an embankment simultaneously inclined under the fictitious angle φ'_d (fig. R 198-2).

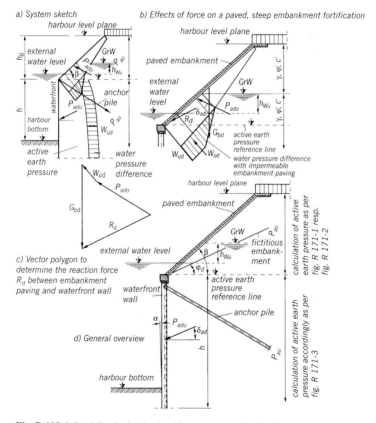

Fig. R 198-1. Partially sloping bank with a steep, paved embankment

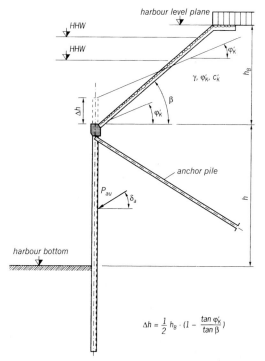

Fig. R 198-2. Approximation formulation for determining P_{au}

In the case of preponderant cohesion

$$\frac{c'_d}{\gamma \cdot h} \leq 0.1$$

calculation with straight failure planes in accordance with fig. R 171-2 resp. R 171-3 results in too small active earth pressure P_{ak}. In this case it is recommended for the active earth pressure to be determined in accordance with DIN 4085-100, both for the section above and section below the active earth pressure reference line with curved or broken failure lines.

2.6 Determination of Active Earth Pressure in Saturated, Non- or Partially Consolidated, Soft Cohesive Soils (R 130)

2.6.1 General

Use of the characteristic value φ_k produces the characteristic value of the active earth pressure for verification in the context of LS 1B; use of the design value φ_d produces the design value of the active earth pressure for verification in the context of LS 1C. The essential factor in determining the magnitude of the active earth pressure is the magnitude of the shear strength which develops at failure in the applicable failure plane. It is expressed by the shear equation for cohesion and friction:

$$\tau = c' + \sigma' \cdot \tan \varphi'.$$

The symbols have the following meaning:
- τ = sliding shear strength (critical shear strength) [kN/m²]
- c' = cohesion of the drained soil in sliding state [kN/m²]
- φ' = angle of slide = friction angle of the drained soil in sliding state [degrees]
- σ' = effective normal stress [kN/m²]

The values for c' and φ' are determined by testing undisturbed soil samples.

2.6.2 Determination of Active Earth Pressure for the Case of a Suddenly Applied Additional Load

2.6.2.1 Calculation with Total Stresses

As $\Delta\sigma = \Delta u$, τ_u becomes $= c_u$.
c_u is the shear strength of undrained, saturated cohesive soils (DIN 18 137, part 1, section 2.7.2).

The shear strength c_u is determined by tests on undisturbed soil samples or by field tests as per R 88, section 1.4. At a vertical wall ($\alpha = 0$) and horizontal ground ($\beta = 0$) with a wall friction angle $\delta_a = 0$, the horizontal component of the entire active earth pressure stress at any considered horizon is:

$$c_{ah} = \sigma \cdot K_{agh} - 2c_u \cdot \sqrt{K_{agh}}.$$

As $\varphi_u = 0$, K_{agh} becomes $= 1$ and thus

$$c_{ah} = \sigma - 2c_u.$$

This simple approach is however not acceptable if the soil is not saturated and if no c_u values have been determined for the project concerned by tests. In such cases, the active earth pressure can be determined according to section 2.6.2.2.

2.6.2.2 Determination of Active Earth Pressure with the Effective Stresses

If c_u is not known, the effective shear stresses as calculated from the general shear equation

$$\tau_k = c' + \sigma' \cdot \tan\varphi'$$

must be used (fig. R 130-1).
The excess pore water stress $\Delta u = \sigma - \sigma'$ does not increase the shear strength. It acts normal to the failure plan in each case and is transmitted in the magnitude existing there, to the retaining wall through the sliding wedge.

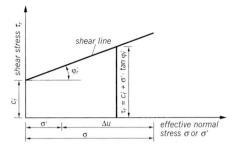

Fig. R 130-1. Depiction of the shear diagram with introduction of the effective shear strength τ for a non- or partially consolidated state with excess pore water stress Δu

The horizontal component of the total active earth pressure distribution thereby occurring is depicted in the example shown in fig. R 130-2 where a uniformly distributed surcharge Δp extends indefinitely towards the land side, i.e. a plane state of stress and deformation exists. In the soft cohesive layer, every active earth pressure co-ordinate is produced as a result of the effective stress σ' and the cohesion c', increased by the neutral overstress $\Delta u = \Delta p$ prevailing there, where Δp signifies the additional loading on the hitherto existing ground surface.

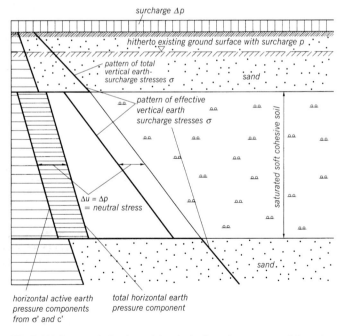

Fig. R 130-2. Example for determining the horizontal component of the active earth pressure distribution for the initial state with shear parameters of the drained soil

Accordingly, in the cohesive layer, not consolidated for Δp the initial state is:

$$e_{ah} = \sigma' \cdot K_{agh} - c' \cdot K_{ac} + \Delta p$$

K_{agh} and K_{ac} to DIN V 4085-100.
If the structure lies partially in groundwater, the pattern of the total vertical surcharge stresses σ is determined taking account of the density under uplift. Differing levels between outer water and groundwater create a pressure difference which must also be taken into account.
In difficult cases, for example with sloping strata, irregular ground surface, irregular surcharges and the like, it is possible to work with a graphical method with variations of the failure surface inclination ϑ_a. In doing so, it must be observed that the surcharges for which the soil in the failure plane section has not been consolidated, create a pore water pressure force which is normal to the failure plane (fig. R 130-3, e.g. layer 2).

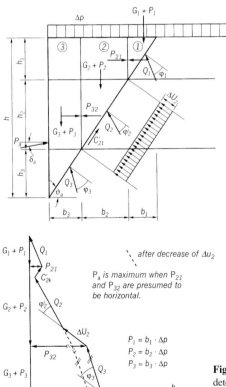

Fig. R 130-3. Example for graphical determination of active earth pressures for a non-consolidated layer under surcharge

$P_1 = b_1 \cdot \Delta p$
$P_2 = b_2 \cdot \Delta p$
$P_3 = b_3 \cdot \Delta p$

$\Delta U_2 = \Delta u_2 \cdot \dfrac{h_2}{\sin \vartheta_a}$

P_a is maximum when P_{21} and P_{32} are presumed to be horizontal.

If the soil is not saturated or if no plane state of deformation exists, or if an appreciable period of time has passed since the load was placed, the additional active earth pressure from Δp is less than Δp. In this case it can only be determined appropriately by more accurate investigations and calculations. Its lower limit value is the additional active earth pressure in the fully consolidated state.

2.7 Effect of Artesian Water Under the Harbour Bottom or River Bed, on Active and Passive Earth Pressures (R 52)

If the harbour bottom or river bed consists of a cohesive surface layer of low permeability, which lies on a ground-water bearing non-cohesive layer, with free low water levels below the simultaneous level of the hydraulic head of the groundwater, the effects of this artesian water must be taken into account.

The artesian level difference will then act on the surface layer from below and thereby reduce its effective weight load. As a result, the passive earth pressure decreases not only in the surface layer, but also in the non-cohesive soil, because of the reduction of the surcharge. At the same time, the safety against slope failure and foundation failure is decreased. Given a level difference $h_{wü}$ between the hydraulic head in the groundwater and the free water level, and the weight density of water is γ_w this results in an excess artesian pressure of $h_{wü} \cdot \gamma_w$.

2.7.1 Influence on the Passive Earth Pressure

If excess artesian pressure acts under a surface layer of thickness d_s and submerged weight density γ', the passive earth pressure is calculated as follows:

2.7.1.1 Case with Preponderant Weight Load on the Surface Layer

$(\gamma' \cdot d_s > h_{wü} \cdot \gamma_w)$ (Fig. R 52-1)

(1) Assuming a linear reduction of the excess artesian pressure through the surface layer, the passive earth pressure introduced into the surface layer by soil friction is calculated according to the reduced density $\gamma_v = \gamma' - h_{wü} \cdot \gamma_w / d_s$.
(2) The passive earth pressure in the surface layer as a result of cohesion is not reduced by the artesian pressure.
(3) $\gamma_v \cdot d_s$ acts as surcharge for the passive earth pressure under the surface layer.

2.7.1.2 Case with Preponderant Artesian Pressure

$(\gamma' \cdot d_s < h_{wü} \cdot \gamma_w)$.

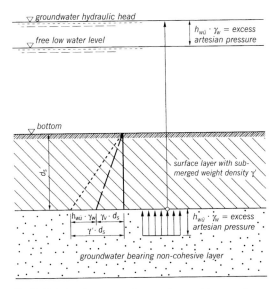

Fig. R 52-1. Artesian pressure in groundwater with preponderant weight load of surface layer

This case can occur in tidal areas, for example. At low water, the surface layer lifts from the non-cohesive subsoil and slowly begins to enter a state of suspension as the groundwater influx takes effect.

At the ensuing high water, it is again pressed down onto its bed. This process is generally not dangerous with thick surface layers. However, if the surface layer is weakened by dredging or the like, boiling-type eruptions of the groundwater can occur, leading to local disturbances in the vicinity of the eruption. It can also lead to easing of the artesian pressure.

Similar conditions can also occur in excavations involved with sheet piling.

The following design principles are then applicable:

(1) Passive earth pressure due to soil friction may not be taken into account in the surface layer.

(2) Passive earth pressure in the surface layer due to cohesion c' of the drained soil may only be used if bottom eruption cannot occur (e.g. as a result of structural countermeasures, such as surcharges).

(3) Passive earth pressure under the surface layer is to be calculated for a load free upper surface, lying in the bottom face of the surface layer. A reduction as a result of flow pressure is only necessary when vertical flow through the non-cohesive strata is possible.

In this context, reference is made to R 114, section 2.9.

2.7.2 Scope
The assumed passive earth pressure dealt with in section 2.7.1 is applicable to sheet pile calculation and to investigations of ground and foundation failures.

2.7.3 Influence on Active Earth Pressure
The influence of artesian pressure on active earth pressure is generally so slight that it can be neglected, particularly in view of the fact that it is on the safe side.

2.7.4 Influence on Excess Water Pressure
In permeable subsoil under the surface layer, the excess water pressure can be taken as zero. In the cohesive surface layer there is a linear transition from the artesian to the groundwater pressure level. In the case of waterfront walls ending in the surface layer, special investigations are required to determine the passive earth pressure at the foot of the wall. This can include determining a flow and potential line network according to R 113, section 4.7.

2.8 Action of Active Earth Pressure and Water Pressure Difference and Construction Hints for Waterfront Structures with Soil Replacement and Fouled or Disturbed Dredge Pit Bottom (R 110)

2.8.1 General
When waterfront structures are constructed with soil replacement as per R 109, section 7.9, the effects of a fouled dredge pit bottom and the non-consolidated conditions there, as well as the inshore dredge pit slope in the existing soft soil, must be accurately taken into account in the design, calculation and dimensioning of the waterfront structure, especially in silt-laden water. In this connection, the time factor is also to be considered as it may have affected the consolidation of the intrusive layer.

2.8.2 Calculation for Determining Active Earth Pressure
Besides the customary calculation of the structure for the improved soil conditions and the ground failure investigations as per DIN V 4084-100, the edge and disturbing influences arising from the failure plane created by the dredging as per fig. R 110-1 must also be taken into account.
In this context, the waterfront structure is to be designed using the partial safety factors of the limit state 1B to DIN V 1054-100. Verification of ground failure is to be based on limit state 1C with the partial safety factors for influence and resistances contained in DIN V 1054-100.
For the active earth pressure P_a acting on the structure down to the dredge pit bottom, the following are the most important factors:

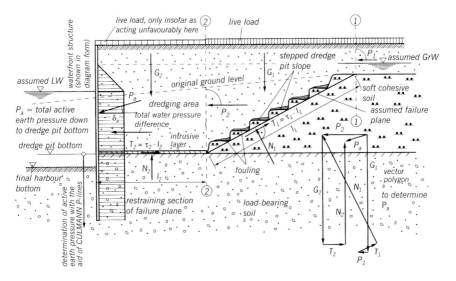

Fig. R 110-1. Determination of active earth pressure P_a on the waterfront structure

(1) Length, and insofar any exists, inclination of the restraining section l_2 of that part of the dredge pit bottom which acts as a failure plane.
(2) Thickness, shear strength τ_2 and effective soil surcharge load of the intrusive layer on l_2.
(3) The possibility of dowelling section l_2 by means of piles and the like.
(4) Thickness of the adjacent soft cohesive soil along the inshore edge of the dredge pit, its shear strength, together with nature and inclination of the dredge pit slope.
(5) Sand surcharge and live load, especially on the dredge pit slope.
(6) Characteristics of the fill soil.

The distribution and action of the active earth pressure P_a down to the dredge pit bottom depend on the deformations on the basis of the statical system and the design of the waterfront structure.

The earth pressure and its distribution below the dredge pit bottom can be determined for example with the aid of CULMANN P-lines. In doing so, the shear strength of section l_2, inclusive of any dowelling, is to be taken into account.

At any stage of construction, including the original dredging of the harbour bottom, or any subsequent deepening of the harbour, the shear stress τ_2 existing at the time in section l_2 of the intrusive layer can be determined for the material then in the layer using the formula:

43

$$\tau = (\sigma - u) \cdot \tan \varphi' \approx \sigma' \cdot \tan 20°,$$

where σ' is the effective vertical surcharge stress transmitted at the place and time of the investigation. The final shear strength after consolidation is then

$$\tau_2 = \sigma'_A \cdot \tan 20°,$$

where σ_A' represents the effective surcharge stress of the investigation area of section l_2 at full consolidation ($u = 0$).

Special calculations are required to consider the effects of dowelling section l_2 to the load bearing soil by piles [11]. If dredging of slopes in soft soil has been properly done in larger steps, the acting failure plane passes through the rear corners of the steps and thus lies in undisturbed soil (fig. R 110-1). Due to the thickness of the soft cohesive soil and its long lasting consolidation, the shear strength is $\tau_d = c_{ud}$, which is the design shear strength of the saturated, undrained soil before excavation. If the soft cohesive soil proves to have layers of different initial shear strength, these different c_d values must be taken into account.

Should the dredge pit slope in soft soil be severely disturbed, excavated in small steps or unusually fouled, the lesser c_{ud} values of the disturbed slip layer before placing the fill soil, which are then to be determined by additional laboratory tests, must be used in calculations instead of the c_{ud} values of the natural soil.

On account of the slow consolidation of the soft cohesive soil underneath the dredge pit slope, it is only worth considering the gradually improving shear strength values of the saturated, undrained soil when the soft soil is drained by closely spaced sand drains. This can also allow for the favourable flattening of the dredge pit slope caused by the induced settlement.

2.8.3 Calculation for Determining the Water Pressure Difference

The total difference in elevation between the theoretical groundwater table at reference line 1-1 (fig. R 110-1) and the lowest attendant calculated outer water level, is to be taken into account. Permanently effective backwater drainage behind water front structures can lower the theoretical groundwater level in the catchment area, and thus reduce differences in levels.

The water pressure difference can be applied in the usual approximated trapezoidal form (fig. R 110-1). However it can be determined more accurately using an equipotential flow net because the actual pore pressures derived from the flow net are available for use in the calculations in the investigation joints (R 113, section 4.7 and R 114, section 2.9).

2.8.4 Hints for the Design of Waterfront Structures

2.8.4.1 Investigations at comparable construction projects have shown that an intrusive layer of up to around 20 cm thick deposited on the restraining section l_2 of the failure plane, even if only partly drained, will have become sufficiently consolidated to support its surcharge, during the construction period, including dredging the harbour bottom. If the intrusive layer is thicker, the value of τ_2 in this layer used in calculations for various construction stages must be the most unfavourable one which exists at the time. This may determine the scheduling of certain operations, such as initial dredging or later deepening of the harbour.

2.8.4.2 Anchoring forces are best transmitted fully into the load-bearing soil through the dredge pit bottom by means of piles or other load-bearing elements. Supportive forces introduced above the dredge pit bottom place additional loads on the slip wedge.

2.8.4.3 Part from its static functions, section l_2 is to be designed with the greatest possible length to accommodate all structural piles. Thus, providing that the sand has been properly placed, the bending stresses in the piles will be kept to a minimum.

2.8.4.4 Brittle fracture must be avoided by using only piles of double killed steel when silt deposits are so heavy that thick intrusive layers of soft cohesive material or very loose layers of sand cannot be avoided in spite of observing all precautions for soil replacement as per R 109, section 7.9, thus possibly causing severe deflection of piling accompanied by stresses in the yield point range, (R 67, section 8.1.6.1 and R 99, section 8.1.18.2).

2.8.4.5 If foundation piles have been driven so as to produce dowelling of section l_2 of the failure plane [11] to verify the stability of the overall system as per DIN V 4084-100, the maximum working stress in the stress verification for these piles due to axial loads, shear or bending must in no case exceed 85 % of the yield point.

The dowelling calculation only allows for such pile deflections which are compatible with the other movements of the structure, i.e. only a few centimetres. Effective dowelling is therefore not possible in the yielding, soft cohesive soil of the dredge pit slope (fig. R 110-1).

Piles with anticipated possible stresses close to the yield point due to settlements of the subsoil or fill soil may not be used for dowelling.

2.8.4.6 In order to prevent the presence of silt deposits and weak soils resulting in an unnecessarily large structure, it is essential for the dredge pit bottom to be as clean as possible, with section l_2 of the failure plane of adequate length and/or a suitably flat dredge pit (see effects in the vector polygon in fig. R 110-1).

If only a relatively thin intrusive layer is expected, the shearing strength of section l_2 can be substantially improved if it is cleaned and then covered with a layer of rubble.

If sufficient time is available, closely spaced sand drains in the soft cohesive soil up to the top of the dredge pit slope can also reduce the load on the structure.

A temporary reduction of the live load on the dredge pit slope, and/or a temporary lowering of the groundwater table to a point behind reference plane 1-1 can also be used to overcome unfavourable initial conditions.

2.8.4.7 If the restraining section l_2 is to be omitted in areas with clay soils without imposing additional loads on the structure, the dredge pit slope may only be about 1 : 4, provided that the soil replacement has been efficiently and carefully carried out in all respects. However, calculated verification is also required as well because the c_u values in the dredge pit slope are the controlling factors together with the effective water pressure difference.

2.9 Effect of Percolating Groundwater on Water Pressure Difference, Active and Passive Earth Pressures (R 114)

2.9.1 General

If water percolates around a structure, the flowing groundwater exerts seepage pressure of differing magnitudes and in differing directions on the masses of the slide wedges which relate to the active and passive earth pressures, this changing the magnitude of these forces.

The active earth pressures for verification of LS 1B to DIN V 1054-100 are determined using the characteristic values of the soil properties, and using the design values for verification of LS 1C to DIN V 1054-100. For this reason, the corresponding indices have been omitted in the following drawings.

The total effect of the groundwater flow on P_a and P_p can be determined with a flow net as described in R 113, section 4.7.7 (fig. R 113-2). All other water pressures acting on the boundary surfaces of the sliding sections of earth are determined, and are taken into account in the COLOUMB vector polygon for the active earth pressure (fig. R 114-1a)) and the passive earth pressure (fig. R 114-1b)). These figures show the forces which are to be included for the general case of a straight failure line. G_a and G_p are the dead loads of the sliding wedges for the saturated soil. W_1 is the corresponding free water surcharge on the sliding wedges, W_2 the water pressure acting directly on the structure in the sliding wedge area, W_3 the water pressure acting in the failure plane, determined according to the flow net (R 113, section 4.7.7, fig. R 113-2). Q_a and Q_p are the soil reactions at angle φ' to the sliding surface normal and P_a resp. P_p the total active and passive earth pressures acting at the wall friction δ_a resp. δ_p to the wall normal, taking into account all the flow effects. In this approach, the water pressure difference to be taken into account is the difference between the water pressures acting directly on the structure from inside and from outside. The result is all the more applicable, the better the flow net coincides with natural conditions.

As the solution according to fig. R 114-1 supplies only the total values of P_a and P_p but not their distribution, separate consideration of the horizontal and vertical seepage pressure effects is recommended in practice. In this case, the horizontal effect is added to the water pressure difference. For this, the water pressure difference relates to the respective failure plane for the active or the passive earth pressures (fig. R 114-2). The vertical seepage pressure effects are added to the vertical soil pressures from the dead weight load of the soil, neglecting uplift, or are approximated by assuming a changed effective soil density. These methods of calculating are dealt with in the following.

2.9.2 Determination of the Theoretical Water Pressure Difference

In order to explain the calculation, the flow net as per R 113, section 4.7.7 fig. R 113-2 is used, and the calculation sequence is shown in fig. R 114-2. Accordingly, the first requirement is to determine the water pressure distribution in the failure planes for the active and passive earth pressures. This is shown in fig. R 114-2 only for the applicable active earth pressure failure plane. It is determined in each case at the points of intersection of the equipotential lines with the pertinent failure plane. The water pressure corresponds in each case to the product of the weight density of the water and the height of the water column, which occurs in the standpipe (fig. R 114-2, right side) used at the investigation point. If the water pressure in the considered points of intersection is plotted horizontally from a vertical reference line, the result is the horizontal projection of the water pressure acting in the investigated failure plane. The horizontal water pressure difference acting on the structure, in which the seepage pressure influences are already contained, results through superposing the horizontal water pressure acting from outside and inside. A good approximation solution can also be found as per section 2.9.3.2. This takes account of the fact that percolation around the wall results in decreased potential on the active side and increased potential on the passive side, which can be theoretically equated with a reduction or increase of the weight density γ_w of the water by $\Delta\gamma_w$.

These $\Delta\gamma_w$ values are calculated with inverted signs with the same formulas as the $\Delta\gamma'$ values according to section 2.9.3.2. They lead to a reduction of the hydrostatic water pressure distribution on the land side

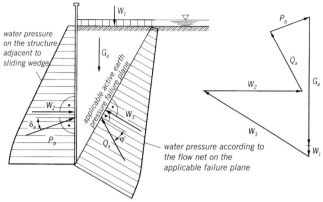

a) Determination of active earth pressure P_a

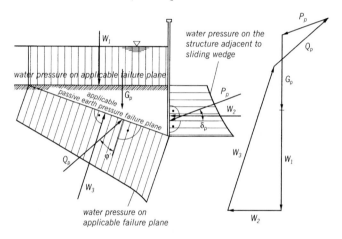

b) Determination of passive earth pressure P_p

Fig. R 114-1. Determination of active and passive earth pressures P_a and P_p, taking into account the effects of flowing groundwater

and a corresponding increase on the water side. The difference of the water pressure altered in this way then gives a highly relevant water pressure difference acting on the percolated sheet piling.

In most cases however, improved calculation of water pressure difference is not necessary if calculated according to R 19, section 4.2, or, if the inflow is mainly horizontal, according to R 65, section 4.3. However, the influence can be considerable for larger water pressure differences.

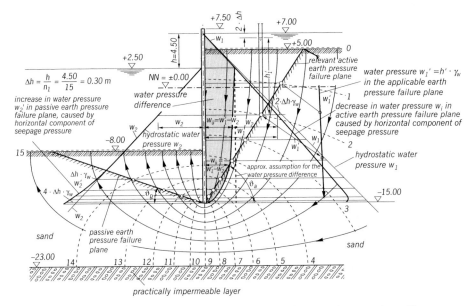

Fig. R 114-2. Determination of the water pressure difference acting on a sheet piling structure, with the flow net as per R 113, section 4.7.7

2.9.3 Determination of the Effects on Active and Passive Earth Pressures when the Flow is Mainly Vertical

2.9.3.1 Calculation Using the Flow Net

The flow net described in R 113, section 4.7.7, fig. 113-2 is used to explain the calculation.

The calculation process is shown in detail in fig. R 114-3. Here it is observed that the drop in the respective standpipe water level per net field takes place because a vertical seepage pressure, corresponding to this drop in head, is transmitted into the section of earth. These effects increase with the downward flow in fig. R 114-3 on the active earth pressure side and decrease with the upward flow on the passive earth pressure side.

In this investigation, in the interest of simplicity, the ordinates are also referred to the points of intersection of the equipotential lines with the applicable failure planes. If Δh is the hydrostatic level difference between adjacent equipotential lines in the flow net, and n is the number of fields beginning with the applicable edge equipotential line, the result on the active earth pressure side, according to Krey is an increase in the horizontal component of the earth pressure stress of

$$\Delta p_{\text{ahn}} = +n \cdot \Delta h \cdot \gamma_w \cdot K_{\text{agh}} \cdot \cos \delta_a$$

49

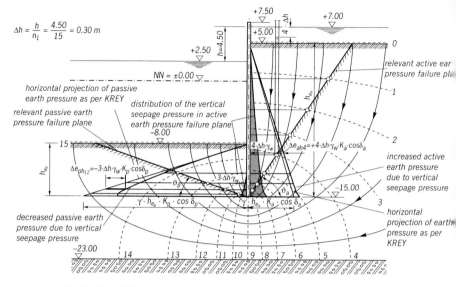

Fig. R 114-3. Influence of the vertical seepage pressure on the active and passive earth pressure with chiefly vertical flow, determined with the flow net for the sheet piling structure, as per R 113, section 4.7.7

for the vertical supplement stress $n \cdot \Delta h \cdot \gamma_w$ at $\gamma_w = 10$ kN/m^3 and on the passive earth pressure side, a corresponding decrease of the horizontal component of the passive earth pressure stress of

$$\Delta p_{\text{phn}} = -n \cdot \Delta h \cdot \gamma_w \cdot K_{\text{pgh}} \cdot \cos \delta_p.$$

In contrast to the decreased water pressure, the result on the active earth pressure side is an increase in the earth pressure; in sandy soil by about one third of the water pressure decrease. As a result of the substantially larger K_p value, the seepage pressure acting from below has a strongly reducing effect on the passive earth pressure. As the greatest part of the reduction takes place in the vicinity of the point of the sheet piling, the influence on the entire structure is as a rule not critical, even here. However, this must be verified theoretically if there are larger water table differences.

The effect of the horizontal component of the seepage pressure on the active and/or passive earth pressures is taken into account by determining the water pressure difference according to section 2.9.2, fig. R 114-2 with the water pressure applied to the active and/or passive earth pressure failure planes.

2.9.3.2 Approximate Solution Based on the Assumption of Changed Effective Density of the Soil on the Active and Passive Earth Pressure Sides

When water percolates around a sheet pile wall, the increase in the active earth pressure or the decrease in the passive earth pressure due to vertical components of the seepage pressures can be approximated by assuming suitable changes in the weight density of the soil.

The increase $\Delta\gamma'$ of the weight density on the active earth pressure side and its decrease on the passive earth pressure side can be determined approximately as per [12] by the following equations:

– on the active earth pressure side:

$$\Delta\gamma' = \frac{0.7 \cdot h_{wü}}{h_w + \sqrt{h_w \cdot d}} \cdot \gamma_w,$$

– on the passive earth pressure side:

$$\Delta\gamma' = -\frac{0.7 \cdot h_{wü}}{d + \sqrt{h_w \cdot d}} \cdot \gamma_w.$$

The symbols herein and in fig. R 114-4 mean:

$h_{wü}$ = hydrostatic difference in level [m]
h_w = vertical depth of soil flowed through on land side of sheet piling down to the pile point [m]
d = driving depth [m]
γ' = submerged weight density of soil [kN/m^3]
γ_w = weight density of water [kN/m^3]

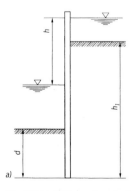
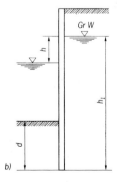

Fig. R 114-4. Dimensions for the approximate determination of the effective specific weight of the soil in front of and behind a sheet piling structure, as changed by seepage pressure

Otherwise, the general remarks made in section 2.9.3.1 are similarly applicable.

2.10 Determination of the Amount of Displacement Required for the Mobilisation of Partial Passive Earth Pressures in Non-cohesive Soils (R 174)

2.10.1 General

As a rule, a considerable amount of displacement is required for the mobilisation of full passive earth pressure. The amount is chiefly contingent on the embedded length of the pressing wall area and on the degree of density of the soil, as well as on the ratio of the wall height h to the wall width b.

On the basis of large-scale model tests [13], [14] and [15], it is possible to estimate the amount of displacement required for given wall loads or the magnitude of the mobilised partial earth pressure at a given displacement.

This recommendation is initially restricted to spatial problems, but can also be applied to the passive earth pressure of walls by way of approximation.

2.10.2 Calculation Formula

According to [13], the correlation

$$P_{p(s)} = w_e \cdot P_p$$

is applicable for a parallel displacement (fig. R 174-1).

The following therein mean:

P_p = passive earth pressure according to DIN V 4085-100 [MN/m]
$P_{p(s)}$ = mobilised partial passive earth pressure contingent on displacement s [MN/m]
w_e = displacement coefficient; $w_e = f(s/s_B)$ [1]
s = displacement [mm]
s_B = required displacement of mobilisation of P_p (failure displacement) [mm]
D = degree of density according to DIN 18 125 [1]
h = wall height or embedded length of the wall [m]
b = wall width [m]

The correlation between the failure displacement s_B and the wall height h together with the degree of density D has been adopted in DIN V 4085-100 as per [13] and [14] for compact walls ($h/b < 3.33$) (fig. R 174-2) to:

$$s_B = 100 \cdot (1 - 0.6 \; D) \cdot \sqrt{h^3}.$$

The following is applicable as per [15] to narrow pressure loaded walls ($h/b \geq 3.33$):

$$s_B = 40 \cdot \frac{1}{1 + 0.5 \; D} \cdot \frac{h^2}{\sqrt{b}}.$$

2.10.3 Method of Calculation

The displacement s related to a pressure loaded wall, or the mobilised partial passive earth pressure $P_{p(s)}$ related to the allowable displacement, may be determined directly with figs. R 174-1 and R 174-2 for $h/b < 3.33$, the case with compact wall occurring most frequently.

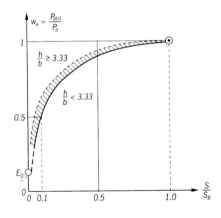

Fig. R 174-1. Passive earth pressure $P_{p(s)}$ contingent on displacement s

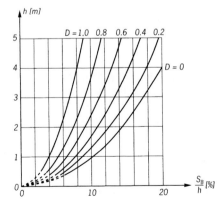

Fig. R 174-2. Failure displacement s_B contingent on the wall height or embedded length h and the degree of density D for $h/b < 3.33$

The value s_B/h and thus the failure deformation s_B result from fig. R 174-2 with the wall height or embedded length h contingent on D. The value s/s_B and thus the displacement s to be expected can then be determined from fig. R 174-1 at given wall compressive force $P_{p(s)}$ or the mobilised partial passive earth pressure

$$P_{p(s)} = w_e \cdot P_p$$

can be ascertained from fig. R 174-1 at given allowable displacement. P_p is to be calculated in accordance with DIN V 4085-100.

2.11 Measures for Increasing the Passive Earth Pressure in Front of Waterfront Structures (R 164)

2.11.1 General

An increase in the passive earth pressure must generally be executed below water. The following measures for example may be considered for this:

(1) Replacement of soft cohesive soil with non-cohesive soil.
(2) Compacting of loosely deposited, non-cohesive soil which may have been encountered, or which may have been placed in case of soil replacement, if need be with added surcharge.
(3) Drainage of soft cohesive soils.
(4) Placing of a fill.
(5) Consolidation of the natural soil.
(6) Combination of measures (1) to (5).

If possible, the respective measure shall be carried out so that later deepenings or forward extensions of the waterfront structure are not obstructed, or only as little as possible.
Reference is made to the details below.

2.11.2 Soil Replacement

When replacing soft cohesive subsoil with non-cohesive material, R 109, section 7.9 is to be observed in so far as the measure itself is concerned. In determining the passive earth pressure, any depositing of intrusive layers in the boundary plane is to be taken into account. The pertinent remarks in R 110, section 2.8 apply accordingly.
The extent of soil replacement in front of the waterfront structure is as a rule determined according to earth statical aspects. Full exploitation of the maximum passive earth pressure which can be achieved with the placed replacement soil means that these measures have to include the entire area of the passive earth pressure slide wedge.

2.11.3 Soil Compaction

Vibroflotation is chiefly used for soil compaction. This is however only possible with non-cohesive soil. The reciprocal spacing of the vibration points (grid widths) depends on the natural subsoil and the desired mean degree of density. The grid width has to be narrower for a greater desired improvement in the prevailing degree of density and with finer grained natural or filled soil. A mean grid width of 1.80 m is an indication. The compaction cores must effectively penetrate the dredge pit bottom through to the dowelling.

The deep compaction shall cover the entire area of the passive earth pressure slide wedge in front of the structure and thereby adequately penetrate the applicable passive earth pressure failure plane emanating from the theoretical sheet piling point. If doubt arises, the adequate extent of chosen compaction measures must be verified by investigating various failure planes, including curved or broken planes.

It should however be considered that compaction with vibroflotation cannot compact a surface zone of 2–3 m depth.

This type of soil compaction can also be used for subsequent strengthening of a waterfront structure. On the other hand, the stability of the structure can be endangered by wide-spread soil liquefaction in particularly loosely deposited, fine-grained, non-cohesive soils and finer sands. Vice versa, in earthquake regions, the danger from soil liquefaction can be eliminated effectively with soil compaction before the structure is erected.

2.11.4 Soil Surcharge

Under special conditions, for example for repairing or modernising an existing waterfront structure, it may be practical to improve the support of the structure by placing a fill with high specific weight and high angle of internal friction. Suitable materials here are steel mill slag, or natural stones. The submerged weight density of the materials differ. Steel mill slag can attain values of $\gamma' \geq 18$ kN/m^3. The characteristic angle of internal friction may be assumed to be $\varphi'_k = 42.5°$.

In the case of existing soft subsoil, care must be taken that the fill material does not sink by suitably grading the grains of the fill material or by installing a filter layer between fill and existing soil.

The material to be placed must be checked constantly for compliance with the conditions and specifications. This applies particularly to weight density.

The remarks in sections 2.11.2 and 2.11.3 apply accordingly.

The soil surcharge can be combined with vertical drainage to accelerate consolidation of soft strata.

2.11.5 Soil Stabilisation

If well permeable, non-cohesive soils are available in the passive earth pressure area (for instance, gravel, gravelly sand or coarse sand), the soil can be stabilised by injecting cement. In less permeable, non-cohesive soils, stabilisation is still possible, using chemicals at increased costs. It must be noted however that in groundwater with greater salt content, stabilisation for example with sodium silicate is not possible because under these conditions, no setting occurs.

Prerequisite for all types of injection is an adequate surcharge, which must be placed in advance when an adequate cover layer is lacking, and then removed again where applicable.

The required dimensions of the stabilisation area can be stipulated in accordance with sections 2.11.2 and 2.11.3. Subsequent placing of driving elements, later dredging for harbour deepening and the like must be taken into consideration. Corresponding adaptations are also possible by adjusting the degree of soil stabilisation.

Core borings and their evaluation are required to verify the success of soil stabilisation measures.

2.11.6 Combined Methods

The methods according to sections 2.11.2 to 2.11.5 can also be combined as required.

2.12 Passive Earth Pressure in Front of Sheet Piles in Soft Cohesive Soils, with Rapid Loading on the Land Side (R 190)

2.12.1 General

The same principles apply to determining passive earth pressure in front of sheet piling with rapidly applied additional loading on the land side as for determining active earth pressure for this load case (R 130, section 2.6.2). The passive earth pressure can be calculated or determined graphically with total stresses when using the shear parameter from the undrained test (c_u, φ_u) or with effective stresses if use is made of the shear parameter from the drained test (c', φ').

Depending on the design case (LS 1B or 1C to DIN V 1054-100), the characteristic values of the soil parameters (LS 1B) or the design values (LS 1C) are to be used.

2.12.2 Procedure in Accordance with DIN V 4085-100

With a rapidly applied additional load on the active side of the sheet piling, a wall bearing pressure is generated which is elevated by the amount ΔP_p. The latter is the wall bearing pressure on the passive earth pressure sliding wedge resulting from rapidly applied additional loading Δp, allowing for the horizontal equilibrium from this additional loading (fig. R 190-1b).

When calculating with c_u and φ_u (generally $\varphi_u = 0$), the calculation in accordance with DIN V 4085-100, section 5.10.2 (fig. R 190-1a) can be performed as a self-contained solution.

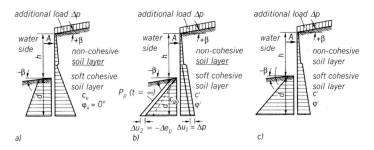

Fig. R 190-1. Active earth pressures on sheet piling in accordance with DIN V 4085-100, in unconsolidated, soft cohesive soil as a result of rapidly applied terrain surcharge of unlimited extent

If only shear parameters from drained tests (c', φ') are known, the passive earth pressure can be calculated in accordance with DIN V 4085-100, section 5.10.3.

Here the total available passive earth pressure $P_{p\,(t=\infty)}$ at a time $t = 0$ (time of application of additional load) is reduced by the uniformly distributed assumed excess pore pressure $\Delta u_2 = -\Delta e_p$. Not before the pore pressure decreases, does the effective passive earth pressure become definitive once more (fig. R 190-1 c)).

The total available passive earth pressure $P_{p\,(t=\infty)}$ is expediently determined at time $t = 0$ in such a way that a uniform distribution of Δu_2 (fig. R 190-1b)) is assumed and the sheet piling is considered as a statically determined system (beam on 2 supports). The magnitude of Δu_2 is determined with the aid of $\Sigma M = 0$ around A. The embedment depth of the wall must initially be assumed here, and subsequently determined iteratively.

The available passive earth pressure $P_{p\,(t=0)}$ with rapidly applied additional loading is then at time $t = 0$ equivalent to the passive earth pressure for the consolidated state, calculated with φ' and c', less $\Delta u_2 \cdot d$:

$$P_{p\,(t=0)} = P_{p\,(t=\infty)} - \Delta u_2 \cdot d.$$

For time $t = \infty$ (following decline of the excess pore pressure) is $\Delta u_2 = 0$ and the passive earth pressure again achieves the value for the fully consolidated state.

2.12.3 Graphical Procedure

If only φ' and c' are known, the passive earth pressure available for absorption of a rapid additional load on the land side ΔP_p can be determined graphically with specification of a failure plan which is inclined at $\delta_p = 0$ and $\alpha = \beta = 0$ under $\vartheta_p = 45° - \varphi'/2$ (fig. R 190-2).

In so doing, we start from the basic premise that the partial passive earth pressure P_{p0} has already been mobilised from the preceding loading, but, in total, the maximum passive earth pressure $P_{p\,(t=\infty)}$ is available from effective parameters φ' and c' following consolidation. At the time of application of the rapid additional loading, as a result of which the pore pressure U_2 (fig. R 190-2) is generated, sufficient reserve ΔP_p must always exist between P_{p0} and P_p. P_{p0} is the point supporting force to balance the earth- and water pressure before the rapid loading, and must be determined from the equilibrium condition ΣH or $\Sigma M = 0$ around the anchor point.

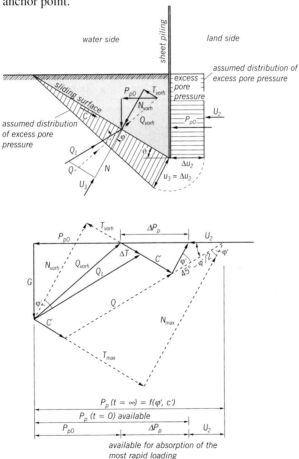

Fig. R 190-2. Determination of available passive earth pressure in unconsolidated, soft, cohesive soil as a result of rapidly applied additional loading on the land side of the sheet piling

Initially, the available sliding surface force Q_{avail} resulting on the failure plane is determined for the assumption $\delta_p = 0$ from the weight of the sliding body G and the passive earth pressure P_{p0} in the sliding surface already used (fig. R 190-2). With the direction (φ' to the normal) of the mobilisable sliding surface force Q_1, the mobilisable friction force ΔT is generated. The rapid additional loading can now only be absorbed by the cohesion force C and ΔT, whereby due to equilibrium the pore pressure U_3 is generated in such a magnitude that the vector of U_3 in the vector polygon ends with the influence line of the passive earth pressure. From case to case, this results in a non-linear distribution of the pore water pressure difference in the passive earth resistance failure line. In the vector polygon this results in the available reserve force ΔP_p for absorbing the rapid loading.

As shown in fig. R 190-2, virtually the same result is achieved with this graphical procedure as with the procedure in accordance with DIN V 4085-100: the maximum passive earth pressure P_p from the shear parameters φ' and c' is reduced by the amount U_2. For $\varphi' = 30°$, U_3 is exactly = U_2. There are slight differences when φ' deviates from this.

2.12.4 Concluding Remarks

Since the calculation of passive earth pressure with effective shear parameters is affected by the uncertainty as to the calculation of excess pore pressure, preference should be given to calculation with shear parameters c_u and φ_u determined from the "undrained tests".

2.13 Effects of Earthquakes on the Design and Dimensioning of Waterfront Structures (R 124)

2.13.1 General

2.13.1.1 If a waterfront structure is to be built in an earthquake region, careful provisions must be made for the effects of a possible earthquake in the area where the site of work is located (DIN V ENV 1998-1).
In practically all countries where earthquakes are to be expected, there are various standards, directives and recommendations specially for buildings in which the prescribed requirements for design and calculation are given in more or less detail. DIN 4149 and [16] are referred to in this matter with regard to the Federal Republic of Germany.

2.13.1.2 The intensity of the earthquakes to be expected in the various areas is generally expressed in the above mentioned publications, by the magnitude of the horizontal seismic acceleration a_h, which occurs during a quake. A possible simultaneously vertical acceleration a_v is generally negligibly low, compared to the acceleration due to gravity g.

2.13.1.3 The acceleration a_h affects not only the structure as such but also the acting active earth pressure, the possible passive earth pressure, the safety against foundation, ground and slope failure and in some cases also the shear strength of the earth masses surrounding the foundation. Under unfavourable circumstances, this shear strength may temporarily disappear completely.

2.13.1.4 The requirements for accuracy of the calculations are correspondingly more stringent when earthquake damage can endanger human lives or cause destruction of supply facilities or the like which are of importance to the population.

2.13.1.5 The static-kinetic problems occurring during an earthquake are taken into account in the actual structure as a rule in such a manner, that the additonal horizontal forces

$$\Delta H = \pm k_h \cdot V,$$

which each act at the centre of gravity of the accelerated masses, are used as being applied simultaneously with the other loads.

The symbols herein mean:

k_h = a_h/g = seismic coefficient = ratio of horizontal earthquake acceleration to acceleration due to gravity g
V = weight load of the considered structural member or sliding wedge including pore water

The magnitude of k_h depends on the intensity of the earthquake, the distance from the epicentre and the foundation soil. The first two factors named are taken into account in most countries by dividing the endangered areas into earthquake zones which have the same k_h value (see also DIN 4149 and [16]). When doubt arises, as far as possible agreement on the magnitude of k_h to be used is to be reached by consulting an experience seismic expert together with the engineer, the building promoter and the construction supervisory agency.

2.13.1.6 In the case of high, slender structures with danger of resonance, i.e. when the natural oscillation period and the earthquake period are nearly the same, the dynamic effects of the earthquake must also be taken into account in the calculations. This however is generally not required in the case of waterfront structures.

2.13.1.7 The principal requirements which must be met in the design of an earthquake-proof waterfront structure is that the additional horizontal seismic forces will be safely absorbed, even with the attendant reduction in passive earth pressure.

2.13.2 Effects of Earthquakes on the Subsoil

2.13.2.1 Waterfront structures in earthquake regions must also take special account of the conditions in deeper subsoil. Thus for example, it should be understood that earthquake vibrations are most severe where loose, relatively thin deposits rest on solid rock (see [7]).

2.13.2.2 The most sustained effects of an earthquake make themselves felt when the subsoil, especially the foundation soil, is liquefied by the earthquake, that is to say, when it loses most, or even all, of its shear strength. This takes place when loosely deposited, fine-grained, non- or lightly cohesive, saturated, only slightly permeable soil (e.g. loose find sand or coarse silt) is transformed into a denser deposit (settlement flow, liquefaction). This condition continues until the resulting superfluous pore water has disappeared. The lower the overburden pressure at the depth in question and the greater the intensity and duration of the shocks, the sooner will liquefaction occur.

2.13.2.3 When the risk of liquefaction cannot be definitely ruled out, it is advisable to consult experienced experts in this particular field.

2.13.2.4 Soil strata tending to liquefaction in the area of planned waterfront structures in earthquake regions, should be thoroughly compacted before starting construction work on the waterfront structure.

2.13.2.5 Cohesive soils do not tend to liquefy.

2.13.3 Statical Determination of the Effects of Earthquakes on Active and Passive Earth Pressures

2.13.3.1 The influence of earthquakes on active and passive earth pressure is also generally determined according to COULOMB, however, the additional forces ΔH (section 2.13.1.5) created by earthquakes must receive added consideration. Moreover, the weight loads of the earth wedges may no longer be applied vertically but rather at a definite angle deviating from the vertical. This is best taken into consideration by referring the inclination of the active and passive earth pressure reference plane and the plane of the ground surface to the new force direction [7]. This results in imaginary changes in the inclination of the reference plane ($\pm \Delta\alpha$) and of the ground surface ($\pm \Delta\beta$).

$$k_h = \tan \Delta\alpha \text{ resp.} = \tan \Delta\beta$$

The active and/or passive earth pressures are then calculated on the basis of the imaginary system rotated by the angle $\Delta\alpha$ or $\Delta\beta$ (reference plane and ground surface).

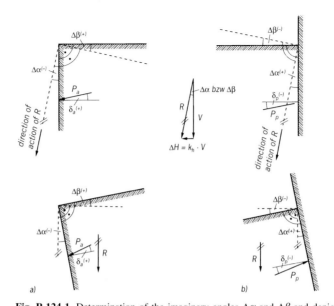

Fig. R 124-1. Determination of the imaginary angles $\Delta\alpha$ and $\Delta\beta$ and depiction of the systems rotated by angle $\Delta\alpha$ or $\Delta\beta$ (signs as per KREY)
a) for calculating the active earth pressure, b) for calculating the passive earth pressure

An equivalent procedure is to calculate on the assumption that the inclination of the wall is $\alpha \pm \Delta\alpha$ and that of the ground surface $\beta \pm \Delta\beta$ (fig. R 124-1).

2.13.3.2 In determining the active earth pressure below the water table, it must be recognised that the mass of the soil and the mass of the water enclosed in the soil pores are accelerated, but that the reduction of the submerged weight density of the soil remains as it is and that the pore water flows downwards of its own accord. In order to take this into account, a larger seismic coefficient – the so-called apparent seismic coefficient k'_h – is used for practical purposes for calculations in the area below the groundwater table.

In the section shown in fig. R 124-2,

$$\Sigma p_v = p + h_1 \cdot \gamma_1 + h_2 \cdot \gamma'_2 \text{ and}$$
$$\Sigma p_h = k_h \cdot [p + h_1 \cdot \gamma_1 + h_2 \cdot (\gamma'_2 + \gamma_w)] = k \Sigma p_v + h_h \cdot h_2 \cdot \gamma_w.$$

The apparent seismic coefficient for determining active earth pressure below the water table thus results in:

$$k'_h = \frac{\Sigma p_h}{\Sigma p_v} = \frac{p + h_1 \cdot \gamma_1 + h_2 \cdot (\gamma'_2 + \gamma_w)}{p + h_1 \cdot \gamma_1 + h_2 \cdot \gamma'_2} k_h.$$

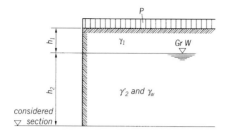

Fig. R 124-2. Sketch showing the arrangement for calculating the value of k'_h

A similar procedure can be followed for passive earth pressure. For the special case where the groundwater level is at the ground surface and where there is no ground surcharge, the result for the active earth pressure side with $\gamma_w = 10$ kN/m³ is:

$$k'_h = \frac{\gamma' + 10}{\gamma'} \cdot k_h = \frac{\gamma_r}{\gamma_r - 10} \cdot k_h \cong 2 k_h.$$

The symbols herein mean:

γ' = submerged weight density of soil
γ_r = weight density of the saturated soil

For the sake of simplicity, the unfavourable value of k'_h determined for the active earth pressure side in this manner is customarily also used as the basis for calculating in other cases, even when the groundwater table is lower and there are existing live loads.

2.13.3.3 In calculating the active earth pressure coefficient K_{ah}, which is determined by k_h and k'_h, a sudden change theoretically takes place in the active earth pressure at the elevation of the groundwater table, as shown in fig. R 124-3. This comes about because the value of k'_h depends on the ratio of the horizontal seismic forces to the existing acting vertical

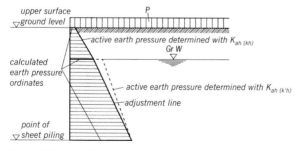

Fig. R 124-3. Simplified disposition of the active earth pressure

force, and hence changes with the depth. If this change in the value of k'_h and the consequent change in K_{ah} can be disregarded, the active earth pressure can be applied in simplified form as shown in fig. R 124-3.

2.13.3.4 In difficult cases for which table figures are not available for calculating the active and passive earth pressures, it is possible to determine the influences of both the horizontal and possibly vertical earthquake accelerations on active and passive earth pressures by an extension of the CULMANN method. The forces resulting from earthquake accelerations and correspondingly acting on the investigated wedges must then also be taken into account in the vector polygons. Such a sophisticated calculation is already recommended for larger, horizontal accelerations, above all if the soil partly lies below the groundwater table.

2.13.4 **Water Pressure Difference**
In the case of waterfront structures in earthquake regions, the application of water pressure difference may be approximated as in the normal case, i.e. according to R 19, section 4.2 and R 65, section 4.3, because the statical-kinetic effects of the earthquake on the pore water have already been taken into account in determining active earth pressure with the apparent vibration coefficient k'_h as per section 2.13.3.2. It must be observed in the earthquake case however, that the applicable active earth pressure failure plane is inclined at a smaller angle to the horizontal than in the normal case. For this reason, increased water pressure difference may act on the failure plane.

2.13.5 **Live Loads**

2.13.5.1 Because the simultaneous occurrence of earthquake, full live load and full wind load is improbable, it suffices to combine the increased loads caused by the earthquake with only half of the live load and half of the wind load (see also DIN 4149, explanations and [16]). The crane wheel loads caused by wind and the component of the line pull from wind should therefore be correspondingly reduced. The loads due to the travel and rotating movement of cranes need not be included in the influence of the earthquake.

2.13.5.2 However, those loads which in all probability remain constant for a longer period of time, such as loads from filled tanks or silos and from bulk cargo storage, may not be reduced.

2.13.6 **Safety Factors**
Earthquake forces may be taken into consideration according to DIN V ENV 1991-1 as extraordinary design situation with the corresponding

partial safety factors for influences and resistances, taking account of cases B and C (LS 1B and 1C) according to DIN V ENV 1997-1. According to DIN V 1054-100, load case 3 applies.

2.13.7 Reference to Consideration of Seismic Influences in Various Waterfront Structures

Taking into account the foregoing remarks and other recommendations by the EAU, waterfront structures can also be calculated and designed systematically and with adequate stability for earthquake regions. Supplementary references to definite types of construction such as sheet pile structures (R 125, section 8.2.18), waterfront structures in blockwork construction (R 126, section 10.8) and walls on pile foundations (R 127, section 11.8) are given in the mentioned recommendations.

Experience gained from the earthquake in 1995 in Japan is dealt with in [198].

3 Overall Stability, Foundation Failure and Sliding

3.1 Relevant Standards

The following standards are recommended for verification of safety from sliding, foundation failure and to provide overall stability:

sliding: DIN V 1054-100
foundation failure: DIN V 4017-100
slope failure: DIN V 4084-100
earth pressure: DIN V 4085-100

The titles of the standards are stated in appendix I, 3.1.
Standard DIN V 4084-100 applies for verification of stability of anchoring for the lower failure plane and safety against eruption of the anchoring soil. Simplifications are contained in R 10, section 8.4.9. The safety factors are determined according to section 0.2.

3.2 Safety Against Failure by Heave (R 115)

Failure by heave occurs when a body of earth in front of a structure is raised by the upward flow force of the groundwater. The passive earth pressure is lost. This occurs when the vertical component W_{st} of the flow force is equal to or greater than the dead load G_{br} of the body of earth under uplift, which lies between the structure and the assumed failure surface used as a basis for the test calculation.

All heaving failure surfaces to be considered in the analysis extend progressively outward from the base of the structure. The assumed surface determined by test calculations with the smallest safety factor is the basis for judging safety.

The required safety factor at wide excavation with large length expansion is:

$$G_{Br} \geq 1.5 \cdot W_{St}.$$

W_{St} can be calculated using a flow net as per R 113, section 4.7.7, fig. R 113-2 or R 113, section 4.7.5. W_{St} is the product from the volume of the hydraulic ground failure body times the weight density of the water γ_w and times the mean flow gradient measured in this body in the vertical.

The same value for W_{St} may be obtained as shown in fig. R 115-1. Here the hydraulic head difference at the considered site still not yet decreased compared to the lower water table $n \cdot \Delta h$ is multplied by γ_w as ideal pressure surface. W_{St} is then the vertical partial force of the contents of this pressure surface.

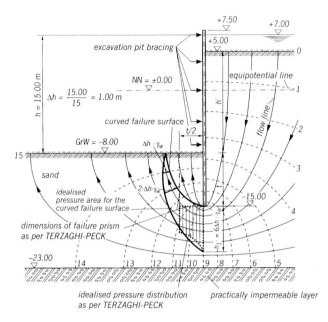

Fig. R 115-1. Safety against failure by heave in a dredge pit bottom, determined with the flow net as per R 113, section 4.7.7

The further evaluation is indicated in fig. R 115-1 for an example of a waterfront structure of larger length expansion. The safety factor is given there for both an optionally selected curved failure plane as well as for the assumption as per TERZAGHI-PECK [17, p. 241]. The latter works on the premise of a rectangular failure body whose width is equal to half the embedment depth t of the sheeting. Approximately the above equation results in

$$\gamma' \geq 1{,}5 \cdot \gamma_w \cdot i\,.$$

With $\gamma' \approx \gamma_w$ and $i = \dfrac{h_r}{t}$,

the tolerable i is zul $I \leq 0.67$.

The effective potential h_r at the point of the sheet piling can be ascertained with a flow net as described in R 113, section 4.7.7, in section 4.7.5 or for the sheet piling structure with a primarily vertical flow of water around it according to the formula by KASTNER [19], expanded by SCHULTZE:

67

$$h_r = \frac{h}{1 + \sqrt[3]{\frac{h'}{t}} + 1}$$

The symbols herein mean:

h_r = difference between the standpipe water level at the point of the sheet piling and the underwater table level [m]
h' = depth of soil flow through on land side of the sheet piling, down to the bottom of the water course [m]
t = mbedment depth of the sheet piling [m]

In contrast to the foregoing formula, a calculation of h_r from the development of the path of flow along the sheet piling furnishes a result of questionable accuracy, e.g. in fig. R 115-2 with an error ≈ 2 · Δh.
According to fig. R 115-2, the cause of this error lies in the non-uniform lowering of the hydraulic head along the sheet piling, as can be seen in the distribution of potential in fig. R 115-1.
The danger of an impending failure by heave in an excavation pit is indicated by a pronounced bulge in front of the base of the sheet piling. If this occurs, the excavation pit should immediately be at least partially flooded. Subsequently, corrective measures can be initiated, by means suggested in R 116, section 3.3, fifth paragraph and following, or, if considered preferably, by placing local earth surcharge in the excavation pit or by suitable means for relieving the flow pressure by lowering the groundwater level.

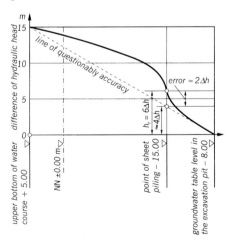

Fig. R 115-2. Lowering of the hydraulic head along sheet piling according to the flow net as per fig. R 115-1

3.3 Foundation Failure Due to Erosion; its Occurrence and its Prevention (R 116)

The danger of foundation failure due to erosion exists when soil in a river or excavation bottom or similar starts to wash out. This is initiated when the exit gradient of the water flowing around the waterfront structure is capable of moving soil particles upwards. Caused by the retrogressive erosion, a canal in the shape of a pipe (piping) forms in the ground in connection with the hydraulic gradient which constantly increases there. This canal develops along the flow lines in the direction of the elevated water level. If it extends to reach a free elevated pool of water, water bursts through the initially small canal and erodes its walls. After a short period of time, large quantities of soil are washed out and foundation failure by erosion occurs and can lead to collapse of the structure being percolated around.

This erosion foundation failure depends on the presence of loose zones in the foot support area (e.g. old funnels, inadequately treated bore holes) or loose zones in direct contact with the wall and at relatively high hydraulic gradients, i.e. low safety from failure by heave.

Slope erosion failure occurs in a similar manner.

The occurrence of a foundation failure due to erosion in homogeneous non-cohesive soil is depicted in fig. R 116-1 with the aid of the flow diagram according to R 113, section 4.7.7, fig. R 113-2. A possible foundation failure due to erosion is first indicated on the lower water table or on the excavation bottom by swelling of the ground and the ejection of soil particles. At this stage, the impending failure can still be brought under control be depositing adequately thick graded or mixed gravel filters to prevent further soil from being washed out.

If an advanced stage has already been reached, and the danger of a breakthrough to the elevated pool is acute, immediate equalisation of the water tables must be brought about by raising weir gates, flooding the excavation or similar measures. Only after this has been accomplished is it possible to undertake permanent remedial measures, such as placing a sturdy filter on the low side, grouting the eroded pipes from the lower end, deep vibration of the soil in the endangered area, lowering the groundwater table, or placing a dense covering on the elevated pool bottom to extend well beyond the endangered area.

The danger of foundation failure by erosion cannot generally be determined by calculation, nor is detailed statistical information available because of the diversity of the designs and marginal conditions. Other conditions being the same, the danger of failure increases in proportion to the increase in head difference between the upper water and the lower groundwater, as well as with an increase in the presence of loose, find grained, non-cohesive or weakly cohesive material in the subsoil, particu-

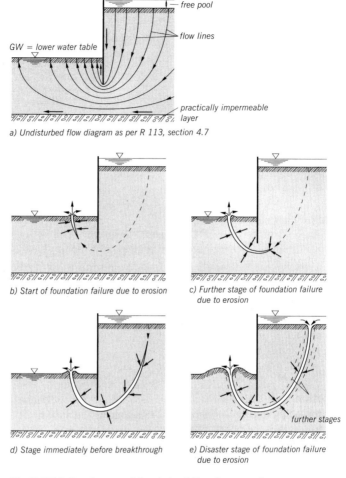

Fig. R 116-1. Development of foundation failure due to erosion

larly when there are embedded sand lenses or veins. In strongly cohesive soil, there is generally no danger of foundation failure due to erosion.

Even if there is no free elevated pool, a piping effect may still begin from the low water table side. In general however, a disaster stage is not reached because the eroded pipe shaped cavity dies out in the subsoil or because there is insufficient free water available for a disastrous erosion effect, unless by chance an extraordinarily capacious water-bearing stratum is encountered.

If conditions prevail which appear to make foundation failure by erosion possible, precautions for its prevention are to be planned from the very beginning at the site in order to take appropriate countermeasures immediately if this should prove necessary.

The minimum embedment depth to prevent foundation failure due to erosion can be determined according to R 113, section 4.7.7.

3.4 Verification of Overall Stability of Structures on Elevated Pile-founded Structures (R 170)

3.4.1 General

Verification of overall stability of structures on elevated pile-founded structures can be carried out using DIN V 4084-100.

3.4.2 Data

The following must be available for verification:

(1) Data on the design and dimensions of the pile-founded structure, applicable loads and internal forces, most unfavourable water levels and live loads
(2) Soil mechanics characteristic of the subsoil, especially the weight densities (γ, γ') and the shear parameters (φ', c') of the encountered soil types, which are also to be determined for the initial state (φ_u, c_u) in cohesive soils. If required, time-dependent influences of surcharges or excavations on the shear strengths are also to be taken into consideration for cohesive soils.

3.4.3 Application of the Loads (Influences)

The following loads in most unfavourable combination are to be taken into account:

(1) Loads in or on the sliding body, especially individual loads from live loads or other external loads.
(2) Dead weight of the sliding body and the soil surcharge thereon, taking into account the groundwater table (weight densities γ and γ').
(3) Water pressure loads on the sliding surfaces of the failure mechanisms being examined from pore water pressure, determined as per R 113, section 4.7.

3.4.4 Application of the Resistances

(1) The resistances (axial and dowelling forces) of the rows of piles within the pile bent plane are distributed along the equivalent length according to fig. R 170-1. They are calculated with the least favourable values as per DIN V 4084-100, section 6.
(2) The passive earth pressure may be calculated with the kinematically compatible wall friction angle δ_p. In special cases the question arises

whether the displacement required for mobilising the passive earth structure does not already impair the utilisation of the structure. But this is generally not necessary when using rating values for the shear parameters. Exceptions are stated in DIN V 4084-100, section 10. In addition, the supporting section of soil must be permanently on hand. Any possible future harbour dredging must therefore be taken into account in advance.

(3) The favourable effect of deeper embedded stabilising walls or aprons may be taken into account.

(4) Effects of earthquake action are to be covered in accordance with R 124, section 2.13. For verification with composed failure mechanisms with straight failure lines as per DIN V 4084-100, the earthquake force may be calculated as horizontal mass force.

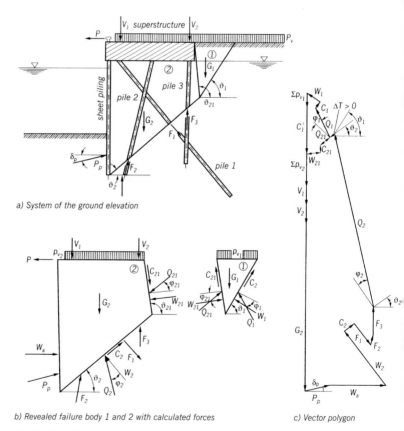

a) System of the ground elevation

b) Revealed failure body 1 and 2 with calculated forces

c) Vector polygon

Fig. R 170-1. Sketch for determining overall stability of an elevated pile-founded structure

4 Water Levels, Water Pressure, Drainage

The loads determined in this section are allocated to the load cases as per R 18, section 5.4. Section 5.4.4 is to be observed regarding the partial safety factors.

4.1 Mean Groundwater Level (R 58)

The allocation of the relevant hydrostatic load situation resulting from changing free and groundwater tables requires an analysis of the geological and hydrological conditions of the area concerned. Where available, long years of observations are to be evaluated to obtain knowledge about the sinking velocity of the water tables. The groundwater table behind a waterfront structure is governed decisively by the soil strata and the design of the waterfront structure. In tidal areas, the groundwater head for permeable soils follows the tide in a more or less attenuated manner. The presumption of a theoretical groundwater head of 0.3 m above $T\,\frac{1}{2}\,W$ in tidal areas respectively MW in non-tidal areas only applies to preliminary drafts.

In areas with stronger groundwater flow, the mean water level is higher. If this flow is also impeded by an extended waterfront structure, the groundwater head may rise considerably. Soil layers with low permeability can cause water-bearing layers at various levels.

4.2 Water Pressure Difference in the Water-side Direction (R 19)

The magnitude of water pressure difference is influenced by fluctuations in the free water level, location of the structure, groundwater flow, permeability of the foundation soil, permeability of the structure and the efficiency of available backfill drainage.

The water pressure difference $w_ü$ comes to

$$w_ü = h_{wü} \cdot \gamma_w.$$

at a difference in height $h_{wü}$ between the applicable free water level and the corresponding groundwater level at the specific weight of water γ_w. The water pressure difference may be assumed as shown in fig. R 19-1 and R 19-2 in the case of weepholes or permeable soil and unhindered circulation around the foot – if the waterfront structure is extended and plane flow can be assumed and without appreciable wave action. It is then allocated to the load cases 1–3.

The water levels stated in figs. R 19-1 and R 19-2 are nominal values (design values). Reference is made to section 8.2.0 as regards the safety factors.

For strong horizontal water flow the water pressure difference is to be increased correspondingly, the same if a waterfront structure is hydraulically backfilled or heavy wave action occurs in front of the structure.

Non-tidal area

Situation	Figure	Load cases as per R 18		
		1	2	3
1 Minor water level fluctuations ($h < 0.50$ m) with weepholes or permeable soil and structure		$h_{wü} = 0.50$ m	$h_{wü} = 0.50$ m	–
2a Major water level fluctuations ($h > 0.50$ m) with weepholes or well permeable soil and structure		$h_{wü} = 0.50$ m in frequent elevation	$h_{wü} = 1.00$ m in unfavourable elevation	$h_{wü} \geq 1.00$ m max. drop in outer water level over 24 h and least favourable elevation
2b Major water level fluctuations without weepholes		$h_{wü} = a + 0.30$ m $a = \dfrac{MHW - MLW}{2}$	$h_{wü} = a + 0.30$ m	–

Fig. R 19-1. Water pressure difference at waterfront structures for permeable soils in non-tidal area

Special investigations are required where flooding of the banks, stratified soils, highly permeable sheet pile locks or artesian pressure occur (R 52, section 2.7.4).

The relieving action of drainage installations according to R 32, section 4.5, R 52, section 4.4 and R 53, section 4.6 may only be considered if its effectiveness can be constantly monitored and the drainage installation can be set up again at any time.

Non-tidal area

Situation	Figure	Load cases as per R 18		
		1	2	3
3a Major water level fluctuations without drainage – normal case		$h_{wü} = a + 0.30$ m $a = \dfrac{MHW - MLW}{2}$ $d = MLW - MLWS$	–	–
3b Major water level fluctuations without drainage – limit case extreme low water level		–	$h_{wü} = a + 2b + d$ $a = \dfrac{MHW - MLW}{2}$ $b = \dfrac{MHWS - LLW}{2}$ $d = MLW - MLWS$	
3c Major water level fluctuations without drainage limit case falling high water		–	–	$h_{wü} = 0.30$ m $+ 2a$
3d Major water level fluctuations with drainage		$h_{wü} = 1.00$ m for outer water level in MLWS	$h_{wü} = 0.30$ m $+ b + d$	–

Fig. R 19-2. Water pressure difference at waterfront structures for permeable soils in tidal area

4.3 Water Pressure Difference on Sheet Piling in Front of Built-over Embankments in Tidal Areas (R 65)

4.3.1 General

In the case of built-over embankments, a partial equalisation of water levels is possible where the flow is predominantly horizontal. No water pressure difference occurs at the surface of the embankment. In the ground further back, a water pressure difference exists with respect to the free outer water level, which depends on the position of the point under consideration, soil conditions, magnitude and frequency of water level fluctuations and flow from land. It is a component of the flow pressure due to percolation through the ground in front.

The water pressure difference should be related to the relevant earth pressure failure plane. For this, knowledge of the variation of the groundwater level is necessary.

4.3.2 Approximation

The approximation shown in fig. R 65-1, which incorporates recommendations R 19, section 4.2 and R 58, section 4.1, can be taken to show conditions generally experienced in the North German tidal area, with fairly uniform sandy subsoil and with negligible groundwater flow. It shows load case 2, but can be similarly used for the other load cases.

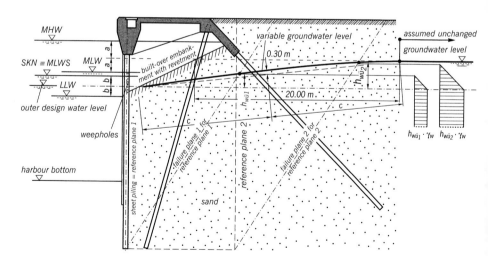

Fig. R 65-1. Assumption of water pressure difference at a built-over embankment for load case 2

4.4 Design of Filter Weepholes for Sheet Piling Structures (R 51)

Filter weepholes should only be used in silt-free water and groundwater of sufficiently low iron content. Otherwise they would quickly become clogged. Weepholes should not be used where there is a danger of heavy barnacle growth. When very high outer water levels occur, filter weepholes should not be used because there is a risk of penetrating water causing damage to the underground systems on the inner side or other damage from subsidence of non-cohesive soil.

The filter weepholes must be placed below mean water level so that they do not become encrusted. They are constructed with graded gravel filters as per R 32, section 4.5.

Slots 1.5 cm wide and approx. 15 cm high are burned in the sheet pile webs for drainage (fig. R 51-1). The slot ends are to be rounded off. In contrast to round holes, these slots cannot be blocked by pebbles. Otherwise reference is made to R 19, section 4.2, last paragraph.

Filter weepholes are substantially cheaper than flap valves (R 32, section 4.5). However, experience shows that they only produce a slight reduction of the water pressure difference in tidal areas, as too much water flows behind the sheet piling through the weepholes during the high water hours. Filter weepholes should also be avoided for quay sheet piling which also serves as flood protection.

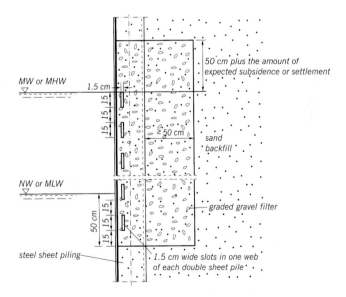

Fig. R 51-1. Filter weepholes for wave-shaped steel sheet piling

Filter weepholes are particularly effective in locations where there is no tide, and where a sudden drop of the free water level, strong groundwater or bank water flow or flooding of the structure may occur. Sheet piling must normally be designed for complete failure of the drainage. In this case the sheet piling may be designed for load case 3 according to R 18, section 5.4.3. This verification can be omitted when redundant drainage systems are provided, for example pump system with two redundant pumps.

4.5 Design of Flap Valves for Waterfront Structures in Tidal Areas (R 32)

4.5.1 General

Effective drainage is possible only in non-cohesive soils. If drainage is to remain effective in silt-laden harbour water and is to lower the water pressure difference where there is a considerable tidal range, it must be equipped with collector pipes and reliable flap valves which permit the outflow of water from the collector in the harbour water, but hinder the backflow of muddy water.

Simple backwater drainage, preferably with special types of drainage chambers, can be considered.

4.5.2 Flap Valves

Flap valves must be positioned to be readily accessible at mean low water for checking during periodic structure inspections and for easy maintenance and repair. They should be checked at least twice a year and also before each dredging operation, as well as, for example, when the transit of heavy goods may cause unusually severe surcharge, and also after heavy wave action.

The flap valves must be designed for maximum reliability. Manholes are to be provided at appropriate intervals (max. 50 m) for maintenance. The usual spacing of simple flap valves is 7 to 8 m for sheet piling drainage systems.

In silt-laden water, only fault-free double secured backwater drainage systems can be used. Efficient drains in the rear part of the backfill soil, which lead to the flap valves, can improve the drainage considerably.

Primarily plastic drain pipes are used as collectors. For further details see figs. R 32-1 and R 32-2.

Fig. R 32-2 shows a modern groundwater drainage system for a quay structure in a tidal area. Two full drain pipes DN 350 made of PE-HD (DIN 19 666) run through the whole quay structure and provide groundwater relief. The pipes are embedded in a graded gravel filter as per section 4.5.3; the filter is separated from the surrounding soil by a

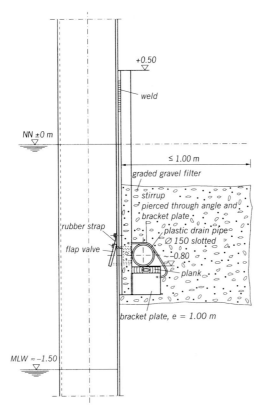

Fig. R 32-1. Example of flap valve drainage of steel sheet pile wall with plastic drain pipe as collector

nonwoven geotextile. The depth was selected to keep the full drain pipes permanently in the groundwater; this rules out practically any risk of clogging because there is no air access to the filter. The outlet depth of −4.20 mean sea level results from the position of the filter and the gradient of the outlet pipe to the outside water.

Filter wells are positioned at intervals of 75 m to deduct the groundwater from deeper strata. The collected groundwater is led off to the outer water in outlet pipes (steel pipe ∅ 609.6 x 20) at intervals of about 350 m. Two flap valves with adjustable float hollow covers prevent any flow of silt-laden river water. One flap is positioned immediately at the outlet into the outer water and the second flap is located in an outlet structure, protected from any damage.

79

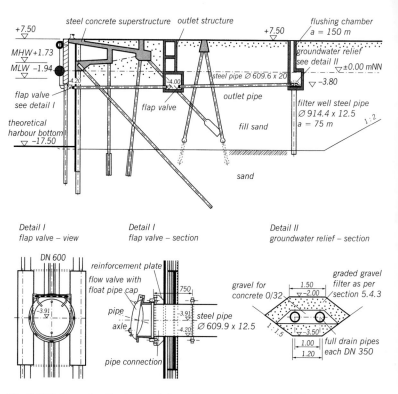

Fig. R 32-2. Example of backwater drainage for a quay structure in tidal area

4.5.3 Filter

Each collector must be separated from the soil to be drained by means of a carefully constructed gravel filter or geotextile filter. Mixed gravel filters should only be used when segregation cannot occur during installation.

4.6 Relieving Artesian Pressure Under Harbour Bottoms (R 53)

4.6.1 General

Relief is best accomplished by means of efficient relief wells of ample capacity. Their effectiveness is independent of the use of mechanical equipment or their power supply. The outlets of the wells should always be placed below LLW. Since there must be a pressure difference when the relief wells are discharging water, it follows that residual artesian pressure will remain under the confining layer, even under favourable conditions with fully effective, closely spaced wells. The residual pressure difference shall be taken as 10 kN/m^2 in calculations according to R 52, section 2.7.

4.6.2 Calculation

The design of the relief wells must always be verified by a lowering calculation. This must work on the premise that a residual pressure difference of 10 kN/m^2 may occur only outside the passive earth pressure wedge. Within the wedge, the residual pressure is thus correspondingly smaller, providing a desirable safety margin.

In exceptional circumstances, for example during alterations to a structure, when the outlet lies above LLW, a residual pressure level of 1.0 m above the outlet must be expected.

4.6.3 Layout

The relief wells are best placed in steel box and steel pipe piles, forming part of the outboard face of the sheet piling. This not only simplifies construction but also positions the well where it is protected and most effective for relief.

In tidal areas, the harbour water level at high tide generally lies above the artesian pressure level of the groundwater. In simple relief wells, the harbour water then flows into the wells and into the subsoil. This leads to rapid silting up of the relief well in silt-laden water, because the flushing force at the bottom of the well at low tide is not sufficient to remove a silt deposit. In such cases, relief wells must be equipped with fault-free backwater valves. Ball valves have proven ideal for this purpose.

The valves must be easily removed for inspection of the wells, and reinstalled again without damage to the water-tight seal.

In addition, the filter zone of the wells is fitted with an insert which forces the groundwater flowing in the well through a narrow space between the insert pipe and the well bottom. This has the effect of forcefully flushing out any sediment which might accumulate (fig. R 53-1). Dredge cuts in the bottom surface layer are not sufficient for permanent relief of artesian pressure in silt-laden water. They silt up again, as do wells without flap valves.

4.6.4 Filter

In order to obtain maximum capacity, the filters of the relief wells should be placed in the most permeable layer. The best filters must be used with complete freedom from corrosion and clogging. The wells must be properly installed by an experienced contractor.

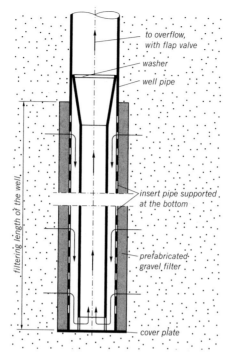

Fig. R 53-1. Flow into a relief well

4.6.5 Checking

The effectiveness of the installation must be checked frequently through observation wells which lie inboard of the quay and which extend below the confining layer.

If the required relief is not achieved, the wells must be cleaned and, if necessary, additional wells installed. Therefore a sufficient number of steel box piles must be installed to be accessible from the deck of the quay.

In other respects, reference is made to R 19, section 4.2, last paragraph.

4.6.6 Radius of Action

The area affected by a relieving installation is generally so small that harmful long-distance effects are not a problem, at least not in tidal areas. In special cases however, this effect should also be investigated. The relieving wells must be omitted if there are harmful effects.

4.7 Assessment of Groundwater Flow (R 113)

4.7.1 General

In order to make a correct design with refined calculations of a quay wall and other marine structures and their parts, to be built where flowing groundwater occurs, the designer must be thoroughly familiar with the essential characteristics of the flowing groundwater and with the corresponding flow net. Only then can he recognise and avoid dangers and, conversely, arrive at the best technical and most economical solutions by a more accurate determination of the acting loads. To accomplish this, only a few fundamental facts need to be observed.

In an adequately uniform, not overly coarse-grained subsoil, the groundwater flow follows the well-known DARCY law

$$v = k \cdot i$$

wherein

v = flow velocity [m/s]
k = permeability coefficient [m/s]
i = hydraulic gradient [1]

4.7.2 Prerequisites for the Determination of Flow Nets

If the DARCY law and the continuity condition have been fulfilled, the groundwater flow follows the potential theory. Its solution is depicted by two sets of curves which intersect at right angles, whose net widths have a constant ratio (fig. R 113-2).

In this flow net, the one set of curves represents the flow lines and the other the equipotential lines.

The flow lines are the paths of the water particles, whereas the potential lines are those corresponding to equal standpipe water heads (fig. R 113-2). Appropriate definition of the boundary conditions is prerequisite for correct determination of the flow net; examples are shown in fig. R 113-1.

4.7.3 Determining the Boundary Conditions for a Flow Net

The boundary of a flow net can be a flow line or an equipotential line or, if the groundwater discharges unconfined into the air, a seepage line. The boundary conditions relevant for the flow net are the appropriate flow and the equipotential lines. They are determined by the structure and the local conditions (water levels and soil conditions).

The following can be boundary flow lines: the surface of an impermeable layer of soil, the surfaces of an impermeable structure, a phreatic line, if it has a sloping, convex surface (seepage line), etc. (Fig. R 113-1).

The following can be boundary equipotential lines: a horizontal groundwater table, a river bottom, a submerged sloping bank, etc. For clarification, fig. R 113-1 shows the boundary conditions for several characteristic situations. For convex sloping phreatic lines, technical literature is available giving comparatively simple and clear methods for making an approximate calculation or construction of seepage lines.

4.7.4 Graphic Procedure for Determining the Flow Net

One method for determining a flow net is the so-called graphic procedure which still provides a rapid overview of critical flow zones, in spite of all groundwater models which can be used easily today for simple cases and steady flow conditions. It is generally presumed that the soil is homogeneous through which the water flows. When stratified soil is encountered or the single strata subsoil is anisotropic, the graphic method can still be used.

Once the flow field and all boundary conditions have been determined, the flow net is depicted according to the following rules:

– Flow lines are vertical to potential lines, i.e. flow lines move away from boundary potential lines vertically or move vertically toward them, or vertical potential lines move vertically from all boundary flow lines, including the free surface.
– The entire potential difference Δh between the highest and lowest hydraulic potential is divided into equal (equidistant) potential steps (in example fig. R 113-2 in 15 steps = 4.50 m/15 = 0.3 m), in some cases Δh is also 100 % and the potential lines are drawn at 10, 20, 30 ... 90 %.
– All flow lines (in example fig. R 113-2, 8 each) lead through the available flow cross section, i.e. they are closer together in narrower passages and are further apart in wider sections.
– Once a grid has been selected (number of potential steps/flow lines), the procedure continues by presuming that the rectangles formed by neighbouring potential and flow lines (ideally squares) remain geometrically similar, i.e. are uniformly larger or smaller.

The procedure continues by trial and error until the requirement of orthogonal lines in the whole net is fulfilled with adequate accuracy together with the boundary conditions and a constant net width ratio; here it is generally beneficial to aim for an orthogonal net from the very start. This task can be solved with sufficient precision in relatively short time by an experienced engineer who has an appropriate flow net in his mind's eye from the start.

Some details for determining a flow net are indicated in section 4.7.7 using a simple example (fig. R 113-2), which would apply among others to dams or excavations in open water.

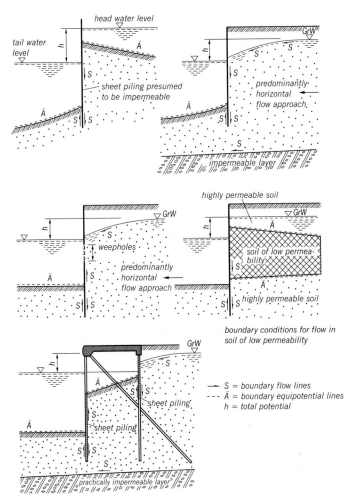

Fig. R 113-1. Boundary conditions for flow nets, characteristic examples with flow around the bottom of sheet piling

4.7.5 Use of Groundwater Models

4.7.5.1 Physical and Analog Models

Physical models use natural media (water, sand gravel, clay) on a model scale to be selected. They are mainly used today for research purposes in the 3-D problem and are no longer relevant for practical applications.

Analog models use media whose movement is also covered by the LAPLACE differential equations. Examples here are the movement of viscous materials through two close plates (gap models), the passage of electric current through conductive paper or a network of electrical resistances (paper models). Another example is the deformation of thin skins by point load (seepage line when lowering wells). In any case, kinematic similarity factors are required beside geometric factors in order to convert a potential (e.g. electric voltage) into the hydraulic potential (water table height). However, all these procedures have disappeared from practical use today since groundwater models now run on efficient personal computers.

4.7.5.2 Numerical Groundwater Models

Numerical groundwater models are not really models but calculation methods in which the entire potential field results from the (calculated) potential heights of individual points. It goes without saying that an adequately large quantity of points must be available. These points are the corner points of individual small but finite surfaces (finite elements) or the gravity points of small rectangular surfaces (finite differences). In both cases, the flow field is divided into individual elements and the geometry-dependent flow equation is calculated at the mentioned points (nodes), whereby the potential height h of the considered node depends on the potential heights of the neighbouring nodes (linear or also non-linear in some programmes). The most frequent element form is the triangle for the FE method; the geohydraulic and geometric properties of the elements are processed in the individual node point equations, whereby both the DARCY law and the continuity conditions are to be satisfied. Any action in the flow field must take place at node points or element lines, i.e. a node must be provided at all relevant points (wells, sources, drains, slope, seal, etc.).

All nodal equations represent an equation matrix, whereby the number of equations is identical to the number of unknown nodal hydraulic heads. After solution of the equation system (directly by an equation solver or indirectly by iteration), the hydraulic heads of all nodes are known and other variables of the field can be calculated (part of the programmes).

The user of the programmes must determine the flow field with all boundary conditions and set up the element mesh, whereby he can be assisted by mesh generators. There have to be instructions as to how his case must be schematised. Although all existing numerical problems are already solved, enough problems remain when selecting the boundary conditions and the geohydraulic parameters, particularly when using programmes which take account of unsteady flow. Basic knowledge of the solution procedure is essential in order to be able to assess how exact the result is and whether or how it can be improved.

4.7.6 Calculation of Individual Hydraulic Variables

Whereas the groundwater model calculates the entire hydraulic potential field with distribution of gradients, velocities, discharge, etc., there are procedures which only determine individual variables.
These are either hydraulic heads at specific points (e.g. bottom of sheet piling) or discharges (e.g. underseepage of structures).

Here are some examples:
- Flow pipe procedure for determining flow (example see 4.7.7).
- Resistance coefficient method according to CHUGAEV for determining gradients and discharges of underseepage [168].
- Fragment procedure according to PAVLOVSKY for calculating underseepage [167].
- Diagrams by DAVIDENHOFF & FRANKE for calculating hydraulic heads for excavation pits with sheet piling [166].

4.7.7 Evaluation of Examples

4.7.7.1 Sheet Piling with Underseepage in Homogeneous Subsoil

The example in fig. R 113-2 (compiled with the graphic method) shows the flow net. The potential difference from MSL + 7 m minus MSL + 2.50 m = 4.50 m = h is divided into n_1 = 15 potential steps of 0.30 m each, the flow lines form n_2 = 8 flow pipes.

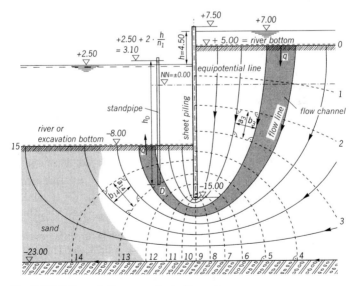

Fig. R 113-2. Example for a groundwater flow net

The following characteristics for the potential field are apparent:

– hydraulic head in point D:

$h_D = 7.00 - 13/15 \cdot 4.50$ m $= 3.10$ m $(= 2.50 + 2/15 \cdot 4.50$ m$)$

– hydraulic head at the foot point of the sheet piling:

$h_{FP} = 7.00$ m $- 9/15 \cdot 4.50$ m $= 4.30$ m $(= 2.50$ m $+ 6 \cdot 0.30$ m$)$

– hydraulic gradients (examples):

$i_3 = \Delta h/a_3 = 0.30/4.00 = 0{,}075$; $i_{14} = \Delta h/a_{14} = 0.30/3.60 = 0.083$;

the directions are marked by the a arrows, a_3 downwards, a_{14} slanting upwards.

– discharge: $q = v \cdot A = k \cdot i \cdot A$
The discharge is the same in every flow pipe, as all rectangles are mathematically similar and as large as between two potential lines in a partial section. The following equation applies to each flow pipe:

$q_i = k \cdot i \cdot b_i = k \cdot \Delta h/a_{14} \cdot b_{14} = k \cdot \Delta h/a_3 \cdot b_3 = k \cdot \Delta h \cdot b/a$.

For 8 flow pipes, this results in

$q = 8 \cdot k \cdot \Delta h \cdot b/a$ or generally $q = k \cdot h \cdot n_2/n_1 \cdot b/a$.

b/a is a factor from the drawing and $= 1$ in the orthogonal net, otherwise $\neq 1$. In this case, $b/a = 0.75$, the discharge therefore with $k = 10^{-4}$ m/s: $q = 8 \cdot 10^{-4} \cdot 4.5/15 \cdot 0.75 = 1.8$ m³/s/m.
This value coincides well with a groundwater model calculation (fig. R 113-3) which gives $q = 1.91 \cdot 10^{-4}$ m³/s/m.

The calculation of the failure by boiling at the toe of the sheet piling is carried out according to TERZAGHI (see R 115, section 3.2) with the head difference $h_{FP} = 4.3$ m and 7.0 m embedment with the presumption $\gamma' = \gamma_w$.

exist. $i = \dfrac{4.3 - 2.5}{7.0} = 0.26 <$ allow $i = 0.67$ as per section 3.2

4.7.7.2 Sheet Piling with Underseepage in Stratified Subsoil

The example shown in fig. R 113-2 resp. R 113-3 is retained, only the originally homogeneous subsoil ($k = 10^{-4}$ m/s) is interrupted by a 2 m thick horizontal layer at differing depths, with a permeability $k_s = 10^{-5}$ m/s (fig. R 113-4) resp. $k_s = 10^{-6}$ m/s (fig. R 113-5). Flow and potential lines are calculated with a groundwater model ($n_1 = 15, n_2 = 8$), together with the discharges q_i.

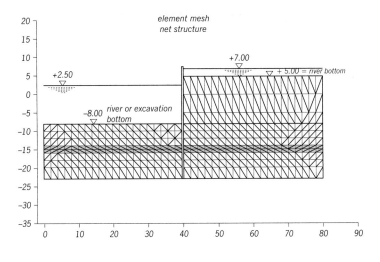

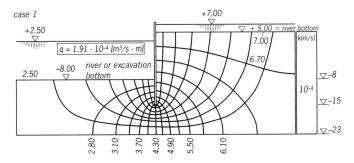

Fig. R 113-3. Element mesh and stratification for the calculation with a groundwater model. Potential net of the homogeneous case ($k = 10^{-4}$ m/s, case 1)

Case 2 indicates the concentration of potential lines at the less permeable layer, which reduces the safety against failure by boiling considerably in case 2a, but increases it in case 2b. However, the ruling potential at the sheet piling for calculating failure by boiling is now to be considered at the lower edge of this layer (case 2a) or at the upper edge (case 2b). The potentials can be read with 3.80 m or 3.40 m from the potential net for strata thicknesses of 2 m down to the bottom (potential here 2.50 m) (less permeable layer), respectively 4 m (other subsoil). This results in ($\Delta h = 3.80$ m $- 2.50$ m $= 1.3$ m resp. $\Delta h = 3.40$ m $- 2.50$ m $= 0.90$ m and the safety factors in case 2a $\eta = 2/1.30 \approx 1.5$ resp. $\eta = 4/0{,}90 \approx 4.45$ in case 2b.

89

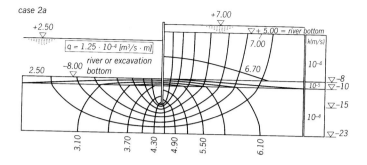

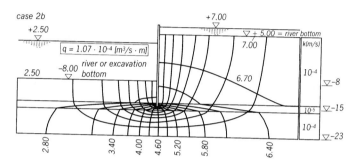

Fig. R 113-4. Potential nets of the stratified subsoil ($k_s = 10^{-5}$ m/s, case 2). Less permeable strata at the top (case 2a) or bottom (case 2b)

In addition, it may be necessary to consider raise of the whole body of soil, consisting of the less permeable layer and the soil above. In this consideration the weights are compared with the uplift pressure. In case 3a (fig. R 115-5; permeability ratio 100 : 1) it is even more evident that the uplift pressure increases at the bottom of the impermeable layer.

The comparison of cases 2 and 3 also shows a decrease in the discharge q compared to the homogeneous case and the effect of the presumptions regarding the boundary potential: for a horizontal flow of groundwater (case 3b, vertical boundary potential line at the edge of the model; simulates groundwater at head MSL + 7 m instead of surface water), the risk of boil at the sheet piling is greater than for vertical flow (case 3a, pure underseepage, presumed an infinite layer of low permeability).

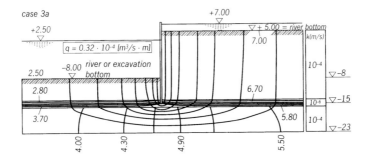

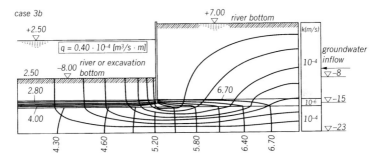

Fig. R 113-5. Potential nets of stratified subsoil ($k_s = 10^{-6}$ m/s) for horizontal (case 3a) or vertical (case 3b) presumed boundary potential line of free water respectively groundwater level

4.8 Temporary Stabilisation of Waterfront Structures by Groundwater Lowering (R 166)

4.8.1 General

The stability of a waterfront structure can be increased for a limited period of time until the definitive stabilisation measures are completed, by lowering the groundwater on the land side of the waterfront structure. First of all, studies are necessary to ensure that the structure itself or other structures in the area influenced by lowering the groundwater will not be endangered. Here reference is also made to a possible increase of the negative skin friction at pile foundations.

The increase of the stability is based on:

- the decrease in water pressure difference, whereby even a supporting effect can be attained from the water side, and

- the increase of the effective mass of the passive earth pressure wedge through reducing the seepage pressure from below or vice versa through a water surcharge and seepage pressure from above.

These influences are counteracted by the following negative factors:

- the increase of active earth pressure on the structure as a result of increased weight of the soil through the loss of uplift in the area where the groundwater has been lowered, and
- from case to case, an increase of active earth pressure through seepage pressure acting from top to bottom.

4.8.2 Case with Uniformly Permeable Soil

If the soil around the structure is of uniform permeability, there are no difficulties in the calculations. The water pressure difference is decreased corresponding to the lowering Δh. The additional active earth pressure behind the structure increases linearly from zero at the original groundwater table, till the lowered table linearly to the value

$$\Delta p_{ah} = \Delta h \cdot (\gamma - \gamma') \cdot K_{agh} \cdot \cos \delta_a$$

and remains constant downward provided there is no vertical groundwater flow. Otherwise, the increase of specific weight from seepage pressure and a possible change in K_{agh} are to be taken into consideration.

No change is generally taken into account on the passive earth pressure side, although the favourable change in specific weights could be included in the investigation.

4.8.3 Case with Soft, Cohesive Soil Near the Ground Surface

If less permeable soft soil exists from surface to a greater depth, which is underlain by well permeable, non-cohesive soil (fig. R 166-1), the soil is not consolidated for the time being under the additional weight due to the lost uplift in the range of the lowering depth Δh. Since in this state, the active earth pressure coefficient is $K_{agh} = 1$ for the additional load, and since with cohesive soil $\gamma - \gamma' = \gamma_w$, the additional active earth pressure at the start of the consolidation results in $\Delta p_{ah} = \Delta h \cdot \gamma_w \cdot 1$ at the level of the lowered groundwater table.

This level corresponds to the hydraulic head of the groundwater of the lower permeable layer provided there is negligible water flow from above. The reduced water pressure difference is therefore fully compensated in the initial state by the increased active earth pressure in soft soil. With increasing consolidation however, the additional active earth pressure drops to the value

$$\Delta p_{ah} = \Delta h \cdot \gamma_w \cdot K_{agh} \cdot \cos \delta_a .$$

On the passive earth pressure side, an increase of the water surcharge caused by groundwater lowering, with the addition of seepage pressure, has a favourable effect, mainly with non-cohesive sublayers (fig. R 166-1). In overlying cohesive soil, the state of consolidation must be taken into consideration accordingly.

4.8.4   Case According to Section 4.8.3 but with Stronger Upper Water Flow

If deviating from fig. R 166-1, a strongly water-bearing non-cohesive stratum exists above the soft cohesive soil behind the waterfront structure, a predominantly vertical potential flow to the lower permeable stratum takes place in the uniformly cohesive sublayer during groundwater lowering. In this case, the hydraulic head of the water in the overlying, highly permeable layer is applicable for the water pressure at the upper surface of the cohesive layer, and the hydraulic head of the groundwater

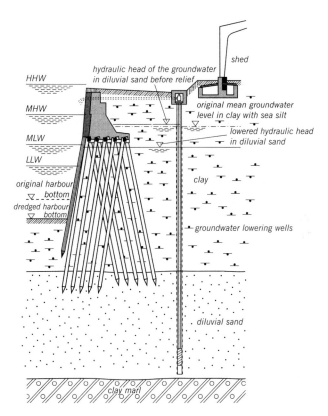

Fig. R 166-1. Executed example of quay wall stabilisation by groundwater lowering

in the lower, non-cohesive layer corresponding to the groundwater lowering is applicable for the water pressure at the lower surface of the cohesive layer. The changes in active and passive earth pressures depend on the respective flow conditions and/or water surcharge, in which case the state of consolidation according to section 4.8.3 must be taken into account.

4.8.5 Conclusions for Structure Stabilisation

The success of stabilisation through groundwater lowering is always ensured in the final state, but in the initial state is strongly dependent on the soil conditions. When this measure is used, the initial state and intermediate conditions must also be carefully considered. The method may then be used successfully for overloaded waterfront structures and above all for compensating the dredging of the harbour bottom in front of a waterfront structure. As a result, the ultimately required final strengthening of the structure, which is generally substantially more expensive, may be postponed to a future, economically more favourable time.

4.9 Flood Protection Walls in Seaports (R 165)

4.9.1 General

Flood protection walls generally have the function of shielding harbour terrain from flood waters. They are also utilised, however, when flood protection cannot be attained with earth dikes.

The special additional demands for such types of walls are commented on in the following sections.

4.9.2 Design Water Levels

4.9.2.1 Design Water Levels for High Tide

(1) Outer water level and theoretical level

The theoretical height of a flood protection wall is determined according to the design still water level (design water level according to the HHW) plus supplementary freeboard as local influences (waves) and, if necessary, wind congestion.

Due to the higher wave run-up on walls, the tops of flood protection walls are placed higher than those of dikes from case to case, unless brief overflowing of the walls can be accepted.

In any case, structural measures must be provided to make allowance for at least slight quantities splashing over the wall without causing damage.

Walls are generally less vulnerable to overflow than dikes. However, verification is required that waves splashing over the walls will not cause damage.

(2) Corresponding inner water level
The corresponding inner water level is to be taken generally at the ground surface, unless other possible water levels are less favourable (fig. R 165-1), such as embankments, for example.

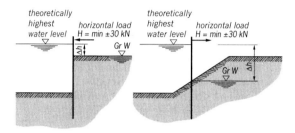

Fig. R 165-1. Design water levels at high water

For water levels as per section 4.9.2.1 (1) and (2), verification of stability for the wall is provided according to load case 3, as long as special forces according to section 4.9.5 are taken into consideration.

4.9.2.2 Design Water Levels for Low Tide

(1) Outer water levels
The mean tidal low water (MLW) is to be taken into account for normal load water in load case 1.
Extraordinarily low water levels, which occur only once a year, are to be classified in load case 2.
The lowest tidal low water (LLW) ever measured or a lower outer water level yet to be expected in future is to be considered as load case 3.

(2) Corresponding inner water levels
As a rule, the inner water level is to be taken at the ground surface (fig. R 165-2), unless a lower water level can be allowed due to more accurate seepage flow investigations, or can be permanently ensured through structural measures such as drains. However, in case of drain failure, a safety factor of ≥ 1.0 (hazard case) must still be available. In individual cases, the design inner water level – if local circumstances are precisely known – can also be determined on the basis of observations of groundwater levels.

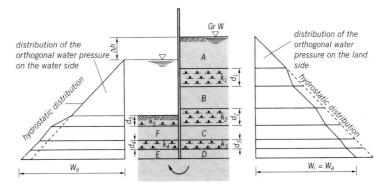

Fig. R 165-2. Example for the reduction of Δh (A, B, C, D, E, F are highly permeable layers)

4.9.3 Water Pressure Difference and Specific Weight for the Design of Flood Protection Walls

4.9.3.1 Applications with Almost Homogeneous Soils
Progression of the water pressure difference ordinates can be determined with the aid of a potential flow net according to R 113, section 4.7, or R 114, section 2.9.2. Changes in effective specific weight due to the groundwater flow can be taken into account as per R 114, section 2.9.3.2.

4.9.3.2 Applications with Stratified Soils
Usually, horizontal or only slightly sloped layer boundaries occur, so that the following remarks will remain confined to these. The decrease in water pressure due to flow occurs almost entirely in the layers with lower permeability, provided that substantial disturbances do not prevail and that safety against failure by boiling is ensured. Flow resistance in the relatively permeable layers can be neglected. In approximately horizontal layers with strongly differing permeability, a purely vertical flow may be applied to determine the water pressure ordinates.

The percentual decrease of Δh in the less permeable strata ΔW_i is proportional to the respective layer thickness d_i and inversely proportional to the permeability k_i [m/s].

$$\Delta W_i = \Delta h \cdot \gamma_w \cdot \frac{d_i}{k_i} \cdot \frac{1}{\sum \frac{d_i}{k_i}}.$$

Under these prerequisites, the effective specific weight of the soil with water flowing through it changes practically only in the layers of low permeability, namely by the amount

$$|\Delta \gamma_i| = \Delta h \cdot \gamma_w \cdot \frac{1}{k_i} \cdot \frac{1}{\sum \frac{d_i}{k_i}}.$$

The following signs apply to $\Delta \gamma_i$:

flow from top to bottom: +
flow from bottom to top: –

If a cleavage occurs between the wall and a less permeable layer as a result of wall flexure, this layer is to be regarded as fully permeable. Relief of γ deserves special attention, if on the land side the uppermost layer of low permeability is very close to or forms the land side ground surface. Thereby $\Delta \gamma_i$ can be equal to γ_i, which leads to buoyancy of the top stratum. For support of the flood protection wall, passive earth pressure only from cohesion may then be considered for the upper stratum, but should preferably be fully neglected to increase safety. Inversely, the effective external water pressure load would then be reduced by higher land sided water pressure.

4.9.4 Minimum Embedment of the Flood Protection Wall

The minimum embedment of the flood protection wall results from the static calculation and the required verification of landslide safety. Attention must also be paid to the following:

- The subsoil risk and the execution risk are to be considered with regard to possible leaks (lock damages), where even only one defect in the flood protection wall can lead to failure of the entire structure, and
- A qualification test for the design load case high tide (or flood water) is not possible.
- A supplementary driving depth is to be determined, taking account of a possibly unfavourable slope as per R 56, section 8.2.9.

Therefore, in the flood load case, the flow path in the ground should not fall below the following values:

- in homogeneous soils with relatively permeable soil structure and in the case of cleavage following wall flexure: the 4-fold difference between the design water level and the land-sided ground surface (independent of the actual inner water level),
- in inhomogeneous soils with less permeable strata with differences in permeability of at least 2 tenth powers: the 3-fold difference between the design water level and the ground surface (independent of the actual inner water level). Horizontal seepage paths may not be included when voids can occur.

4.9.5 Special Loads Acting Against a Flood Protection Wall

Apart from the usual live loads, loads of at least 30 kN deriving from the bumping of drifting objects at high tide and the impact of land vehicles are to be considered (see fig. R 165-1). At risky locations with unfavourable current and wind conditions, respectively good accessibility, the impact loads must be taken to be substantially higher. Distribution of the loads through suitable structural measures is allowed, if the functional capability of the flood protection wall is not impaired as a result.

In case of special loads, stability verification is allowed as per load case 3.

4.9.6 Structural Measures – Requirements

4.9.6.1 Surface Stabilisation on the Land Side of the Flood Protection Wall

Surface stabilisation must be provided to prevent land-sided scouring from splashing water in the case of flooding. The width of the stabilisation should match at least the land-sided height of the wall.

4.9.6.2 Defence Road

The installation of a flood protection wall defence road with asphalt roadway, close to the wall, is recommended. It should be at least 2.50 m wide and can serve at the same time as surface protection as per section 4.9.6.1.

4.9.6.3 Stress Relief Filter

Directly at the flood protection wall, a stress relief filter should be installed on the land side, approx. 0.3–0.5 m wide, so that no major bottom water pressure can develop below the defence road.

In the case of sheet piling structures, it is sufficient for the land-side valleys to be filled with corresponding filter material (e.g. steel mill slag 35/55).

4.9.6.4 Water Tightness of the Sheet Piling

An artificial lock sealing according to R 117, section 8.1.20 should be generally provided for the section of the sheet piling surrounded by air and water.

4.9.7 Hints for Flood Protection Walls in Embankments

In general the low water levels of the outer water are applicable for the set-up and design of flood protection walls in or in the vicinity of embankments.

The passive earth pressure outside decreases simultaneously with the increase of loads from inner water pressure difference and increased specific weight, also from seepage pressure. The changed water levels also frequently lead to a decrease in slope failure safety.

In the case of stratified soils (cohesive intermediate layers) which are not dowelled by adequately long sheet piles, usual verification of slope failure must also be accompanied by verification of the stability of the soil wedge in front of the flood protection wall (barnacle failure, slide safety).

The outer slope is to be protected from scouring by hand-set stone pitching or equivalent measures. Safety from ground or slope failure is to be verified as per DIN V 4084-100 at least for load case 2. Regular checks of these embankments are to be arranged.

4.9.8 Conduits in the Area of Flood Protection Walls

4.9.8.1 General

Conduits running in the area of flood protection walls can constitute weak points for several reasons, in particular:

- Leaking liquid conduits reduce internal erosion through the existing seepage path.
- Excavations to replace damaged conduits reduce the supportive influence of the passive earth pressure and also shortens the seepage path.
- Decommissioned conduits can leave uncontrolled cavities.

4.9.8.2 Conduits Parallel to a Flood Protection Wall

Conduits running parallel to a flood protection wall should not be installed within an adequately wide protective strip on both sides of the wall. Existing conduits should be moved or decommissioned. Resulting cavities or remaining cavities must be securely filled.

Allowance must be made for conduits remaining in protective strips as follows:

- A trench reaching to the bottom of the pipeline is to be considered for conceivable work on the conduit, when determining the passive earth pressure and in particular the seepage paths.
- Trench bracing is to be rated for the passive earth pressure applied in the wall calculation.
- It must be possible to seal off conduits carrying liquids by means of suitable shut-off gear on entering and leaving the protective strips.
- Conduit work should not be carried out in the season at risk from storm tides, if this can be avoided.

4.9.8.3 Conduits Crossing a Flood Protection Wall

The passage of conduits through a flood protection wall always constitutes potential weak points and should be avoided where possible. Therefore:

- The conduits should be routed over the wall, particularly high pressure or high voltage conduits.
- Single conduits in the ground outside the protective strip should be grouped together and routed through the flood protection wall as collective conduit or conduit bundles, and
- Conduit crossings should be approximately at right angles.

Structural measures are to be provided to take account of the differing settlement behaviour of conduits and the protection wall (flexible passages, articulated pipes). Rigid passages are not allowed.

A conduit crossing is rated individually depending on the type of medium concerned.

- Cable crossings
  Communications cables and power cables may not be routed through the flood protection wall directly but protected in a casing. The cables are to be sealed against the casing in a suitable manner.

- Pressure conduit crossings
  Pressure conduits (gas, water ...) are to be protected by a casing in the whole protective strip in such a way that on failure of the medium conduit, they can be replaced without excavations in the protective strip. The casing should be able to withstand the operating pressure with the same safety factors as the medium conduit. The same applies to the sealing of the medium conduit.

- Canal or sewer crossings
  If there is danger that water will be forced into the polder through canals or sewers in the case of flooding, suitable double shut-off devices must be provided. Here either a shaft with gate valve or sluice valve is to be installed on both sides of the flood protection wall, or both installations and a part of the protection wall are combined in a structure with double gate or sluice valve safeguard. When the risk is less, one of both valves can also be installed as a backflow valve.

- Within the protective strip, supply and disposal conduits are to be rated with loads as per load case 3. Stability verification is to be provided for load case 2. Resistance to abrasion, corrosion and other chemical attacks is to be given special attention.

- Dike openings
  Dike openings in connection with the flood protection wall are rated appropriately according to the proven design principles of sea dike openings. The load cases named above also apply in this case.

- Decommissioned conduits
  Decommissioned conduits are to be removed from the protective strip. If this is not possible, the conduit voids are to be securely filled.

5 Ship Dimensions and Loading of Waterfront Structures

5.1 Ship Dimensions (R 39)

The following exemplarily listed average ship dimensions may be used for the calculation and design of waterfront structures and in the design and layout of fenders and dolphins:

5.1.1 Seagoing Vessels

5.1.1.1 Passenger Vessels (table R 39-1.1)

Tonnage	Carrying capacity	Displacement G	Overall length	Length between perps	Beam	Draft
GT	DWT	t	m	m	m	m
80 000	–	75 000	315	295	35.5	11.5
70 000	–	65 000	315	295	34.0	11.0
60 000	–	55 000	310	290	32.5	10.5
50 000	–	45 000	300	280	31.0	10.5
40 000	–	35 000	265	245	29.5	10.0
30 000	–	30 000	230	210	28.0	10.0

5.1.1.2 Bulk Carriers (table R 39-1.2) (oil, ore, coal, grain, etc.)

–	450 000	524 000	424	404	68.5	25.0
–	420 000	490 000	418	398	67.0	24.5
–	380 000	445 000	407	386	64.5	24.0
–	365 000	428 000	404	383	63.5	23.0
–	340 000	400 000	398	378	62.5	23.0
–	300 000	356 000	385	364	59.5	22.0
–	275 000	326 000	376	355	57.5	21.5
–	250 000	300 000	367	346	55.5	20.5
–	225 000	270 000	356	336	53.5	20.5
–	200 000	240 000	345	326	51.0	19.5
–	175 000	212 000	330	315	48.5	18.5
–	150 000	180 000	315	300	46.0	16.5
–	125 000	155 000	295	280	43.5	16.0
–	100 000	125 000	280	265	41.0	15.0
–	85 000	105 000	265	255	38.0	14.0
–	65 000	85 000	255	245	33.5	13.0
–	45 000	60 000	230	220	29.0	11.5
–	35 000	45 000	210	200	27.0	11.0
–	25 000	30 000	190	180	24.5	10.5
–	15 000	20 000	165	155	21.5	9.5

5.1.1.3 Mixed Cargo Freighters (Full Deck Construction) (table R 31-1.3)

Tonnage	Carrying capacity	Displacement G	Overall length	Length between perps	Beam	Draft
GT	DWT	t	m	m	m	m
10 000	15 000	20 000	165	155	21.5	9.5
7 500	11 000	15 000	150	140	20.0	9.0
5 000	7 500	10 000	135	125	17.5	8.0
4 000	6 000	8 000	120	110	16.0	7.5
3 000	4 500	6 000	105	100	14.5	7.0
2 000	3 000	4 000	95	90	13.0	6.0
1 500	2 200	3 000	90	85	12.0	5.5
1 000	1 500	2 000	75	70	10.0	4.5
500	700	1 000	60	55	8.5	3.5

There appears to be no trend towards construction of larger cargo freighters. If necessary, the dimensions used in section 5.1.1.2 may be used accordingly.

5.1.1.4 Fishing Vessels (table R 39-1.4)

2 500	–	2 800	90	80	14.0	5.9
2 000	–	2 500	85	75	13.0	5.6
1 500	–	2 100	80	70	12.0	5.3
1 000	–	1 750	75	65	11.0	5.0
800	–	1 550	70	60	10.5	4.8
600	–	1 200	65	55	10.0	4.5
400	–	800	55	45	8.5	4.0
200	–	400	40	35	7.0	3.5

5.1.1.5 Container Ships (table R 39-1.5)

Carrying capacity	Displacement G	Overall length	Length between perps	Beam	Draft	Number of containers	Generation
DWT	t	m	m	m	m	circa	
75 000	90 000	350	335	45.0	14.0	6 000	6th
66 300	80 000	275	262	40.0	14.0	4 800	5th
64 500	77 500	294	282	32.2	13.5	4 400	5th
55 000	77 000	275	260	39.4	12.5	3 900	4th
50 000	73 500	290	275	32.4	13.0	2 800	3rd
42 000	61 000	285	270	32.3	12.0	2 380	3rd
36 000	51 000	270	255	31.8	11.7	2 000	3rd
30 000	41 500	228	214	31.0	11.3	1 670	2nd
25 000	34 000	212	198	30.0	10.7	1 380	2nd
20 000	27 000	198	184	28.7	10.0	1 100	2nd
15 000	20 000	180	166	26.5	9.0	810	1st
10 000	13 500	159	144	23.5	8.0	530	1st
7 000	9 600	143	128	19.0	6.5	316	1st

5.1.1.6 Car transport Ships (table R 39-1.6)

Carrying capacity	Displacement G	Overall length	Length between perps	Beam	Draft	No. of cars
DWT	t	m	m	m	m	approx.
28 000	45 000	198	183	32.3	11.8	6 200
26 300	42 000	213	198	32.3	10.5	6 000
17 900	33 000	195	180	32.2	9.7	5 600

5.1.1.7 Ferries and Ro-Ro Ships (table R 39-1.7)

Carrying capacity	Displacement G	Overall length	Length between perps	Beam	Draft
DWT	t	m	m	m	m
106 400	115 000	253.00	238.00	40.00	15.10
64 400	76 100	225.00	215.00	34.00	13.00
42 500	53 000	182.50	173.00	32.30	12.00
27 750	39 800	177.30	158.10	27.30	11.55
18 000	32 650	181.20	165.00	30.40	9.30
16 000	23 400	178.10	164.00	26.80	7.60
14 000	21 500	163.80	148.60	23.50	8.80
12 000	20 000	190.90	173.00	26.00	7.18
10 000	23 410	192.50	181.00	27.30	6.75
8 000	16 000	156.00	137.00	22.60	7.30
6 000	20 750	179.40	170.00	27.80	6.27
4 000	17 500	163.40	150.00	27.00	6.20
2 000	10 800	164.70	159.60	17.70	5.90

The data in the table vary according to type of load (cars, trucks, trailers, waggons, passengers) and load shares.

5.1.2 River-sea Ships (table R 39-2)

Tonnage	Carrying capacity	Displacement G	Overall length	Beam	Draft
GT	DWT	t	m	m	m
999	3 200	3 700	94.0	12.8	4.2
499	1 795	2 600	81.0	11.3	3.6
299	1 100	1 500	69.0	9.5	3.0

The length, width and draft of all types of freighters depend on the ship's construction and the country of origin. The dimensions can be expected to vary by up to 5 % (see also [197], [199] and [200]).

The gross tonnage (GT) is taken as the dimension-less gross space number [201]. The carrying capacity is stated in deadweight tons (DWT), namely the weight of provisions, supplies, fresh water, crew, reserves of boiler water, fuel, freight and passengers, measured in English tons (long tons) at 2240 lbs = 1016 kg.

5.1.3 Inland Vessels (table R 39-3)

Designation	Carrying capacity	Displacement G	Length	Beam	Draft
	t	t	m	m	m
Motor freighters:					
Large Rhine ship	4 500	5 200	110.0	11.4	4.5
2600-ton class	2 600	2 950	110.0	11.4	2.7
Rhine ship	2 000	2 385	95.0	11.4	2.7
"Europe" ship	1 350	1 650	80.0	9.5	2.5
Dortmund-Ems-Canal ship	1 000	1 235	67.0	8.2	2.5
Large-Canal-Class ship	950	1 150	82.0	9.5	2.0
Large-"Plauer"-Class ship	700	840	67.0	8.2	2.0
BM-500 ship	650	780	55.0	8.0	1.8
Kempenaar	600	765	50.0	6.6	2.5
Barge	415	505	32.5	8.2	2.0
Peniche	300	405	38.5	5.0	2.2
Large-Saale-Class ship	300	400	52.0	6.6	2.0
Large-Finow-Class ship	250	300	41.5	5.1	1.8
Push lighters:					
Europe IIa	2 940	3 275	76.5	11.4	4.0
	1 520	1 885			2.5
Europe II	2 520	2 835	76.5	11.4	3.5
	1 660	1 990			2.5
Europe I	1 880	2 110	70.0	9.5	3.5
	1 240	1 480			2.5
Carrier ship lighters:					
Seabee	860	1 020	29.7	10.7	3.2
Lash	376	488	18.8	9.5	2.7
Push tows:					
with one lighter Europe IIa	2 940	3 520[1]	110.0	11.4	4.0
	1 520	2 130[1]			2.5
with 2 lighters Europe IIa	5 880	6 795[1]	185.0	11.4	4.0
	3 040	4 015[1]	110.0	22.8	4.0
					2.5
with 4 lighters Europe IIa	11 760	13 640[2]	185.0	22.8	4.0
	6 080	8 080[2]			2.5

[1] Push vessel 1 480 kW; approx. 245 t displacement
[2] Push vessel 2963–3333 kW; approx. 540 t displacement

According to ECE resolution no. 30 dated 12.11.1992 – TRANS/SC 3R.153, the following classification applies to European waterways:

Type of inland waterway	Class of inland waterway	Motor vessels and barges in tow — Type of vessel: general features					Push tow — Type of pushed lighter: general features					Vertical clearance under a bridge [m] [2]	Graphical symbol on the map
		Designation	Max. length L [m]	Max. beam B [m]	Draft d [m] [7]	Tonnage T [t]	Formation	Length L [m]	Beam B [m]	Draft d [m] [7]	Tonnage T [t]		
1	2	3	4	5	6	7	8	9	10	11	12	13	14
of regional significance — west of the Elbe river	I	Peniche	38.5	5.05	1.8–2.2	250–400							
	II	Kempenaar	50–55	6.6	2.5	400–650						4.0	
	III	Gustav Konigs	67–80	8.2	2.5	650–1000						4.0–5.0	
of regional significance — east of the Elbe river	I	Large Finow	41	4.7	1.4	180						4.0–5.0	
	II	BM-500	57	7.5–9.0	1.6	500–630						3.0	
	III	[6]	67–70	8.2–9.0	1.6–2.0	470–700		118–132[1]	8.2–9.0[1]	1.6–2.0	1000–1200	3.0	
of international significance	IV	Johann Welker	80–85	9.5	2.5	1000–1500		85	9.5[5]	2.50–2.80	1250–1450	4.0	
	Va	large Rhine ship	95–110	11.40	2.50–2.80	1500–3000		96–110[1]	11.40	2.50–4.50	1600–3000	5.25 or 7.00[4]	
	Vb							172–185[1]	11.40	2.50–4.50	3200–6000	5.25 or 7.00[4]	
	VIa							95–110[1]	22.80	2.50–4.50	3200–6000	7.00 or 9.10[4]	
	VIb	[3]	140	15.00	3.90			185–195[1]	22.80	2.50–4.50	6400–12000	7.00 or 9.10[4]	
	VIc							270–280[1]	22.80	2.50–4.50	9600–18000	9.10[4]	
								195–200[1]	33.00–34.20[1]		9600–18000		
	VII							285	33.00–34.20[1]	2.50–4.50	14500–27000	9.10[4]	

Foot notes for the classification table:

[1] The first number considers the current situation, whereas the second shows both future developments and, in some cases, the existing situation.
[2] Considers a safety clearance of approx. 30 cm between the highest fixed point of the ship or its cargo and a bridge.
[3] Considers the dimensions of vessels under own power expected in Ro-/Ro- and container traffic. The stated dimensions are approximate values.
[4] Rated for transporting containers:
    – 5.25 m for ships with two layers of containers,
    – 7.00 m for ships with three layers of containers,
    – 9.10 m for ships with four layers of containers.
    – 50 % of the containers can be empty, otherwise ballast is required.
[5] Some existing waterways can be allocated to class IV on account of the greatest permissible length of ships and barges, although the greatest beam is 11.40 m and the largest draft 4.00 m.
[6] Vessels used in the region of the Oder and on the waterways between Oder and Elbe.
[7] The draft for a specific federal waterway is to be ascertained according to the local conditions.
[8] On certain sections of waterways in class VII, push tows can be used consisting of a larger number of lighters. Here the horizontal dimensions can exceed the values stated in the table.

Table R 39-3.1. Classification of the European inland waterways

5.1.4 Displacement

The displacement G [t] is the product of the length between perpendiculars, the width, the draft, the block coefficient c_B and the mass density ρ_w [t/m^3] of the water. The block coefficient varies from 0.50 to 0.80 for seagoing vessels, from 0.80 to 0.90 for inland vessels, and from 0.90 to 0.93 for push lighters.

5.2 Assumed Berthing Pressure of Vessels at Quays (R 38)

In preparation of the design, accidental impacts need not be taken into consideration but only the usual berthing loads. The magnitude of these berthing loads depends on the ship's dimensions, the berthing velocity, the fenders and the deformation of the ship's hull and the structure.

In order to give the quay sufficient stability against normal berthing loads, but on the other hand to avoid unnecessarily large dimensions, it is recommended that the front wall be so designed that at any position of a section, a concentrated impact load in the magnitude of the relevant line pull force can act, without the total stresses exceeding the permissible limits. Berthing impact for quay walls in seaports according to R 12, section 5.12.2 with the values in table R 12-1, and for quay walls in inland harbours 100 kN according to R 102, section 5.13.2.

This concentrated force may be distributed over a square area 0.50 m on a side. In sheet pile walls without solid superstructures, only the wales and bolts need be designed for this force.

The berthing loads on dolphins are dealt with in R 128, section 13.3.

5.3 Berthing Velocities of Vessels Transverse to Berth (R 40)

When vessels make their approach transverse to a berth, it is recommended that the following berthing velocities be taken into consideration when designing the corresponding fendering:

Condition	Approach	Berthing velocity transverse to berth (m/s)			
		up to 1000 DWT	up to 5000 DWT	up to 10000 DWT	Larger ships
		corresponding to approx. 1500 t | 6500 t | 13000 t displacement			
Strong wind and heavy sea	difficult	0.75	0.55	0.40	0.30
Strong wind and heavy sea	favourable	0.60	0.45	0.30	0.20
Moderate wind and heavy sea	moderate	0.45	0.35	0.20	0.15
Protected	difficult	0.25	0.20	0.15	0.10
Protected	favourable	0.20	0.15	0.10	0.10

Table R 40-1. Berthing velocity transverse to berth

5.4 Load Cases (R 18)

The following load cases (combination of actions) are considered as a general principle for the statical calculations and allocation of the partial safety factors:

5.4.1 Load Case 1

Loads due to active earth pressure (in unconsolidated, cohesive soils, separately for both initial and final states) and to water pressure differences where unfavourable outer and inner water levels frequently occur (see R 19, section 4.2), Earth pressures resulting from the normal live loads, from crane tracks and pile loads, directly acting surcharges from dead weight and normal live load.

5.4.2 Load Case 2

Same as load case 1, but with restricted scour from flow or from ship's screw action, and together with the following, insofar as they can occur simultaneously: water pressure difference according to R 19, section 4.2, wave loads from frequent waves ("design wave" as per R 136, section 5.6.5), water pressure difference caused by regularly anticipated flooding of the waterfront structure, the suction effect of passing ships, loads and active earth pressure from unusual local surcharges, hawser pull on bollards, recess bollards or mooring hooks, and the impact of vessels; the effect of temporarily unfavourable loads during construction and the protection afforded by any existing piling should be neglected.

5.4.3 Load Case 3

Same as load case 2, but taking into consideration additional surcharges not previously allowed for on larger areas, or the possibility that elements which help to stabilize the structure in general may fail because of unfavourable circumstances. Examples of these contingencies are the complete failure of the drainage system, an unusual slumping of an underwater slope in front of the sheet piling, unusual scouring due to current or ship's screws, water pressure difference after extreme water level situations and wave loads from rare waves ("design wave" as per R 136, section 5.6.5), unexpected flooding of the banks or a severe groundwater rise due to an ice jam with subsequent sudden drop of the outer water after the jammed ice goes out, the bursting of a large water pipe behind the waterfront structure, unforeseen transhipment of unusually heavy goods. The combination of several such unfavourable actions is also to be taken into consideration, as far as this occurrence is possible and probable.

5.4.4 Partial Safety Factors

It must be taken into account whether the acting loads are stated as *"nominal* loads" (i.e. design loads) or *"characteristic* loads". In the latter case, they are to be multiplied by the partial safety factors according to DIN V 1054-100. For load case 3, as a rule the partial safety factor 1.0 is used.

5.5 Vertical Live Loads (R 5)

All quantitative loads (actions) stated in this section are *characteristic values*.

5.5.1 General

Vertical live loads (variable loads in accordance with DIN V ENV 1991-1) are essentially the surcharges resulting from stored material and the loads from vehicular traffic. The load actions of rail-mounted or vehicular mobile cranes must be considered separately, insofar as they exert any effect on the waterfront structures. At waterfront structures in inland ports, the latter is generally only the case for waterfront structures which are expressly intended for heavy load handling with mobile cranes. In seaports, in addition to the rail-mounted quay cranes, mobile cranes are being used increasingly for general cargo handling, that is to say, not only for heavy loads.

A distinction is to be made between three different basic types (table R 5-1) for the live loads:

In *basic type 1*, the bearing members of the structures are driven over directly by the vehicles and/or stressed by the stacked materials, e.g. at pier bridges (table R 5-1a).

In *basic type 2*, the load from vehicles and the stacked material acts on a more or less deep bedding course, which distributes and transmits the loads to the structural members. This type of design is used for example at super-structured slopes with load distributing bedding layer on the pier slab (table R 5-1 b).

In *basic type 3*, the load from vehicles and the stacked goods acts only on the solid mass of earth fill behind the waterfront structure, which consequently is subject only indirectly to additional stress from the live loads as the result of increased active earth pressure. Simple sheet piling bulkheads or partially sloped banks are characteristic for this (table R 5-1c).

Supplementing the three basic types, there are also transitional types, for example pile-founded structures on piles with a short pile cap.

If complete and reliable calculations are available, the live loads should normally be taken at the anticipated magnitude. Any subsequently necessary increases in the live loads can be better accommodated within the tolerable limits, the greater the deadweight share and the better the distribution of loads in the structure. Support systems according to basic type 2 and in particular basic type 3 offer particular advantages in this respect.

Reference is made to R 18, section 5.4 when it comes to allocation of the corresponding loads to load cases 1, 2 and 3.

Basic type	Traffic live loads[1]				Storage area outside the waterfront cargo-handling area
	Railroad	Roads			
		Vehicle	Road-bound cranes	Light-weight traffic	
a) BT 1	Issue B.3 dated 08.03.93 (DS 804) Regulations for railroad bridges and other engineering structures (VEI)	Load assumptions as per DIN 1072 (road and foot bridges – load assumptions)	Fork lift loads as per DIN 1055, Claw loads of 550 kN for mobile cranes	5 kN/m^2	Loads according to the use actually anticipated in accordance with section 5.5.6
		Impact factor: The parts exceeding 1.0 can be decreased by half			
b) BT 2	As 1, but further reduction of the impact factor to 1.0 at bedding layer thickness $h = 1.00$ m. For bedding layer thickness $h \geq 1.50$ uniformly distributed surface load of				
	20 kN/m^2	33.3 kN/m^2			
c) BT 3	Loads as in BT 2 with a bedding layer thickness of more than 1.50 m				

[1] Crane loads are to be taken as stipulated in R 84, section 5.14

Table R 5-1. Vertical live loads

5.5.2 Basic Type 1

Railroad live loads correspond to the load diagram UIC[1] 71 of the Regulation for Railroad Bridges and other Engineering Structures (VEI), issue B3 of 8.4.1993 (DS 804). The load assumptions according to DIN 1072 are to be applied for road traffic. Bridge class 60/30 is to be adopted in general. In indicated impact factors (DS 804) and vibration coefficients (DIN 1072), with which the live loads of the main track are to be multiplied, the parts exceeding 1.0 can as a rule be decreased by half, because of the slow speed. For piers in seaports, loads from fork lifts are to be taken according to DIN 1055 and claw pressures for mobile cranes of 550 kN, insofar as higher assumptions are not required in special cases (see table R 84-1, section 5.14.3).

Outside the waterfront cargo handling area, the actually expected surcharge from stored goods is to be taken, but at least 20 kN/m^2 (see section 5.5.6), because of later possible changes in use of the area. A live load of 5 kN/m^2 is adequate if the nature of the facility means that only light traffic is possible or anticipated.

5.5.3 Basic Type 2

Essentially the same as basic type 1. The impact factors and coefficients however may be linearly further reduced according to bedding layer thickness, and completely ignored when the bed is at least 1.00 m thick, for road traffic taken from the top of the road, and when the rails are embedded in the pavement, from the top of the rails. Load by sections is however still to be taken into account.

If the bedding layer thickness is at least 1.50 m, the total live load can be replaced by a uniformly distributed area load corresponding to the actually anticipated live load, but not less than 20 kN/m^2. In cases of light traffic, a live load of 5 kN/m^2 suffices.

5.5.4 Basic Type 3

Load as for basic type 2, with a bedding layer thickness of more than 1.50 m.

5.5.5 Load Assumptions Directly Behind the Head of the Waterfront Structure

When working with heavy vehicular cranes or similar heavy-duty vehicles and heavy construction gear, such as crawler excavators and similar, which drive along directly behind the front edge of the waterfront structure, the following are to be applied for the design of the uppermost parts of the structure, inclusive of an eventual upper anchoring:

---

[1] UIC = Union Internationale des Chemins de Fer.

a) Live load = 60 kN/m² from rear edge of coping, inboard for 2.0 m width, or

b) Live load = 40 kN/m² from rear edge of coping, inboard for 3.50 m width.

In a) and b), effects from a claw end load P = 550 kN are covered insofar as the distance between the axis of the waterfront structure and the axis of the claw is at least 2.0 m.

5.5.6 Loads Outside the Waterfront Cargo Handling Area

Outside the waterfront cargo handling area, the following live loads are taken as the basis in accordance with [140], working on the basis of 300 kN gross load for 40' containers and 200 kN for 20' containers.

- Light traffic (cars)   5 kN/m²
- General traffic (trucks)   10 kN/m²
- General cargo   20 kN/m²
- Containers:
  - empty, stacked 4 high   15 kN/m²
  - full, stacked 2 high   35 kN/m²
  - full, stacked 4 high   55 kN/m²
- Ro-Ro loads   30–50 kN/m²
- Multi-purpose facilities   50 kN/m²
- Offshore feeder bases   55–150 kN/m²
- Paper
- Timber products
- Steel
- Coal
- Ore

depending on the bulk/stacking height, calculating values of the weight density according to DIN1055, part 4

Further details regarding the material properties of bulk and stacked goods are to be found in the tables of ROM 02.-90 [197].

When calculating the active earth pressure of retaining structures, as a rule the differing loads in the cargo handling and container area can be grouped together to produce an average surface load of 30 to 50 kN/m².

5.6 Determining the "Design Wave" for Maritime and Port Structures (R 136)

5.6.1 General

In order to rate the wave loads acting on maritime and port structures, the sea conditions in the planning area should be analysed and studied with regard to probabilities. This includes an investigation of the wave data, such as wave heights, periods, lengths and directions in connection with wind conditions, tides and currents, including their seasonal frequency. It is then possible to determine the applicable wave value as so-called "design wave" taking account of the damage risk for the structure. This is the *characteristic* value. The resulting wave pressures are to be multiplied by the partial safety factors according to DIN V 1054-100 of the load cases 2 respectively 3 according to R 18, section 5.4.

A comprehensive presentation of wave determination is not possible here. Scientifically sound procedures other than the simple method described here can be used.

It is recommended that an institute or engineering bureau experienced in coastal engineering should be consulted when it comes to investigating the wave conditions in the planning area.

5.6.2 Description of Waves and Statistical Conditions

5.6.2.1 Definitions and Designations

Among others, the following types of seas are recognised:

- Wind sea = very steep, short crested waves, constantly influenced by the wind,
- Swell = long crested waves of lesser steepness which have advanced into weaker wind regions,
- Deep water waves = waves in which the ratio water depth d/wave length L ≥ 0.5,
- Waves in the transition zone = waves where $0.5 > d/L > 0.04$,
- Shallow water waves = waves where $d/L \leq 0.04$,
- Breaking waves = waves which fall forward.

The regularly advancing gravity wave is illustrated in fig. R 136-1 with the most important designations.

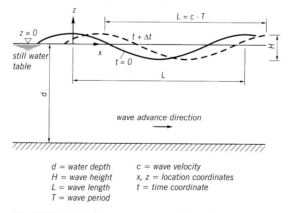

d = water depth c = wave velocity
H = wave height x, z = location coordinates
L = wave length t = time coordinate
T = wave period

Fig. R 136-1. Advancing gravity wave, designations

5.6.2.2 Wave Theories

When deep water waves enter shallow water, they are affected by topographic and geomorphologic influences, as well as those which are caused by structures, which have to be taken into account when determining the wave characteristics in the planning region.

The theories regarding the description of regular waves can generally be divided into the following two classes according to [27]:

- Theories for waves of small amplitude,
- Theories for long waves.

For further information on wave theories and physical relations, see [24], [27], [28] and [46].

The applicable scope of various theories is indicated quantitatively in fig. R 136-2.

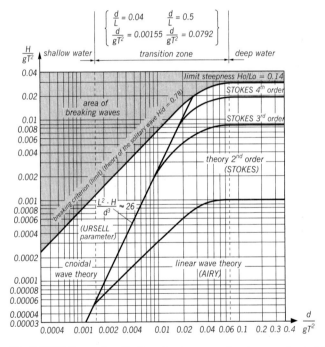

Fig. R 136-2. Scope of application of various wave theories according to [27] and [21], depicted on double logarithmic scale

5.6.2.3 Description of the State of the Sea

The state of the sea can be shown in two ways:

(1) By making a graph of wave characteristics (wave heights and wave periods) defined as arithmetical means.
(2) By graphic presentation of a wave spectrum, which shows the energy content of the state of sea as a function of the wave frequency. This spectrum can be made as "one-dimensional spectrum" with joint presentation of all wave directions or as "directional spectrum" with separate presentation of the various wave directions.

Presentations in line with (1) are mainly used for determining the applicable waves in statical calculations, whereas the spectral, energetic presentation as per (2) is used particularly in model tests because these allow more detailed conclusions.

5.6.2.4 Determination of the State of the Sea not Influenced by the Structure

It is determined by:

(1) Direct measurements away from the sphere of influence of the structure, over the longest possible period of time. The measurements are generally made at intervals of 3 or 6 hours.
(2) Determination of the distinguished values in accordance with a wave forecast. Customary methods for practical wave forecasting are stated in [21], [27] and [46].

The zero transit method according to [24] is recommended for evaluation of wave records (fig. R 136-3).

The most essential influences for a wave forecast are:
- Wind force, direction and duration,
- Wind field range,

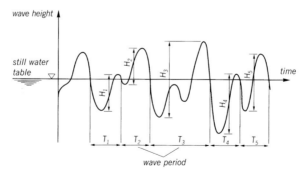

Fig. R 136-3. Determination of the state of the sea using the zero transit method [24]

- Effective fetch,
- Water depth.

A distinction is to be made between the deep and shallow water area. Diagrams are available for both cases [21]. A critical analysis of the various forecasting methods is to be found in [24] and [29]. The selection of a suitable method should be made as the occasion requires on the basis of local conditions. Its results should be verified by supplementary wave calculation and, if possible, by actual wave measurements.

The wave forecast is to be carried out for a definite period of time, e.g. determining the maximum for one year or several years (frequently 50 or 100 years). The chosen period of time in such cases need not coincide with, but should not be less than, the service life of the structure according to R 46, section 14.1.

As far as offshore platforms are concerned, the design period is usually taken to be 50 years.

5.6.2.5 Graphic Presentation of the State of the Sea Showing Relationship Between Wave Heights and Frequency

Fig. R 136-4 shows the normal presentation.

The symbols are defined as follows:

n = Frequency of the wave heights H in the observed period, expressed as a percentage,

H_m = Arithmetical mean value of all wave heights, recorded during a period of observation [m],

H_d = Most frequent wave height [m],

$H_{1/3}$ = Significant wave height,
= Arithmetical mean value of the one-third highest waves [m],

$H_{1/10}$ = Arithmetical mean value of the one-tenth highest waves [m],

$H_{1/100}$ = Arithmetical mean value of the one-hundredth highest waves [m],

$\max H$ = Maximum wave height [m].

For practical evaluation of state of sea measurements, it is advisable to plot the frequency distributions on suitable logarithmic paper so that the measurements lie on a straight line. By means of extrapolation it is possible to find the highest wave from a certain number of waves (e.g. from 1000 waves) or specific period of time (e.g. in 50 or 100 years).

The measurements of an irregular state of sea are frequently electronically evaluated by means of a FOURIER analysis which is used to plot a spectrum of the energy density (see section 5.6.2.3 (2)).

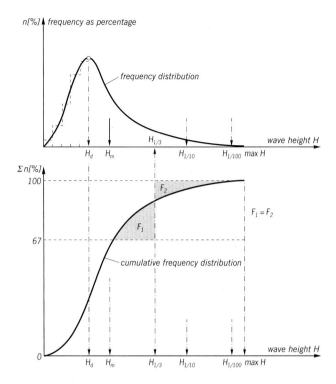

Fig. R 136-4. Frequency and cumulative frequency distribution of wave heights in % in accordance with [25]

5.6.2.6 Statistical Conditions in State of Sea

The following approximations result from the frequency distribution shown in fig R 136-2 according to [24] respectively [26]:

$$H_m = 0.63 \cdot H_{1/3},$$
$$H_{1/10} = 1.27 \cdot H_{1/3},$$
$$H_{1/100} = 1.67 \cdot H_{1/3},$$
$$\max H = 2 \cdot H_{1/3}.$$

5.6.3 Reflections Caused by Structures

Non-breaking waves of height H are reflected by the shore and by structures. The reflection is described by the reflection coefficient $\kappa_R = H_R/H$, with H_R = height of the reflected wave. The reflection coefficient κ_R is highly dependent on the steepness of the wave and therefore subject to change with the waves contained in the wave spectrum.

A vertical wall throws a normally arriving wave back practically at full height, theoretically forming a standing wave with twice the height of the incoming wave.

For a wall inclination of 1 : 1, κ_R is between 0.7 and 1.0; for an inclination of 1 : 4 it equals 0.2, but can increase to 0.7 depending on the steepness of the waves. Further information is available in [29].

The reflection coefficient is also influenced by the type of contact surface between wave and structure, e.g. by a perforated face of a wall. It also depends on the direction of the wave approach.

Reference is made to [30] and [31] when it comes to the steepening of the waves at an oblique wave attack, known as the "Mach reflection". Accordingly, it is advisable to calculate with $\kappa_R = 1.0$, i.e. twice the height of the arriving wave for oblique waves at a vertical wall.

5.6.4 Transformation of the State of Sea when Waves Enter Shallow Water

(1) Shoaling effect

When a wave touches bottom, its velocity and wave length are thereby decreased. However, since the energy equation must be satisfied, after a local insignificant diminution, the height increases steadily as the wave approaches the shore, caused by equilibrium of energy. This is referred to as the shoaling effect.

The shoaling factor can be calculated with sufficient accuracy according to the linear wave theory (e.g. as per [28]).

(2) Bottom friction and percolation

The wave height is decreased by frictional losses and by any tendency of the water to seep into the bottom. These losses have been partially included in the methods for wave forecasting. For details see [23].

(3) Refraction and diffraction

Refraction occurs at a rising bottom when the waves do not approach at right angles to the contour lines of the bottom. In such cases, the wave crests tend to turn parallel to the coast line. The wave energy is changed thereby. For details see [28] for example.

Diffraction occurs when waves encounter obstacles (islands, spits or structures). The waves then run into the wave shadows, generally at reduced height. At certain points beyond the wave shadow however – caused by superposition of diffraction waves of closely positioned obstacles, e.g. harbour moles on both sides – increased wave heights may occur.

The calculation methods now in use are applicable if the marginal conditions are very simple. They are based in general on the linear wave theory. Diagrams for ascertaining the influence of diffraction are found in [21].

(4) Breaking waves
The height of the deep water waves running into shallow water is limited by the breaker conditions (index b). According to the solitary wave theory, we have:

$H_b/d_b = 0.78$ (breaker criterion).

However, the following should be assumed in engineering practice:

$H_b/d_b = 1.0$.

At upward sloping tidal flats however, values $H_b/d_b > 1.0$ can also occur [33].

The ratio of the breaker height H_b to water depth d_b is a function of the beach slope α and the steepness of the deep water wave H_0/L_0. Both influences also determine the type of breakers as surging/collapsing breaker, plunging breaker or spilling breaker. Further details can be found in [34] and [46]. The following terms and relationships are used:

α = angle of slope of the bottom,
H/L_0 = wave steepness,
H = respective wave height,
L_0 = length of the incoming deep water wave,

$\xi = \dfrac{\tan \alpha}{\sqrt{H/L_0}}$ = refraction coefficient.

For a given H and L_0, the slope of the bottom is near the critical point if the computed reflection factor $\xi = 2.3 = \text{crit } \xi$.

ξ can be used to state the type of breaker, with reference to the nomenclature of GALVIN [35]. ξ can be defined for the deep water height H_0 (ξ_0) or for the wave height at breaking point H_b (ξ_b) (see table R 136-1):

Designation of breaker	ξ_0	ξ_b
Surging/collapsing breaker	> 3.3	> 2.0
Plunging breaker	0.5 to 3.3	0.4 to 2.0
Spilling breaker	< 0.5	< 0.4

Table R 136-1. Determination of types of breakers

These values are derived from investigations according to BATTJES [34] with bottom slope of 1 : 5 to 1 : 20. It must be considered however that underwater beaches are frequently substantially flatter than this.

White cappings occur only in deep water of the open sea and are therefore of no significance for waterfront structures.

According to [36], the type of breaking can also be indicated by the breaker index $\beta = L_H / L_B$. The symbols mean:

L_H = Distance of the breaking point from the point at which the surging wave has lost half its height,
L_B = Wave length at the breaking point.

High breaker indexes ($\beta > 1$ to 100 and more) occur in surface breakers with spilling breakers, low breaker indexes ($\beta < 1$) in line breakers with plunging breakers. The latter lead to high energy transmission on waterfront structures.

Very low breaker indexes ($\beta < 0.1$) occur without exception at dike and bank protection embankments. The breaker forms are then essentially plunging breakers at flat embankments and surging breakers at steep embankments.

Plunging breakers create high compressive stresses on the embankments. Surging breakers lead to an especially high wave run-up, which increases with the embankment slope. For further information see [36] and [37]. The height of the breaker can be determined approximately, relative to ξ, from table R 136-2:

ξ_0	H_b/d_b
< 0.3	0.8 ± 0.1
0.3 to 0.5	0.9 ± 0.1
0.5 to 0.7	1.0 ± 0.1
0.7 to 2.2	1.1 ± 0.2

Table R 136-2. Determination of the height of breakers

5.6.5 Determining the "Design Wave"

In order to determine the statistical conditions for the state of the sea, regular waves are idealised resulting in the wave characteristics stated in section 5.6.2.5. On this basis, the height of the "design wave" H_{des} can be determined contingent on the tolerable risk for the structure being built. Recommendations for various structures are stated in table R 136-3:

Structure	$H_{des}/H_{1/3}$
Breakwater	1.0 to 1.5
Sloped moles	1.6
Vertical moles	1.8
Quay walls with warehouses	1.9
Excavation enclosures	1.5 to 2.0

Table R 136-3. Determining the heights of the "design wave"

For high safety requirements, the ratio of the "design wave" height H_{des} to the significant wave height $H_{1/3}$ should be taken as 2.0.

As a rule, the partial safety factor 1.0 can be used, if the *characteristic* values for the height of the "design wave" were determined with high risk addition according to this section.

Structures which are essentially insensitive to overloads can be designed for a lower wave height according to the permissible risk of flooding or destruction.

For fatigue strength studies with models, it is advisable to use the actual frequency distribution of the wave height in the period of observation, instead of the "design wave" height.

5.7 Wave Pressure on Vertical Waterfront Structures in Coastal Areas (R 135)

5.7.1 General

The wave pressure and/or wave movement on the front of a waterfront structure is to be taken into account for:

- blockwork walls in uplift and in joint water pressure,
- built-over embankments with non-backfilled forward wall, taking into account the effective water pressure difference on both sides of the wall,
- non-backfilled sheet piling walls,
- flood protection walls,
- stresses during construction,
- backfilled structures in general, also because of the lowered outer water level in the wave trough,
- assessment and elimination of scour damage in front of a waterfront structure.

Furthermore, the waterfront structures are loaded by waves through line pulls, impacts of vessels and fender pressures from ship movements.

When taking wave pressure on vertical waterfront structures into account, a distinction is to be made between three loading modes as follows:

(1) The structure is subject to load from waves which are fully or partially reflected by the structure,
(2) The structure is subject to load from waves breaking on it,
(3) The structure is subject to load from waves which have already broken at some distance from the wall.

Which of these three loading modes applies depends on the water depth, on the motion of the sea, and on the morphological and topographic conditions in the area of the planned structure.

5.7.2 Determination of Loads of Waves Reflected by the Structure

A structure with a vertical or approximately vertical front wall, in water of such a depth that the highest incoming waves do not break, is stressed by the increased water pressure difference on the water side at the crest of the reflected wave, and on the landward side at the wave trough.

When incoming waves are superimposed on the backwash, standing waves are formed. In reality, ideal standing waves never occur: the irregularity of the waves creates certain wave impact loads, but their impulse is generally negligible compared to the following load assumptions, so that these are considered to be practically static. The wave height doubles as a result of reflection when the waves meet a vertical or approximately vertical wall and if no losses are incurred (reflection coefficient $\kappa_R = 1.0$). A reduction in the wave height from partial reflection ($\kappa_R < 1.0$) at vertical walls should only be taken into consideration with verification provided by large-scale model tests. Otherwise, reference is made to the reflection coefficients in R 136, section 5.6.3.

For calculating wave action when the wave approach is at right angles to the wall, SAINFLOU's method [20] is recommended, as shown in fig. R 135-1. However, more recent investigations have shown that the loads calculated by this method are too large if the waves are steep. Particulars and detailed procedure are given in [21].

The symbols in fig. R 135-1 mean:

H = height of incoming wave [m],
L = length of incoming wave [m],
h = difference in level between still water level and mean water level in the reflected wave on the face of the wall,

$$= \frac{\pi \cdot H^2}{L} \cdot \cot h \frac{2 \cdot \pi \cdot d}{L} \text{ [m]},$$

Δh = difference between the still water level in front of the wall and the ground water respectively rearward harbour water table [m],
d_s = depth of groundwater or rearward harbour table [m],
γ = weight density of the water [kN/m^3],
p_1 = pressure increase (wave crest) or decrease (wave trough) at the foot of the structure due to wave effect

$$= \gamma \cdot H / \cos h \frac{2 \cdot \pi \cdot d}{L} \text{ [kN/m}^2\text{]},$$

p_0 = maximum water pressure difference ordinate at level of the landside water table (according to fig. R 135-1c)

$$= (p_1 + \gamma \cdot d) \cdot \frac{H + h - \Delta h}{H + h + d} \text{ [kN/m}^2\text{]},$$

p_x = water pressure difference ordinate at level of the wave trough according to fig. R 135-1d)
 = $\gamma \cdot (H - h + \Delta h)$ [kN/m²].

Application of this procedure to the case of oblique wave approach is dealt with in [30]. Accordingly, the assumptions for right-angled wave approach should also be used for small angled wave approach, particularly for long structures.

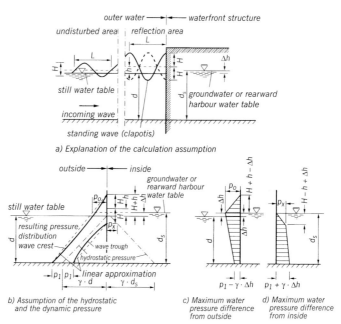

Fig. R 135-1. Dynamic pressure distribution on a vertical wall at total wave reflection, in accordance with SAINFLOU [20], as well as water pressure difference at wave crest and wave trough

5.7.3 Wave Loads at Breaking Waves

Waves breaking on a structure can exert impact pressures of 10 000 kN/m² and more. These pressure peaks however are limited to local effects and last for only a very brief period (1/100 s to 1/1000 s).

The structure should be suitably arranged to ensure that high waves do not break immediately at the structure, because of the huge pressure impulses which can occur. If this is not possible, model studies on the largest possible scale are recommended for definitive design.

The most common calculation procedure for preliminary designs is illustrated in fig. R 135-2 according to MINIKIN [22].

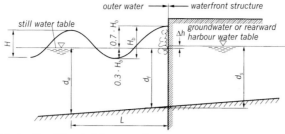

a) Explanation of the calculation assumption

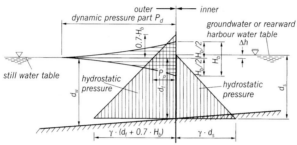

b) Assumption of the hydrostatic and dynamic water pressure

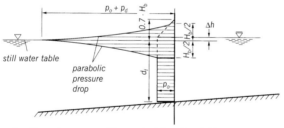

c) Resulting water pressure difference load from the outside

Fig. R 135-2. Wave action and dynamic and hydrostatic water pressure distribution, and resulting water pressure difference at a vertical wall at the moment when the wave breaks, according to MINIKIN [22]

The total water pressure is composed according to fig. E 135-2b) from the superimposition of a hydrostatic pressure distribution, and a dynamic water pressure distribution resulting from the wave impact. The approximate assumption of the maximum dynamic water pressure at the level of the still water head with parabolic pressure drop to zero in the area of the wave height is the closest approximation to measurement results obtained in nature and on large models, although in [21] the numerical factor in the equation for p_d (see below) was kept the same when the

125

transition to the decimal system was made, resulting in a theoretically too large value.

The symbols in fig. R 135-2 mean:

H_b = height of wave at moment of breaking [m],
d_w = water depth at a full wave length from structure [m],
d_f = water depth at foot of structure [m],
d_s = water depth at groundwater or rearward harbour water table [m],
L = wave length corresponding to d_w [m],
Δh = difference between the still water level in front of the wall and the groundwater or harbour water table [m],
p_0 = hydrostatic pressure ordinate at level of the land-sided groundwater or harbour water table
  = $\gamma \cdot (0.7 \cdot H_b - \Delta h)$ [kN/m^2],
p_d = highest dynamic water pressure at level of the still water table
  = approx. $100 \cdot \gamma \cdot \dfrac{H_b}{L} \cdot \dfrac{d_f}{d_w} \cdot (d_w + d_f)$ [kN/m^2].

The diagrams in [21] are recommended for determining the water pressure ordinates indicated in figs. R 135-2 b) and c), as well as the resulting forces and moments as functions of the wave parameters.

Section 5.7.2 is applicable for the outward water pressure difference at the wave trough.

An assumption to GODA for load from standing or breaking waves of a natural sea state is described in [26] and [46], although this does not take any account of pressure impact load.

5.7.4 Wave Load by Broken Waves

An approximate determination of the forces of a broken wave is possible according to [21]. In this case, it is assumed, that the broken wave runs on at the same height and velocity which it had at breaking (fig. R 135-3).

The symbols in fig. R 135-3a) mean:

H_b = wave height at moment of breaking [m],
d_b = water depth at point of breaking [m],
h_c = $0.7\, H_b$ [m],
Δh = difference between the still water level and the groundwater table [m].

The pressures are calculated according to the following formulas:

dyn $p \approx 1/2 \cdot \gamma \cdot d_b$ [kN/m^2],
$p_s = \gamma \cdot (d_s + h_c - \Delta h)$ [kN/m^2],
$p_0 = \gamma \cdot (h_c - \Delta h)$ [kN/m^2].

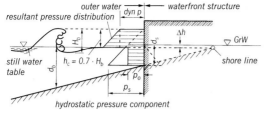

a) Waterfront structure seaward of the shore line

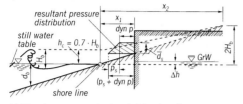

b) Waterfront structure landward of the shore line

Fig. R 135-3. Resultant dynamic and hydrostatic pressure distribution on a vertical wall with already broken waves, corresponding to CERC [21]

The formulas appertaining to fig. R 135-3b) read:

$$d_s = h_c \cdot \left(1 - \frac{x_1}{x_2}\right) \; [\text{m}],$$

$$\text{dyn } p = 1/2 \cdot \gamma \cdot d_b \cdot \left(1 - \frac{x_1}{x_2}\right)^2 \; [\text{kN/m}^2],$$

$$p_s = \gamma \cdot d_s = \gamma \cdot h_c \cdot \left(1 - \frac{x_1}{x_2}\right) \; [\text{kN/m}^2].$$

Sample calculations are stated in [21].

5.7.5 Additional Loads Caused by Waves

If the structure standing on a permeable bedding course does not have a watertight face, e.g. in the form of an impervious diaphragm, additional bottom water pressure from the effects of the waves must also be taken into account along with the water pressure on the wall surfaces. This also applies to the water pressure in block joints.

5.8 Loads Arising from Surging and Receding Waves Resulting from the Introduction or Extraction of Water (R 185)

5.8.1 General

Surging and receding waves arise in bodies of water as a result of temporary or temporarily increased water introduction or extraction. However, surging and receding waves manifest themselves significantly only with wetted cross-sections of bodies of water which are small in comparison with the volume introduced or extracted per second. Therefore, great importance is generally attached to allowing for surging and receding waves and their effects on waterfront structures only in navigation canals. In such cases, the effects of changes in water levels on embankments, the linings of bodies of water, waterfront revetments and other facilities must be taken into account.

5.8.2 Determination of Wave Values

Surging and receding waves are shallow-water waves in the range:

$$\frac{d}{L} < 0.04$$

(see R 136, section 5.6.2.1). The wave length depends on the duration of the water introduction or extraction. The wave propagation speed can be roughly calculated as:

$$c = \sqrt{g \cdot (d \pm 1.5 \ H)} \ \text{[m/s]} \quad \begin{cases} + \text{ for surging} \\ - \text{ for receding} \end{cases}$$

in which:

- g = acceleration due to gravity [m/s^2],
- d = depth of water [m],
- H = rise in the case of surging or fall in the case of receding as compared with still water level [m].

If the H/d ratio is small,

$$c = \sqrt{g \cdot d}$$

can be used.

The water level rise or fall is roughly:

$$H = \pm \frac{Q}{c \cdot B},$$

in which:

- Q = volume of water introduced or extracted per second [m^3/s],
- B = mean breadth of water level [m].

The wave height can increase or reduce as a result of reflections or subsequent surging or receding waves. Particularly in the case of uniform canal cross-sections and smooth canal linings, the wave attenuation is small so that the waves can run backwards and forwards several times, specially with short reaches.

In navigation canals, the most frequent cause of surging and receding phenomena is the introduction of extraction of lockage water. In order to prevent extreme surging and receding phenomena, the volume of lockage water is generally restricted to 70 to max. 90 m^3/s.

5.8.3 Load Assumptions

Load assumptions for waterfront structures must take account of the hydrostatic load arising from the height of the surging or receding wave and its possible superimposition by reflected or subsequent waves, as well as simultaneously possible fluctuations in the water table, e.g. from water accumulation by wind, ship's waves etc., in the least favourable constellation. Owing to the long periodicity of the shape of surging and receding waves, the effect of the flow gradient of the groundwater produced thereby must be examined simultaneously in permeable revetments. The dynamic effects of surging and receding waves can be ignored because of the mostly low flow rates caused by such waves.

The resulting loads are *characteristic* values which are to be multiplied with the partial safety factors of load case 2 (see R 18, section 5.4.2) according to DIN V 1054-100.

5.9 Effects of Waves from Movements of Ships (R 186)

5.9.1 General

Waves of different types are constantly generated by moving vessels. Depending on the local circumstances, these cause different stresses on waterfront structures.

Additionally, a water accumulation, the headwater wave, is created in front of the moving vessel as a result of the water displacement, and a drop in water level, which generally exceeds the accumulation many times over, arises as a result of the backflow below and beside the vessel. The headwater wave and the drop in level manifest themselves particularly clearly in restricted waterways and must be allowed for as hydrostatic loads on waterfront structures and comparable structural elements, e.g. lock gates.

5.9.2 Wave Sizes

Bow waves spread out from the ship's bow at a specific angle. Because of the relationships between wave speed and wave group-speed, for ships' waves under deep water conditions the direction of dispersion does not

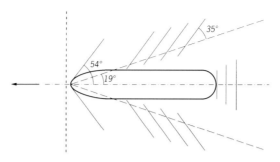

Fig. R 186-1. Ship's waves

depend on the speed of travel, and lies at approx. 19° to the direction of travel. The wave crest direction is inclined at approx. 35° in relation to the direction of dispersion, the wave crests being inclined at approx. 54° in relation to the direction of travel.

The symbols below are used in the following text:

d = water depth [m],
L = wave length [m],
g = acceleration due to gravity [m/s^2],
v_s = relative speed of vessel compared with water [m/s],
L_H = length of stern wave [m],
L_B = length of bow wave [m],
H_1 = height of solitary wave above calm water level [m],
c = propagation speed of headwater wave [m/s].

In conditions $d/L < 0.5$, the direction of the bow waves becomes dependent on the travelling speed. Under shallow-water conditions ($d/L \leq 0.04$), it reaches the value

$$\sin \alpha = \frac{\sqrt{g \cdot d}}{v_s}.$$

The stern waves have the propagation direction of the ship. Their wave crests therefore run at right-angles to the direction of travel. According to the speed of travel, the wave length of the stern waves can be calculated as approx.:

$$L_H = v_s^2 \cdot \frac{2 \cdot \pi}{g}.$$

The wave length of the bow wave is:

$$L_B = 2/3 \, L_H.$$

The wave heights depend on the shape of the vessel and its travelling speed, and generally do not exceed 0.6 m. The headwater wave can be thought of as a "solitary wave", i.e. a wave with only one vertex of height H_1 above the calm water level.

It generally precedes the vessel with a propagation speed:

$$c = \sqrt{g \cdot (d + H_1)}$$

so that a very protracted headwater wave of several hundred metres up to 1 km is created as a result of continuous new accumulations. The headwater height is generally small and seldom exceeds 0.2 m.

If the depth of water is relatively great in comparison with the water level breadth, i.e. in conditions such as exist in navigation canals, the hydraulic radius F_w/U_w must be used instead of the water depth. Since H_1/d is small at the same time, the equation for wave propagation speed becomes:

$$c = \sqrt{g \cdot \frac{F_w}{U_w}}$$

F_w = wetted cross-section of body of water [m^2],
U_w = wetted perimeter (bottom and embankments) of body of water [m].

The water level drop corresponds with the backflow below and beside the submerged body of the vessel, and its form and size depend on the shape of the vessel, the vessel's propulsion, the travelling speed of the vessel and the waterway conditions (relationship n of wetted cross-section of body of water to submerged midship frame cross-section of vessel, proximity and form of waterfront). The drop seldom exceeds 0.8 m.

5.9.3 Load Assumptions

Particular allowance must be made for the effects of the bow and stern waves on embankments and waterfront revetments in restricted waterways. The applicable *design* loads arise from the increase and decrease in pressure of the wave height and the breaking of the wave on transition into the shallow-water area at the embankment, this depending on the direction in which the waves travel. The effects of the headwater wave and the water-level drop on waterfront structures with the hydrostatic pressure change must be taken into account. In the case of possible reflections, e.g. in short branches with vertical termination (outer lock harbours), the height of accumulation or drop can increase to twice its value. More precise values can be determined in model tests. If applicable in the case of permeable waterfront structures, the time sequence of the accumulation or drop must be taken into account, with its effect on groundwater movement. Attention is drawn to the possible effects on automatically operating closures, dike locks for example (opening and closing of the gates as a result of the sudden pressure changes), and to lock gates.

Reference is made to R 106, section 6.4.3 in respect of the water level drop arising from movements of vessels in the case of canal quay walls. Further details and wave values for particular cases can be found in [123], [124] and [125], for example.

The maximum values determined in this way can be taken as *design* values.

5.10 Wave Pressure on Pile Structures (R 159)

5.10.1 General

In the calculation of pile structures, the loads originating from the wave motion are to be taken into account in respect of both the loads on the individual pile as well as on the entire structure, insofar as this is required by local conditions. The superstructures should be located above the crest of the design wave if possible. Otherwise, large horizontal and vertical loads from the direct wave action can affect the superstructures, whose determination is not an object of this recommendation, as reliable values for such cases can only be obtained from model studies. The elevation of the crest of the design wave is to be determined in consideration of the simultaneously occurring highest still water level, where applicable also taking into account the wind-raised water level, the influence of the tides and the raising and steepening of the waves in shallow water.

The superposition method according to MORISON, O'BRIEN, JOHNSON and SCHAAF [38] is used for slender structural members, whereas the calculation method for wider structures is based on the diffraction theory [39]. The subject of this recommendation is the superposition method according to MORISON [21], which is applicable for non-breaking waves. Due to lack of accurate calculation methods for breaking waves, a makeshift method is proposed in section 5.10.5.

The method according to MORISON furnishes useful values, if there is

$$\frac{D}{L} \leq 0.05$$

for the individual pile.

The symbols mean:

D = pile diameter or for non-circular piles, characteristic width of the structural member (width transverse to direction of wave propagation) [m],

L = length of the "design wave" [m] in accordance with R 136, section 5.6, in conjunction with table R 159-1, no. 3.

This criterion is mostly fulfilled.

For determination of the wave loads, reference is made to [42] and [21], which contain tables and diagrams for execution of the calculation. The diagrams in [21] are developed on the stream-function theory and may be applied to waves of differing steepnesses up to the breaking limit, whereas the diagrams in [42] are applicable only under the prerequisites of the linear wave theory.

Other design methods are valid for offshore structures, e.g. according to API (American Petroleum Institute).

5.10.2 Calculation Method According to MORISON [38]

The wave load on an individual pile consists of the components
- force due to the water particle velocity (drag force), and
- force due to the water particle acceleration (inertial force)

which must be separately determined and superimposed according to phases.

According to [41], [43] and [21], the horizontal total load per unit of length for a vertical pile comes to:

$$p = p_D + p_M = C_D \cdot \frac{1}{2} \cdot \frac{\gamma_w}{g} \cdot D \cdot u \cdot |u| + C_M \cdot \frac{\gamma_w}{g} \cdot F \cdot \frac{\partial u}{\partial t}.$$

For a pile with circular cross section, there is accordingly

$$p = C_D \cdot \frac{1}{2} \cdot \frac{\gamma_w}{g} \cdot D \cdot u \cdot |u| + C_M \cdot \frac{\gamma_w}{g} \cdot \frac{D^2 \cdot \pi}{4} \cdot \frac{\partial u}{\partial t}.$$

The symbols mean:

p_D = pressure due to the water particle velocity caused by the flow resistance per unit length of pile [kN/m],
p_M = inertial pressure due to instationary wave movement per unit length of pile [kN/m],
p = total load per unit length of pile [kN/m],
C_D = drag coefficient taking into account the resistance of the pile against the flow pressure [1],
C_M = inertia coefficient taking into account the resistance of the pile against the acceleration of water particles [1],
g = gravity acceleration [m/s²],
γ_w = weight density of the water [kN/m³],
u = horizontal component of the velocity of the water particles at the studied pile location [m/s],
$\frac{\partial u}{\partial t} \approx \frac{du}{dt}$ = horizontal component of the acceleration of the water particles at the studied pile location [m/s²],

	Shallow water $\frac{d}{L} < \frac{1}{25}$	Transition area $\frac{1}{25} < \frac{d}{L} < \frac{1}{2}$	Deep water $\frac{d}{L} > \frac{1}{2}$
1. Profile of the free surface	General equation $\eta = \frac{H}{2} \cdot \cos\vartheta$		
2. Wave velocity	$c = \frac{L}{T} = \frac{g}{\omega} kd = \sqrt{gd}$	$c = \frac{L}{T} = \frac{g}{\omega} \tanh(kd) = \sqrt{\frac{g}{k}\tanh(kd)}$	$c = \frac{L}{T} = \frac{g}{\omega} = \sqrt{\frac{g}{k}}$
3. Wave length	$L = c \cdot T = \frac{g}{\omega} kdT = \sqrt{gd} \cdot T$	$L = c \cdot T = \frac{g}{\omega} \tanh(kd) \cdot T = \sqrt{\frac{g}{k}\tanh(kd)} \cdot T$	$L = c \cdot T = \frac{g}{\omega} \cdot T = \sqrt{\frac{g}{k}} \cdot T$
4. Velocity of the water particles			
a) Horizontal	$u = \frac{H}{2} \cdot \sqrt{\frac{g}{d}} \cdot \cos\vartheta$	$u = \frac{H}{2} \cdot \omega \cdot \frac{\cosh[k\ (z+d)]}{\sinh(kd)} \cdot \cos\vartheta$	$u = \frac{H}{2} \cdot \omega \cdot e^{kz} \cdot \cos\vartheta$
b) Vertical	$w = \frac{H}{2} \cdot \omega \cdot \left(1 + \frac{z}{d}\right) \sin\vartheta$	$w = \frac{H}{2} \cdot \omega \cdot \frac{\sinh[k\ (z+d)]}{\sinh(kd)} \cdot \sin\vartheta$	$w = \frac{H}{2} \cdot \omega \cdot e^{kz} \cdot \sin\vartheta$
5. Acceleration of the water particles			
a) Horizontal	$\frac{\partial u}{\partial t} = \frac{H}{2} \cdot \omega^2 \cdot \sqrt{\frac{g}{d}} \cdot \sin\vartheta$	$\frac{\partial u}{\partial t} = \frac{H}{2} \cdot \omega^2 \cdot \frac{\cosh[k\ (z+d)]}{\sinh(kd)} \cdot \sin\vartheta$	$\frac{\partial u}{\partial t} = \frac{H}{2} \cdot \omega^2 \cdot e^{kz} \cdot \sin\vartheta$
b) Vertical	$\frac{\partial w}{\partial t} = -\frac{H}{2} \cdot \omega^2 \cdot \left(1 + \frac{z}{d}\right) \cos\vartheta$	$\frac{\partial w}{\partial t} = \frac{H}{2} \cdot \omega^2 \cdot \frac{\sinh[k\ (z+d)]}{\sinh(kd)} \cdot \cos\vartheta$	$\frac{\partial w}{\partial t} = \frac{H}{2} \cdot \omega^2 \cdot e^{kz} \cdot \cos\vartheta$

Table R 159-1. Linear wave theory. Physical relationships [28]

D = pile diameter or (for non-circular piles) characteristic width of the structural member [m],

F = cross sectional area of the flowed-around pile in the studied area of flow direction [m^2].

The velocity and acceleration of the water particles are taken from the wave equations. Different wave theories may be used, whose areas of application are stated in R 136, figure R 136-2, section 5.6.2.2. For the linear wave theory, the relations required for the calculation of velocity and acceleration have been compiled in table E 159-1. Reference is made to [21], [44] and [46] for the use of theories of a higher order.

The symbols used in table R 159-1 mean:

$$\vartheta = \frac{2\pi \cdot x}{L} - \frac{2\pi \cdot t}{T} = kx - \omega t \quad \text{(phase angle)},$$

$$k = \frac{2\pi}{L}, \quad \omega = \frac{2\pi}{T}, \quad c = \frac{\omega}{k},$$

t = time duration [s],
T = wave period [s],
c = wave velocity [m/s],
k = wave number [1/m],
ω = wave angular frequency [1/s].

Otherwise, see fig. R 159-1.

5.10.3 Determination of the Wave Loads on a Vertical Individual Pile

Since the velocities and, accordingly, accelerations of the water particles are a function, among others, of the distance of the studied location from the still water level, the wave load diagram according to fig. 159-1 results from the calculation of the wave pressure load for various values of z.

The coordinate zero point lies at the level of the still water table, but can be fixed optionally in the abscissa.

z = ordinate of the investigated point ($z = 0$ = still water table),
x = abscissa of the investigated point,
η = temporal changeable elevation of the water table, referred to the still water table (water surface displacement) [m],
d = water depth below the still water table [m],
D = pile diameter [m],
H = wave height [m],
L = wave length [m].

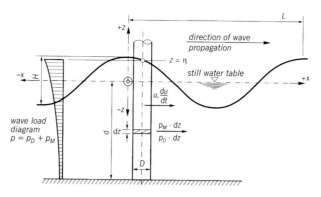

Fig. R 159-1. Wave action on a vertical pile

Please note that the components of the wave load max p_D and max p_M occur phase-displaced. The calculation is therefore to be executed for differing phase angles and the maximum load determined by phase-adjusted superposition of the components from flow velocity and flow acceleration. Thus for instance, the acceleration force is phase-displaced by 90° ($\pi/2$) when the linear wave theory is applied, as compared to the flow velocity compressive force, which lies phase-equal to the wave profile (fig. R 159-2).

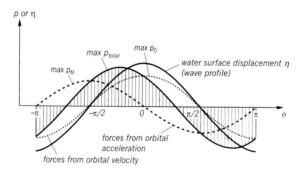

Fig. R 159-2. Variation of forces from velocity and acceleration over one wave period

5.10.4 Coefficients C_D and C_M

5.10.4.1 Drag Coefficient for Flow Pressure C_D

The drag coefficient C_D taking into account the resistance against flow pressure is determined from measurements. C_D depends on the shape of the flowed-around body, the REYNOLDS number Re, surface roughness of the pile and the initial degree of turbulence of the current [42], [43] and [47].

The location of the separation point of the boundary layer is decisive for flow pressure. The C_D value is practically steady for piles where the separation point is indicated by corners or break-off edges (fig. R 159-3).

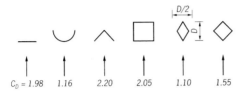

$C_D = 1.98 \quad 1.16 \quad 2.20 \quad 2.05 \quad 1.10 \quad 1.55$

Fig. R 159-3. C_D values of pile cross-sections with stable separation points [41]

In the case of piles without stable separation points, for instance circular cylindrical piles, distinctions are made between a subcritical range of the REYNOLDS number with a laminar boundary layer, and a supercritical range with turbulent boundary layer.

However, in general under natural conditions high REYNOLD numbers prevail, so that in the case of smooth surfaces, a uniform value of $C_D = 0.7$ is recommended, [21] and [42]. For further details, please consult [162].

Larger C_D values are to be used for rough surfaces, see e.g. [48].

5.10.4.2 Inertia or Mass Coefficient C_M for Accelerated Flow

With the potential flow theory, the value $C_M = 2.0$ is obtained for the circular cylindrical pile, whereas C_M values up to 2.5 have also been ascertained on the basis of tests for the circular cross-section [49].

Normally, the theoretical value $C_M = 2.0$ can be used. Otherwise, reference is made to [21], [48] and [162].

5.10.5 Forces from Breaking Waves

At present there is no usable calculation method available for suitable determination of the forces from breaking waves. The MORISON formula is therefore used again for these waves, but under the assumption that the wave acts as water mass with high velocity on the pile, without acceleration. Thereby the inertia coefficient C_M is set to 0, whereas the drag coefficient C_D is increased to 1.75 [21].

5.10.6 Wave Load with Pile Groups

In determining the wave loads on pile groups, the phase angle ϑ applicable for the respective pile location is to be taken into account.

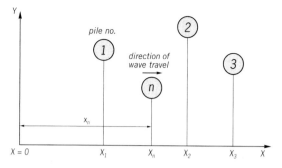

Fig. R 159-4. Definitions for a pile group (layout) (according to [21])

With the designation according to R 159-4, the horizontal total load on a pile structure of N piles comes to:

$$\text{total } P = \sum_{n=1}^{N} P_n(\vartheta_n).$$

The symbols mean:

N = number of piles,
$P_n(\vartheta_n)$ = wave load of the individual pile n, taking into account the phase angle $\vartheta = k \cdot x_n - \omega \cdot t$ [kN],
x_n = distance of the pile n from the y-z plane [m].

It should be noted that for piles standing closer than about four pile diameters, an increase in load occurs for piles standing side by side in the wave direction, and a reduction of load for piles standing one behind the other.

The correction factors compiled in table R 159-2 are proposed for this load [49]:

Pile centre-to-centre distance e / Pile diameter D	2	3	4
for piles in rows parallel to the wave crest	1.5	1.25	1.0
for piles in rows vertical to the wave crest	0.7[1]	0.8[1]	1.0

[1] Reduction does not apply to the front pile directly exposed to wave action.

Table R 159-2. Multiplier for small pile distances

5.10.7 Inclined Piles

In the case of inclined piles, it should also be noted that the phase angle ϑ for the local coordinates x_0, y_0, z_0 differs for the individual pile sections d_s.

The pressure on the pile at the considered section is to be determined with the coordinates x_0, y_0 and z_0 according to fig. R 159-5.

The local force $p \cdot d_s$ due to velocity and acceleration of the water particles on the pile element d_s ($p = f[x_0, y_0, z_0]$) can be equated with the horizontal force on an assumed vertical pile at (x_0, y_0, z_0) according to [21]. However, when the pile slopes at a steeper angle, it should be checked whether the determination of load furnishes more unfavourable values, taking into account the components of the resulting velocity acting vertically to the pile axis

$$v = \sqrt{u^2 + w^2}$$

and the resultant acceleration

$$\frac{\partial v}{\partial t} = \left(\frac{\partial u}{\partial t}\right)^2 + \left(\frac{\partial w}{\partial t}\right)^2.$$

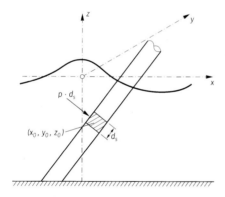

Fig. R 159-5. For calculating the wave forces on an inclined pile [21]

5.10.8 Safety Factors

The design of pile structures against wave action is strongly dependent on the selection of the "design wave" (R 136, section 5.6 in conjunction with table E 159-1, no. 3).

Also of influence are the wave theory used and the coefficients C_D and C_M assigned to it. This applies particularly to pile structures in shallow water. In order to allow for such uncertainties, it is recommended to multiply the calculated loads with increased partial safety factors according to [21].

Consequently, when the "design wave" occurs only rarely, i.e. in normal cases with deep water conditions, the resulting wave load on piles is to be increased by a partial safety factor $\gamma_d = 1.5$ When the "design wave" occurs frequently, which is usually the case in shallow water conditions, the partial safety factor $\gamma_d = 2.0$ is recommended.

Reference is made to [162] and [46] regarding the possibility of using the coefficients C_D and C_M depending on the REYNOLDS- and KEULEGAN-CARPENTER-number and a corresponding reduction of the partial safety factor.

Critical vibrations may occur occasionally at pile structures, especially when separation eddies act transverse to the flow direction, or the inherent frequency of the structure is close to the wave period resulting in resonance phenomena. In this case, regular waves smaller than the "design wave" can be less favourable. In such cases, special investigations are required.

5.11 Wind Loads on Moored Ships and Their Influence on the Dimensioning of Mooring and Fendering Facilities (R 153)

5.11.1 General

This recommendation is applicable as supplement to the proposals and instructions dealing with the planning, design and dimensioning of fender and mooring facilities, especially to

R 12, section 5.12, R 111, section 13.2 and R 128, section 13.3.

The loads for mooring installations, such as bollards or quick release hooks with the appertaining anchorages, foundations, retaining structures etc., which result according to this recommendation, replace the load values in R 12, section 5.12, only when the influence of swells, waves and current on ship berths can be neglected. Otherwise, these latter must be specially checked and additionally taken into consideration

R 38, section 5.2, is not affected by this recommendation. In determining the "normal berthing loads" dealt with therein, reference to R 12 section 5.12.2 therefore remains applicable without restriction.

5.11.2 Applicable Wind Velocity

If there is no other specific data available on the wind conditions for the area of the ship berth, the values given in DIN 1055, part 4, section 4.3 are to be used as applicable wind velocities v for all wind directions at comparatively sheltered location. In open coast regions however, increased wind velocities are to be expected.

If the maximum gust velocities (peak values max v from many years of local measurements) are known,

$$v = \frac{\max v}{1.10}$$

may be taken as applicable wind velocity.
This basic value can be differentiated according to wind directions, insofar as more detailed data is available.

5.11.3 Wind Loads on the Moored Vessel

The loads quoted are *characteristic* values.

Wind load components:

$$W_t = (1 + 3.1 \sin \alpha) \cdot k_t \cdot H \cdot L_{\ddot{u}} \cdot v^2$$
$$W_l = (1 + 3.1 \sin \alpha) \cdot k_l \cdot H \cdot L_{\ddot{u}} \cdot v^2.$$

Equivalent loads for $W_t = W_{tb} + W_{th}$:

$$W_{tb} = W_t \cdot (0.50 + k_e)$$
$$W_{th} = W_t \cdot (0.50 - k_e).$$

The symbols herein mean:

H	=	greatest freeboard height of ship (ballasted or empty) [m],
$L_{\ddot{u}}$	=	overall length [m],
v	=	applicable wind velocity [m/s],
W_i	=	wind load components [kN],
k_t and k_l	=	wind load coefficients [kNs^2/m^4],
k_e	=	coefficient of eccentricity [1].

As international experience has shown, the load and eccentricity coefficients may be applied in accordance with tables R 153-1 and 153-2.

Forces diagram:

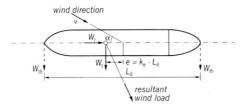

Fig. 153-1. Application of wind loads on the moored vessel

Ships up to 50 000 dwt			
α^0	k_t [kN · s²/m⁴]	k_e [I]	k_l [kN · s²/m⁴]
0	0	0	$9.1 \cdot 10^{-5}$
30	$12.1 \cdot 10^{-5}$	0.14	$3.0 \cdot 10^{-5}$
60	$16.1 \cdot 10^{-5}$	0.08	$2.0 \cdot 10^{-5}$
90	$18.1 \cdot 10^{-5}$	0	0
120	$15.1 \cdot 10^{-5}$	−0.07	$-2.0 \cdot 10^{-5}$
150	$12.1 \cdot 10^{-5}$	−0.15	$-4.1 \cdot 10^{-5}$
180	0	0	$-8.1 \cdot 10^{-5}$

Table R 153-1. Load and eccentricity coefficients for ships up to 50 000 dwt

Ships over 50 000 dwt			
α^0	k_t [kN · s²/m⁴]	k_e [I]	k_l [kN · s²/m⁴]
0	0	0	$9.1 \cdot 10^{-5}$
30	$11.1 \cdot 10^{-5}$	0.13	$3.0 \cdot 10^{-5}$
60	$14.1 \cdot 10^{-5}$	0.07	$2.0 \cdot 10^{-5}$
90	$16.1 \cdot 10^{-5}$	0	0
120	$14.1 \cdot 10^{-5}$	−0.08	$-2.0 \cdot 10^{-5}$
150	$11.1 \cdot 10^{-5}$	−0.16	$-4.0 \cdot 10^{-5}$
180	0	0	$-8.1 \cdot 10^{-5}$

Table R 153-2. Load and eccentricity coefficients for ships over 50 000 dwt

5.11.4 Loads on Mooring and Fender Facilities

In order to determine the mooring and fender forces, a static system is to be introduced formed by the ship, the hawsers and the mooring and fender structures. The elasticity of the hawsers, which is contingent on material, cross-section and length, is to be taken into account just as much as the inclination of the hawsers in horizontal and vertical direction, under variable loads and water level conditions. The elasticity of the mooring and fender structures is to be ascertained at all support and bearing points of the static system. Anchored sheet pile walls and structures with inclined pile foundations may therefore be considered as rigid elements. Please note that the static system can alter if individual lines fall slack or fenders remain unloaded under specific load situations. All *characteristic* mooring and fender loads determined on the basis of the wind loads given in section 5.11.3 are to be multiplied by a partial safety factor of $\gamma_d = 1.25$ to cover dynamic and other non-ascertainable influences.

The wind-shielding effect of structures and facilities may be taken into account in a reasonable manner.

5.11.5 Safety
The load case 2 is applicable for the design of structures taking into account the wind loads acting on the ship.

5.12 Layout and Loading of Bollards for Seagoing Vessels (R 12)
5.12.1 Layout
For simple and straightforward structural treatment, bollard spacing of 30 m is recommended, i.e. equal to the normal length of a wall section (see R 17, section 10.1.5) for quays and walls on piles with superstructures of mass concrete or reinforced concrete. Generally, the bollard should be placed in the centre of a section. If there are two bollards in each section, they should be placed symmetrically at about the centre of the section at the outer quarter points. Proceed accordingly for shorter section lengths. The distance of the bollard from the face of the wall should be as recommended in R 6, section 6.1.2.

The bollards can be designed as a simple bollard or as a double bollard. They can hold several hawsers simultaneously. They should be designed to allow for easy repairs or replacement.

5.12.2 Loads
The laid-on hawsers are generally not fully stressed at the same time, and the hawser forces partially cancel each other out in action, so that the line pull forces according to table R 12-1 may be applied for both single and double bollards:

Displacement t	Line pull force kN
up to 2 000	100
up to 10 000	300
up to 20 000	600
up to 50 000	800
up to 100 000	1 000
up to 200 000	1 500
> 200 000	2 000

Table R 12-1. Line pull forces

The stated loads are *characteristic* values which are to be multiplied by the partial safety factor of 1.3 for the design of the structure elements.
At quays and berths for larger vessels where there is a strong current, the line pull forces as per table R 12-1 should be increased by 25 % beginning with ships of 50 000 t displacement.
Main bollards at the ends of the individual large vessel berths at river structures are designed for 2 500 kN line pull forces under unfavourable

conditions for ships up to 100 000 t displacement, and for double the value in table R 12-1 for larger ships, insofar as more accurate values are not available from determining the wind and current influences.

5.12.3 Direction of the Line Pull Force

The line pull force may occur at any angle toward the waterside. An inboard line pull force is not assumed, unless the bollard serves a waterfront structure lying in that direction, or has a special purpose as corner bollard. When designing the waterfront structure, the line pull force is usually assumed to act horizontally.

When designing the bollard itself and its connection to the waterfront structure, upward slanting inclines of up to 45° with corresponding line pull forces are to be taken into consideration.

Load case 3 with the partial safety factor 1.0 is applicable to designing the bollards.

5.13 Layout, Design and Loading of Bollards in Inland Harbours (R 102)

This recommendation has been brought into line with DIN 19 703 "Locks for waterways for inland navigation – principles for dimensioning and equipment" insofar as the principles of this standard can be applied to waterfront structures.

The term bollard has been used to cover all mooring facilities, and includes edge bollards, recess bollards, dolphin bollards, mooring hooks, mooring lugs, mooring rings and similar.

5.13.1 Layout and Design

In inland harbours, ships should be moored to the shore by three hawsers, namely with a bow line, a breast line and a stern line. An appropriate number of bollards should be provided on the bank.

Bollards must be arranged on and above the level of the port ground surface, with their upper surface above HHW (fig. R 102-1). The bollard diameter must be more than 15 cm. If the bollard does not project sufficiently above HNW, slippage of the line must be prevented with a crosspiece.

In addition to bollards along the top of the bank, in river ports other bollards must be installed at various elevations according to local water level fluctuations. Only then can the ship's crew moor the ship without any difficulty at every water level and every freeboard height.

In the case of vertical waterfront structures, the bollards lie at differing heights in rows perpendicular to each other. The position of the rows depends on the position of the rising ladders. To avoid undue tension on the ladders, a row of bollards is located to the left and right of every rising ladder at a distance of approx. 0.85 m to 1.00 m to the axis of the

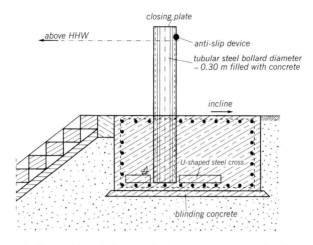

Fig. R 102-1. Foundation for a bollard on the port ground surface (design to statical requirements)

ladder. A further row of bollards is positioned in the middle between the ladders. The axis distance of the row of bollards is around 15 m for a distance of around 30 m between the ladders. In the case of steel piling walls, the precise axis dimension is determined by the system dimension of the planks, and by the section lengths in the case of solid walls.

The lowest bollard is located at approx. 1.50 m above LLW, in tidal areas over MLWS. The perpendicular distance between this and the upper edge of the waterfront structure is divided by further bollards at a distance of 1.30 to 1.50 m (up to 2.00 m in borderline cases).

In the case of waterfront structures of reinforced concrete, the bollards are positioned in recesses, the housings of which are anchored and concreted in position. In the case of steel piling walls, the bollards can be screwed or welded in position. The front edge of the bollard cone should be 5 cm behind the front edge of the waterfront structure. Appropriate clearance is to be left at the side behind and over the bollard cone so that the ships' hawsers can be looped over it and removed again easily. The edges to the waterfront structure are to be rounded off to prevent any damage to the hawsers and waterfront structure.

In the case of partially sloping and sloping banks, the bollards are positioned on both sides next to the steps (fig. R 102-2). The steps are located as extension of the ladders.

For this arrangement, the bollard foundation is best located under the steps for both bollards.

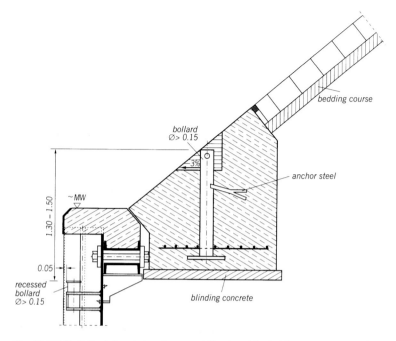

Fig. R 102-2. Bollard foundation for a partially sloped bank (shown as an example, design to statical requirements)

5.13.2 Loads

The occurring line pull forces depend essentially on the ship's size, the speed of and the distance from passing ships, the velocity of water currents at the berth and the quotient of water cross section to the submerged ship's cross section.

100 kN load per bollard is to be taken. A line pull force of 200 kN is to be used for berths and transhipment sites with very heavy use. The stated loads are *characteristic* values which are to be multiplied by the partial safety factor of 1.3 for designing structural components.

Ships in motion are not allowed to brake at bollards so that the corresponding effect is not taken into consideration in the load assumptions (actions).

5.13.3 Direction of Line Pull Forces

Line pull forces can only be applied only from the water side, mostly at an acute angle and only rarely at right angles to the bank.

The calculations must however take account of every reasonable horizontal and vertical angle to the bank.

5.13.4 Calculation

Stability verification is to be provided for a single line pull force acting at the most unfavourable angle. Stability verification can also be provided by means of test loadings.

If the bollard has its foundations immediately behind or in a solid waterfront structure, this structure must be capable of assuming the line pull force as per load case 2. The bollards themselves and their connections are rated to load case 3 with a partial safety factor of 1.0.

5.14 Quay Loads from Cranes and Other Transhipment Equipment (R 84)

The following loads are *characteristic* values which are to be multiplied with the partial safety factors of the relevant load cases (see R 18, section 5.4) according to DIN V 1054-100.

5.14.1 Usual General Cargo Harbour Cranes

5.14.1.1 General

The customary general cargo harbour cranes constructed in Germany are chiefly full portal level-luffing cranes spanning 1, 2 or 3 railway tracks.

Occasionally however, half portal cranes are manufactured. The bearing capacity varies between 7 and 50 t at a working radius of 20 to 45 m.

The axis of rotation of the crane superstructure should lie as close as possible to the outboard crane rail in the interest of the working radius from the centre of rotation. However, to avoid any collision between crane and heeling ship, care must be taken that neither the crane operator's cabin nor the rear counterweight project out over an inboard inclined plane, toward the top by 5°, proceeding from the quay edge.

The distance between the outboard crane rail and the face of the quay wall shall be in accordance with R 6, section 6.1. The corner spacing of the small cranes is approx. 6 m. Minimum corner spacing should not fall below 5.5 m, as otherwise excessive corner loads occur and the crane must be equipped with much too high central ballast. The length between buffers is approx. 7 to 22 m, depending on the size of the crane. If this results in too high a wheel load, lower wheel loads can be achieved by increasing the number of wheels. Today however there are general cargo handling facilities whose craneways have been built for especially high wheel loads.

General cargo harbour cranes are as a rule classified into lifting class H 2 and load group B 4 or B 5 in accordance with DIN 15 018, part 1. Additional, reference is made to the F.E.M. 1001 [154].

The vertical wheel loads from dead weight, live load, inertia forces and wind forces are to be applied in the calculation of the craneway (DIN

15 018, part 1). Vertical inertia forces from the travel motion or from the lifting or setting down of the live load are to be allowed for through use of impact factor, which comes to about 1.2 in lifting class H 2. The foundation of the craneway may be dimensioned without taking an impact factor into account. All crane booms can slew through 360°. The appertaining corner load changes accordingly. With higher wind loads and crane not in operation, calculations may be used with load case 3, if necessary, for dimensioning the quays and craneways.

When dealing with preliminary designs for quays and craneways with appropriate prerequisites, line load can be taken simply as 500 kN/m, in some cases restricted to the length of the chassis, or 550 kN respectively 700 kN as wheel load for wheel spacing of 1.10 m respectively 1.40 m, with one crane rail taking the full load and the other 0.2 times the load. The horizontal forces per wheel are to be taken into account as follows with impact coefficient:

in rail direction 1/7 each of the wheel load of the braked wheels, transverse to rail direction 1/10 each of the wheel load to allow for inertia effects, angular deviation and wind. For heavy luffing jib cranes, the horizontal transverse force in preliminary design calculations is to be taken as 1/8 and not 1/10 of the wheel load. For cranes in operation with operating winds substantially above the value as per DIN 15 018, even higher values are to be assumed where applicable. It may be taken into account however when horizontal forces from side thrust acting in opposite directions act simultaneously on the same structural member. The final structural calculations are always to proceed on the basis of the vertical and horizontal corner or wheel loads indicated by the crane manufacturer.

5.14.1.2 Full Portal Cranes

The portal of light harbour cranes with low lifting capacities has either four or three supports, of which each has one to four wheels. The number of wheels in each case depends on the allowable wheel load. General cargo heavy-lift cranes have at least six wheels per support. On straight quay stretches, the centre-to-centre spacing of the crane rails is at least 5.5 m, in general however 6, 10 or 14.5 m, depending on whether the portal spans one, two or three tracks. The dimensions 10 m and 14.5 m result from the theoretical minimum dimension of 5.5 m for a track, to which one or two times the track spacing of 4.5 m is to be added.

5.14.1.3 Half Portal Cranes

The portal of these cranes has only two supports which run on the outboard crane rail. On the inboard side, it is supported by a short leg on an elevated craneway, thus providing free access to any section of the quay area. The remarks in section 5.14.1.2 apply to the number of wheels under both supports and the leg.

5.14.2 Container Cranes

The actual container cranes are constructed as full portal cranes with cantilever beams and trolley (loading bridges), whose supports have eight to ten wheels as a rule. The crane rails of existing container cargo handling facilities are in general 15.24 m (50') or 18 m on centres. A track width of 30.48 m (100') is often selected for new facilities. The clear support spacing = free space between the corners longitudinally is 17 m to 18.5 m at an overall spacing between the buffers of about 27 m (fig. R 84-1). If the handling of 20' containers requires use of the smaller overall spacing between the buffers, a smallest corner spacing up to 12 m is possible. The overall spacing between the buffers is then 22.5 m. The corner spacing in this case is not the same as the portal support spacing. The lifting capacity of the cranes is selected between 45 t and 55 t including spreader. The maximum corner load is affected in particular by the design and length of the jib. The working range usual hitherto (38 m to 41 m, corresponding to the vessel width of Panmax vessels) is insufficient for the "post Panmax" vessels recently put into service, which cannot pass through the Panama Canal owing to their width. Jib lengths of at least 44.5 m are necessary for this type of vessel. The maximum corner loads for container cranes in operation range up to 4500 kN for Panmax ships and up to 8000 kN for post Panmax ships.

5.14.3 Loads for Harbour Cranes

	Rotating cranes	Container cranes and other transhipment gear
Bearing capacity [t]	7–50	10–80
Dead weight [t]	180–350	200–1200
Portal span [m]	6–19	9–45
Clear portal height [m]	5–7	5–13
Max. vertical corner load [kN]	800–3000	1200–8000
max. vertical wheel surcharge load [kN/m]	250–600	250–700

Horizontal wheel load
  transverse to the direction of the rail
  in the direction of the rail
	up to approx. 10 % of vertical load
	up to approx. 15 % of the vertical load of the braked wheels	

Claw load [1] [kN]	Mobile crane up to 4800	

[1] Prerequisite is a zone of 40 m^2 subject to no other loads; the claw load can be taken as distributed over 4 m^2.

Table R 84-1. Dimensions and characteristic loads of rotating and container cranes

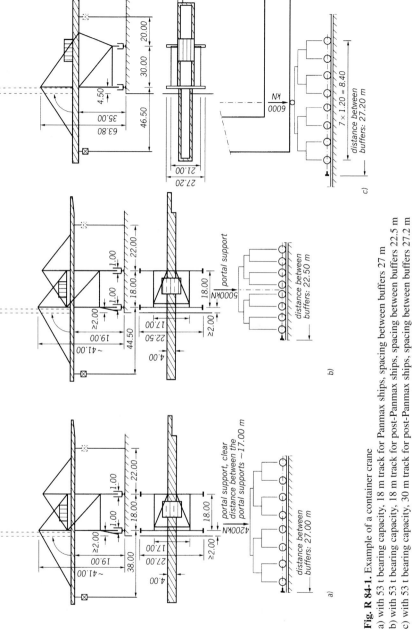

Fig. R 84-1. Example of a container crane
a) with 53 t bearing capacity, 18 m track for Panmax ships, spacing between buffers 27.00 m
b) with 53 t bearing capacity, 18 m track for post-Panmax ships, spacing between buffers 22.50 m
c) with 53 t bearing capacity, 30 m track for post-Panmax ships, spacing between buffers 27.20 m

The support structure always consists of a portal-type substructure, either with rotating or rigid beam which can be hinged up when decommissioned. The portal usually stands on four corner points with several wheels arranged in rockers, depending on the magnitude of the corner load. The corner load is distributed as evenly as possible among all wheels of the corner point. Table R 84-1 summarises general loads and dimensions to supplement the details given in section 5.14.1 and 5.14.2.

5.14.4 Notes

Further details about harbour cranes are to be found in the AHU recommendations and reports [185] E 1, E 9, B 6 and B 8, in the ETAB [45] recommendation E 25 and in the VDI directive 3576 [184].

5.15 Impact and Pressure of Ice on Waterfront Structures, Fenders and Dolphins in Coastal Areas (R 177)

5.15.1 General (see also [46], [148])

Loads on hydraulic engineering facilities due to the effects of ice can occur in various ways:

a) as ice impact from collisions with ice flows carried along by the current or by wind,
b) as ice pressure exercising its effect through ice thrusting against an ice layer adjacent to the structure or through vessel movements,
c) as ice pressure exercising its effect on the structure through an unbroken ice layer as a result of thermal expansion,
d) as live ice loads on the structure when ice forms on the structure or as live or lifting loads when the water levels fluctuate.

Among others, the magnitude of possible load effects depends on:

- the shape, size, surface condition and resilience of obstacles against which the ice mass collides,
- the size, shape and propagation speed of the ice mass,
- the nature of the ice and ice formation,
- the salt content of the ice and the ice strength which depends on this,
- the angle of impact,
- the applicable strength of the ice (compression, bending and shear strength),
- the loading rate,
- the ice temperature.

Whenever possible, it is recommended that the applicable load values for waterfront structures including pile-founded structures be checked with the assumptions for constructed facilities, which have proven their worth, or with local ice pressure measurements.

The ice loads determined below are *characteristic* values. A partial safety factor of 1.0 can be used because of the usually slight probability of occurrence.

5.15.2 Ice Loads on Waterfront Structures

In the northern German coastal region, an ice thickness of 50 cm and an ice compression strength of $\sigma_0 = 1.5$ MN/m^2 with temperatures around freezing point can generally be assumed as the starting point for determination of the horizontal ice loads on surface structures. This gives rise to the assumption:

a) 250 kN/m as the mean horizontally acting linear load at the least favourable altitude of the water levels under consideration, it being a prerequisite that the max. load calculated from the ice compression strength of 750 kN/m on average has an effect of only 1/3 of the length of the structure (contact coefficient $k = 0.33$).
b) 1.5 MN/m^2 as local surface load.
c) 100 kN/m as the mean horizontally acting linear load at the least favourable altitude of the water levels under consideration for groynes and waterfront revetments in tidal regions if a broken ice layer is created as a result of water level fluctuations.

The simultaneous action of ice effects with wave loads and/or vessel impact is not to be assumed.

5.15.3 Ice Loads on Piles of Pile-founded Structures or on Individual Piles

5.15.3.1 Principles of Determination of Ice Load

The ice loads acting on piles depend on the shape, inclination and layout of the piles and on the ice compression, bending and shear strength applicable for the fracture of the ice. The magnitude of the load also depends on the nature of the load – whether primarily a dead load or impact load from colliding ice floes.

In the case of North Sea ice, it can generally be assumed that the mean compression strength does not exceed the value $\sigma_0 = 1.5$ MN/m^2, for Baltic Sea ice the figures are $\sigma_0 = 1.8$ MN/m^2 and for fresh-water ice, $\sigma_0 = 2.5$ MN/m^2. The values apply to a specific expansion rate $\dot{\varepsilon} = 0.003$ s^{-1}, at which the ice compression strength achieves its maximum value according to tests [108].

To the extent that more accurate ice strength investigations are not available, the bending tensile strength σ_B can be assumed to be approx. 1/3 σ_0 and the shear strength τ as approx. 1/6 σ_0. Standard values in accordance with table R 177-1 apply for the ice thicknesses h for the German North Sea and Baltic coasts.

North Sea	max h (cm)	Baltic	max h (cm)
Heligoland	30–50	Kiel Canal	60
Wilhelmshaven	40	Flensburg (outer fjord)	32
"Hohe Weg" lighthouse	60	Flensburg (inner fjord)	40
Büsum	45	Schlei mouth	35
Meldorf (harbour)	60	Kappeln	50
Tönning	80	Eckernförde	50
Husum	37	Kiel (harbour)	55
Wittdünn harbour	60	Bay of Lübeck	50
		Wismar harbour	50
		Wismar – bay	60
		Rostock – Warnemünde	40
		Stralsund – Palmer Ort	65
		Saßnitz – harbour	40
		Koserow – Usedom	50

Table R 177-1. Measured maximum ice thicknesses as standard values for measurement

On the German North Sea coast, the ice load is frequently taken at a height of 0.5 m to 1.5 m height above MHW for free-standing piles. The assumptions below apply to slim components up to 2 m wide with flat ice. In the event of the formation of compression ice ridges, the ice loads listed below must be doubled.

5.15.3.2 Ice Load on Vertical Piles

Irrespective of the form of the pile cross-section, the horizontal ice load from the effect of drift ice is calculated on the basis of the investigations in accordance with [108] to be:

$$P_p = 0.36\, \sigma_0 \cdot d^{0.5} \cdot h^{1.1}.$$

in which:

σ_0 = ice compression strength in MN/m² at specific expansion rate $\dot{\varepsilon} = 0.003$ s⁻¹,
d = width of individual pile [cm],
h = thickness of ice [cm],
P_p = ice load [kN].

If the case of incipient ice movement with tightly adjacent ice is to be considered, the following load assumptions are applicable:

with a round or half-round pile:

$$P_i = 0.33\, \sigma_0 \cdot d^{0.5} \cdot h^{1.1} \text{ [kN]},$$

with a square pile:

$$P_i = 0.39\, \sigma_0 \cdot d^{0.68} \cdot h^{1.1} \text{ [kN]},$$

or with a wedge-shaped pile:

$$P_i = 0.29\, \sigma_0 \cdot d^{0.68} \cdot h^{1.1} \text{ [kN]}.$$

5.15.3.3 Ice Load on Inclined Piles

With inclined piles, the fracture of the ice flow can occur earlier than the crushing of the ice, as a result of shearing off or bending. In accordance with [109], the smaller ice load is applicable in each case. In piles with an inclination more pronounced than 6 : 1 ($\beta \geq$ approx. 80°), the ice load must be calculated in accordance with section 5.15.3.2.

In the case of shear fracture, the horizontal ice load is:

$$P_s = c_{fs} \cdot \tau \cdot k \cdot \tan\beta \cdot d \cdot h \text{ [kN]}.$$

in which:

P_s = horizontal load on shear fracture [kN],
τ = shear strength [MN/m^2],
c_{fs} = shape coefficient as per table R 177-2 [1],
k = contact coefficient, generally approx. 0.75 [1],
β = angle of inclination of pile against horizontal [°],
d = pile width [cm],
h = thickness of ice [cm].

In bending fracture, the horizontal ice load is:

$$P_b = c_{fb} \cdot \sigma_B \cdot \tan\beta \cdot d \cdot h \text{ [kN]}$$

in which

P_b = horizontal ice load for bending fracture [kN],
σ_B = bending tensile strength [MN/m^2],
c_{fb} = shape coefficient as per table R 177-3 [1].

5.15.3.4 Horizontal Load on Pile Groups

The ice load on pile groups arises from the sum of the ice loads on the individual piles. Generally, assumption of the sum of the ice loads which act on the piles facing the ice drift will suffice.

5.15.4 Live Ice Loads

The live ice load must be assumed in accordance with local conditions. Without more detailed checks, a minimum live ice load of 0.9 kN/m^2 can be regarded as sufficient [110]. In addition to the live ice load, as-

Edge angle 2α [°]	Shape coefficient c_{fs}
45	0.29
60	0.22
75	0.18
80	0.17 (= round pile)
90	0.16
105	0.14
120	0.13
180	0.11 (= square pile)

Table R 177-2. Shape coefficient c_{fs} for round pile, square pile or wedge-shaped edge with 2α = edge angle, measured in the horizontal plane

Edge angle 2α [°]	Shape coefficient c_{fb}				
	at inclination angle β [°]				
	45	60	65	70	75
45	0.019	0.024	0.029	0.037	0.079
60	0.017	0.020	0.022	0.026	0.038
from 75	0.017	0.019	0.020	0.021	0.027

Table R 177-3. Shape coefficient c_{fb} for round pile, square pile or wedge-shaped edge with 2α edge angle, measured in the horizontal plane

sumption of the usual snow load at 0.75 kN/m² also comes into consideration. Conversely, traffic loads which have no effect with thick ice formation, generally do not need to be assumed at the same time.

5.15.5 Vertical Loads with a Rising or Falling Water Level

With rising or falling water levels, vertical supplementary forces from immersing or projecting ice act on frozen structures or piles. Ice adhering to the side of the structure with a strip width $b = 5$ m and ice thickness h and any ice present below the structure in its full volume can be included for rough calculations. With the weight density of the ice γ_E = approx. 9 kN/m³ and a falling water level, the ice volume thus determined, V_E generates the load $P = V_E \cdot \gamma_E$ acting vertically downwards, and with the difference between the weight densities of water and ice $\Delta\gamma_E = 1$ kN/m³, it generates the load $P = V_E \cdot \Delta\gamma_E$ acting vertically upwards.

5.15.6 Supplementary Information

The above recommendations for ice loads on structures are rough assumptions which apply to German conditions, i.e. not to Arctic regions. Considerably reduced values apply to protected areas (bays, harbour basins, etc.) and in seaports with clear tidal action and considerable traffic.

Corresponding reductions of the assumed loads also apply insofar as measures are introduced to reduce the ice load, such as timely breaking or exploding of the ice, influencing the current, use of air bubbling facilities, heating or other thermal influence etc., or when the ice field is only small in size.

Ice formation and ice loads also depend extensively on the wind direction, flow and shear zone arrangement in the ice. This must be taken into consideration particularly in the layout of harbour entrances and the alignment of harbour basins, for example. In narrow harbour basins, considerable ice loads arising from deformation can occur as a result of temperature changes in the ice. In accordance with [111], it can be assumed, in view of the ice temperatures in the North German coastal region, which are generally not very low, that the thermal ice pressure does not exceed 400 kN/m^2.

In individual cases, where it is important to determine the ice loads more precisely, experts should be consulted and, if applicable, model tests should be carried out.

If the ice loads in the case of dolphins significantly exceed the loads from vessel impact or bollard tension, an investigation should be made as to whether such dolphins should be designed for the higher ice loads, or whether rarely-occurring overstresses can be simply accepted for economic reasons.

5.16 Impact and Pressure of Ice on Waterfront Structures, Piers and Dolphins in Inland Areas (R 205)

5.16.1 General

The data stated in the recommendation R 177, section 5.15, is generally applicable to inland areas. This is valid for the general statements and for the loads, as they depend on the construction dimension, ice thickness and strength properties of the ice.

As the heat balance of the inland waterways is nowadays in most cases influenced by the inlet of cooling and sewage water, it can be assumed that as far as inland waterways and inland harbours are concerned, situations of extreme cold are seldom and thus the probability of ice-formation and ice thickness is considerably reduced.

The determined ice loads are *characteristic* values in accordance with R 177, section 5.15.1, to be multiplied by a partial safety factor of 1.0.

5.16.2 Ice Thickness

The ice thickness can be deducted according to [109] from the sum of daily degrees below zero during an ice period – the so-called "cold sum". So for example according to BYDIN [109], $h = \sqrt{\Sigma|t_L|}$ in which h is the ice thickness in cm and $\Sigma|t_L|$ the sum of the absolute values of the negative mean diurnal air temperature in °C for the considered period of time.

If no more detailed statistics or measurement results are available, in general an arithmetical ice thickness of $h \leq 30$ cm can be presumed, the conditions stated under section 5.16.1 being prerequisite.

5.16.3 Ice Strength

The ice strength depends on the ice temperature t_E, the average ice temperature being half the ice temperature at the surface, as 0°C can always be achieved at the underside of ice.

For moderate ice temperatures, the compression strength of fresh water ice can be assumed to be $\sigma_0 = 2.5$ MN/m^2 according to R 177, section 5.15.3.1.

If the average temperature of the ice falls under –5°C, the compression strength increases according to [148] by approx. 0.45 MN/m^2 per below zero degree.

At ice temperatures above –5°C, the compression strength of the ice can also be determined according to [109] to be

$$\sigma_0 = 1.1 + 0.35 \; |t_E| \; [\text{MN/m}^2].$$

5.16.4 Ice Loads on Waterfront Structures and Other Structures of Larger Extension

In general according to R 177, section 5.15.2,

$$p_0 = 10 \cdot k \cdot \sigma_0 \cdot h \quad [\text{kN/m}] \text{ is valid}$$

in which

p_0 = ice load [kN/m],
k = contact coefficient, generally approx. 0.33,
σ_0 = compressive strength of the ice [MN/m^2],
h = thickness of the ice [cm].

On sloped surfaces, the horizontal ice load according to [109] can be taken to be:

$$p_h = 1.0 \cdot h \cdot \sigma_B \cdot h \cdot \tan \beta \quad [\text{kN/m}]$$

with

σ_B = bending tension strength of the ice [MN/m^2],
$\tan \beta$ = slope inclination [1].

5.16.5 Ice Loads on Small Structures (Piles, Dolphins, Bridge- and Weir Piers, Ice Repellers)

The loads for vertical or inclined piles according to R 177, section 5.15.3.2 and 5.15.3.3 are valid in the same manner in consideration of the respective ice strength for the inland area. They are also applicable for pier construction and ice control devices in consideration of the form of the cross-section and the surface as well as the inclination.

5.16.6 Ice Loads on Groups of Structures

The indications in R 177, section 5.15.3.4 are valid.

For structures built in water, a distance of at least

$$l = \frac{1.1 \cdot \sigma \cdot d}{v^2}$$

is recommended according to [109] in order to avoid obstructions when hauling away the ice,

in which

l = distance of piers [m],

σ = $10 \cdot \dfrac{P}{d \cdot h}$ for an ice load P [kN] according to section 5.16.5,

v = drift speed of the ice [m/s],

d = pier diameter [cm].

Otherwise, the possibilities of the formation of pack ice can be assumed according to [148].

Pack ice does not necessarily result in an increase in the ice load when the failure conditions of the packing ice are applicable; changes to load distribution and height of load application must also be considered, together with additional loads from congested water together with changes in current by means of cross section restrictions.

5.16.7 Vertical Loads with a Rising or Falling Water Level

The indications in R 177, section 5.15.5 are valid.

5.16.8 Additional Indications

The information in R 177, section 5.15.6 are to be observed. Approximate values for the thermal ice pressure dependent on the initial temperature and the hourly temperature increase are stated in [148]. The thermal ice pressure remains under 200 kN/m^2 for moderate temperatures.

6 Configuration of Cross-section and Equipment of Waterfront Structures

6.1 Standard Dimensions of Cross-section of Waterfront Structures in Seaports (R 6)

6.1.1 Walkways

The walkway space in front of the outboard crane rail is necessary to provide room for bollards, storing of gangways, as a path and working space for line handlers, for access to the berths and for the accommodation of the outboard portion of the gantry. Consequently, it is of special importance in harbour operations. In the selection of its width, accident prevention regulations must also be considered.

The greater the width required, the farther the crane must be moved back from the edge of the quay. This in turn requires a greater length of the jib. Although a longer jib makes cargo handling more expensive, on the other hand ample clearance is becoming increasingly desirable because of the growing number of ships which are being built with superstructure hanging over the hull. This overhang is especially hazardous to crane operation when the ship is listing. For this reason, all positions of the crane operator's cab must clear the vertical plane through the face of the quay by at least 1.00 m and preferably 1.50 m (fig. R 6-1). If need be,

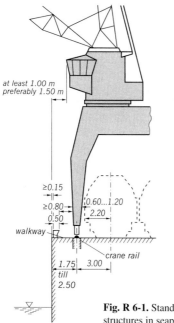

Fig. R 6-1. Standard cross sections of waterfront structures in seaports (supply channels are not illustrated)

159

this dimension is measured from the face of the dolphin, fender or fender system.

Railings are not required at the edges of quays for mooring and cargo handling services. However, the edges of such quays should be provided with adequate nosing or slipping protection as per R 94, section 8.4.6. The edges of quays with public access which do not serve for mooring or cargo handling should be equipped with a railing.

6.1.2 Edge Bollards

Edge bollards which are placed flush with the face of the wall have caused difficulties in handling heavy rope hawsers when the ship is lying close to the quay. Therefore, the face of the bollards must lie at least 0.15 m behind the front face of the quay wall. The width of the head of a bollard may be taken as 0.50 m. The trucks of modern harbour cranes may be assumed to be about 0.60 to 1.20 m wide (fig. R 6-1).

6.1.3 Other Equipment

The dimensions given in fig. R 6-1 are recommended for the installation of new harbours and reconstruction of existing harbours, taking all relevant factors into account. The dimension of 1.75 m between the crane rail and the edge of the wall is recommended as a minimum. It is better to allow 2.50 m for the new construction and deepening of berths, particularly when very wide outboard crane trucks affect safety aspects for mooring or for access to and from the gangway.

Railway traffic requires adherence to the safety standards regarding clearances at the crane, even when the outboard trucks are independent; therefore, the centre line of the first track must lie at least 3.00 m inboard of the outer crane rail.

6.2 Top Elevation of Waterfront Structures in Seaports (R 122)

6.2.1 General

The elevation of the top of waterfront structures in seaports is determined by the working ground level of the port. When fixing the working ground level, the following principal factors are to be observed:

(1) Water levels and their fluctuations, especially heights and frequencies of possible storm tides, wind-raised water levels, tidal waves, the possible effects of flow from upstream water and other actions mentioned in section 6.2.2.2.

(2) Mean level of the groundwater table including frequency and magnitude of fluctuations.

(3) Ship traffic, harbour installations and cargo handling operations, live loads.

(4) Ground conditions, subsoil, fill material and possible mass compensation.

(5) Available options for designing waterfront structures.
(6) Environmental concerns.

According to the requirements of the harbour with respect to operations, economy and construction methods, the relative importance of these criteria must be adjusted in making decisions, in order to arrive at the optimum result.

6.2.2 Levels and Frequency of High Water

A fundamental distinction must be made here between wet dock harbours and open harbours with or without tide.

6.2.2.1 Wet Dock Harbours

In flood-free wet dock harbours, the port operation range must be designed as high above the officially determined *mean working water level* as is required

(1) for the prevention of flooding of the port terrain at the *highest possible working water level*,
(2) for an adequate height above the highest groundwater table in the port terrain which is associated with the *mean working water level*, and
(3) for practical handling of general cargo, bulk cargo and containers.

The ground surface level should generally be 2.00 to 2.50 but at least 1.50 m above the *mean working water level*.

6.2.2.2 Open Harbours

The height and frequency of high tides are decisive when it comes to selecting the suitable port operation range.

Planning tasks must use frequency lines for surpassing the *mean high tide level* as far as possible. In addition to the main influencing factors stated in section 6.2.1 (1), the following influences must be taken into consideration:

- wind-raised water level in harbour basin,
- oscillating movements of the harbour water due to atmospheric influences (seiche),
- wave run-up along the shore (so called Mach effect),
- resonance of the water level in the harbour basin,
- secular rise of the water table, and
- long-term coastal lifting or settlement.

If inadequate water level measurement records are available, as many readings as possible must be made on the spot during the design period and integrated with existing record frequency lines of high tide levels in adjacent areas.

6.2.3 Effects of Elevation and Changes of the Groundwater Table in the Port Terrain

The mean elevation of the groundwater table and its local seasonal and other changes as well as their frequency and magnitude, must be taken into consideration, particularly where the construction of proposed pipelines, cables, roads and railways or anticipated live loads etc. may be affected. In this connection, the need for drainage means that the course of the groundwater table to the harbour water must also be given attention.

6.2.4 Port Operation Range Depending on Cargo Handling Procedures

(1) General cargo and container handling
In general, a flood-free operation range is essential. Exceptions should be allowed only in special cases.

(2) Bulk cargo handling
Because of the diversity of cargo handling methods and types of storage, as well as the sensitivity of the goods and susceptibility of the handling gear to damage, it is not possible to give a general guiding principle here. An effort should nevertheless be made to provide a flood-free surface, particularly in view of the environmental problems involved.

(3) Special cargo handling equipment
For ships with side doors for truck-to-truck handling, bow or stern flaps for roll-on/roll-off handling or other special types of equipment, the top elevation of the waterfront structure must be compatible with the type of ship and equipped with either a fixed or movable transition ramp. In this case, the top of the waterfront structure need not be at the same level as the general ground surface.

(4) Cargo handling with ship's gear
In order to have adequate working clearance under the load hooks even for low lying vessels, the height of the quay must generally be designed lower than one where cargo is handled by quay cranes.

6.3 Standard Cross-sections of Waterfront Structures in Inland Harbours (R 74)

6.3.1 Port Operation Range

The port operation range in inland harbours should normally be arranged over the level of the highest water table. In the case of flowing waters with high water table fluctuations, this is frequently only possible at considerable expense. As far as handling sites for solid bulk goods are concerned, occasional flooding can be taken into account. The risk of contamination of the water during such flooding must be taken into consideration. In cargo handling ports on inland canals with lesser fluctuations in the water level, the port operation range should be at least 2.00 m above the normal canal water table.

6.3.2 Waterfront

As far as possible, the waterfronts in inland ports should be straight with a smooth face (R 158, section 6.6). Wave-shaped sheet piling is ideal as waterfront structure apart from a few exceptions (R 176, section 8.4.16). It is important that the outermost outboard structural part of the crane does not protrude in the alignment of or over the front edge of the waterfront structure. A crane support width of 0.60–1.00 m is assumed.

6.3.3 Clearance Profile

When designing crane tracks and cargo handling cranes, care must be taken to comply with the required lateral and upper safety clearances as stipulated in valid specifications (EBO, BOA, UVV German Federal Railways, TAB.-E25), see fig. R 74-1.

As far as roadways under crane portals are concerned, the recommendation given in fig. 74-2 applies (EBO, BOA, UVV; EVV, E 25 in [45]).

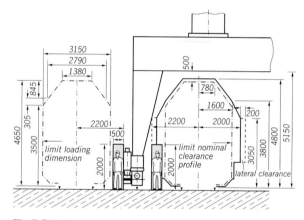

Fig. R 74-1. Lateral and upper safety clearances for the railways

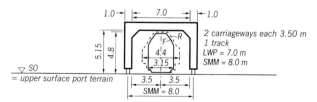

Fig. R 74-2. Recommended track middle dimension (SMM) and width clearance (LWP) for crane portals over roadways and covered tracks

6.3.4 Layout of the Outboard Crane Rail

The aim is to place the outboard crane rail as close as possible to the waterfront edge. This reduces the crane jib to a minimum and saves valuable storage space near the waterfront. The necessary walkway is to be arranged inboard of the crane portal support (fig. R 74-1). Otherwise a walkway of 0.80 m is to be provided between crane portal and waterway edge. The vertical waterfront structure can consist of

- reinforced concrete support wall
- sheet piling
- combination of sheet piling and reinforced concrete support wall on bored piles

Reinforced concrete support walls produce no problems in compliance with the specifications in section 6.3.2.

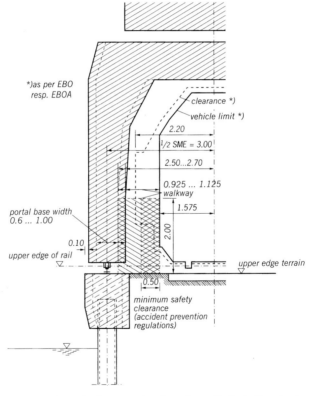

Fig. R 74-3. Standard cross section dimensions with sheet piling structures in inland harbours (crane rail in axis line of sheet piling)

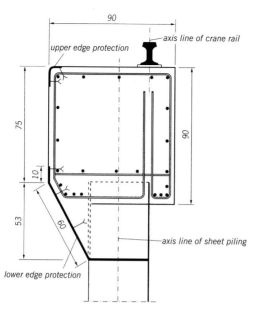

Fig. R 74-4. Crane rail out-of-centre to axis line of sheet piling, example

In the case of sheet piling, the crane rail should be planned in the axis line of the sheet piling wall (fig. R 74-3).
The required geometrical requirements (6.3.2) can make it necessary to support the crane rail out-of-centre (fig. R 74-4).
The combined solution (steel sheet piling with mooring piles/reinforced concrete wall) offers the advantage of separated introduction of load. In addition, the crane rail can be routed close to the edge without vessel impacts having any influence on the crane. Ladders can be placed in ideal positions for access to the ship, fig. R 74-5 and recommendation R 42 in [45].

6.3.5 Mooring Facilities

Sufficient mooring facilities for the ships are to be installed on the outboard side of the waterfront structure (R 102, section 5.13).

6.4 Sheet Piling Waterfront on Canals for Inland Vessels (R 106)

6.4.1 General

When canals are to be constructed or extended in areas where only limited space is available, waterfront structures of anchored steel sheet pil-

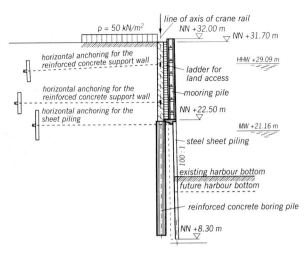

Fig. R 74-5. Anchored sheet piling with mooring piles/reinforced concrete wall, example

ing are frequently the best technical solution, and considering the reduced land purchase and maintenance costs, also the most economical. This is true especially for stretches required to be made impervious. The sheet piling interlocks can be sealed to improve watertightness, see R 117, section 8.1.20.

Fig. R 106-1 shows a typical example of such a structure.

If operating conditions allow, the upper edge of the sheet piling should remain under the water table for reasons connected with corrosion protection and landscape protection. Reference is made to [52] as regards cross-section design.

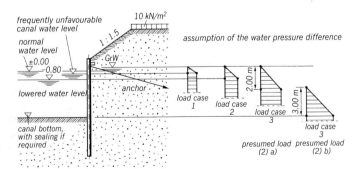

Fig. R 106-1. Cross-section for the sheet piling waterfront of the normal stretch of an inland waterway canal with the most essential load assumptions

Prof. Dr. Lackner & Partners
Consulting Engineers GmbH

Independent consulting engineers since 1936

Activities

- Port operations, organisation and management
- Offshore and coastal marine structures
- Ports and harbours
- Shipyards and dry docks
- Inland waterways
- Navigation locks, barrages, flood control structures
- Shore / bank protection
- Aids to navigation
- Specialised heavy foundation designs
- Industrial estate development
- Traffic infrastructure
- Environmental engineering
- Power supply and distribution
- Water, sewerage, waste disposal

Shipyard in Wismar / Germany

Port of Conakry / Guinea

Services

- ✔ Traffic analyses and forecasts
- ✔ Technical and financial feasibility studies
- ✔ Master planning (operational and development projections)
- ✔ Preliminary and detailed designs, working documents
- ✔ Tender documents, bidding procedures
- ✔ Field investigations (subsoil, surveying on- / off-shore, structure condition)
- ✔ Special studies (oceanographic, vessel manoeuvring, mooring and berthing)
- ✔ Expert opinions, advisory
- ✔ Project management (financial and technical)

Zifta Lock / Egypt

Head Office:
Lindenstrasse 1A • D - 28755 Bremen • Germany
Telephone: ++49 - 421 - 65 84 10 • Telefax: ++49 - 421 - 65 84 110
e-Mail: lacknerconsult@t-online.de

Branch Office:
Rosa-Luxemburg-Strasse 16/18 • D – 18055 Rostock • Germany
Telephone: ++49 - 381 – 4 56 78 0 • Telefax: ++49 - 381 - 4 56 79 19

Represented in Bangladesh, Chile, Czech Republic, Ghana, Guinea, India, Indonesia, Morocco, Namibia, Pakistan, Philippines, Tanzania, Thailand, Togo

Concrete Structures Euro-Design Handbook

Editor: Josef Eibl
1st edition 1994/96. Approx. 800 pages, numerous illustrations and tables. Format: 14,8 x 21 cm.
Hb DM 168,-/öS 1226,-/sFr 156,-
ISBN 3-433-01429-9

As a result of the creation of the European Economic Community, concrete construction products and components can be freely exchanged among various users and countries. This development now calls for the design and construction of civil engineering works and components which comply with uniform European regulations.

We at Ernst & Sohn wish to give this development due consideration by publishing for the first time an English edition of the Beton-Kalender, one of the best known standard texts for civil engineers.

This Handbook presents essential European standards, practical examples for their application, and detailed accounts of the most recent developments in this field.

Highlights from the table of contents: Concrete - Action on Structures - Design of Concrete Structures to ENV 1992-Eurocode 2 - Design of Reinforced and Prestressed Concrete Structures for Bending with Axial Forces, Shear and Torsion - Dimensioning of Slender Elements Related to Ultimate Limit States Influenced by Structural Deformations, Stability check - European Codes and Standards.

Ernst & Sohn
Verlag für Architektur
und technische Wissenschaften GmbH
Bühringstraße 10, D-13086 Berlin
Tel. (030) 470 31-284
Fax (030) 470 31-240
E-Mail: mktg@ernst-und-sohn.de
www.ernst-und-sohn.de

total support

Corus is dedicated to a total support package. Based in Dusseldorf, the steel sheet piling team supported by the UK Sales Office and Design Team, offers all Civil Engineering Contractors and Consultants, Local Authorities, Universities, Port and Harbour Operators the ultimate in quality and service. Able to meet all European Standards including the 'U' mark, Corus are proud of their record as a supplier of steel sheet piling. Support when and where you need it.

Corus Europe
Gartenstraße 2
40479 Düsseldorf
Germany
Tel   +49 211 4926-0
Fax   +49 211 4926-282

Glass for Structural Design

Gerhard Sedlacek, Haegh Gulvanesian, Kurt Blank, and Wilfried Laufs
2000. Approx. 220 pages.
Format: 17 x 24 cm.
Pb approx. DM 118,-/öS 861,-/sFr 105,-
ISBN 3-433-01736-X
Publication date: April 2000

To the structural designer, glass poses new problems relating to its qualities and behaviour as a construction material, especially its structural strength. There is a special need for scientifically founded design methods/processes for dimensioning and construction with various kinds of glass and load-bearing glass elements according to state-of-the-art technology.

Ernst & Sohn
Verlag für Architektur
und technische Wissenschaften GmbH
Bühringstraße 10, D-13086 Berlin
Tel. (030) 470 31-284
Fax (030) 470 31-240
E-mail: mktg@ernst-und-sohn.de
www.ernst-und-sohn.de

6.4.2 Calculation

The waterfront structure and its components are calculated and rated according to the pertinent recommendations. Special reference is made to R 9, section 4.2 and R 18, section 5.4. In the case of vertical live loads, in deviation from R 5, section 5.5, the *characteristic* value is taken to be a uniformly distributed ground live load of 10 kN/m^2 (fig. R 106-1). Reference is also made to R 41, section 8.2.10 and R 55, section 8.2.8.

6.4.3 Load Assumptions

The loads attributed to the load cases are *characteristic* values which are to be multiplied by the partial safety factors of DIN V 1054-100.

The water pressure difference is to be calculated with in load case 1, as this can be expected to occur frequently due to unfavourable canal and groundwater levels. The groundwater table is often assumed at the elevation of the top of the sheet piling.

If the groundwater table is inclined towards the sheet piling, the water pressure difference is to be referred to the applicable active earth pressure failure plane (R 65, section 4.3, fig. 65-1 and R 114, section 2.9, fig. R 114-2).

Load case 2 takes account of a lowering of the canal water in front of the sheet piling by 0.80 m due to passing ships.

Load case 3 takes the following loads into consideration:
(1) In canal sections which are purposely emptied at times (e.g. between two gates, the canal water level is to be assumed to coincide with the elevation of the canal bottom, while the groundwater level will depend on local conditions.
(2) In the other areas (normal stretches), it need not be assumed that the canal will be completely dewatered while the surrounding groundwater table remains at its normal level.
   If in exceptional circumstances local conditions are such that a rapid and severe drop of the canal water level can be expected in case of serious damage to the canal, the following two load cases must be investigated:
   a) the canal water level is assumed to be 2.00 m below the groundwater table,
   b) the canal water level is assumed to be at the canal bottom while the groundwater table is 3.00 m above the canal bottom.
(3) In locations where the failure of the canal wall would destroy the stability of bridges, loading facilities or other important installations along the canal, the sheet piling must be designed for the load case "emptied canal", or must otherwise be adequately reinforced by means of special structural supplements.

In the statical investigations, the planned canal bottom may be assumed to be the actual one for calculating purposes. Overdepth dredging up to

0.30 m below the nominal bottom is generally admissible without special calculations, if the EAU are observed and when the wall is fully fixed in the ground (see R 36, section 6.7). This does not apply in any case to unanchored walls and anchored walls with free toe support. If in exceptional cases, greater deviations are to be expected and severe scour damage due to ship's screw action is probable, the calculations should assume a bottom elevation which is at least 0.50 m below the nominal bottom.

6.4.4 Embedment Depth

If impervious soil is encountered at attainable depth in dam stretches which are to be made watertight, the sheet piling is driven to a uniform depth at which it is firmly embedded in the impervious layer. In this way, the sealing of the canal bottom is not necessary.

6.5 Partially Sloped Waterfront Construction in Inland Harbours with Extreme Water Level Fluctuations (R 119)

6.5.1 Reasons for Partially Sloped Construction

The mooring, tying up, lying off and departure of unmanned vessels must be possible at every water level without the use of anchors, and port and operations personnel should at all times have safe access to moored vessels. These conditions exist only if the vessels are moored alongside a vertical wall. Fully sloped banks are *not suitable* as berths or cargo handling sites.

Vertical banks are required at cargo handling areas with flood-free working ground level. As far as the handling of bulk goods at the upper part of the waterfront is concerned, a vertical structure is not necessary and frequently also not wanted.

The partially sloped bank is therefore ideal for inland harbours with extreme water fluctuations. It consists of a vertical quay wall for the lower part and an adjoining upper slope (figs. R 119-1 and 119-2 as examples).

6.5.2 Design Principles

In the planning of a partially sloped bank, special importance is attached to correct determination of the elevation of the toe of the sloped portion and to the point of transition from the vertical waterfront structure to the sloped structure. In any case, it must be located above the mean water level recorded over a long period of years.

For embankments, it is generally acceptable if the slope is not flooded for more than about 60 days per year on average, based on recorded observations of many years. On the lower Rhine for example, this corresponds to a toe elevation of about 1 m above mean water level (MW) (fig. R 119-1).

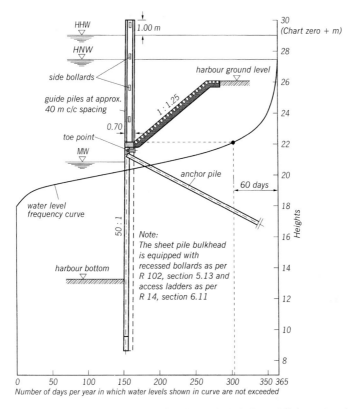

Fig. R 119-1. Partially sloped bank of berths, particularly for push lighters where harbour ground level is subject to flooding

For embankments with higher harbour ground surface, the elevation of the top of the sheet piling should be selected so that the slope height is limited to a maximum of 6 m, to avoid difficult operating and unfavourable statical conditions (fig. R 119-2).

Within the entire harbour basin, the elevation of the toe of the slope should always be uniform.

Guide piles at intervals of about 40 m are advisable along the vertical bank section at berths and coupling points without cargo handling operations for unmanned vessels in river ports with extreme water level fluctuations. These serve for marking, safe mooring and as slope protection. They should extend 1.00 m above HHW without projecting over the water side (fig. R 119-1).

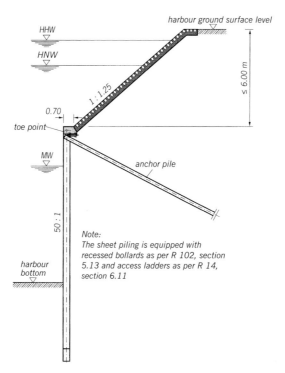

Fig. R 119-2. Partially sloped bank with flood-free port ground level

The vertical bank section is generally constructed of single-anchored sheet piling with fixed earth support.

The coping could consist of a 0.70 m wide steel or reinforced concrete capping beam (figs. R 119-1 and 119-2) which also suffices as a berm, which can be walked on safely in the area of the ladder recesses. With this width there is still no danger of ships grounding at falling water levels, provided the vehicles are properly maintained.

In the area of guide piles, the berm lying behind must be constructed continuously.

The waterside edge of the reinforced concrete capping beam is to be protected from damage with a steel plate as described in R 94, section 8.4.6.

The slope must not be steeper than 1 : 1.25 so as to permit the installation of safe, properly designed stairways. Gradients of 1 : 1.25 to 1 : 1.5 are chiefly used.

Bollards on partially sloped banks are constructed according to R 102, section 5.13.

6.6 Design of Waterfront Areas in Inland Ports According to Operational Aspects (R 158)

6.6.1 Requirements

The requirements for the development of the waterfront areas result principally from navigation and cargo handling conditions, but partially also from railroad and road operation, as far as the design of the upper part is concerned. In order to ensure duly proper shipping operations, it must be possible for ships to moor on the waterfront and to cast off safely and easily and to lie calmly, so that even passing ships or formations of ships do not create disadvantageous effects, and the mooring wires or lines can be well slacked off at water level changes. The waterfront structure shall also act as a guide when berthing the ship. There must be the possibility of direct access for passengers between shore and ship, or for the safe placing of a gangway (E 42 in [45]).

Operations with push lighters move relatively large masses; in addition, push lighters are box-shaped with angular boundaries. This results in increased requirements for a waterfront structure face which is as plane as possible.

For cargo handling operations, conditions are to be created which enable the ship to be loaded and unloaded rapidly and safely. During these operations, the ship should move as little as possible. On the other hand, if necessary it should be possible to move or shift the vessel without difficulty. Economic and operational aspects determine the waterfront cross-section. The crane operator must have a clear view. Together with vertical embankments, partially sloped banks can also be recommended for cargo handling. If the crane is located on the upper surface of the slope however, the crane boom requires a greater reach. At sloped banks, safe manoeuvring of the ship is only ensured in connection with mooring dolphins at adequately close intervals.

6.6.2 Planning Principles

Long, straight stretches of the waterfront structures are required because of the great length of the ships and ship formations, but also because of the better guidance of the ship. If changes in direction are unavoidable, they should be designed in the form of bends (polygons) and not continuously (circular arcs). The bends are to be spaced so that the intermediate straight stretches are adapted to the ship or formation lengths.

The shape of the ships and their operation compel the design of the smoothest possible structures without protruding installations and recesses in which the ships can catch. The front face can be sloped, partially sloped, inclined or vertical. In longitudinal direction however, it should be as smooth as possible.

6.6.3 Waterfront Cross-sections

(1) Embankments

Embankments should be designed as plane as possible. Intermediate landings should be avoided if possible. Stairs should be installed at right angles to the shore line. Bollards and mooring rings may not protrude beyond the embankment surface. If intermediate berms are unavoidable for high embankments, they must not be arranged in the area of frequent water level fluctuations but rather placed in the high water zone. The breakpoint at the transition from the sloped to the vertical bank is to be positioned in the same manner (see R 119, section 6.5.2).

(2) Vertical waterfronts

Vertical or slightly inclined waterfront structures of solid construction are chiefly suitable for waterfront facilities to be newly constructed in a dry excavation. They present a smooth forward face. When erected in or at water, diaphragm or bored diaphragm walls may be used, for instance. The shaping of their forward face does however generally not meet operational requirements after dredging. Measures for bringing about a smooth surface in the ship contact area are then required according to R 176, section 8.4.16.

The sheet piling method of construction represents a proven and economical solution for a waterfront structure. However, given a pile system dimension in the magnitude of half a metre and an opening width of the sheet piling troughs exceeding 0.7 m, the limit of deviation of the corrugated surface from the smooth surface is already reached in some cases. Where special loads from shipping operations are concerned, it may even be necessary to require a smooth surface instead of the corrugated one (see R 176, section 8.4.16).

6.7 Nominal Depth and Design Depth of Harbour Bottom (R 36)

6.7.1 Nominal Depth in Seaports

The nominal depth is the water depth below a definite reference height which should be observed.

When stipulating the nominal depth of the harbour bottom at quay walls, the following factors should be considered:

(1) The draft of the largest, fully loaded ship berthing in the port, taking account of salinity of the harbour water and heeling of the ship.
(2) The safety clearance between ship's keel and the nominal depth should generally have a minimum depth of 0.50 m.

The harbour depth is measured from low water (LW) and in tidal areas from mean low water spring tides MLWS which approx. corresponds to chart zero. Under special circumstances, it may be necessary to work on the basis of an even lower theoretical low water level.

6.7.2 Nominal Depth in Inland Harbours

The nominal depth of the harbour and the harbour entrance is to be selected so that the ships can travel on the waterway with the greatest possible unloading depth. In inland harbours on rivers, the water depth should as a rule be 0.30 m deeper than the nominal bottom of the adjoining waterway, to avoid any dangers for ships at all water levels.

6.7.3 Design Depth at Quay Walls

If dredging is to be carried out in front of quay walls because of silt, sand, gravel or rubble deposits, the dredging must be deeper than the planned nominal depth of the harbour bottom stipulated in sections 6.7.1 and 6.7.2 (fig. R 36-1).

The design depth consists of the nominal depth of the harbour bottom, the maintenance dredging zone down to the planned dredging depth, plus dredging tolerance and other surcharges for special conditions. The dredging depth is influenced by the following factors, with reference to recommendation R 139, section 7.2:

(1) Extent of the silt mass, sand drift, gravel or rubble deposits per dredging period.
(2) Depth under the nominal depth of the harbour bottom, to which the soil may be dredged or disturbed.
(3) Costs of every interruption to cargo handling operations caused by dredging works.
(4) Constant or only occasional availability of the required dredging apparatus.
(5) Costs of dredging work with regard to the level of the maintenance dredging zone.
(6) Extra costs of a quay wall with deeper harbour bottom.

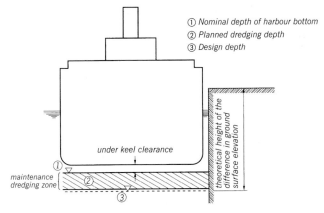

Fig. R 36-1. Calculating the design depth

Due to the importance of all afore-mentioned factors, the scope for dredging at quay walls must be stipulated with care. On the one hand, inadequate scope results in high costs for maintenance dredging and more interruptions to operations, on the other hand excess scope causes higher construction costs and possibly also additional sedimentation.

It is practical to attain the harbour bottom depth first in at least two dredging cuts executed at time intervals. A maximum cut thickness of 3 m must be observed.

Table R 36-1 shows the thickness of the maintenance dredging zones under the nominal depth of the harbour bottom for various water depths, with the corresponding minimum tolerances, only as a general orientation.

Water depth m	Thickness of the maintenance dredging zone m	Minimum tolerance[1] m	Theoretical total depth below the nominal depth of the harbour bottom m
5	0.5	0.2	0.7
10	0.5	0.3	0.8
15	0.5	0.4	0.9
20	0.5	0.5	1.0
25	0.5	0.7	1.2

[1] depending on the dredging equipment

Table R 36-1. Maintenance dredging depths and minimum tolerances: indicative values in [m]

Stipulation of the theoretical total depth already takes account of the surcharges required in DIN V ENV 1997-1 (see also section 2.0).

In case of greater bottom erosion, the design depth is to be enlarged, or suitable measures are to be taken to prevent erosion.

6.8 Reinforcement of Waterfront Structures to Deepen Harbour Bottoms in Seaports (R 200)

6.8.1 General

Developments in vessel dimensions mean that occasionally it is necessary to deepen the harbour bottom in front of existing quay walls. Greater crane and live loads are often an additional factor.

The possibility of deepening the harbour bottom in such cases depends on:
a) the design of the quay wall,
b) deformation in the wall since it was built,

c) the structural condition of the quay wall,
d) the degree of necessary deepening, particularly in respect of the design depth of the harbour bottom,
e) the possibility of reducing the allowed live loads behind the quay wall,
f) the anticipated service life of the quay wall after any reinforcement,
g) the availability of the static calculations performed earlier with all associated loads, theoretical soil values and water levels and the design drawings,
h) the costs of reinforcement in comparison with the costs of different solutions (e.g. new construction elsewhere).

For a) and b) reference is made in particular to R 193, section 15.1. With respect to g), it may be useful to perform new soil investigations to establish, for example, the consolidation figure for the cohesive soil, and to check and reduce or increase the theoretical soil values.

With the new loads, water levels, soil values and the increased theoretical depth, a static calculation can then be performed for the design of a reinforced quay wall.

If no old calculations and design drawings are now available, it is recommended that the consequences of a necessary bottom deepening be combined with a reduction in live loads. The deformation behaviour of such a quay wall plays an important role in this case and special account must therefore be taken of this.

6.8.2 Design of Structure Reinforcements

There are numerous possibilities for reinforcing quay walls for deepening the harbour bottom. In the case of combined walls, care must be taken that the filling piles also have sufficient embedment length. The following sections illustrate a few exemplary solutions, depending on the factors stated in section 6.8.1.

6.8.2.1 Measures to Increase the Passive Earth Pressure

Reference is made to R 164, section 2.11.

a) Replacement of soft, cohesive soil with non-cohesive material of high specific weight and shear strength in front of the quay wall (fig. R 200-1). The change must be executed so as to be filter-stable. Construction work must be regulated with especially stringent requirements, using verification measurements, if appropriate, in order to prevent or restrict the wall from giving way on the water side during dredging. The quay wall will then generally be relieved by partial reclamation of the backfill.
Subsidence of the soil directly in front of the wall must be anticipated specially if no "prestressing" took place.

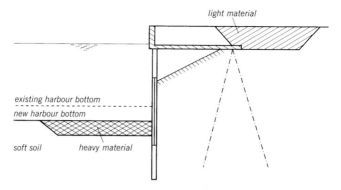

Fig. R 200-1. Soil replacement in front of and/or behind the structure

b) Soil compaction for non-cohesive soil (fig. R 200-2).
c) Soil stabilisation for permeable, non-cohesive soil by injections (fig. R 200-2).

6.8.2.2 Measures to Reduce Active Earth Pressure
a) Create a reinforced concrete relieving platform on piles (fig. R 200-3).
b) Replace the backfill with a lighter material; reference is made to R 187, section 7.11 in this respect.
c) Stabilisation of non-cohesive backfill with good permeability by injections (fig. R 200-2).

6.8.2.3 Measures Involving the Quay Wall
a) Use of supplementary anchors, inclined or horizontal (fig. R 200-4).
b) Drive more deeply and raise the existing waterfront structure (fig. R 200-5).

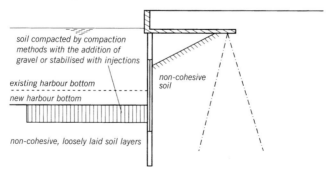

Fig. R 200-2. Soil stabilisation or compaction in front of the structure

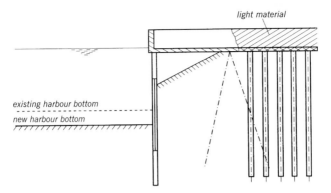

Fig. R 200-3. Stabilisation with a relieving construction on piles

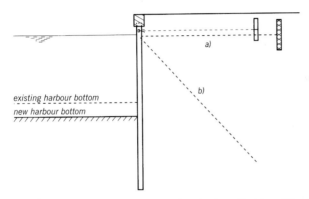

Fig. R 200-4. Use of supplementary anchors horizontally (a) or obliquely (b)

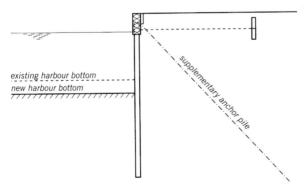

Fig. R 200-5. Drive more deeply and build up existing water front structure plus supplementary anchoring

c) Drive new sheet piling directly in front of the quay wall. The sheet piling can then be anchored in different ways:
- by means of a new superstructure on piles over the existing relieving platform (fig. R 200-6),
- by means of obliquely anchor piles or horizontal anchors (fig. R 200-7).

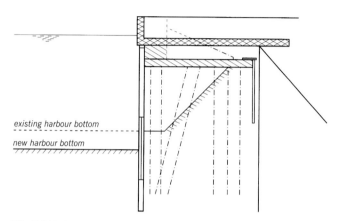

Fig. R 200-6. Forward extension of sheet piling and new superstructure

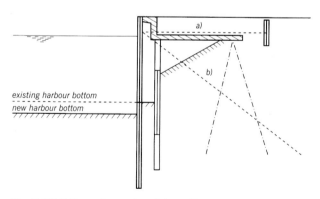

Fig. R 200-7. Forward extension of sheet piling and supplementary anchoring a) or b)

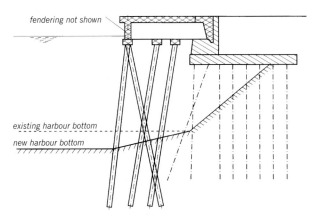

Fig. R 200-8. Forward extension on piles with underwater embankment

d) Forward extension with resilient reinforced concrete relieving platform on piles, provided sufficient space is available. Reference is made particularly to R 157, section 11.5 (fig. R 157-1). An additional bonus with this method is that the quay surface area is increased, to the advantage of cargo handling operations (fig. R 200-8).

6.9 Re-design of Waterfront Structures in Inland Harbours (R 201)

6.9.1 General

Initially, everything said in R 200, section 6.8 about quay wall reinforcement in seaports generally applies accordingly here. However, the reasons for re-designing waterfront structures in inland harbours are frequently different. It is mostly the erosion of the river bottom which gives rise to the need for deepening the harbour bottom in lateral side basins. In canals and rivers regulated by damming, extension to accommodate a greater unloading depth can also necessitate deepening. In individual cases, increases in the crane and live loads can lead to re-designing.

In the case of sloping banks, a deepening of the river bottom results in a reduction in the harbour basin width and water cross-section. Together with the increasing length of the crane jib, this results in a need for a partially sloping or vertical extension. Harbour bottom deepening or increased live loads result in higher stresses on individual components which are then no longer adequately rated.

6.9.2 Possibilities for Re-design

It is generally possible to construct a new waterfront structure from, or instead of, the old one. However, it is often sufficient to renew or reinforce certain parts of the waterfront, or to implement other design measures. Thus, for example, sheet piling can be driven deeper with a new superstructure. Increased anchoring forces can be absorbed by supplementary anchoring. In non-cohesive soils, compaction of the harbour bottom leads to an increase in passive earth pressure. The stability of an embankment can be improved by needling with the subsoil by embedding driving components.

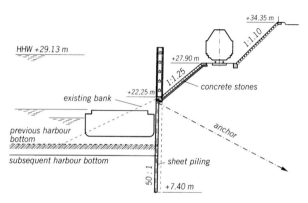

Fig. R 201-1. Bank construction by replacing a sloped bank with a partially sloped bank

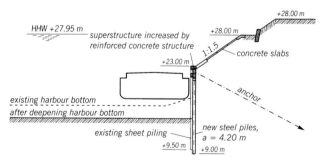

Fig. R 201-2. Bank construction by driving more deeply and building up the superstructure on the existing bank sheet piling

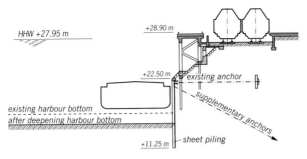

Fig. R 201-3. Bank construction by supplementary anchoring of remaining sheet piling

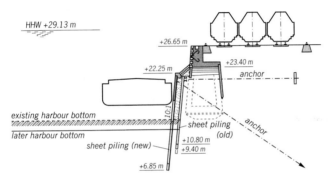

Fig. R 201-4. Bank construction by pre-driving new sheet piling

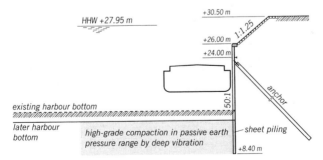

Fig. R 201-5. Bank construction by vibration of non-cohesive soil in passive earth pressure range in front of sheet piling

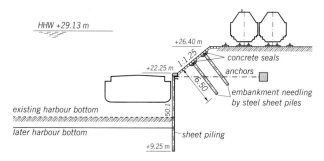

Fig. R 201-6. Bank construction with bank stabilisation by needling

6.9.3 Construction Examples

Figs. R 201-1 to R 201-6 show typical examples of re-designs for waterfront structures in inland harbours:

Bank construction by replacing the sloped bank with a partially sloped bank (fig. R 201-1).

Bank construction by driving more deeply and increasing the superstructure on existing bank sheet piling (fig. R 201-2).

Bank construction by supplementary anchoring of temporarily remaining sheet piling (fig. R 201-3).

Bank construction by pre-driving new sheet piling (fig. R 201-4).

Bank construction by compaction of non-cohesive soil to increase the passive earth pressure in front of the sheet piling (fig. R 201-5).

Bank construction by needling the slope (fig. R 201-6).

6.10 Equipping of Berths for Large Vessels with Quick Release Hooks (R 70)

Instead of bollards, heavy-duty quick release hooks are used for loads of 30–3000 kN with manual or oil-hydraulic release device and remote control, to guarantee easy mooring and quick release of hawsers even when heavy steel hawsers are used.

Fig. R 70-1 shows an example of a quick release hook for 1250 kN maximum load capacity with manual triggering device. It can be used by several hawsers and releases them at full load and also at lesser load by operating the handle with little effort.

The quick release hooks are attached to a quick release hook bearing by a Cardan universal joint. The number of quick release hooks depends on the hawser tension given in R 12, section 5.12, and on the directions from which the principal hawser pulls may occur simultaneously. Several quick release hooks can be installed on one quick release hook bearing. The range of movement must ensure that the hook can meet all anticipated operational requirements without jamming. The swivel range extends from 180° in the horizontal to 45° in the vertical.

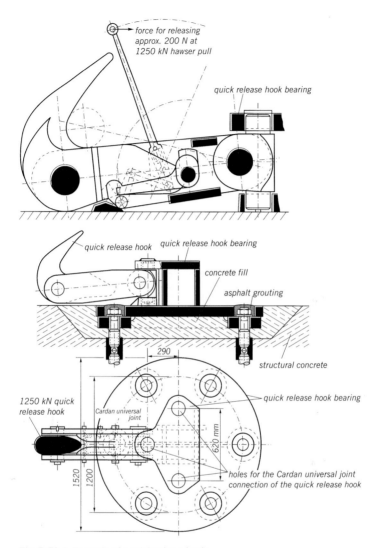

Fig. R 70-1. Example of a quick release hook

It is easier to fasten heavy-duty hawsers when the quick release hook is combined with a capstan. Quick release hooks are ideal for mooring large vessels at special berths where the range of movement can be determined according to the mooring plan.

6.11 Layout, Design and Loading of Access Ladders (R 14)

6.11.1 Layout

Access ladders are intended above all to provide access to the mooring facilities, and, in emergencies, to make it possible for someone who has fallen into the water to regain the shore. Qualified and practised shipping and cargo handling staff can also be expected to use the access ladders even when there are greater differences in water table levels (up to approx. 4 m).

Access ladders in waterfront structures of reinforced concrete should be placed at 30 m intervals, so that there is one ladder for each normal section of the quay. The position of the ladder in a normal section depends on the position of the bollard, since the ladders must not be obstructed

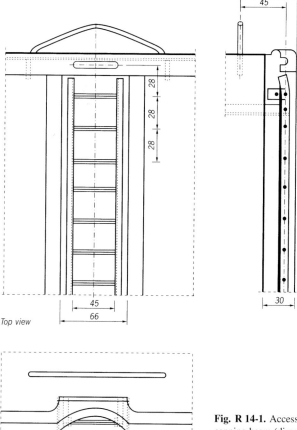

Fig. R 14-1. Access ladder in steel capping beam (dimensions in cm)

by mooring lines. As a general rule, it is advisable to place the ladders near the section joints. Proceed in a similar manner for shorter sections lengths as per R 17, section 10.1.5.

In the case of waterfront structures of steel piling, it is recommended to position the access ladders in the pile troughs.

Mooring facilities are to be installed on both sides of each ladder (R 102, section 5.13.1).

6.11.2 Practical Dimensions

In order to be accessible from the water at all times, even under LLW, the ladders must extend down to 1.00 m below LLW. For easy installation and replacement of the ladders, the lowest ladder mounting is designed as a plug device into which the side pieces can be inserted from the top. The transition between the top of the ladder and the deck of the quay must ensure that ascent and descent of the ladder can be accomplished safely. At the same time, the ladder must not be a hazard to traffic on the quay. These two requirements can be met best by dishing in the nosing of the wall at the ladders and pulling it back to a distance of about 15 cm. In addition, a hand grip should be placed at the head of the

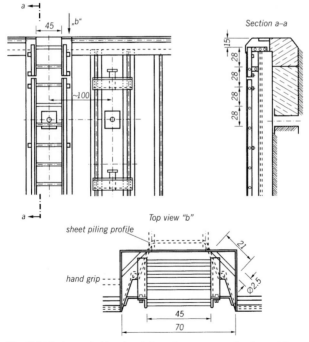

Fig. R 14-2. Access ladders in reinforced concrete capping beam (dimensions in cm)

ladder, at least for flood-free waterfront structures. This hand grip should be of material 40 mm in diameter with the top 30 cm above the deck of the quay and its longitudinal axis 55 cm from the face of the wall (fig. R 14-1). If the hand grips should present an obstacle during cargo handling, other suitable aids to climb the ladder must be provided. Fig. R 14-2 shows a proven design of this type. The top rung of the ladder in this solution is 15 cm below the deck of the quay.

The ladders should be installed with centre of rungs about 10 cm behind the face of the wall, consisting of square steel bars 25/25 mm installed edge upwards. This reduces the danger of slipping due to ice or dirt. The rungs are fastened to the side pieces at a centre-to-centre distance of 28 cm. The clear width between the side pieces of the ladder should be 45 cm.

Otherwise reference is made to DIN 19 703.

6.12 Layout and Design of Stairs in Seaports (R 24)

6.12.1 Layout

Stairs are used in seaports when required by public traffic. They must be safe for use by persons not acquainted with the conditions in harbours. The upper exit of the stairs should be placed so that there is little or no interference with foot traffic and harbour cargo handling. The approach to the stairs must be clearly visible and thus permit the smooth flow of foot traffic. The lower end of the stairs is to be positioned so that ships can berth easily and safely with safe passage between ship and stairs.

6.12.2 Practical Dimensions

Stairs should be 1.50 m wide so that they are clear of the outer crane track on seagoing vessel quays and do not conflict with the fastenings of the crane rails at a distance of 1.75–2.50 m from the edge of the quay. The inclination of the stairs should be determined by the well-known formula $2s + a = 59 - 65$ cm (DIN 18 055) (height of step s, breadth of step a). Concrete stairs should have a rough, hard concrete overlay and the edges of the stairs should have a steel edge protection.

6.12.3 Landings

For tidal ranges of greater magnitude, landings should be installed at 0.75 m above MLW, MW and MHW, respectively. Additional landings may be required, depending on the height of the structure. Intermediate landings are to be positioned after max. 18 stairs – DIN 18 065; the length of the landing should be 1.50 m or the same as the stair width.

6.12.4 Railings
The stairs should be fitted with a handrail of 1.10 m height above the front edge of the tread. When harbour operations permit, the stairs should be enclosed by a 1.10 m high railing, which could be removable if necessary.

6.12.5 Mooring Equipment
The quay wall next to the lowest landing is equipped with cross poles (R 102, section 5.13). In addition, a recessed bollard or mooring hook is positioned below each landing. Recessed bollards are used for solid quay walls or quay wall members, mooring hook generally for steel piling structures.

6.12.6 Stairs in Sheet Piling Structures
These are frequently made of steel. The sheet piling wall is driven with a recess large enough to contain the stairs.

The stairs must be protected by suitable means against underrunning (fender piles).

6.13 Equipment of Waterfront Structures in Seaports with Supply and Disposal Facilities (R 173)

6.13.1 General
The supply facilities serve to supply the required matter, energy etc. to existing public installations and facilities, but also to firms located in the port, as well as to the docked ships and the like. The disposal facilities serve to drain any water.

In the planning of such facilities in port areas, consideration must be given to the fact that they must also be located in the immediate vicinity of the waterfront structures and in part, placed directly in these structures themselves.

Adequate openings for all lines are to be foreseen in structural members situated below the ground surface, such as craneway beams and the like. Therefore, in order to avoid unnecessary costs, consultation of all participants must take place in time during the early planning of such structural members. Reserve openings are to be included to allow for any later expansion.

The supply facilities include:

- water supply facilities
- electric power facilities
- communications and remote control facilities
- other facilities

and the discharge facilities:

- pure water drainage
- sewage disposal and
- fuel and oil interceptors

The legal regulations are to be observed.

6.13.2 Water Supply Facilities

The water supply facilities provide potable and raw water, and can also be used for extinguishing purposes in case of fire.

6.13.2.1 Potable and Raw Water Supply

In order to safeguard the potable and raw water supply system in the port, at least two feed-in points are required independent of each other for each port section, whereby the lines are laid out as ring nets.

Hydrants are installed at approx. 100–200 m intervals. Underground hydrants are placed on quay walls and in paved crane and rail areas so as not to hinder operations. The hydrants are to be arranged so that there is no danger of being crushed by rail-bound cranes and vehicles, even when standpipes have been fitted. When using underground hydrants, special attention is to be paid to the fact that the connection coupling is protected from impurities even in case of any possible flooding of the quay wall. An additional shut-off valve is required to separate the hydrant off from the supply line. The hydrants must always be accessible. They must be situated in areas where storage of goods is not possible for operational reasons.

Under Central European conditions, the pipe lines will be generally laid with an earth cover of 1.5 to 1.8 m, with spacing of at least 1.5 from the front face of the quay wall for protection from frost. In loaded areas with tracks of the port railway, the lines are to be placed in protective pipes. In quay walls with concrete superstructures, the lines may be placed in the concrete construction. Here differing deformation behaviour of the individual construction sections must be taken into account, together with the differing settlement behaviour of deep or shallow founded structures. For the potable water supply, crossings with tracks must be accepted. When ring mains are laid on the land side of the superstructure, drainable connection lines are to be installed between the ring main and the hydrants lying on the front face quay wall. Lines without a constant flow of water pose a risk to potable water hygiene.

In order to avoid major breakage/demolition work at operation surfaces in case of a pipe failure, as far as possible the mains are to be laid not under reinforced concrete surfaces but under strips reserved for the pipe lines.

6.13.2.2 Fire-fighting Water

When there is a high fire risk in a certain section of the port, it is frequently recommended to supplement the potable and raw water supply system with an independent fire fighting supply system. The water is taken directly from the harbour basin with pumps. The necessary pump rooms can be located in a chamber of the quay wall below ground so as not to disturb cargo handling.

It is also possible to feed the fire-fighting water supply system through pumps of the fire-fighting boats at special connection points, owned by the fire department.

In sheet piling quay walls, the suction pipes may be placed in the sheet piling troughs, whereby the fire-fighting water can be obtained through mobile pumps belonging to the fire department. These suction pipes are adequately protected from the impact of a vessel. The same is also possible in recessed slits of concrete superstructures.

The fire-fighting pipeline system has to satisfy the same requirements as for the potable and raw water supply.

6.13.3 Electric Power Supply Facilities

The electric power supply facilities provide power to the administration buildings, port installations, crane installations, lighting installations of the rail areas, roads, operations areas, sites, quays, berths and dolphins, etc.

Only cables will be used for the high and low voltage supply system of the port, except for provisional states of construction. The cables will be laid in the ground with an earth cover of approx. 0.80–1.00 m, in quay walls and operational areas in a plastic pipe system with concrete cable draw pits which can be driven over. The advantage of such pipe system is that the cable installations can be increased or expanded without interrupting port operations.

Where there is a risk of frequent flooding of the quay walls, the power connection points are to be installed in high, flood-free stands.

Power connection points are generally installed in the quay wall coping at intervals of 100–200 m. They must be capable of being driven over and have a drainage pipe. These connection points are used among others to provide the power connection for welding generators carrying out minor repairs on ships and cranes, as well as the power connection for emergency lighting.

Contact line ducts, cable channels and crane power feeding points must be provided in the quay areas for power supply to the crane systems. The drainage and ventilation of these facilities is particularly important. In quay walls with concrete superstructures, these facilities can be included in the concrete construction.

Special attention is drawn to the fact that the electric power supply networks must be provided with potential compensation facilities . This is to prevent unduly high contact potentials from occurring through a fault in the electrical facilities of a crane in crane rails, sheet piling or other conductive members of the quay wall area. Such potential compensation systems should be installed about every 60 m.

In the case of craneways integrated in the quay wall superstructure, the potential compensation lines are as a rule concreted in during construction of the superstructure, for reasons of costs. These protective pipes must however be laid in areas in which differential settlements may be expected.

6.13.4 Other Facilities

These include all supply facilities not mentioned in sections 6.13.2 and 6.13.3 as required for example at shipyard quay walls. These include: gas, oxygen, compressed air and acetylene lines, as well as steam and condensate lines in channels. The layout and installation of such facilities must comply with the pertinent regulations, particularly the safety regulations.

6.13.5 Disposal Facilities

6.13.5.1 Rain Water Drainage

The rain water falling in the quay wall area and also on its land side will be drained in the harbour directly through the quay wall. For this, the quay and operations areas are equipped with a drainage system consisting of inlets, cross and longitudinal culverts and a collecting main with outlet into the harbour. The catchment drainage areas depend on local characteristics. An attempt should be made to install the minimum possible number of outlets in the waterfront structure. Such outlets can suffer damage from the impact of vessels.

Suitable slide valves are to be provided in the drainage system in case of dike drainage to impair flood protection in the case of extremely high water levels. The outlets are to be provided with slide valves in quay and operational areas with a risk of leakage of dangerous or toxic substances or fire-fighting water into the drainage system so that the valves can be closed to prevent pollution of the harbour water.

6.13.5.2 Sewage Disposal

The sewage occurring in the port is fed through a special sewage disposal system into the municipal sewer system. It may not be drained into the harbour water. A sewage main is therefore only found in waterfront structures in exceptional cases.

6.13.5.3 Fuel and Oil Interceptors
Fuel and oil interceptors are placed wherever that they are also required at facilities outside the port area.

6.13.5.4 Disposal Regulations for Ship Waste
According to the MARPOL convention, ports should provide facilities for the disposal of ship waste, such as liquids containing oils and chemicals, solid ship waste (galley waste and packaging refuse) together with sanitary sewage.

6.14 Fenders for Berths for Large Vessels at Quays (R 60)

6.14.1 General

6.14.1.1 Purpose
In order to provide safe berthing for large vessels at quays which are exposed to unfavourable wind conditions, strong currents, bad approaches etc., these walls must be equipped with fenders. They absorb the impact of vessels when berthing and avoid damage to ship and structure during laydays.

6.14.1.2 Design
The fenders can consist of steel, timber, plastic, rope and in special cases brushwood.
Suitable designs include scraper beams, fender timbers, fender piles, buffer fenders, fender walls with fendering by means of piles, suspended fenders, weighted fenders, torsion fenders, floating fenders, etc.
Large fenders are generally placed in the centre of normal sections depending on section length as per R 17, section 10.1.5, smaller fenders at the quarter points of the section.

6.14.1.3 Economy
The maintenance costs of fenders are generally quite high. It is therefore advisable to make a careful review before installing waterfront structure fenders to determine the nature and extent of the hazards to which the ship or waterfront structure will actually be exposed. It may be possible to omit the waterfront structure fenders if these are exposed to wear.

6.14.2 Plastic Fenders

6.14.2.1 Elastomer Fenders

General

Elastomer elements are used in many ports for fending off impacts of vessels and for absorbing berthing pressure. These elements are generally resistant to sea water, oil and ageing (see under d) and are not destroyed by occasional overloads, so that they enjoy a long service life. Their use is therefore generally economical, despite their comparatively high procurement cost.

Elastomer fendering elements of various shapes, dimensions and specific performance characteristics are manufactured by the industry, so that it is possible to meet every requirement, from simple fendering for small vessels to fender structures for large tankers and bulk cargo freighters. Special attention is directed to the fendering stresses required for ferry terminals, locks, dry docks and the like.

Elastomers are used either alone as fender material, against which the ships berth directly, or as suitably designed buffers behind fender piles, fender walls or berthing contact panels. Occasionally, both types of usage are combined. In such cases, it is possible to attain those spring constants best suited to any specific requirement using elements made from commercially available elastomers (see R 111, section 13.2).

In all elastomer structures used against a rigid port structure, the force/deformation characteristic of the elastomer elements must be given special attention. If the energy being absorbed increases over the working capacity A on which the design was based, the then occurring force impact P increases progressively to infinity. A rigid structure must therefore be rated to comply with the required safety from the fender reaction forces. This results from the position of the structure and the local conditions of the specific purpose. A rough indication is to rate the structure for an approaching energy of the ship which is twice the fender dimension. This requirement can be considerably reduced in the case of resilient piers or approach dolphins additionally equipped with elastomer fenders.

a) Tube fenders

Thick-walled elastomer tubes are frequently used (fig. R 60-1). They can have various diameters from 0.125 m to more than 2 m. They have variable spring characteristics depending on use. Tubes with smaller diameters are placed with ropes, chains or rods in a horizontal, vertical or, where applicable, diagonal position. In this case they are frequently suspended as a "garland" in front of a quay wall, mole head etc.

Large tube fenders are usually installed in a horizontal position (fig. R 60-2). They may not be suspended directly on the quay wall with ropes or chains because of the risk of deflection and tearing under load. They are drawn onto rigid steel pipes or steel pipe trusses etc. These are then suspended at the quay wall with chains or steel cables, or placed on steel brackets located next to the fenders (fig. R 60-2).

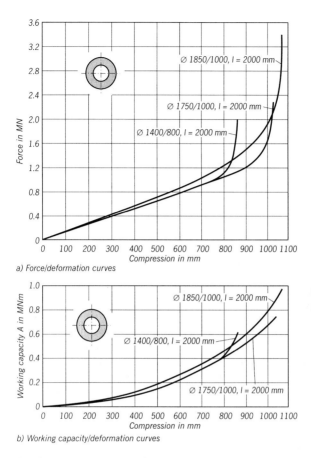

Fig. R 60-1. Example for the force/deformation curves and working capacity/deformation curves of large round fenders

Besides round tubes, square tubes are also used, although only for smaller dimensions. They can have either round or polygonal inner openings. They are generally only used as buffers.

When feeder ships or inland vessels are also handled at the berths of larger vessels, there is risk of them hooking under fixed fenders. In addition, the heeling of small ships resulting from cargo handling procedures at low water can result in damage to the superstructures and loads by the upper fender facilities. In new developments to the container quay in Bremerhaven, floating fenders have been fitted in front of the fender panels which move up and down with the tide with the aid of lateral

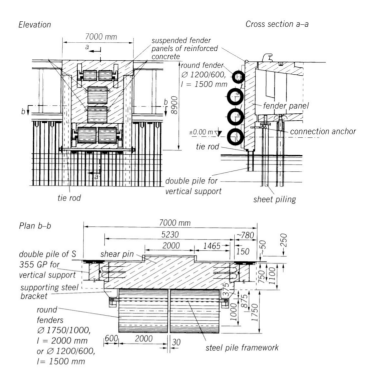

Fig. R 60-2. Example of a large diameter tubular fender facility at a large vessel berth

guide pipes. This solution is shown as an example in fig. R 60-2a. The fender structure here consists of a fixed fender facility with roll fenders ∅ 1.75 m and a floating fender ∅ 2.0 m in front of a fixed fender panel. The diameters are selected to ensure that adequate heeling is still possible even at low water levels.

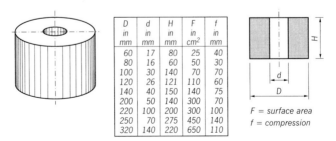

Table R 60-1. Dimensions of elastomer round fenders

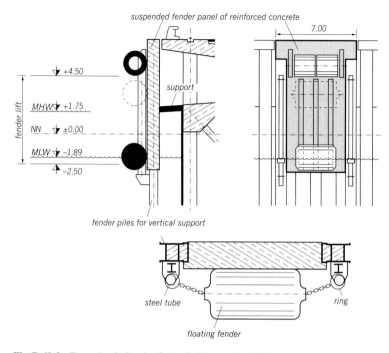

Fig. R 60-2a. Example of a floating fender facility at a berth for large vessels with berthing possibility for feeder and inland vessels

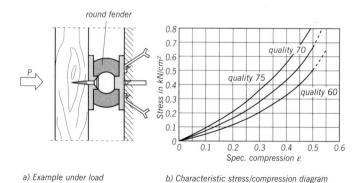

a) Example under load

b) Characteristic stress/compression diagram

Fig. R 60-3. General data for round fenders loaded in longitudinal direction made of elastomer qualities with 60, 70 and 75 (ShA) to DIN 53505

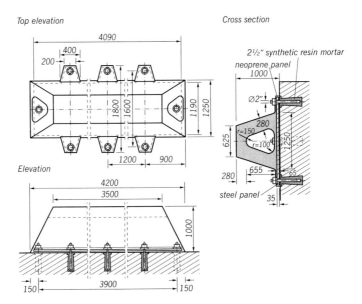

Fig. R 60-4. Example of a trapezoidal fender

Reference is made to manufacturer publications as regards dimensions and properties of the various elastomer fender elements, together with the force and working capacity curves. However, special caution is urged that the curves stated there only apply when the fenders cannot buckle to the side, and when extreme creep movement cannot occur under permanent load (see d).

When designing a quay wall or pier etc. together with the fender brackets, it is not only the berthing pressures which must be taken into account. The horizontal and vertical movements of the ships on mooring and departing, unloading and loading procedures, swell or water table fluctuations etc. can cause friction forces in a vertical and/or horizontal direction, unless these movements are absorbed by the rolling action of suitable round fenders. If lower values cannot be verified, dry elastomer fenders should be rated with a friction coefficient of $\mu = 0.9$ to be on the safe side.

Round fenders can be installed to bear loads either crosswise or lengthways. However, in the latter case only short lengths can be considered because of the buckling risk (table R 60-1 and fig. R 60-3).

If the spring paths are then not adequate under compression, several elements can be used in succession. Steel plates with suitable guides can be arranged between the individual elements to prevent such a structure from buckling.

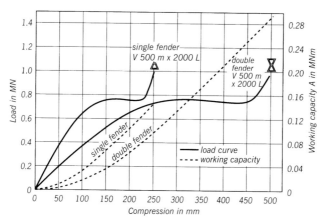

Fig. R 60-5. Example for the relationship between compression and loading as well as the working capacity A for individual and double trapezoidal fenders

b) Trapezoidal fenders
In order to obtain a more favourable working line, special shapes have been developed using special inlays, for example fabric, spring steel or steel panels vulcanised into the elements. These inlays must be blasted with a metallic bright surface and completely dry when vulcanised into the structure. These elements are frequently made in a trapezoidal form with a height of 0.2 to approx. 1.3 m. They are fastened to the quay wall with dowels and screws (fig. R 60-4).
It is possible to double the compression and hence the capacity for the same impact force by installing fenders with working diagrams as per fig. R 60-5. Particularly soft fender facilities can be created in this way, as required particularly for fendering large ships, where the impact forces would be excessive if the pairs of fenders were installed parallel instead of in series.
Here again, any buckling of the double fenders must be prevented. To do so, a guided steel panel is positioned between the two fenders; the steel plate can only shift along the centre line of the fenders but not be pressed out or turned to the side.

c) Special forms of fender elements
Fender elements are also used with unsymmetrical shapes in the load direction, with steel plates or enclosing steel frames vulcanised to both sides (fig. R 60-6a). They can be bolted to the quay wall and the structure member being fendered through these steel components. The panels buckle under load because of the non-symmetry, resulting in good performance characteristics. Such elements can also be installed in series, if

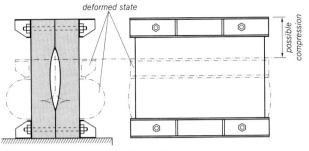

a) Fender element of two asymmetrical panels

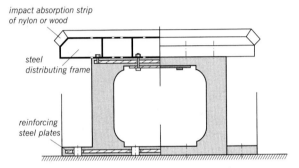

b) Special fender elements or fender for large working capacity

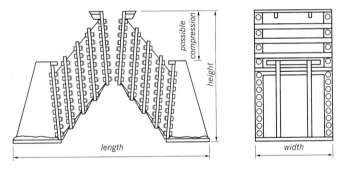

c) Shear deformation fender

Fig. R 60-6. More examples of fender elements or fenders

proper structures are installed to prevent lateral buckling and twisting of the intermediate points.

The side walls of the special fenders shown in fig. R 60-6b) also buckle under load. They are available in specially large dimensions up to a maximum height of 2 m and length of 4 m. Fenders with these dimensions according to Fig. R 60-6b) have a working capacity A of around 3 MNm with an impact force P of around 3.3 MN.

d) Properties of fender elastomers

Elastomers for fenders must provide the following characteristics:

Watertightness shown by freedom from pores and cracks (visual check)	
Tensile strength in accordance with DIN 53 504	≥ 15 N/mm^2
Elongation at rupture in accordance with DIN 53 504	≥ 300 %
Hardness in accordance with DIN 53 505	between 60 and 75 Shore A depending on requirements with a delivery tolerance of ± 5, but within the nominal values
Limit temperature range for use in Central Europe	$-30/+70$°C
Tear resistance in accordance with DIN 53 507	≥ 80 N/cm
Resistance to sea water in accordance with DIN 86 076 (pre-standard), section 7.7: – change in hardness – change in volume	max. ± 10 Shore A max. $+ 10$ % max. $- 5$ %
Tested to section 8.8 for 28 days in artificial sea water at 95 ± 2°C	
Abrasion in accordance with DIN 53 516	≤ 100 mm^3
Ozone resistance in accordance with DIN 53 509, 24 hrs., 50 pphm	crack formation level 0
After kiln-ageing in accordance with DIN 53 508, 70 °C, 7 days: – relative change of tensile strength – relative change in ductile yield	< -15 % referred to < -40 % the value in delivered state

When the above conditions have been fulfilled, adequate light resistance is ensured.

Deviations from the above conditions may be required from case to case but only permissible following previous agreement with the authorities responsible for the design and construction supervision.

When fender elastomers of commercially available shape and quality are used for special structures which are exposed to permanent loads under pressure, tension and/or shear, for example by pre-tension or only slowly changing loads, the creep behaviour of the material must be taken into consideration.

6.14.2.2 Sliding Battens and Panels of Polyethylene

In addition to other friction components such as scraper beams, timber piles, fender piles etc., sliding battens or panels made of plastic and frequently polyethylene (PE) are used in order to reduce the friction stresses at waterfront structures when vessels are berthing and lying up. Their properties, requirements and features of application and design are described below; depending on the conditions of application and types of stress, this does not rule out the use of other substances or structural elements. These components must absorb the loads arising from pressure and friction without fracture and be capable of transmitting them into the harbour structure via their mountings. In certain cases, they must be supported by supplementary carrying members for this purpose. In order to keep the friction forces small, sliding battens should be made from a material with the smallest possible friction coefficient, e.g. ultrahigh molecular-weight polyethylene (UHMW-PE). Low abrasion and wear rates of the component are also of particular importance here.

Polyethylene compounds of medium density with the requirements and properties according to DIN 16 776 have proven suitable for use as sliding battens in hydraulic engineering and seaport construction.

Normal delivery forms are rectangular full profiles with cross sections of 50 x 100 mm up to 200 x 300 mm, and profile lengths of up to 5500 mm. Special profile cross sections and lengths are also available. The shaped parts must always be free of voids and produced and processed in such a way that they are free of distortion and stress-free. The quality of processing can be checked by acceptance tests to verify the property values and by additional hot storage tests of samples cut from the sections, e.g. in accordance with DIN 16 925. After hot storage for 90 minutes at 105°C, a change in mass of 3 % may not be exceeded.

Regenerated PE waste may not be used because of the reduced material properties.

PE sliding battens have to fulfil the following requirements for physical and mechanical properties in accordance with DIN 16 776:

Apparent density	DIN 53 479	0.93 g/cm³ (0.92 to 0.94 g/cm³)
Water absorption capacity	DIN 53 894	< 0.5 % by weight
Strength, tensile and bending	DIN 53 455	10 N/mm² (8 to 12 N/mm²)
Extension at tear	DIN 53 452	380 % (200 to 400 %)
Modulus of elasticity	DIN 53 455	300 MN/m² (200 to 400 MN/m²)
Impact strength (ductility)	DIN 53 453	no fracture down to –40°C
Ball hardness (surface pressure)	DIN 53 456	≥ 15 N/mm²
Shore hardness "D"	DIN 53 505	50 (40 to 60)
Friction coefficient between PE batten and steel in accordance with	DIN 53 375	0.2 to 0.5

If the above requirements are fulfilled, the allowable surface pressure may be 5.3 MN/m². The temperature application range is between –40°C and +80°C.

Figures R 60-7 and 60-8 show fastening and construction examples. The heads of the fastening bolts should stop at least 40 mm behind the contact surface of the sliding battens. Replaceable screws should be at least 22 mm thick and concrete-embedded screws at least 24 mm thick, and galvanised.

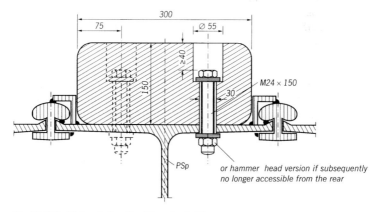

Fig. R 60-7. Sliding batten fixed to a sheet piling

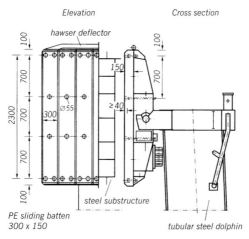

Fig. R 60-8. Equipping the fender apron of a tubular steel dolphin with sliding battens

6.14.2.3 Fenders of Rubber Waste

Various seaports use old car tyres filled with rubber waste as fenders, suspended flat against the face of quays. They have a cushioning effect. They do not have any appreciable working capacity.

Frequently several stuffed truck tyres – usually 5 to 12 – are placed on a steel shaft fitted at both ends with a welded-on pipe collar for attaching the guy rope and holding rope. The fender is placed with these ropes and held so that it can rotate on the face of the quay wall. The tyres are filled with diagonally placed elastomer slabs which brace the tyres against the steel shaft. The remaining voids are filled with elastomer material (fig. R 60-9). Fenders of this type, occasionally of even a simpler design with wooden shaft, are economical. They have generally performed well in circumstances where the requirement to absorb impact stresses from berthing ships has not been severe, even through their working capacity and thus the anticipated berthing pressure cannot be reliably determined. For this reason, the reproduction of characteristic force/deformation and working capacity/deformation has been dispensed with here.

Not to be confused with these improvisations are the accurately designed fenders, freely rotating on an axle. These fenders are fabricated mostly of very large special tyres which are either stuffed with rubber waste or inflated with compressed air. Fenders of this type are successfully used at exposed positions, such as at the entrances to locks or dry docks, as well as at narrow harbour entrances in tidal areas. They are suspended horizontally and/or vertically as guidance for ships, which must always navigate with caution here.

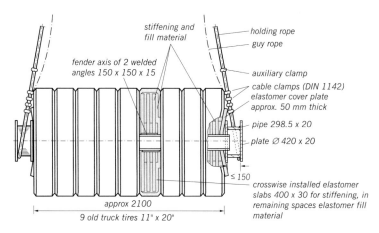

Fig. R 60-9. Example of a truck tyre fender

6.14.3 Fenders of Natural Materials

Suspended brushwood fenders can also be considered as fendering for the berths of large vessels along river banks where sizeable waves strike the bank constructions at rectangles. The dimensions are adapted to the largest vessel berthed. Unless special circumstances require larger dimensions, fender sizes may be chosen in table R 60-2:

Size of vessel DWT	Fender length m	Fender diameter m
up to 10 000	3.0	1.5
up to 20 000	3.0	2.0
up to 50 000	4.0	2.5

Table R 60-2. Fender sizes

Suspended brushwood fenders are subject to natural wear and tear due to ship operations, drifting ice, waves etc. Although they cause higher investment and maintenance costs than elastomer fenders, their use may be considered particularly in countries where suitable raw materials are available and foreign exchange resources should be economised. Fig. R 60-10 shows an example.

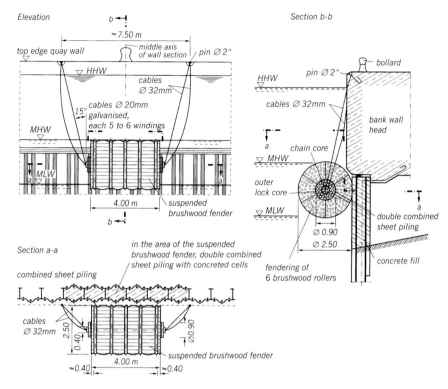

Fig. R 60-10. Example of a suspended brushwood fender at a quay for large ships

6.15 Fenders in Inland Harbours (R 47)

Modern inland vessels have a smooth shell plating. The manoeuvring capability of ships and groups of ships is so good that during berthing operations, the crew only has to place about 1 m long timber fenders horizontally on short wire ropes between the shore and the ship's side. While the vessel is lying at the berth, these timber fenders are suspended between ship and shore.

Experience has shown that these measures are adequate for protection of the ship's hull and the waterfront structure. It is therefore recommended not to equip inland ports with vertical waterfront structures with fender piles, timber fenders or other fendering. Safety measures by means of fendering are only required at pier structures and special installations in the port, e.g. stairs, projecting ladders, etc.

In specially endangered waterfront sections, sheet steel piling is provided with steel plating as per R 176, section 8.4.16.

6.16 Foundation of Craneways at Waterfront Structures (R 120)

6.16.1 General

The type of craneway foundation to be constructed along a waterfront structures depends above all on the prevailing local subsoil conditions. For craneways with a wider track gauge, the subsoil is also to be investigated along the line of the inboard crane rail (see R 1, section 1.2). In many cases, especially for heavy structures in seaports, structural requirements make it practical to construct a deep foundation for the outboard crane rail as an integral part of the quay wall, whereas the foundation for the land-side crane rail is generally independent of the waterfront structure, except at embankments with superstructured slope, on pier slabs and the like.

By contrast, in inland ports the outboard crane rail is frequently also on a foundation independent of the waterfront structure. This facilitates later modification which could be required, for example, for changed operating conditions due to new cranes, or alterations to the waterfront structure.

If the quay wall, craneway and crane are owned by different authorities, it may be necessary to disperse ownership of the structure. The best total solution should always be aimed at with respect to technical and economic aspects.

6.16.2 Design of Foundations/Tolerances

The craneway foundations may be shallow or deep, depending on local subsoil conditions, sensitivity to settlement and displacement of the cranes, occurring crane loads, etc.

The allowable dimensional differences of the craneway must be taken into account here, distinguishing between differences in manufacture (assembly tolerances) and differences during the course of operation (operational tolerances).

Whilst the assembly tolerances in harbour cranes mainly relate to displacement and fixing of the crane rails, the allowable operational tolerances must be taken into account in selecting the type of foundation depending on the subsoil.

Depending on the design of the crane portal, the following reference values can be taken as the basis for the operational tolerances:

- height of a rail (gradient)   2 ‰ to 4 ‰
- height of rails in relation to each other (camber)   max 6 ‰ of track width
- inclination of rails in relation to each other (offset)   3 ‰ to 6 ‰

These operational tolerances include any assembly tolerances.

Reference is made to [53] regarding the relationship between craneway and crane system.

6.16.2.1 Shallow Craneway Foundations

(1) Strip foundations of reinforced concrete
In settlement-sensitive soils, the crane rail beams may be constructed as shallow strip foundations of reinforced concrete. The crane rail beam is then calculated as a resilient beam on a continuous resilient support. In doing so, the maximum allowable soil pressures according to DIN V 1054-100 for settlement-sensitive structures are to be observed. Furthermore, a soil settlement calculation must confirm that the maximum allowable non-uniform settlements for the crane stipulated by the crane manufacturer are not exceeded.

DIN 1045 is applicable for dimensioning the beam cross section. The stresses from vertical and horizontal wheel loads are to be verified, including longitudinal loads due to braking. The beam cross sections should be divided by joints which are keyed both vertically and horizontally (concrete hinges). The length of the beam sections depends on the condition of the existing subsoil, the customary length being 30 m. Construction joints are to be avoided as far as possible.

In craneways with relatively narrow gauge widths, e.g. for gantry cranes spanning only one track, the design gauge of the tracks is maintained by tie-beams or tie-bars installed at intervals equal to about the gauge width. With wider gauges, both crane rails are designed separately on individual foundations. In this case, the cranes must be equipped with socketed stanchions on one side.

Reference is made to R 85, section 6.17 for the design of the rail fastening.

Settlements of up to 3 cm can still be absorbed generally by installing rail bearing plates or special rail chairs. Where larger settlements are probable, a deep foundation will be more economical as a rule, because later realignment and the consequent shut-down of cargo handling operations will cause much lost time and great expense.

(2) Tie foundations
Crane rails on ties in a ballast bed are used primarily in mining subsidence regions, because they are comparatively easy to realign. Even massive movements in the subsoil can be corrected quickly by realignment of elevation, lateral position and gauge, to avoid major damage to craneway and cranes. Ties, tie spacing and crane rails are calculated according to the theory of the resilient beam on a continuous resilient support, and according to the standards for railway structures. Wooden ties, steel ties, reinforced concrete ties and prestressed concrete ties can be used. Wooden ties are preferred at facilities for loading lump ore,

scrap and the like, because of the reduced risk of damage from falling pieces.

6.16.2.2 Deep Craneway Foundations

Deep foundations are practical in settlement-sensitive soils or deep fills, insofar as soil improvement is not carried out by means of replacement, vibration and the like. An adequately deep foundation also reduces stresses on the waterfront structure.

Basically, all customary types of piles can be used for deep foundations of craneways. However, especially in the area of the outboard crane track, horizontal bending of piles caused by deflection of the waterfront structure must be taken into consideration. Furthermore, considerable horizontal loading of the piles can occur in unconsolidated soils, due to one-sided larger live loads.

All horizontal forces acting on the craneway are to be absorbed by partially developed passive earth pressure in front of the craneway beam by sloping piles or by effective anchoring.

For deep foundations on piles, the craneway beam is to be calculated as a resilient beam on a continuous resilient support.

6.17 Installation of Crane Rails on Concrete (R 85)

The crane rails are to be installed with longitudinal movement and free of stress.

The following possibilities can be used for perfect installation of crane rails on concrete:

6.17.1 Bedding the Crane Rail on a Continuous Steel Plate over a Continuous Concrete Base

In this method, the bedding plate is grouted in a proper manner, or rests on slightly moist stone-chip compressed concrete. The rail fastenings are designed to permit the rail to move longitudinally, but to restrain it from vertical motion so securely that even negative bending stresses resulting from interaction between the concrete base and the rail will be satisfactorily absorbed. In calculating maximum moment, anchoring force and maximum concrete compression stress, the method of the subgrade reaction coefficient may be used.

Fig. R 85-1 shows a typical example for a heavy crane rail. The embedding concrete is tamped in between the steel angles, levelled and given a levelling coat ≥ 1 mm of synthetic resin or a thin bituminous coating.

If an elastic intermediate course is laid between the concrete and the bedding plate, rail and anchoring are to be calculated for this softer bedding, which can result in larger dimensions. The rails are to be welded to avoid joints as far as possible. Short rail bridges are to be used at expansion joints between quay wall sections.

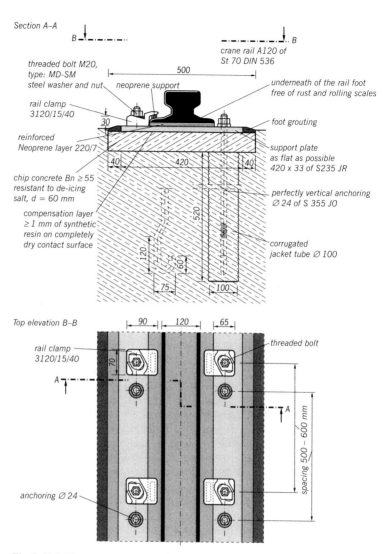

Fig. R 85-1. Heavy craneway on continuous concrete base (example)

6.17.2 Bridge Type with Crane Rail Supported at Intervals on Bedding Plates

In this case, bedding plates of a special type are used which assure a well-centred introduction of the vertical forces into the points of support. They must also guide the rails, which have freedom of movement in longitudinal direction. Furthermore, they must prevent tilting of the rail which of necessity is quite high because it must act as a continuous beam. They must absorb the lifting forces resulting from both the negative bearing force as well as from the direct horizontal forces acting on the rail.

Light-weight crane rails of this type are used for the normal general cargo cranes, preferably in inland ports, and for bulk cargo cranes. In heavy-duty design, they are to be recommended above all for the runways of heavy lift cranes, very heavy short unloaders, unloader bridge cranes and the like. Rail sections S 49 and S 64 are used in light installations; in heavy installations, PRI 85 or MRS 125 as per DIN 536 are used, or very heavy special rails made of St 70 or St 90.

A typical example of a light installation is shown in fig. R 85-2. Here, rail S 49 or S 64 according to the K-type of the railway structures of the German Railway AG rests on horizontal bedding plates. The rail, bedding plates, anchors and special dowels are placed completely assembled on the formwork or on a special adjustable steel support, which can be securely mounted. The concrete is then so placed with the aid of vibration that complete contact is made with bedding plates. Occasionally, an intermediate course of approx. 4 mm thick plastic is also placed between bedding plates and upper surface of concrete (fig. R 85-2). With

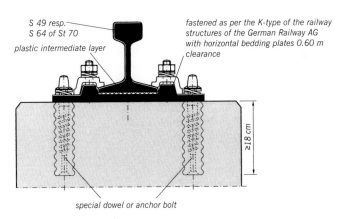

Fig. R 85-2. Light craneway on individual supports

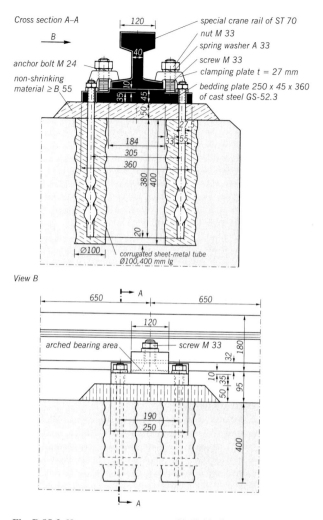

Fig. R 85-3. Heavy craneway on tamped individual supports

arched bedding plates, design measures must ensure that the plastic intermediate layer cannot slip off.

Fig. R 85-3 shows a heavy craneway in which the chair for the rail is arched upwards so that the rail is supported tangentially on the arching. The bedding plate is supported or packed with a non-shrinking material. The bedding plates are also provided with slotted holes in transverse direction, so that gauge changes can be corrected if need be. This bedding must be provided above all for long-stem rails.

Intermittent support can also be provided with very high wheel loads. For crane rails with small section moduli, e.g. A 75 to A 120 or S 49, however, continuous support is recommended for loads above approx. 350 kN, since otherwise the spacing between the panels or chairs becomes too small.

6.17.3 Bridge Type Crane Rails with Support on Rail Bearing Elements

When rail bearing elements or so-called rail chairs are used, the rail is a continuous beam with an infinite number of supports. In order to make full use of the resilience of the rail, an elastic slab is inserted between rail and chair, e.g. for bearing pressures up to 12 N/mm^2 this could be made of neoprene or similar, or for higher pressures rubber fabric in thicknesses up to 8 mm. This also tends to cushion impacts and shocks on the crane wheels and chassis.

The top of the rail chair is arched, thus causing the bearing force to be introduced centrally into the concrete. This arched bearing lies some distance above the concrete, and a certain amount of yield occurs in the lock washers of the mounting hardware so that the rail will be free to move longitudinally. In this way, any temperature changes and rocking movements can be absorbed (fig. R 85-4). Through flexible shaping, the rail chairs can be adapted to any desired requirements. For example, the chairs offer the facility of subsequent rail regulation of:

$$\Delta s = \pm\, 20 \text{ mm and } \Delta h = +\, 50 \text{ mm}$$

or of fitting lateral pockets to absorb edge protection angles for traversible rail sections.

The rail chairs are mounted together with the rails, whereby after adjustment and locking in position, additional longitudinal reinforcement is drawn through special openings in the chairs and connected with the projecting bars of the substructure (fig. R 85-4). The concrete quality depends on statical requirements. However, at least B 25 is required. Since the height of the rail cannot be adjusted easily by tamping, the design should be used only when any settlement worth mentioning can be ruled out.

If settlement and/or horizontal displacement of the crane rail necessitating re-alignment of the rail must be anticipated, it must be taken into account right at the early planning stage by means of appropriate design-dependent selection of the type of construction, e.g. special chairs.

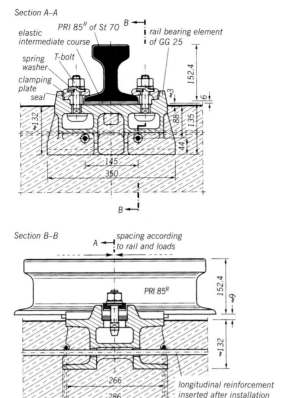

Fig. R 85-4. Example of a heavy craneway on rail bearing elements

6.17.4 Traversible Craneways

The demands of port operations frequently need the crane rails to be installed sunk into the quay surface so that they can be crossed without difficulty by vehicular traffic and port cargo handling gear. At the same time, all other demands made of craneways must also be met.

(1) Traversible construction of a heavy craneway

Fig. R 85-5 shows a proven construction example here. The bedding of the rails on the craneway beam section completed first, consists of stone-chip concrete > B 55, levelled off horizontally by means of a flat steel bar (ladder gauge). For load distribution, the rail with a bottom surface of thin bitumen rests on a bedding plate fixed on a > 1 mm synthetic resin levelling course. This bedding plate is not connected to the fixing system to prevent loads from the longitudinal movements of the rails and bedding plates being transferred to the bolts. Subsequent installation of the bolts is preferable to be better able to ensure the exact position of the bolts. However, this approach must be allowed for when reinforcing the craneway beam by leaving adequate space between the reinforcing bars for the plate or plastic pipe to be concreted in. If necessary, the bolt holes can also be drilled subsequently. In order to compensate for the transfer of horizontal forces transfer and hold the rail in an exact position, approx. 20 cm wide brackets of synthetic resin mortar will be inserted between the foot of the rail construction and the bordering lateral edges of the top course concrete at approx. 1 m centre spacing.

It is expedient to use permanent-elastic two-component filler as mastic filler in the upper 2 cm of the head area of the reinforced top course concrete jointed with stirrups to the remaining craneway beam.

Further details can be taken from fig. R 85-5.

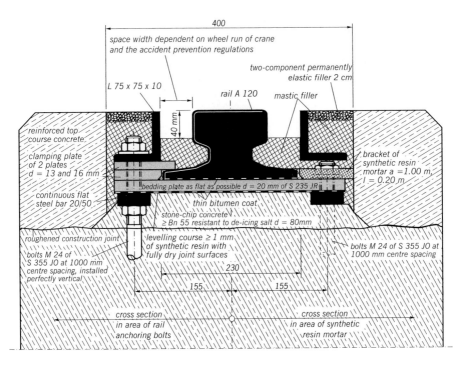

Fig. R 85-5. Execution example of traversible heavy craneway (the reinforcement is not shown)

213

(2) Traversible construction of a light craneway

A proven execution example of this is shown in fig. R 85-6.

Horizontal ribbed slabs are fastened onto the plane levelled reinforced concrete craneway beam with dowels and screw spikes, at centre spacing of approx. 60 cm. The crane rail, for example S 49, is connected to the ribbed slabs with clamping plates and T-bolts in accordance with German Railway AG regulations. Levelling slabs such as steel plates, plastic slabs and the like can be installed beneath the rail foot to correct slight differences in level in the concrete surface.

A continuous steel closure is to be installed as abutment for laterally adjoining reinforced concrete raft foundations. This closure consists of an L angle 80 x 65 x 8 of S 235 JR running parallel to the rail head, beneath which 80 mm long U 80 sections are welded at intervals of three ribbed slabs each. These sections have oblong holes on the bottom for fastening to the T-bolts. In the area of the intermediate ribbed slabs, the angle is stiffened by 8 mm thick steel plates. Recesses will be made in the horizontal leg above the fastening nuts to be covered by 2 mm thick plates after the nuts have been tightened. A bar will be welded on along the rail head at the foot end of the angle to give the subsequent mastic filler an adequate hold below.

To pave the port ground surface, reinforced concrete slabs are placed loosely against the steel closure. In this case, rubber mats should be laid underneath to prevent tilting, and at the same time to create a slope for drainage from the crane rails.

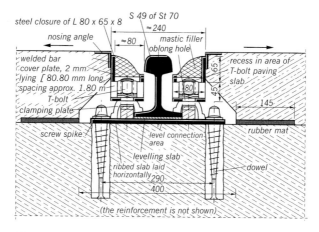

Fig. R 85-6. Execution example of a traversible light craneway

6.17.5 Notes on Rail Wear

The wear to be expected for the foreseen service of life of all crane rails must already be taken into account in the design. As a rule, a height deduction of 5 mm with good rail support is adequate. Furthermore, more or less frequent maintenance and, depending on the type, checks of the fixing devices, are recommended during operation to prolong the service life.

6.18 Connection of Expansion Joint Seal in a Reinforced Concrete Bottom to Bearing External Steel Sheet Piling (R 191)

Expansion joints in reinforced concrete bottoms, e.g. in a dry dock or similar, are protected against massive mutual displacement in the vertical direction by means of denticulation in the form of a dam cover. Thus only slight mutual vertical displacement is possible. The transition between the bottom and the vertically bearing connected steel sheet piling is via a relatively slim reinforced concrete beam connected securely to the sheet piling, to which the bottom plates separated by the expansion joint are flexibly connected, also interleaved with a dam cover.

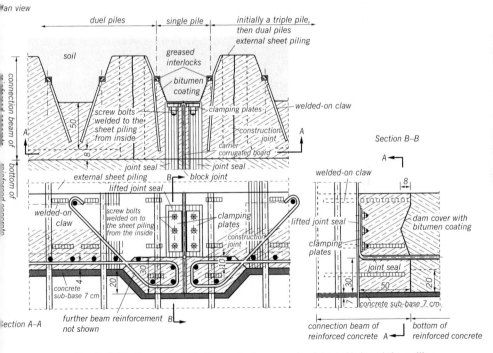

Fig. R 191-1. Connection of bottom seal of an expansion joint to U-shaped sheet piling, example

215

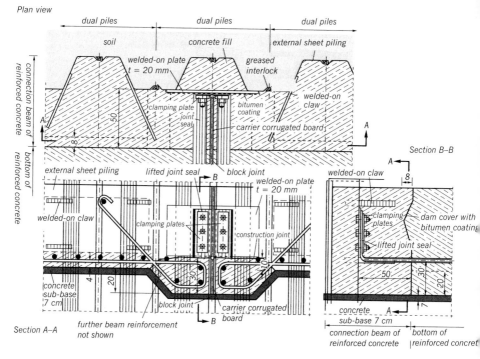

Fig. R 191-2. Connection of bottom seal of an expansion joint to Z-shaped sheet piling, example

The bearing plate joint with denticulation is also constructed in connection beams. The expansion joint of the bearing plate is sealed from below with a joint seal with loop. In U-shaped sheet piling in accordance with fig. R 191-1, this seal ends at a specially installed sheet piling pillar. In Z-shaped sheet piling, a connection plate is welded on to the sheet piling trough as shown in fig. R 191-2, and the joint seal is lifted and clamped to this.

The round locks of the connection piles (single piles for U-profile, dual piles for Z-profile) are to be generously greased with a lubricant before installation.

Further details are noted in figs. R 191-1 and R 191-2.

6.19 Connection of Steel Sheet Piling to a Concrete Structure (R 196)

The connection of steel sheet piling to a concrete structure should be as close as possible and allow mutual vertical movement of the structures. Normally the simplest possible solution should be aimed for. Fig. R 196-1

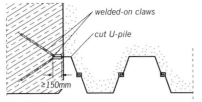

a) Connection to a subsequently produced concrete structure

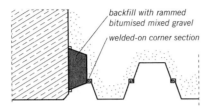

b) Connection to an existing concrete structure

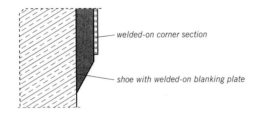

Fig. R 196-1. Connection of a U-shaped sheet piling to a concrete structure

shows such construction examples for a U-shaped sheet piling. Fig. R 196-1 a) shows connection to a previously constructed concrete structure. Here, the necessary connection is provided by an individual pile inserted through the formwork and then equipped with welded-on claws. It is embedded in the concrete structure during its construction to a depth sufficient for the necessary connection. The connecting interlock must be previously treated in a suitable way, see R 117, section 8.1.20). Fig. R 196-1b) shows a recommended method for connecting steel sheet piling as closely as possible to an already existing concrete structure.
Similar construction examples for Z-shaped steel piling are shown in fig. R 196-2.
If stringent requirements are made for the watertightness and/or mobility of the connection, e.g. if seepage in dammed sections of waterways could prejudice stability, special joint designs with joint seals must be provided, these being connected with clamping plates to the sheet piling

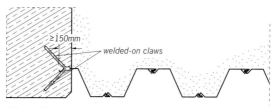

a) Connection to a subsequently produced concrete structure

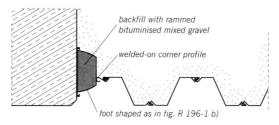

b) Connection to an existing concrete structure

Fig. 196-2. Connection of Z-shaped sheet piling to a concrete structure

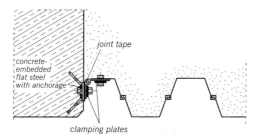

Fig. R 196-3. Connection of U-shaped sheet piling to a concrete structure with stringent requirements for the watertightness of the interlocks

and to a fixed flange in the concrete construction (fig. R 196-3). The embedment of the sheet piling must be stipulated on the basis of the existence of low-permeability soil strata or of the allowable seepage path length. Reference is made in particular to DIN 18 195 parts 1 to 4 and 6, 8, 9 and 10.

6.20 Floating Wharves in Seaports (R 206)

The specification "floating wharves" of the Fed. Ministry of Transport applies to wharves on federal waterways. It can be applied correspondingly to seaports as well, with reference to the following.

6.20.1 General

In seaports, floating facilities are used for passenger ferry transport, as berths for port vehicles and for leisure vehicles, consisting of one or several pontoons and connected to the shore by means of a bridge or permanent stairs. The pontoons are generally held by rammed guide piles, while the access bridge to the land is fastened to a fixed bearing on land and a mobile bearing on the pontoon.

If the facility consists of several pontoons, transition flaps ensure that it is possible to move from one pontoon to the next.

6.20.2 Design Principles

Stipulation of the location for a floating facility must take account of current directions and speeds together with wave influence.

In tidal areas, HHW and LLW should be used as nominal water levels. The incline of the access jetty should not be steeper than 1 : 6 under normal tidal conditions and not steeper than 1 : 4 for extreme water levels.

Especially when used for public transport, the facility must comply with stringent requirements, even for example under ice conditions. Suitable structural and organisational measures must be provided here.

The bulkhead divisions of the pontoons must be rated in such a way that failure of one single cell through an accident or other circumstances will not cause the pontoon to sink. Pore-free foaming of the cells can also be considered. In individual cases it may be practical to use alarm systems which draw attention to undetected water leaks.

A camber of beam (raised section in the pontoon surface) is required for water drainage.

It is advisable to provide a disconnecting possibility for the access bridge to guarantee that the pontoon can float away rapidly in the event of an accident, for example by 2 piles rammed in next to the bridge with suspended cross member.

The minimum required freeboard of the pontoon depends on the permissible heeling, anticipated wave height and anticipated use. For smaller facilities, e.g. for leisure vehicles, a minimum freeboard of 0.20 m is adequate for half-sided use, whereas large steel pontoons have far greater freeboard heights. Indicative values are stated below:

- steel pontoon 30 m long,
  3 to 6 m wide   freeboard height 0.8 to 1.0 m
- steel pontoon 30–60 m long,
  12 m wide   freeboard height 1.2 to 1.45 m

The freeboard heights must be adjusted to the embarking and disembarking heights of the ships particularly when the facility is used for public transport.

6.20.3 Load Assumptions and Design

As a basic rule, the position of the pontoon is to be verified with an even keel, with ballast balance being provided where necessary.

The following stipulations apply to principal and live loads:

A live load of 5 kN/m^2 is to be taken for floating stability and heeling calculations (half-sided load here).

The floating stability tests are to be confirmed in addition by a hydrodynamic approach with tests where applicable. Heeling causes, e.g. congested water pressure, current pressure and waves are to be taken into consideration.

Verification is to be provided for heeling of the pontoon and the incline of the transition points between connected pontoons, together with the incline of the transition flaps. Depending on the pontoon dimensions, heeling acceleration and mutual offset of several pontoons, heeling may not exceed 5°; the upper slewing width limit is 0.25 to 0.30 m. Greater heeling angles are to be checked in the consideration of individual cases.

The ship's mooring impact as load from mooring ships is fundamentally to be taken as 300 kN and 0.3 m/s, or 300 kN and 0.5 m/s for larger facilities (more than 30 m pontoon length).

A cushioning effect to reduce the ship's mooring impact for pontoons can be applied and verified by:

- fendering on the outer surface
- spring brackets and sliding battens and
- "soft guide dolphins" (e.g. coupled tubular piles)

7 Earthwork and Dredging

7.1 Dredging in Front of Quay Walls in Seaports (R 80)

This section deals with the technical possibilities and conditions to be taken into account when planning and executing port dredging work in front of quay walls.

A distinction must always be made between new dredging and maintenance dredging.

Dredging down to the designed depth as per R 36, section 6.7 should be carried out among others by grab dredges, hydraulic dredges, bucket-chain dredges, cutter-suction dredges, cutter-wheel suction dredges, ground suction dredges or hopper suction dredges. When using cutter-suction dredges, ground suction dredges or hopper suction dredges, these dredges are to be equipped with devices which ensure exact adherence to the planned dredging depth. Cutter-suction dredges with high capacity and high suction force are however unsuitable, because of the danger of creating overdepths and disturbances of the soil below the cutter head. Dredging by suction dredges without cutter head must be avoided in any case.

It should be noted that neither the bucket-chain dredge nor the cutter-suction dredge, nor the hopper suction dredge cannot create the exact theoretical nominal depth immediately in front of a vertical quay wall when dredging the last metres, even under favourable dredging conditions and corresponding equipment, because a wedge of earth 3 to 5 m wide remains, unless the earth is able to slide down. The need to remove this residual wedge depends on the kind of fendering on the quay and on the shape of the hull of the ships which will berth there. The residual material can be removed only by clam-shell or hydraulic dredges. Under certain circumstances, the sheet piling troughs must be flushed free of cohesive soil.

When dredging a harbour with floating equipment, the work is usually divided into cuts between 2 and 5 m, depending on the type and size of the dredging equipment. The intended utilisation of the removed soil can also be relevant to the choice of dredging equipment in the case of changing soil types.

It is advisable to survey the front face of the quay accurately not just after dredging is completed but also at intervals throughout the operation, for early detection of any incipient possible excessive movement toward the water side. If necessary, the sheet piling is inspected after every dredging cut.

Reference is made to R 73, section 7.4.4 regarding inspection by diverse of areas on the underwater face of quays which have been exposed by dredging, to detect any damage as may have occurred to sheet piling locks.

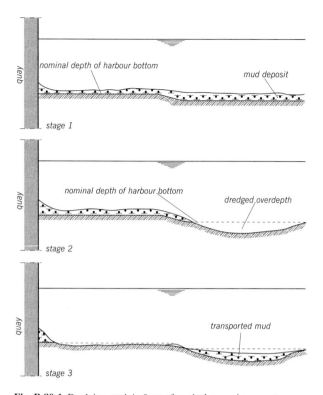

Fig. R 80-1. Dredging work in front of vertical quays in seaports
Stage 1: prevailing situation
Stage 2: situation after dredging with a hopper suction dredge
Stage 3: situation after work with a harrow or grab dredge

An approach such as that depicted in fig. R 80-1 can be more economical and less disturbing to port operations.

After making an overdepth (stage 2) by a wet dredge, the mud lying in front of the quay is moved into the overdepth of the harbour bottom with grab dredges or a harrow (stage 3), which reduces the production costs. The overdepth should be created as far as possible during the new dredging stage.

Before every dredging measure which may exploit the theoretical total depth under the nominal depth of the harbour bottom, the stability of the quay wall must be checked, particularly when equipped with drainage devices, and restored if necessary. In addition, the behaviour of the wall must be monitored before, during and after dredging.

7.2 Dredging and Hydraulic Fill Tolerances (R 139)

7.2.1 General

The specified depths and filled heights are to be produced within clearly defined permissible tolerances. If the difference in elevation between individual points of the produced excavation bottom or filled-in area exceed the permissible tolerances, then supplementary measures must be introduced. Tolerances which are too small can lead to expensive extra work at a later date. The client must therefore carefully consider what value he places on a certain accuracy. The stipulation of tolerances for dredging and hydraulic filling work is thus not merely a question of type and size of equipment but first and foremost a question of costs.

In addition to these vertical tolerances, there are also horizontal tolerances when trenches are to be dredged for such purposes as soil replacement, inverted siphons and tunnels. Here too, it is almost always necessary to reach an optimum balance between the added costs involved in more comprehensive dredging and filling work with more generous tolerances, and the added costs resulting from reduced performance in the equipment from more precise working, together with the costs for any possible additional measures.

The depth tolerances for inland waterways can generally be kept smaller than in waterways for ocean-going vessels, where tide, the waves, shoaling by sand and/or silt deposits play an important part.

Maritime authorities normally demand minimum depths in waterways.

7.2.2 Dredging Tolerances

The main factors to be considered in stipulating dredging tolerances are:

(1) The types of soil and quantities to be dredged.
(2) The type and size of the dredges.
(3) The dredging depths in connection with the optimum position of the bucket or suction ladder.
(4) Instruments on board the dredge (position finding, depth sounding, performance measurement etc.).
(5) Magnitude of performance loss in the dredging equipment because of tolerances.
(6) Current and wind.
(7) Possible tide with its vertical and horizontal effects.
(8) Swell and in general, wave height, length, frequency, action and duration.
(9) Depth below the planned harbour bottom to which the subsoil may be disturbed.
(10) Stability of the underwater slopes, breakwaters, quays and the like in the near vicinity.
(11) Thickness of the layer of dropped and overflow soil.

(12) Possible silt and/or sand deposits or erosion which may develop during dredging work.
(13) Rising of the soil as the result of the removal of loads.

The selection of optimal dredging tolerances is a complicated problem. The many factors influencing the decision mean that utmost caution must be exercised when considering the selection of tolerances from those customarily applied in practice. Before major dredging operations commence, it is imperative to weigh up the many factors and their effects carefully. At the time of inviting bids, it is often not certain which dredge will be used so that it is wise to ask bidders to submit their priced based on the specified tolerances along with their price for tolerances which they themselves propose and will guarantee to meet (special proposal and special bid for dredging work).

For general guidance, reasonable dredging tolerances for various dredge types are given in table R 139-1 in cm; T_v in vertical and T_h in horizontal direction. These are based especially on experience gained in the Netherlands [55]. The soundings should be made with instruments which measure the actual depth to the bottom, and do not give a false reading to the top of an overlying suspended layer. A suitable frequency is to be selected for echo sounders. A 2-frequency sounding is preferable.

7.2.3 Hydraulic Fill Tolerances

7.2.3.1 General Remarks

For hydraulic fill, the tolerances depend largely on the accuracy with which the settlement of the subsoil and the settlement and subsidence of the fill material can be predicted. For this reason too, satisfactory subsoil and soil mechanics investigations are very important. Of course, final grading of the fill, which is usually done by bulldozers in the case of sand, is always necessary. When a shallow hydraulic sand fill is to be placed over a yielding subsoil, the tolerance should be selected under consideration of whether or not the fresh fill will be required to support the traffic of construction equipment immediately. A tolerance does not apply until a hydraulic fill level which can support construction equipment traffic. Otherwise, the purpose of the tolerance is lost because it has to give way to practical difficulties.

7.2.3.2 Tolerances Taking Settlement into Account

If only minor settlement of the subsoil and of the hydraulic fill is expected, a plus tolerance is generally required referred to a specific fill height.

When settlement is expected, the estimated amount of settlement should be indicated in the tender and taken into account in the hydraulic fill specification.

Dredges	Size of equipment	Non-cohesive soils		Cohesive soils							Increment per m tidal range up to	Increment for cross current per 1.5 m/s		Increment for heavy waves	
		Sand		Peat		Silt		Soft clay		Hard clay					
		T_h	T_v	T_h	T_v	T_h	T_v	T_h	T_v	T_h	T_v	T_v	T_h	T_h	T_v
Clam shell dredge	Clam shell capacity in m³														
	0.5–2	100	50	75	40	–	–	50	30	50	10	5	25	25	15
	2 –4	200	75	150	75	–	–	150	75	75	15	5	50	50	25
	4 –7	300	100	250	125	–	–	250	100	100	20	5	75	75	35
Dipper bucket dredge	Bucket capacity in l														
	50–200	100	30	75	25	75	15	75	15	50	10	5	50	25	15
	200–500	150	50	100	35	125	25	125	25	75	15	5	75	50	25
	500–800	200	60	125	45	150	30	150	30	100	20	5	100	75	35
Cutter suction dredge	Cutter ⌀ in m														
	0.75–1.50	200	40	100	30	150	30	100	25	75	15	5	50	50	25
	1.50–2.50	250	50	125	40	200	40	150	40	100	20	5	75	50	25
	2.50–3.50	300	60	175	60	250	50	200	50	150	30	5	100	75	35
Cutter-wheel suction dredge	Cutter wheel ⌀ in mm														
	1.25–2.00	200	40	100	30	150	30	100	25	75	15	5	50	50	25
	2.00–3.50	250	50	125	40	200	40	150	40	100	20	5	75	50	25
	3.50–5.00	300	60	175	60	250	50	200	50	150	30	5	100	75	25
Dipper dredge	Dipper capacity in m³														
	1.8– 4.5	100	25	70	20	75	15	70	15	50	10	5	25	25	15
	4.5– 7.5	125	35	85	30	100	25	85	25	75	20	5	50	50	25
	7.5–12.0	150	50	100	40	125	35	100	30	100	25	5	75	75	35

Table R 139-1. Indicative values of positive and negative dredging tolerances of various dredges in cm for normal conditions [55]

7.3 Hydraulic Filling of Port Areas for Planned Waterfront Structures (R 81)

7.3.1 General

R 73, section 7.4, deals with the direct backfilling of waterfront structures.

In order to obtain good, usable port areas behind planned waterfront structures, non-cohesive material, where possible with a wide range of grain size, should be pumped in. In hydraulic filling above water, a greater degree of density is generally achieved without additional measures than underwater (R 175, section 1.7).

In all hydraulic filling, but especially in tidal areas, measures must be taken to ensure the run-off of the filling water, as well as the water flowing in with the tide.

The fill sand should contain as little silt and clay as possible. The allowable volume is not only dependent on the foreseen waterfront structure and on the required quality of the planned port area, but also on the time before the terrain must be suitable for further earth and construction work. In this respect, the dredging and hydraulic filling processes are of essential significance. Further determining factors can be pollution contamination in the filling material and resulting impairment of the groundwater from leaking pore water. The upper two meters must be ideal for compaction to safeguard adequate distribution of loads under roads etc.

If the port area is intended for settlement-sensitive facilities, silt and clay deposits are to be avoided. A content of fine particles < 0.06 mm of max. 10 % should be allowed for the sand fill. It is frequently economical to use material obtained in the immediate vicinity, for example from dredging work. In so doing, the dredge spoil is often loosened by means of suction dredges with cutting heads or suction dredges and hydraulically pumped directly onto the planned port area. Particularly in such a case, previous technically sound soil investigations at the source are indispensable. Continuous soil profiles are to be taken in the dredging area by means of core borings, whereby deposits of thin cohesive silt or clay strata are to be determined. In connection with static
penetrometer tests, a very good picture of the variations in silt and clay content may be obtained (R 1, section 1.2).

If the available sand fill contains larger silt and clay components, the flushing process must be co-ordinated accordingly. It is important to consider whether the sand fill is hydraulically pumped, dredged and hydraulically filled with a bucket-chain or suction hopper dredge onto the future port area directly from the borrow point, or obtained with a suction dredge and first loaded into scows. This method also enables some cleaning of the sane in the case of scow loading, because mud and clay drain off when the scows overflow. If mud or clay has been deposited locally in the upper surface area of the completed hydraulic fill, this material is to be removed to a depth of 1.5 to 2.0 m and replaced with good material (see R 175, section 1.7). This is necessary because if silt or clay inclusions are left in place, it may take a long time for the surplus pore water to flow off, and permit the cohesive layers to become consolidated. Appropriate maintenance of the flushing field or suitable control of the flushing current can prevent the creation of cohesive strata in the flushing field. Vertical drains can be used to accelerate consolidation (R 93, section 7.7). It is recommended that drainage ditches be dug as soon as possible after the hydraulic fill is completed.

Without special auxiliary means, it is not possible to produce certain types of slopes with hydraulic fill under water, nor truly horizontal areas. The natural slope of medium sand in still water appears to be from 1 : 3 to 1 : 4, in depths from 2 m below water table, from case to case also up to 1 : 2. The slope is even flatter when currents exist.

7.3.2 Hydraulic Filling of Port Areas on Existing Subsoil Above the Water Table

Reference is made to fig. 81-1. The width and length of the flushing field and the layout of the outlets (so-called monks) are of particular importance, particularly when the sand fill is contaminated by silt or clay.

Width, length and layout must be defined so that water carrying fines is kept in motion for as long as possible. To achieve this, hydraulic filling operations must continue without interruption. After every necessary stop, such as weekends, it must be determined if a fine grain layer has settled anywhere: if it has, it must be removed before filling is resumed. If flushing water containing suspended particles and fines is to be returned to the water, the corresponding regulations from the authorities are to be observed. If necessary, settlement basins are to be used and the separated silt particles disposed of separately.

If the hydraulic fill dike is to be the later short line, it is recommended above all to have a sand dike with foil covering. In order to have sand with the coarsest possible grain size in the immediate vicinity of the

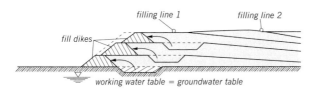

Fig. R 81-1. Hydraulic filling of port areas on a ground surface above the water table

short line, it is recommended to place a shore pipe outlet on or in the immediate vicinity of the hydraulic fill dike so that the dredge discharge will run along the dike (fig. R 81-1). The risk of possible foundation failure must be taken into account.

7.3.3 Hydraulic Filling of Port Areas on Subsoil Below the Water Table

7.3.3.1 Coarse-grained Hydraulic Fill Sand (fig. R 81-2)

Coarse-grained sand can be used for filling without further measures. The inclination of the natural fill slope depends on the grain coarseness of the fill sand and on prevailing water current. The fill material outside the theoretical underwater slope line will be dredged away later (R 138, section 7.5).

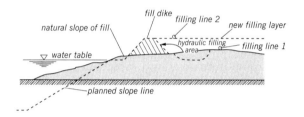

Fig. R 81-2. Hydraulic filling of port areas on a ground surface below water level

The rolling soil filled in the first stage should reach to a level of about 0.50 m above the relevant working water table for coarse sand and at least 1.0 m for coarse to medium sand. Above this, work continues between fill dikes. Tide-dependent filling may be necessary in tidal areas.

7.3.3.2 Fine-grained Hydraulic Fill Sand (fig. R 81-3)

Fine-grained fill sand is placed by filling or dumping between underwater dikes of rock fill material. This method can also be recommended when for instance insufficient space is available for a natural hydraulic fill slope because of ship traffic. Rock fill material is however not suitable when a quay is planned in the final state and the rock fill material would then have to be removed again.

It is also possible to build up the shore in advance with dumped sand (fig. R 81-4) which is back-filled. The coarsest sand should be used for this method. Strong currents can nevertheless lead to difficulties.

The dumped sand outside the theoretical underwater slope line will be dredged away later (R 138, section 7.5).

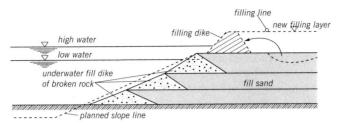

Fig. R 81-3. Underwater fill dike of broken rock. The fine-grained fill sand is flushed in or dumped

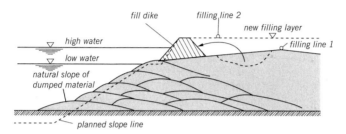

Fig. R 81-4. Underwater filling of dikes of coarse sand by dumping

Some hopper suction dredges are able to fill the sand in the so-called rainbow method. In this method, the fill material is filled through a jet. Currents and fines can cause considerable problems.

7.3.3.3 Filling Harbour Basins Without Large-scale Removal of Existing Sediment Deposits

If existing idle harbour basins are to be filled as part of restructuring work and the resulting surface used for harbour operations, it can be sensible and economical for larger magnitudes of existing sediment deposits to be left in the port and deliberately covered by the fill material. In such cases, a locally restricted soil exchange down to the bearing bottom can be carried out in the statically required extent as part of the planned final structure (embankment or quay wall). An underwater wall is created from bearing material as coarse as possible.

In the harbour basin behind the wall, the sediment remains on the bottom and is covered by the thinnest possible layers (max. 1 m) uniformly by the hydraulic filling method. Local deformations and ground failure from uneven surcharges are to be avoided.

Special filling pontoons will be required for sand filling, which can be moved and positioned specifically in the vicinity of the filling surface.

Consolidation periods between the individual layers depend on the soil mechanical properties of the sediment deposit. As a general rule, several sand layers are required underneath the water table (under uplift) to consolidate the sediment layer to such an extent that further hydraulic filling stages are possible. This requires careful soil mechanical studies and accompanying measurements. Settlement can be accelerated by additional measures in the form of vertical drains and pre-loading.

7.4 Backfilling of Waterfront Structures (R 73)

7.4.1 General

In order to provide a firm support for the backfill, and thereby prevent extensive future settlement and the imposition of heavy earth pressure on the structure, it may be advantageous where practicable to remove those layers of cohesive soil of poor bearing value occurring near enough to the structure to have an effect on it. This should be carried out before any pile driving or work of similar importance. If such material is not removed, its effect on the structure must be taken into consideration. For thick layers of poor soil, the effect in the unconsolidated state must also be taken into consideration (R 109, section 7.9).

7.4.2 Backfilling in the Dry

Waterfront structures erected in the dry shall also be backfilled in the dry as far as possible.

The backfill shall be placed in horizontal layers, adapted to the compaction gear used, and well compacted. If possible, sand or gravel is to be used as fill material.

Non-cohesive backfillings must have a minimum density of $D \approx 0.5$ if excessive maintenance of pavements, tracks and similar work is to be avoided. Density D is to be determined in accordance with R 71, section 1.6.

If the backfilling is executed with non-uniform sand in which the component of fines less than < 0.06 mm in diameter is less than 10 % by weight, a point resistance of at least 6 MN/m^2 shall be required at depths from 0.6 m in static penetrometer tests. With satisfactory backfill and compaction, at least 10 MN/m^2 can in general be expected from 0.6 m depths on. If possible, the pressure resistance measurements should be carried out regularly during the backfill work.

For backfilling in the dry, cohesive types of soil such as boulder clay, sandy loam, loamy sand and in exceptional cases even stiff clay may be considered. Cohesive backfill must be uniform throughout, in thin layers, and especially well compacted to form a uniform solid mass without cavities. This can be achieved without difficulty with suitable modern compacting gear, e.g. vibration rollers. An adequate cover of sand must be provided over the cohesive layers. Any possible increases in earth pressure must be taken into account in the design and dimensioning of the backfilled structure if load-relieving movements are not possible. special investigations are required when there is any doubt.

7.4.3 Backfilling Under Water

Only sand and gravel or other suitable non-cohesive soil may be used as fill under water.

A medium degree of density may be achieved if very non-uniform material is hydraulically pumped in such a manner that it is deposited like sedimentary material. But higher degrees of density can generally be achieved only by using special construction measures, such as deep vibration. When using uniform material as fill, hydraulic pumping along can generally produce only loose density.

Reference is made to R 81, section 7.3.1 and R 109, section 7.9.3 for the quality and procurement of the hydraulic fill sand.

The dredge water must flow off quickly and completely. Otherwise, a strongly increased water pressure difference can cause unacceptable greater structure loads and movements.

Above all, when polluted material or silt and mud are to be expected, backfilling must proceed in such a manner that no sliding surfaces are formed. These lead to a decrease in the passive earth pressure or to an increase in the active earth pressure. Special reference is made to R 109, section 7.9.5.

The drainage system of a waterfront structure may not be used to withdraw the dredge water when backfilling or it will become clogged up and damaged.

In order to enable the backfill to settle uniformly and let the existing subsoil adapt itself to the increased surcharge, there should be a suitably long interval between completion of backfilling and the beginning of dredging.

7.4.4 Additional Remarks

Sheet piling occasionally suffers driving damage at the intervals through which a heavy flow of water occurs when there is a difference in water pressure. Backfill is then washed out and soil in front of the sheet piling piled up or carried off, resulting in cavities behind the wall and shoaling or scour in front. Hydraulic backfilling considerably increases the risk of such damage. These defects can be recognised through settlement of the ground behind the wall. Even in non-cohesive soils, greater cavities can be produced in time which fail to be recognised in good time in spite of careful inspection. Such cavities often do not cave in until after many years of operation, after already causing considerable damage to persons and property.

In consideration of other actions, such as active earth pressure redistribution, consolidation of the backfill and the like, it is recommended at first to backfill the quay walls and then subsequently dredge them free in steps at adequate time intervals (R 80, section 7.1).

During water-sided dredging, the structure should be inspected as early as possible by divers between water level and dredging pit bottom or harbour bottom for defects in the wall. Otherwise, reference is made to the possibilities of inspection during the driving of sheet piling, according to R 105, section 8.1.13.

7.5 Dredging of Underwater Slopes (R 138)

7.5.1 General

In many cases, underwater embankments are constructed as steep as possible considering stability requirements. The inclination of the slope is stipulated above all on the basis of equilibrium investigations. Consideration is taken of choppy seas and currents as well as the dynamic effects of the dredging work itself. Experience has shown that slope failures frequently take place just at this stage. The high costs then required for the restoration of the planned slope, justify extensive initial soil exploration and soil mechanics investigations as a basis for the planning and execution of this kind of dredging.

Stability in non-cohesive soils can be increased during dredging by withdrawing groundwater by means of wells installed immediately behind he slope.

7.5.2 Effects of the Soil Conditions

When choosing the type and extent of the soil investigations, those soil characteristics influencing the dredging operation must also be taken into account. Exact knowledge of these characteristics is required for:

- correct selection of the dredge type,
- deciding on the best mode of operation with the selected equipment,
- estimating the attainable dredge output.

The following parameters are of special significance:

in non-cohesive soils:	in cohesive soils:
grain-size curve,	grain structure,
specific weight,	specific weight,
porosity,	cohesion,
critical degree of density,	angle of internal friction,
permeability,	undrained shear strength,
angle of internal friction,	consistency index,
point resistances of penetrometer tests or SPT values,	point resistances of penetrometer tests or SPT values,

geo-electric resistance measurement,

point resistances with measurement,

of friction and pore water pressure.

point resistances with measurement,

of friction and pore water pressure,

in-situ density and moisture from nuclear geophysical measurements.

Adequate knowledge of the soil strata structure can be obtained by hose core borings. The records of soil findings obtained in this way can be supplemented and enhanced by colour photos taken immediately after removing the samples or after opening the hoses.

Special problems may arise when dredging in loosely deposited sand, if its density is less than the critical density. Large quantities of sand may become liquefied and "flow" due to minor causes, such as vibration or small disturbances and stress changes in the bottom during dredging. Flow sensitivity of the soil must be detected in ample time to take countermeasures such as compacting the soil in the area influenced by the planned embankment, or at least making the slope flatter. The latter alone however is frequently not sufficient.,

Even a comparatively thin layer of loosely deposited sand in the mass of soil to be dredged can lead to a flow failure during the dredging operation.

7.5.3 Dredges

Underwater slopes are excavated by dredges whose type and capacity depend on:

- type, quantity and thickness of the layer of soil to be dredged,
- dredging depth and disposal of spoil.

The following dredge types may be considered for dredging the slopes:

- bucket-ladder dredge,
- cutter-suction dredge,
- cutter-wheel suction dredge,
- hopper suction dredge,
- grab dredge,
- dipper dredge.

The dredge must be selected according to the operating conditions. Availability of the dredge is also important.

The operating mode of suction dredges means that they are apt to cause uncontrollable slope failures. They are therefore generally not considered qualified to meet the specifications for dredging underwater slopes. undercutting must be strictly avoided.

Slopes to a depth of approx. 30 m can be successfully excavated with large cutter-suction dredges; with large bucket-ladder dredges, slopes to a depth of approx. 34 m are feasible.

Dipper dredges are preferably employed for dredging in heavy soils.

Clam-shell dredges are best suited for dredging only small quantities, or when dredging is to be carried out in accordance with R 80, section 7.1.

7.5.4 Execution of Dredging Work

7.5.4.1 Rough Dredging

Dredging is preferably carried out above to just below the water level, where, for example, the profile of these parts of the slope is properly produced with a grab dredge. Before dredging the remainder of the underwater slope, dredging is carried out at such a distance from the embankment that the dredge can operate as close as possible to full capacity without causing any risk of a slope failure occurring in the planned embankment.

Indications as to the safe distance to be maintained between the dredge and the planned embankment are gained by observing soil slides and the ensuing slopes which occur during the rough dredging work.

After conclusion of the rough dredging work, a strip of soil remains clear of the top of the underwater slope. It must be removed by a method suitable for the purpose (figs. R 138-1 and 138-2).

7.5.4.2 Slope Dredging

Dredging along underwater slopes must be carried out carefully so that slope failures are kept within bounds and under control.

(1) Bucket-ladder dredging

Formerly, bucket-ladder and clam-shell dredges were employed exclusively both for coarser dredging and for dredging slopes. Dredging can be accomplished with small bucket-ladder dredges, starting at a depth of approx. 3 m below the water level.

For practical reasons, the bucket-ladder dredge operates parallel to the slope, generally dredging layer by layer. Full or semi-automatic control of movements of the dredge ladder is possible and is to be recommended. The slope is dredged in steps. The type of soil determines the extent to which the steps may intrude into the theoretical slope line (fig. R 138-1). In cohesive soils, the steps are generally dredged symmetrically to the theoretical slope line. In non-cohesive soils however, intrusion into the slope line is not allowed. The possible removal of the protruding soil depends on the tolerances which are to be stipulated contingent on the soil conditions and the marginal conditions listed under R 139, section 7.2.2.

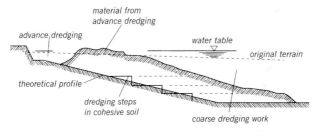

Fig. R 138-1. Dredging an underwater slope with bucket-ladder dredges

The depth of the steps depends on the soil conditions and is generally between 1 and 2.5 m.

The precision with which slopes can be built in this manner depends among others on the planned slope inclination, type of soil, capabilities and experience of the dredge crew.

With slope inclinations 1 : 3 to 1 : 4 in cohesive soils, a tolerance of ± 0.5 m measured vertical to the theoretical slope line may be accepted. The tolerance in non-cohesive soils must be +0.25 to +0.75 m, depending on the dredging depth.

(2) Cutter-suction or cutter-wheel suction dredging

Apart from bucket-ladder dredges, cutter-suction and cutter-wheel suction dredges are also suitable for building underwater slopes, and often do so better and more cheaply.

If a spoil area is not available within a reasonable distance, dredged sand can be loaded into barges, with the help of supplementary equipment. Coarser

sand will settle and remain in the barges. But finer material will flow out with the overflowing water.

When dredging, the cutter-suction dredge preferably moves along the slope, dredging layer by layer, like the bucket-ladder dredge. Computerised control of the dredge and dredge ladder is recommended.

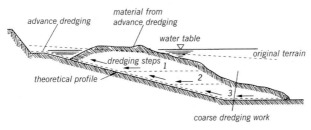

Fig. R 138-2. Dredging an underwater slope with cutter-suction or cutter-wheel suction dredges

Fig. R 138-2 shows how the cutter-head works upward parallel to the theoretical slope line after having made a horizontal cut. Underwater slopes of the greatest precision can be produced in this way. In the case of computerised dredge control, tolerances of +0.25 m measured at right angles to the slope are satisfactory for small cutter suction dredges, and +0.5 m for larger dredges. If dredging is done without special control, the same tolerances as recommended for bucket-ladder dredging apply. A prerequisite is that the soil has no tendency to flow.

7.6 Scour and Scour Protection at Waterfront Structures (R 83)

7.6.1 General

Scour can be caused by natural currents or erosion effects caused by ship movements, such as

- natural gradients or drift currents,
- waves, backflow currents and propeller currents caused by ships.

In port engineering in the vicinity of the ships' berths, the action of the ships' screws is a prime eroding element, with speeds of up to 4 to 8 m/s possible near the harbour bottom. This contrasts with the current speeds of natural river or drift currents and backflow currents caused by the ships of only 1 to 2 m/s. Erosion resulting from waves caused by ship action has an influence on the slopes. A corresponding safeguard is to be included in the principles for coastal construction, see R 186, section 5.9.

Scour caused by the natural current of the water occurs mainly where erosion currents can occur at headlands, narrow passage openings etc.

Special account must be taken of the ship's screw action and its effect on the harbour bottom particularly where ships berth and depart again under their own steam. This includes particularly ferries, Ro-Ro-vessels and container ships.

Scour caused by the drive screw or bow thruster only affects a limited area, but the high and turbulent velocities involved do particularly endanger water structures in the vicinity, such as ferry beds, quay structures, locks etc.

7.6.2 Scour Caused by Ships

7.6.2.1 Scour Caused by Jet Formation by the Stern Screw

The jet velocity caused by the rotating screw, so-called induced jet speed (occurs directly behind the screw), can be calculated to [170]:

$$V_0 = 1.6 \cdot n \cdot D \cdot \sqrt{k_T} , \qquad (1.1)$$

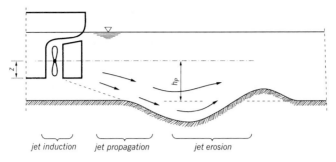

Fig. R 83-1. Jet formation caused by the stern screw

n = speed of the screw [1/s],
D = diameter of the propeller [m],
k_T = thrust coefficient of the screw, $k_T = 0.25 \ldots 0.50$.

The simplified formula for a mean value of the thrust coefficient is:

$$V_0 = 0.95 \cdot n \cdot D. \tag{1.2}$$

If the output of the screw P is known instead of the speed, the induced jet speed can be calculated according to the following assumption:

$$V_0 = C_P \left[\frac{P}{\rho_0 \cdot D^2} \right]^{1/3}. \tag{1.3}$$

P = screw output [kW],
ρ_0 = density of the water [t/m³],
C_p = 1.48 for free screw (without nozzle) [1],
    = 1.17 for screw in a nozzle [1].

As it progresses further, the jet expands cone-shaped from the turbulent exchange and mixing processes loses speed with increasing length. the maximum speed occurring near the bottom which is essentially responsible for scouring, can be calculated as follows, see also [170]:

$$\frac{\max V_{\text{bottom}}}{V_0} = E \cdot \left(\frac{h_P}{D} \right)^a. \tag{2.1}$$

Integration of equation (1.2) then produces:

$$\max V_{\text{bottom}} = 0.95 \cdot n \cdot D \cdot E \left(\frac{h_P}{D} \right)^a, \tag{2.2}$$

E = 0.71 for single-screw vessels with central rudder [170],
   = 0.42 for single-screw vessels without central rudder,
   = 0.42 for twin-screw vessels with middle rudder [170],
   = 0.52 for twin-screw vessels with twin rudders located after the screws,
a = −1.00 for single-screw vessels,
   = −0.28 for twin-screw vessels,
h_p = height of the screw shaft over bottom [m],
   = $z + (h - T)$,
z = $\left(\dfrac{D}{2}\right) + 0.10\ldots 0.15$,
h = water depth [m],
T = draft.

The speed of the screw which is relevant to water jet velocity depends on the power plant output used for berthing and departing. Practical experience has shown that this machine output for port manoeuvres lies between

– approx. 30 % of the rated speed for "slow ahead" and
– approx. 65–80 % of the rated speed for "half-speed ahead".

A speed corresponding to 75 % of the rated speed should be selected for designing the bottom safeguard systems.

The rated speed or, where applicable, increased speeds at maximum power plant output must be assumed for particularly critical local conditions (high wind and current loads for the ship, nautically unfavourable channels), or for basin trials in shipyards. These conditions must be clarified with the future operator of the port facilities, particularly with the port authority.

7.6.2.2 Water Jet Generation by the Bow Thruster

The bow thruster consists of a screw which works in a pipe and is located cross-wise to the longitudinal angle of the ship. It is used for manoeuvring out of a standing position and is therefore installed at the bow – more rarely at the stern. When the bow thruster is used near to the quay, the generated water jet hits the quay wall directly and is diverted to all sides from there. The critical element for the quay wall is the part of the water jet directed at the harbour bottom, which causes scour in the immediate vicinity of the wall on hitting the bottom, see fig. R 83-2.
The velocity at the bow thruster outlet $V_{0,B}$ can be calculated according to [170] as per equation (3):

$$V_{0,B} = 1.04 \left[\frac{P}{\rho_0 \cdot D_B^2} \right]^{1/3}, \tag{3}$$

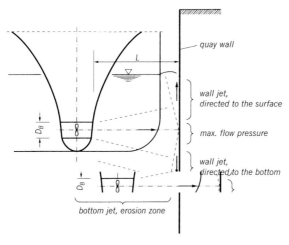

Fig. R 83-2. Jet load on the harbour bottom from the bow thruster

P = output of the bow thruster [kW],
D_B = Inner diameter of the bow thruster opening [m],
ρ_0 = density of the water [t/m³].

Jet velocities of 6.5 to 7.0 m/s must be expected for the bow thrusters of large container ships (P = 2500 kW and $D_B \approx$ 3.00 m).
The velocity of the part of the water jet hitting the bottom max V_{bottom}, which is responsible for erosion, is calculated as follows:

$$\max V_{bottom} = 2.0 \left(\frac{L}{D_B}\right)^{-1.0} \cdot V_{0,B}, \qquad (4)$$

L = distance between opening of the bow thruster and quay wall, see fig. R 83-2, [m].

The bow thruster usually operates at full load.

7.6.3 Scour Protection

The following measures can be considered for averting the dangers to waterfront structures from scour:

(1) Scour surcharge to the structures.
(2) Covering the bottom with a stone fill in loose or as a grouting.
(3) Covering the bottom with flexible composite systems.
(4) Monolithic concrete slabs, e.g. in ferry beds.

To 1):
In this case, a protective layer is not applied and the scour formation is accepted. The structure is secured by calculating the theoretical foundation bottom at a depth which takes account of the corresponding scouring depth (scour surcharge).

This procedure creates problems because it is hard to even approximately estimate the anticipated scour depth and thus accurately calculate the scour surcharge. When this method is used, the bottom should be checked continuously by soundings in the area at risk from scour so that it is possible to react at once when the tolerances are exceeded.

To 2):
A loose stone cover constitutes one of the most frequently used protection systems. The following requirements must be met:

- Stability when exposed to screw action.
- Installation of the stone cover so as to cover the bottom reliably. This means installation in 2 to 3 layers.
- Installation on a grain or textile filter rated for the relevant subsoil, see [128] and [149].
- Connection to the waterfront structure underneath the current to be safe from erosion.

Verification of current stability is provided according to [170] according to the following assumption:

$$d_{req} \geq \frac{V_{bottom}^2}{B^2 \cdot g \cdot \Delta'}, \qquad (5)$$

d_{req} = required diameter of the stones [m],
V_{bottom} = bottom velocity as per equation (2.2) or (4) [m/s],
B = stability coefficient [1] according to [170],
  = 0.90 ship without central rudder (stern screw),
  = 1.25 ship with central rudder (stern screw),
  = 1.20 bow thruster,
g = 9.81 (earth acceleration) [m/s^2],
Δ' = relative density of the stone material under uplift [1],
  = $(\rho_s - \rho_0)/\rho_0$,
ρ_s, ρ_0 = density of the stone material respectively water [t/m^3].

For single-screw vessels with central rudder (e.g. container ships), stone fills using broken rocks with $\rho_s = 2.65$ t/m^3 create problems for bottom speeds of 4 to 5 m/s, as the corresponding diameters with d_{req} = 0.7–1.0 m are so large that they cannot be easily handled.

Stone fills as a grouting are stable up to very high speeds ($V_{bottom} \approx 7$ m/s) as a result of the cramping effect. Materials which adhere well and can be used underwater such as bitumen or colloidal mortar are suitable as grouting material.

The following aspects must be taken into consideration when installing such a solution:

- The grouting should not cover the whole area. A minimum pore volume (continuous from the bottom to the top surface of the fill) of 15–20 % is necessary for pressure compensation.
- The grouting stone fill forms a stable but rigid unit. Scours frequently occurring at the edges cause underwashing and in some cases damage because the grouting cannot react flexibly to these phenomena. A combination with flexible elements at the edges can be beneficial to avoid such damage.
- The grouting thickness, positioning of filters and connection to waterfront structures are rated according to the same aspects as for loose stone fill.

To 3):

Composite systems consists of systems in which various basic elements are combined to create a sheet-type safety mat. The most important principle to be observed here is that the elements are to be connected in a flexible way for good adjustment to and stabilisation of edge scouring. The following technical solutions are known:

– concrete elements connected by ropes or chains,
– mesh containers filled with broken rocks ("gabions"),
– geo-textile mats filled with mortar.

These systems provide excellent stability properties when rated with adequate dimensions. A generalised, current-mechanics design assumption is only available for special cases because of the individual variety of available systems, see [170], so that these systems are frequently dimensioned according to the manufacturer's experience.

When the system is flexible enough, it shows good edge scouring behaviour, i.e. occurring edge scour is stabilised automatically by the system, thus preventing regressive erosion.

The disadvantage of wire rubble mats ("gabions") with equally good stability and edge scour properties is that the wire mesh is liable to corrosion, sand wear and mechanical damage, and the current stability is lost when the wire mesh is destroyed.

To 4):

An underwater concrete bottom offers ideal erosion protection because its depth can be produced with far greater precision than a stone fill. There is no need to fear that individual stones can be dislodged from the system by an anchor or screw action; the thrust force transferred locally to the bottom by the screw is distributed across a wide surface.

The rigid underwater concrete bottom cannot follow uneven settlement, so that special solutions are required for the edges. Underwater concrete bottoms are installed in thicknesses from 0.30 to 1.00 m depending on the installation method. The installation of concrete under water is a technologically complicated, very costly process, which has to take account of diving operations, poor visibility underwater, floating equipment, underwater formwork, special types of concrete, suction of mud on the harbour bottom. A rigid concrete bottom as erosion protection becomes competitive again when it can be installed in the dry, e.g. with the protection of a catchment dam.

The necessary spatial expansion of this type of consolidation depends very much on local conditions and should be rated so that the water jet velocity is essentially reduced at the edges to rule out any risk of the system being underwashed by resulting scour.

The minimum spacing of the system perpendicular to the quay must also be selected so that the area of the passive earth pressure wedge at the foot of the wall is not reduced by scour.

An initial approximation as per Fig. R 83-3 is:

for single-screw vessels:

- perpendicular to the quay: $L_N = 3 \ldots 4 \cdot D$,
  D = screw diameter,
- lengthways to the quay: $L_{L,H,1} = 6 \ldots 8 \cdot D$,
  $L_{L,H,2} = 3 \cdot D$,
  $L_{L,B} = 3 \ldots 4 \cdot D_B$.

for twin-screw vessels:

The above values for L_N and $L_{L,H}$ are to be doubled.

The total expansion of the consolidation layer lengthways to the quay including the intermediate length L_Z depends on the possible variations in the berth positions.

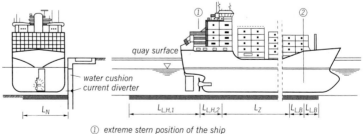

① extreme stern position of the ship
② extreme bow position of the ship

Fig. R 83-3. Expansion of the scour protection areas in front of a quay

The quay wall can be pulled back to create a water cushion between front edge of the quay and ship's wall, where necessary in combination with jet deflectors, offering efficient possibilities for minimising load on the harbour bottom.

In the case of waterfront structures where dredging is necessary because of silt, sand, gravel or rubble deposits, the upper edge of the bottom consolidation layer should correspond to the design depth as per R 36, section 6.7.

7.7 Vertical Drains to Accelerate the Consolidation of Soft Cohesive Soils (R 93)

7.7.1 General

The consolidation (primary settlement) of soft cohesive, relatively impermeable strata, can be generally accelerated by means of vertical drains. This cannot be used as solution for secondary settlements which occur in some soils, caused by creeping of the soil with no change in pore pressure.

Vertical drains are used with good results particularly in soft cohesive soils. Since the horizontal permeability of the soil is generally higher than the vertical, the effectiveness of a vertical drain is increased. In soils with layers of alternating permeability (so-called sandwich structure, e.g. layers of clay and mud-flat sand), the pore water in less permeable layers is not only discharged to the drains directly, but also through adjacent layers of higher permeability.

As a result, the consolidation is additionally accelerated. When secondary settlement accounts for a substantial portion of total settlement under prevailing load conditions – as in peat and certain other soils. The results are as a rule less satisfactory.

7.7.2 Use

Where bulk goods, dams or landfills are dumped on soft cohesive soils, vertical drains are used to shorten the duration of settlement and increase the bearing capacity of the existing foundation soil as quickly as possible. They are also used to prevent sliding of slopes or terraces, to reduce side flow movements, decrease active earth pressure behind quay walls, (increasing cohesion), etc.

When evaluating the utilisation of vertical drains, it must also be investigated to see whether the drains have an unfavourable influence on the surrounding soil, facilitating the transport of any pollution in the soil.

7.7.3 Design

When designing a facility with vertical drains, the following considerations should be taken into account.

- Through the application of an additional load larger than the total of all future loads, including the weight of the material necessary to compensate for settlements, secondary settlements can be additionally accelerated.
- The settlements of drained areas can be calculated according to the theory of consolidation (TERZAGHI), secondary settlements to KEVERLING BUISMAN and combined settlement formula of KOPPEJAN [56]. However, the results of these calculations can only be used to estimate the order of magnitude because of the simplification contained in the calculation method and the inhomogeneity of the foundation soil. Settlement measurements and the measuring of pore water pressure in consolidated soil can be used for more accurate determination of the end of the settlement period and the increase in shear strength from the actual progression of consolidation.
- Greater primary total settlements have been observed in some vertically drained soils than found under similar conditions in undrained soil.
- The results of the use of vertical drains for the hydrological conditions of the area (pore water pressure and effective pressure in soil material) should be carefully analysed.
- When pore water excess pressure is present in a sand layer below cohesive strata to be consolidated, the bottom surface of the drains is to be at least 1 m above this layer.

For optimum spacing of the drains, hose borings, core borings or pressure soundings should be carried out to ascertain the exact position and thickness of the soil layers, and to obtain characteristic samples for determining permeabilities.

The following criteria are to be observed:

(1) The inlet area must be large enough.
(2) The drain must possess sufficient discharge capacity.
(3) Thickness and permeability (horizontal and vertical) of the cohesive soil strata to be dewatered.
(4) Permeability of neighbouring soil strata.
(5) Desired acceleration of the consolidation period.
(6) Costs involved.

In most cases, the criterion according to (1) is decisive. The available consolidation time is also of particular importance.

Early systematic preliminary investigations with settlement, water level and pore water pressure measurements can be used to find the most practicable and economical type, layout and construction of the drains. The effects of later construction work must also be taken into consideration: dirt on the drain walls resulting from construction procedures can decisively hinder the water runoff.

This also applies to runoff down into a water-bearing layer, or to lateral discharge in a sand or gravel layer place on top. In general, sand drains can be installed 0.25 m diameter at 2.5 to 4.0 m centre-to-centre spacing, plastic drains generally have a smaller centre-to-centre spacing.

Plastic drains are supplied in widths of 0.1 to 0.3 m. The design should also take account of the manufacturer's product-specific information.

The choice of drain type (sand or plastic drain) should be based on the following factors:

- During the installation of a plastic drain, kneading and/or measuring of the contact areas soil/drain may occur. This will increase water penetration resistance. The favourable action of thin horizontal sand layers cannot take effect through this smearing etc.
- The vertical drains must experience the same relative compression as the soil strata in which they are located.
- The function of plastic drains can be considerably impaired by settlement.

7.7.4 Construction

Vertical drains can be constructed as bored, jetted or driven sand drains, but also as plastic drains.

Plastic drains are generally used today. They are installed by means of stabbing or vibration.

The timely planning of the drain works and the land fills is especially important.

In order to facilitate the moving of equipment and the transportation of sand over extremely soft terrain, and to prevent the bore spoil from fouling the ground surface, a layer of sand of good permeability at least 0.5 m thick is placed on the ground before work starts. This also acts as drainage layer when dewatering upward. If the sand layer is too thick and compacted, it is much harder to install the plastic drains.

The watery bore spoil is carried off in ditches to locations where deposits can cause no harm.

In the event of contaminated subsoil, avoid embedment procedures in which bored or jetted material is obtained and requires disposal.

In order to gain time for the utilisation of the drained ground, the drains should be installed in good time and the subsoil already pre-consolidated through soil fill on the ground.

7.7.4.1 Bored Sand Drains

For bored sand drains, depending on the type of soil, cased or uncased boreholes are sunk with diameters of about 0.15 to 0.3 m, and filled with sand.

Bored sand drains have the advantage that they cause minimum disturbance of the subsoil and the least reduction in permeability of the soil in the horizontal direction, which is vital for their success. They should be bored swiftly under water pressure difference. When no casing is used, they are to be flushed free of any obstructions so as to provide a perfectly clear hole; when a water pressure difference exists, the holes should be filled without interruption to minimise the danger that the sand will be fouled by crumbling of the walls into the hole. Special care must also be taken that the drains are not severed by soil penetrating from the sides.

The fill material must be chosen so that pore water can flow in and out without hindrance. The proportion of fine sand (≤ 0.2 mm) should not be higher than 20 %.

7.7.4.2 Jetted Sand Drains

Section 7.7.4.1 applies correspondingly to jetted sand drains.

They are cheap to make but have the disadvantage that they are easily clogged by deposits of fines on the bottom and on the drain walls. Special care must be paid to this during their construction. After they are jetted clear and the gear removed, the sand filling is to be started immediately and completed swiftly without interruption.

7.7.4.3 Driven Sand Drains

If displacement of the soil has no harmful effect on the bearing capacity of the subsoil and its permeability in a horizontal direction, driven sand drains may be used. Thus for example, a pipe 0.3 to 0.4 m in diameter is plugged with gravel or sand at its lower end and driven into the soft subsoil until the desired depth is reached using a rammer working inside the pipe. After the bottom closure has been knocked out, suitable graded sand (as with bored drains) is filled into the pipe and driven into the subsoil, while the pipe is simultaneously withdrawn. In this manner, a sand drain is created whose volume is about two to three times that of the volume of the pipe. Zones which may have become stopped up when the pipe was being driven will be opened again, at least partially, by the driving of the sand and the resulting increase in the wall area of the drain.

It is perfectly feasible to direct the drainage both into a lower water-bearing stratum and into a sand fill on the surface by this method.

7.7.4.4 Plastic Drains

Plastic drains are introduced into the soft soil by means of a needle unit. They can be inserted in the soft soil without water, which practically avoids any clogging of the drain. This dry method is also advantageous when passing through sand fills, because if sand drains were used here, much flushing water would flow off through the borehole walls.

7.7.5 Precautions During Construction

The success of vertical drainage depends largely on the care taken during construction. In order to avoid failures, continuous inspection of the sand filling is necessary, and the effectiveness of the drains must be checked in good time by means of fill tests with water, water level measurements in individual drains, pore pressure measurements between the drains, pressure soundings or water pressure measurements and observations of settlement of the ground surface.

7.8 Subsidence of Non-cohesive Soils (R 168)

7.8.1 General

Subsidence is a certain component of the volume reduction of a non-cohesive mass of soil, e.g. of sand.

Altogether, volume reductions are caused by:

(1) Grain readjustments resulting in an increase in density.
(2) Deformations of the grain skeleton.
(3) Grain failure and grain splintering in areas with large local stress peaks at grain contact points.

The indicated effects can even be started simply through load increases. The grain readjustments according to (1) chiefly occur during vibrations, reductions in the structure resistances and/or the friction effects between the grains.

The volume reductions due to the latter two actions are designated as subsidence. It takes place when the so-called capillary cohesion between the grains, i.e. the friction effect due to the surface tension of the residual water in the pore grooves (at the points of grain contact) strongly decreases or disappears completely, at very heavy wetting or drying out of non-cohesive soils. If this occurs for example in connection with a rising groundwater level, vibratory effects due to the changes of the menisci between the grains at the respective groundwater level also act on the grains.

7.8.2 Magnitude of Subsidence

If moist sand is loaded and fully wetted, for example by rising groundwater, a load/settlement diagram as per fig. R 168-1 results.

The subsidence of non-cohesive soils is contingent on the granulometric composition, grain shape, grain roughness, initial moisture content, the state of stress in the ground and, above all, on the initial density. The more loosely the soil is deposited, the greater the subsidence is. In very loosely deposited, uniform fine sands, it reaches 8 %. But even after high-degree compaction it can still reach 1 to 2 %. In general, subsidence is greater in round-grained soils than in angular soils. Uniform sands

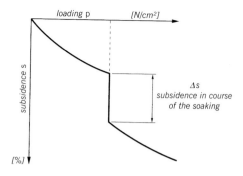

Fig. R 168-1. Load/settlement diagram of moist sand at soaking

show a greater degree of subsidence than non-uniform sands. The difference however is only recognizable at very loose and loose deposit.

In this connection the sudden subsidence of coarse silt when heavily wetted has been known for a long time.

The danger of subsidence of silty sand or clayey sand decreases rapidly with the increase in cohesion.

Tests as per [57] have shown that sands with the same initial moisture content and same initial density experience settlements and subsidence caused by loading and wetting, whose total may be assumed with adequate accuracy, independent of the magnitude of the loading. Accordingly, under small loads, settlement is small and subsidence large.

7.8.3 Effects of Subsidence on Structures and Countermeasures

Subsidence occurs essentially during the first soaking of non-cohesive soils. At further soaking, after repeated groundwater lowering, subsidence is slight.

In order to avoid or reduce subsidence as a result of submerging or wetting at rising groundwater or when fully dried-out, the sand must be extensively compacted. The achieved density can be checked in each case on the basis of undisturbed samples or with the help of dynamic, static or radiometric penetrometers (see R 71, section 1.6).

If necessary, subsidence, which may cause damage, may be effected in advance by adding large quantities of water during the installation or during soil replacement in protection of groundwater lowering, by allowing the groundwater level to rise in the meanwhile.

7.9 Soil Replacement Procedure for Waterfront Structures (R 109)

7.9.1 General

In areas where the site of a proposed waterfront structure, such as quay walls, embankments etc., is covered by thick layers of soft cohesive soils, the replacement of these weak layers with sandy soil may be feasible as well as economically justified under favourable conditions. In this connection, the principal considerations are the height of the earth to be retained by the bulkhead, the volume of fill involved, the live load to be supported by the shore area, the magnitude of the water level fluctuations and the availability of sand for replacement of proper quality and quantity at a favourable price. Soil replacement is also recommended when the soil investigations show that driving obstacles must be expected, which could lead to damage to the sheet piling. The depth of the dredge pit must be sufficient to ensure the stability of the contemplated waterfront structure, even if it becomes necessary to remove all soft cohesive strata down to bearing soil. Unless preventive measures are taken, it will be found that an intrusive layer of sediment has formed on the bottom of the dredge pit. This sediment may be the result of soil spilled by the dredge, possibly of a disturbance of the surface of the bearing soil or of a steady accumulation of silt.

Complete removal of all soft cohesive material is also required when the waterfront structure can only tolerate slight settlement.

The results of adequate exploratory soil investigations and appropriate laboratory tests customarily made in the practice of soil mechanics must be available even before a preliminary study begins, because this information is essential for estimating earth moving costs and for making a cost comparison of designs with and without soil replacement. If the decision is taken in favour of soil replacement, the following can then be determined:

- The dimensions of the dredge pit and the elevation of its bottom,
- The type of dredges to be used,
- The dredge capacities to be required and realized,
- An estimate, as accurate as possible, of the thickness of the intrusive layer of sediment which may form on the dredge pit bottom as the result of soil spilled during dredging, or possibly due to a disturbance of the surface of the dredge pit bottom.

In addition, every effort should be made to become informed about the movement of bed material and sediment, including the kind of material, its proportions and the amount deposited, as it is affected by the velocity of the currents at various tidal ranges, as well as by seasonal influences. This information is necessary to develop a systematic plan for placing the sand fill in such a way as to minimise the deposit of silty layers in the fill.

Before a decision is made in favour of the soil replacement procedure for major structures, a test pit of appropriate size should be excavated and kept under constant observation for an extended period.

Attention is drawn to the vulnerability of sand fill to damage by scour. It follows that construction scaffolding and similar structures placed on the sand must be deeply embedded unless a protective covering has been installed. In order to avoid all effects prejudicial to the stability of the structure, it is essential that workmanship be of a high order of competence and, above all, that the instructions contained herein be observed.

7.9.2 Excavation

7.9.2.1 Selection of the Dredge to be Used

In general, only bucket-ladder dredges, cutter-suction dredges, cutter-wheel suction dredges or dipper dredges can be used for excavating cohesive soil. If a layer with driving obstacles must be dredged (for example, soil interlaced with rubble), a suction dredge may be employed only if it is equipped with special pumps with sufficiently large opening widths, since otherwise the

driving obstacles will remain in place. With all types of dredges, certain losses of soil cannot be fully avoided, leading to the formation of an intrusive layer on the dredge pit bottom, even with special precautionary measures (fig. R 109-2).

When excavating with bucket-ladder dredges (fig. R 109-1), an intrusive layer generally forms on the dredge pit bottom as a result of overfilled buckets, incomplete emptying of buckets in the dumping area, and overflowing of barges. In order to reduce the thickness of this layer, a shallower cut must be used when reaching the dredge pit bottom. Furthermore, at least one cleaning cut should always be made in order to ensure that dropped dredge spoil is largely removed.

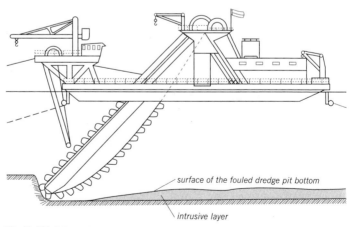

Fig. R 109-1. Intrusive layer formation when excavating with a bucket-ladder dredge

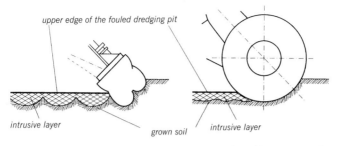

Fig. R 109-2. Intrusive layer formation when excavating with a cutter-suction or cutter-wheel suction dredge

For this, a slack lower bucket chain must be used together with low bucket and cutting speeds. The scows may be fully loaded, but overflowing with soil losses must be absolutely avoided.

Use of cutter-suction and cutter-wheel suction dredges results in an undulating dredge pit bottom as per fig. R 109-2, and the intrusive layer is thicker than that produced by a bucket-ladder dredge.

The intrusive layer thickness can be reduced by using a special cutting head shape, a slower rotation, short thrusts and slow cutting speed. The intrusive layer must be removed immediately before applying the replacement soil.

7.9.2.2 Performance and Control of Dredging Work

In order to guarantee proper dredging, the dredge pit must be boldly dimensioned according to the size of the selected dredge, and appropriately marked. Suitable survey markers and positioning devices must be clearly and unmistakably recognizable by the dredge crew at all times of day or night.

Marking the dredge cutting width solely on the side lines of the dredge is not adequate.

Excavation is performed in steps corresponding to the mean profile inclination at the edge of the dredge pit. The depth of the cuts depends on the type and size of the equipment and on the character of the soil. Strict control over the cut widths must be maintained because cuts that are too wide can cause an excessively steep slope in places with resulting slope slides.

Orderly progress of dredging can be controlled quite effectively by modern surveying methods (e.g. depth sounder in combination with the global positioning system – GPS) which can also be used for early detection of any profile changes caused by slides in the underwater slope. Deformation measurements with inclinometers on the dry edge of the excavator pit have also proven successful for checking the stability of underwater slopes.

The last sounding should be made immediately before the sand fill is placed. In order to obtain information on the characteristics of the dredge pit bottom, soil samples are to be taken here. For this, a hinged sounding tube with a minimum diameter of 100 mm and a gripping device (spring closure) has proven effective. Depending on conditions, this tube is driven 0.5 to 1.0 m or even deeper into the dredge pit bottom. After retrieval and opening, the core in the tube gives a good idea of the soil strata on hand in the dredge pit bottom.

7.9.3 Quality and Procurement of the Fill Sand

It is advisable to investigate the sand procurement areas before starting the actual work on coarse sand. The fill sand should contain only very little silt and clay, and no major accumulations of stones.

In the interests of continuous, rapid and economic filling, adequately large deposits of suitable sand must be available within a reasonable distance. In determining the amount of fill required, soil drift is to be taken into account. The finer the sand, the stronger the current over and in the dredge pit, the slower the rate at which the fill is placed and the greater the depth to the dredge pit bottom, the greater will be the loss of sand due to drift.

Efficient and capable suction dredges are recommended for sand procurement because, along with their high output, they clean the sand at the same time. The cleaning action can be intensified by proper loading of the barges and longer overflow times. Samples of the fill sand are to be taken frequently from the barges and tested for compliance with the requirements in the design, especially for the maximum allowable silt content.

7.9.4 Cleaning of the Dredge Pit Bottom Before Filling with Sand

Immediately before the start of the filling work, especially if silt has been deposited, the dredge pit bottom in the area concerned must be cleaned to the required extent. For this purpose, the use of silt suction equipment is possible if the deposits are not too firm. If however a longer period of time has elapsed between conclusion of the dredging work and the start of suction work on the silt, the silt may already be so solid that another cleaning cut may be necessary.

The water jet or water injection method has proven effective here. Large quantities of water are pumped at approx. 1 bar through nozzles to the dredge pit bottom, using a cross member suspended under a floating machine. The clearance to the bottom is kept to a minimum of between 0.3 m and 0.5 m. This method turns any settled silt back into a full suspension.

The water jet method must be carried out before placing the sand fill until the bottom is proven to be free of silt.

The cleanness of the dredge pit bottom must be checked regularly. The sounding tube described in section 7.9.2.2 can be used for this. If only soft deposits are to be expected, a properly designed grab can also be used for taking the samples; a hand grab may also be considered. A combination of silt soundings and depth soundings with differing frequencies is a good control possibility.

If an adequately clean bottom cannot be ensured, the keying between the natural bearing soil and the fill sand is to be established by other suitable measures. This can best be achieved in cohesive bearing subsoil by a layer of crushed rock of adequate thickness which is to be placed very quickly.

Such a safety measure can be of special importance on the passive earth pressure side. As there is generally no pile driving work in this area, rubble can be used to even better effect than crushed rock.

In non-cohesive soils and only slight depth to the dredging pit, keying between the fill soil and the subsoil can also be achieved by the "dowelling" effect of multiple-vibratory cores (deep vibrating with a unit of 2 to 4 internal vibrators).

7.9.5 Placing the Sand Fill

The dredge pit can be filled hydraulically or by dumping, or even both methods at the same time. Particularly in waters heavily burdened with sediment, this operation must be carried out without interruption around the clock, with carefully co-ordinated equipment, at all times adhering to a plan worked out in every detail from the very beginning. Winter operations with loss of working days to be expected due to cold weather, drifting ice, storms and fog, should be avoided.

Placing the sand fill shall follow dredging of the poor soil as quickly as possible in time and space to minimise the unavoidable deposits of sediment (silt) occurring in the meantime. On the other hand however, mixing of the soil to be replaced with that being filled in may not be permitted due to insufficient separation between dredge and fill operations. This danger exists particularly in waters with strong reversing currents (tidal area) where it must be guarded against with special care.

A certain amount of polluting of the sand fill by continuous depositing of silt cannot be avoided. However, it can be minimised if the fill is made rapidly. The action of the expected pollution on the soil mechanics characteristics of the fill sand must be taken into account accordingly. The sand must be deposited
in a manner which will prevent as far as possible the build-up of continuous silt layers. In case of heavy silt deposits, this can only be achieved by continuous, efficient operations, uninterrupted even at weekends.

It has proven effective to process the sand surface with the water injection method while placing the sand fill. This makes it possible to avoid silt contamination of the sand to a great extent. Sand losses occurring in this way must be taken into account.

Should interruptions and thus larger silt deposits occur despite all efforts, the silt must be removed before further sand filling takes place, or it must be made harmless later by suitable measures. During any interruptions, a check must be made as to whether and where the surface elevation of the fill has changed.

In order to avoid causing active earth pressure on the waterfront structure in excess of the design load, the dredge pit must be filled in such a way that silted-up slopes occurring during the filling will have an inclination opposite to that of the failure plane of the active earth pressure slip wedge, which will later act on the waterfront structure. The same applies similarly to the passive earth pressure side.

Reference is made to the PIANC report of the 3rd Wave Commission, part A, harbour protective structures, for explanations of the construction phases [171].

7.9.6 Control of the Sand Filling

Soundings are to be constantly taken during the sand filling, and the results recorded. In this way, changes in the fill surface from the filling processes itself and those due to the effects of the tidal currents can be determined to a certain extent. At the same time, these records show clearly for about how long a surface has remained unchanged, or if it has had prolonged exposure to the deposit of settling solids, so that timely measures can be initiated for the elimination of intrusive layers, which may have been formed.

The taking of surface samples from the fill area can be dispensed with only when there is speedy uninterrupted hydraulic filling and/or barge dumping.

After completion of the fill and also during the work, if required, the fill must be investigated and tested by means of tube core borings or other equivalent methods. These borings are to be sunk to natural soil at the dredge pit bottom.

An acceptance certificate forms the official basis for final design of the waterfront structure and any measures which may be required to adapt the design to conditions at the site.

7.10 Calculation and Design of Rubble Mound Moles and Breakwaters (R 137)

7.10.1 General

Moles differ from breakwaters particularly as regards the type of use. Moles can support vehicular or at least pedestrian traffic. Its crest is therefore generally higher than that of a breakwater, which may even lie submerged below the still water table. Breakwaters do not always have a shore connection.

When moles and breakwaters are constructed by a rubble mound method, reliable results of subsoil investigations are indispensable, as are careful determinations of the wind and wave conditions, currents and presence of littoral drift. For the sake of simplicity, the remaining discourse refers only to rubble mound breakwaters.

The layout and cross section of large rubble mound breakwaters are determined not only by the purpose but also by the method of execution. Recently the construction of some large breakwaters has had to be changed in retrospect because of special difficulties encountered in the execution, so that particularly detailed requirements are to be satisfied, for example when it comes to the question of whether the rubble required for the structure is delivered by land by trucks or rail, or by sea in "barges", and whether land-bound or floating equipment is to be used. In addition, the anticipated weather conditions during the construction period deserve special attention, because experience shows that the waves are frequently underestimated during this work, and the periods of calm seas are not sufficient for installing the unprotected underneath layers.

If the building work cannot be described exhaustively, suitable requisite items are to be included in the specification. The execution draft depends frequently on the equipment and material being used by the contractor involved, so that a functional work specification can be taken into consideration in order to ascertain the suitable means of execution.

7.10.2 Stability Verification; Settlement and Subsidence; Construction Suggestions

The calculation and design of rubble mound breakwaters must include verification of slope failure and foundation failure, in so far as the construction method does not make these superfluous. The effects of the waves are to be ascertained taking account of the characteristic value of the "design wave", multiplied by the partial safety factor as per R 136, section 5.6. The flow phenomenon following an earthquake in loosely deposited sand strata of the subsoil must also be taken into consideration.

Soil replacement or soil compaction and soil displacement, possibly with the aid of blasting, may be considered for subsoil with lesser bearing capacity.

Even in soft subsoil, it is possible to displace the subsoil by deliberate overload of the soil (increased fill "in front of the head") to such an extent that a state of equilibrium emerges. The resulting mud roll must be removed by careful dredging where necessary; however, mud rolls should be avoided as far as possible.

Contingent on the existing subsoil, larger penetrations, settlement and subsidence are to be expected of up to several metres in some cases. Penetration of the mole rubble due to soil displacement can be roughly estimated through equilibrium investigations or according to DIN V 4017-100 (determination of the penetration depth d in the limit state of bearing capacity with characteristic values). While settlement results from compression of the subsoil and can be partially calculated to some extent at least approximately according to DIN V 4019-100, subsidence also occurs as a result of vibration of the rock fill caused by wave impacts, and by an intrusion of the subsoil into the rock fill. The less the latter occurs, the better a transition in line with the filter rules is ensured by filter mats or by a suitable structure of the core fill on the existing ground.

Suitable choice of the core material, which is finer compared to the cover layers, keeps the permeability of the structure within limits. Breakwaters of single-core coarse material are permeable. The permeability results in through currents which can cause hydraulic foundation failure in the inboard mound of the mole. Filter stability is to be guaranteed in any case. The action of seismic forces in earthquake regions is also to be taken into account (R 124, section 2.13).

Local borings and soundings together with the results of soil mechanics investigations provide an important basis for the mentioned calculations and the resulting conclusions.

The support surface is to be examined by wide-spread depth soundings (e.g. side scanning or sonograph) before construction work starts, and recorded accordingly. Before filter mats are placed, it may be necessary to remove hindrances (larger pieces of rock etc.).

When building a new breakwater, the action of settlement and subsidence is to be compensated by an oversized profile. The method for invoicing the installed quantities should be stipulated in the specification, i.e. invoicing according to quantities used or according to oversize.

7.10.3 Designing of the Cover Layer

The stability of the cover layer, given the prevailing wave conditions, depends on the size, weight and shape of the armour blocks, as well as on the slope of the outer covering. In a series of tests over a period of many years, HUDSON has developed the following equation for the required block weights [21], [172] and [174]. It has proven itself in practice and the equation is as follows:

$$W = \frac{\rho_s \cdot H_{\text{des,d}}^3}{K_D \cdot \left(\dfrac{\rho_s}{\rho_w} - 1\right)^3 \cdot \cot\alpha}.$$

The symbols mean:

W = block weight [t],
ρ_s = mass density of the block material [t/m³],
ρ_w = density of the water [t/m³],
$H_{des,d}$ = height of the "design wave" [m], the characteristic value being multiplied by the partial safety factor,
α = slope angle of the cover layer [°],
K_D = shape and stability coefficient [1].

The preceding equation is applicable to a cover layer composed of rocks of about uniform weight. The most common shape and stability coefficients K_D of rubble and shaped blocks for a sloped cover layer according to [21] have been compiled in table R 137-1. Fig. R 137-1 shows examples of common shaped blocks.

If settlement or subsidence movements in accordance with section 7.10.2 are possible, it should be borne in mind in selecting the type of cover layer components that additional tensile, bending, shear and torsion stresses can occur as a result of the form of the component. Because of the high impact loads, the K_D values must be halved if Dolos is used.

The following modified equation is recommended as per [21] for the design of an outer layer composed of graded rock, for "design wave" heights up to approx. 1.5 m

$$W_{50} = \frac{\rho_s \cdot H_{des,d}^3}{K_{RR} \cdot \left(\dfrac{\rho_s}{\rho_w - 1}\right)^3 \cdot \cot \alpha}.$$

The symbols mean:

W_{50} = weight of a rock of average size [t],
K_{RR} = shape and stability coefficient,
     $K_{RR} = 2.2$ for breaking waves,
     $K_{RR} = 2.5$ for non-breaking waves.

The weights of the largest rocks should be 3.5 W_{50} and that of the smallest, at least 0.22 W_{50}. Because of the complex process involved, the block weights according to [21] should generally not be reduced if the structure will be exposed to waves approaching at an oblique angle.

According to [172], it is recommended that the "design wave" assumed in the Hudson equation should be at least $H_{des,d} = H_{1/3}$ where this value is generally extrapolated with the aid of peak value statistics over a fairly long period (e.g. 100 year repetition). Adequate wave measurement data must be available for the extrapolation. The significance of the design wave is obvious in that the required weight of individual blocks W increases proportionally with the 3rd power of the wave height.

Type of outer covering elements	Number of layers	Type of placing	Breakwater side K_D [1]		Breakwater end K_D		Slope
			Breaking waves	Non-breaking waves	Breaking waves	Non-breaking waves	
Smooth, rounded rocks	2 ≥ 3	random random	1.2 1.6	2.4 3.2	1.1 1.4	1.9 2.3	1 : 1.5 to 1 : 3 1 : 1.5 to 1.3
Angular rubble	2 ≥ 3 2	random random carefully placed[2]	2.0 2.2 5.8	4.0 4.5 7.0	1.9 1.6 1.3 2.1 5.3	3.2 2.8 2.3 4.2 6.4	1 : 1.5 1 : 2 1 : 3 1 : 1.5 to 1 : 3 1 : 1.5 to 1 : 3
Tetrapods	2	random	7.0	8.0	5.0 4.5 3.5	6.0 5.5 4.0	1 : 1.5 1 : 2 1 : 3
Tribar	2	random	9.0	10.0	8.3 7.8 6.0	9.0 8.5 6.5	1 : 15 1 : 2 1 : 3
Dolos	2	random	15.8[3]	31.8[3]	8.0 7.0	16.0 14.0	1 : 2[4] 1 : 3
Tribar	1	uniformly placed	12.0	15.0	7.5	9.5	1 : 1.5 to 1 : 3

[1] for slope of 1 : 1.5 to 1 : 5
[2] longitudinal axis of rocks perpendicular to the surface
[3] K_D values confirmed experimentally only for slope 1 : 2. If requirements are higher (destruction < 2 %), the K_D values must be halved
[4] Slopes steeper than 1 : 2 are not recommended

Table R 137-1. Recommended K_D values for the design of the cover layer at an allowed destruction up to 5 % and only insignificant wave overflow, excerpt from [21]

In the planning of a rubble mound breakwater, economic consideration may make it advisable to depart from the criterion for maximum protection of the outer covering against damage, if extreme loads from wave action occur only very infrequently, or if rapid, extensive land accretion on the sea-side of a stretch of the breakwater towards the shore end makes the outer covering there unnecessary. The most economical design should be chosen if the capitalised maintenance costs together with the antici-

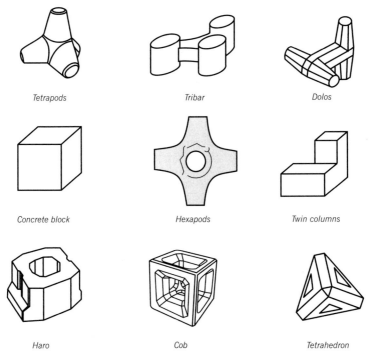

Fig. R 137-1. Examples for common shaped blocks

pated cost of repairing other damage which may occur in the port area, are less than the increase in capital cost which would occur of the block weights were designed to resist an unusually high wave which occurs only infrequently. In this case, however, the general local maintenance possibilities and the expected time required to make repairs should always be taken into account.

Further calculation methods are to be found in [40] and [172].

7.10.4 Breakwater Construction

In actual practice, as recommended in [21] and shown in fig. R 137-2, breakwaters constructed in three graded layers have proven effective. The symbols mean:

W = weight of the individual blocks [t],
H_{des} = height of the design wave [m].

Rubble mound breakwaters should not be constructed as a single layer of rocks.

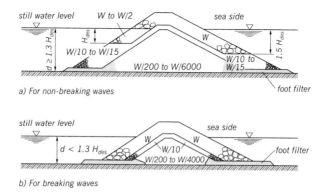

a) For non-breaking waves

b) For breaking waves

Fig. R 137-2. Filter-shaped breakwater construction in three steps

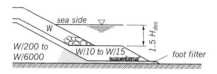

Fig. R 137-2a

It is recommended that the sea side slope be quite generally no more than 1 : 1.5.

Particular attention should be paid to the support for the cover layer, especially if it is not taken down to the base of the slope on the sea side. According to the requirements for the stability of the slope, an adequate retreat must be provided (fig. R 137-2a).

The filter rules are to be obeyed also vis-à-vis the subsoil. It is frequently more practical to do so using a filter mat rather than a filter-shaped structure of the layers between soil and outer layer (particularly at the foot points of the breakwater). Installation safety is to be guaranteed.

7.10.5 Physical Model Tests

Model tests are often a meaningful and valuable aid to accurate dimensioning. However, the results obtained under ideal conditions can supply theoretical safeties which cannot be attained under natural conditions. This is why such model tests should only be carried out by experienced institutes.

7.10.6 Execution of Work and Use of Equipment

7.10.6.1 General

The construction of rubble mound moles and breakwaters often requires the placing of large quantities of material in a comparatively short time under difficult local conditions due to weather, tide, sea state and current. The mutual dependency of the individual operations under such conditions demands especially careful planning of the individual construction steps and use of equipment.

The designing engineer and the contractor should obtain precise information about the wave heights to be expected during the construction period. In order to do this, they require information about the prevailing sea state during the construction period, not only about rare wave events. The duration of the wave heights H_s and H_{max} occurring during twelve months for example can be estimated according to fig. R 137-3 [175].

A reliable description of the wave climate by means of a wave height duration line requires long observation periods.

The design of a breakwater must allow for an execution which avoids major damage even when storms suddenly set in, e.g. layer construction with few steps.

When determining the capacity of the site installation or selecting the capacity of the equipment, realistic assumptions must be taken into consideration for possible work interruptions due to bad weather.

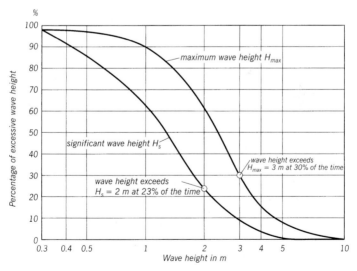

Fig. R 137-3. Wave height duration line. Duration of waves exceeding a certain wave height in a twelve month period, e.g. $H_s = 2$ m; $H_{max} = 3$ m

Depending on their local functions, the fill work is carried out:

(1) with floating equipment,
(2) with land-based equipment when the structure is advanced from the shore,
(3) with fixed scaffoldings, mobile jack-up platforms, etc.,
(4) with cableways,
(5) in combination of (1) to (4).

Construction methods with fixed scaffoldings, mobile jack-up platforms and the like are employed preferably at specially exposed locations with aggravating influence of wind, tide, wave action and current. This applies to an even higher degree if there is no harbour of refuge available at or near the construction site.

7.10.6.2 Supply of Fill and Other Material

The fill material of natural stones is generally procured in rock quarries, less often from the sea. The difficulty lies in obtaining suitable coarse material. The planning and execution must be adapted to the existing possibilities for procurement. Today it is frequently more economical and reliable to arrange for large quantities of the fill to be transported in barges by sea from even more remote sources. In this case, both the cross section of the breakwater and its structure must be designed in line with the method of placing the fill. This approach allows for large block sizes and high transport rates.

7.10.6.3 Placing Material with Floating Equipment

When placing the fill with floating equipment, the cross-section of the breakwater must be rated according to the equipment. Dumping scows always require an adequate depth of water. Deck scows can be used for lateral displacement with lesser water depths. Today's computerised position finding methods mean that floating equipment is now able to achieve the precision which was previously only possible with land-based equipment.

7.10.6.4 Placing the Fill Material with Land-based Equipment

The working plane of the land-based equipment should as a rule lie above the action of normal wave action and surf. The minimum width of this working plane is to be adapted to the requirements of the construction plant being
used. When placing the fill with land-based equipment, the material is brought to the actual site for advance filling using rear dumpers.
This construction method usually requires a core protruding out of the water with an overwide crest. The core which acts as roadway frequently has to be removed again to a certain thickness before the cover layer can be applied to ensure adequate keying and hydraulic homogeneity.

When the working plane is of a narrow width, it is frequently advantageous to use a portal crane for placing the material, as the material for the advancing work can be passed through below the crane.

Rip-rap intended for installation by crane is brought to the site mostly in rock skips on platform trailers, trucks with special tail-gates or low-bed trailers. Where narrower roadways are concerned, trailers are used which can be backed up without turning.

Large rocks and prefabricated concrete elements are placed with multi-bladed grabs or special tongs.

Electrical or electronic instruments in the operator's cab of the placing crane also facilitate true to profile filling under water.

Especially when the breakwater is constructed by advancing its head into the sea, the core fill and cover layer should follow each other rapidly as close as possible, in order to avoid unprotected core material from being washed away, or minimising this danger at least.

For further information, reference is made to [173].

7.10.6.5 Placing the Material with Fixed Scaffoldings, Mobile Jack-up Platforms and the Like

Placing the fill from a fixed scaffolding, mobile jack-up platform and the like or also with a cableway, may be considered above all for bridging a zone with constantly heavy surf.

Otherwise, section 7.10.6.4 applies accordingly.

When using a jack-up platform, the placing progress generally depends on the efficiency of the crane. Therefore, a crane should be used which has high carrying capacity in addition to the required reach.

The design should clarify which parts of the breakwater cross-section should be filled with a very calm sea and which can still be placed at a certain wave action. This applies both to the core material and to the precast concrete elements of the cover layer. When lowering precast concrete elements into position, even a slight swell can cause impact movements under water because of the large weight of the parts, which in turn causes cracks and failure.

7.10.6.6 Settlement and Subsidence

Special reference is made to section 7.10.2. Uniform and relatively slight settlements of moles and breakwaters are compensated for by increasing the height of the fill and an eventual concrete crest. If comparatively large and possibly non-uniform settlement or subsidence is expected, the crest concrete should not be placed until the end of the construction work, in not too long sections, but preferably not until the essential deformations have finished. The absorption of non-uniform settlements is to be verified. If necessary, the length of the construction sections is to be shortened.

In order to obtain reliable bases for evaluation of settlement and subsidence behaviour, and also for the settlement of accounts, settlement gauges are to be placed to the necessary extent.

Differing degrees of settlement frequently impair not the stability of a rubble mound breakwater but its appearance. This is why long, horizontal lines should be avoided in the case of concrete crests.

It is frequently necessary to make minor repairs at least to the roadway just a few years after completion, in order to compensate for subsidence.

7.10.6.7 Settlement of Accounts for the Placed Materials

The settlement and subsidence behaviour of such structures is difficult to predict so that it is recommended to stipulate easily maintained tolerances (±) from the very beginning with respect to the shape of the mole and its individual layers, when accounts are settled according to the drawing.

7.11 Light Backfilling for Sheet Piling Structures (R 187)

If light, durable filling material which is insoluble in water can be procured economically, or if suitable, non-decomposable, non-toxic industrial by-products or similar are available, they can occasionally be used as backfill material. They may make a sheet piling structure more economical if, in addition to the weight, the compressibility and consolidation value are also low, and the permeability will not decrease in time. Depending on the shear strength, the moment stresses and anchoring strength can be reduced as a result of the low weight when installed.

If a construction with light backfill is under consideration, conventional solutions should always be examined by way of comparison and the effects of light backfilling should be examined in both positive and negative respects for the structure as a whole (for example, reduced stability against slippage and slope failure, settlement of the surface, fire risk, etc.).

7.12 Soil Compaction Using Heavy Drop Weights (R 188)

Primarily soils with good water permeability can be effectively compacted with heavy drop weights. This is also possible for weakly cohesive but adequately permeable soils, and for non-water saturated cohesive soils by compression of the gas-containing pore component or by expulsion of water and gas through pores and fissures as a result of percussive effects. Under especially high percussive effects, water expulsion and thus compaction can be induced even in water-saturated, cohesive soils by crack formation in the soil.

In order to ascertain in good time the success of intended compaction measures, the soil to be compacted must be carefully examined, especially for grain distribution and permeability. Zones which can be compacted only with difficulty or not at all should first be excavated and replaced with soil amenable to compaction. The effects of compaction should be ascertain at the outset in a soil mechanics test, where possible in a characteristic test field, in order to establish the success of compaction and its deep action, as well as the economic feasibility and, if applicable, the environmental acceptability of the measures (impact noise and vibration effects on neighbouring structures or adjacent components such as canals, drains, etc.). Values for the minimum degree of density to be achieved for the various soil conditions can be established on the basis of the results in a test field. In doing this, it must be borne in mind that every type of soil requires its own test criteria. Reference is made to R 175, section 1.7.4, for verification of attaining the degree of density. No generally applicable information is available on compaction for soils below the free water level.

7.13 Consolidation of Soft, Cohesive Soils by Preloading (R 179)

7.13.1 General

In new construction areas for harbours, frequently only land with soft, cohesive soil strata with inadequate bearing capacity is available for the creation of sufficiently deformation-free harbour surfaces. However, in many cases the soft soil stratum can be improved by pre-consolidation in such a way that subsequent settlement as a result of raising the terrain, loads from buildings, bulk cargoes, containers etc. does not exceed the allowable tolerances. In addition, the shear strength of the soft strata can be permanently improved by deliberate preloading [112], [113] and [114]. By pre-consolidation of the soft strata, horizontal deformations resulting from vertical live loads can also be anticipated in the same way under certain boundary conditions, so that, for example, pile foundations are no longer subject to impermissibly high lateral pressures and deformations [115] and [116].

7.13.2 Application

In cases in which the soil replacement procedure in accordance with R 109 section 7.9 is too costly, the pre-consolidation of soft, cohesive soil strata offers a proven means of creating harbour terrain of adequate quality. The pre-consolidation of soft, cohesive soil strata can generally be considerably accelerated by incorporating vertical drains in accordance with R 93, section 7.7.

The aim of pre-loading by filling is to achieve in a short time the settlement of soft soil strata which would otherwise take a very long time to achieve its full effect as a result of continuous filling, including the live loads to be absorbed (fig. R 179-1). This time frame depends on the

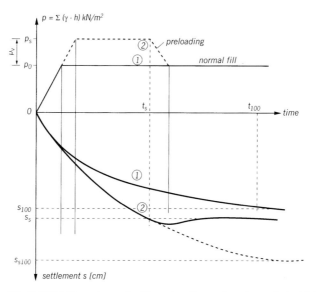

Fig. R 179-1. Dependence of settlement on time and surcharges (principle)

depth of the soft stratum, its water permeability and the level of the preload. However, such preloading by filling can be effectively applied only if sufficient time is available for the consolidation of the soft soil for an adequate deformation stability to be achieved by the soil for the intended purpose. It is expedient to proceed with preloading high enough and early enough for the desired objective for the waterfront structure to be attained before construction work begins. Preloading filling must distinguish between the following in accordance with fig. R 179-1:

a) The part of the fill which, as an earth structure, is to be free of deformation (permanent filling). It generates the surcharge stress p_o.
b) The actual preload fill, which temporarily acts as an additional surcharge with the preload stress p_v.
c) The sum of both fills (overall fill) which yields the overall stress $p_s = p_o + p_v$.

7.13.3 Stability

The prerequisite for applying a fill is its stability on the soft foundation. In many cases, it is therefore only possible to work initially with low fill levels.

The maximum fill height h for avoidance of foundation failure at the edge of the fill can be assumed to be:

$$h = \frac{4c_u}{\gamma} \text{ [m]}.$$

As the fill height and consolidation increase, the shear strength c_u, and thus h, increases.

Occasionally however, especially with filling underwater, the displacement of soft cohesive soil under the fill is desired or accepted, whereby a mud roll is created in front of the head of the fill and, if applicable, must be removed by dredging. Through displacement of this part of the soft soil stratum, its depth is reduced so that consolidation is achieved more rapidly.

7.13.4 Fill Material

The permanent fill material must be filter-stable against the soft subsoil present. If applicable, filter layers or geotextiles should be installed before the permanent fill is added. Otherwise, the required quality of the permanent fill material is governed by the intended use.

7.13.5 Determination of Height of Preload Fill

7.13.5.1 Principle

The requirements in respect of the preload fill arise principally from the available construction time. The dimensioning is based on the principle of the consolidation coefficient c_v. c_v values obtained from compression tests have been shown by experience to lead to grossly false estimates of consolidation times. Therefore, such values should be used only in exceptional cases or in preliminary considerations, an immediate verification by means of settlement measurements then undertaken when filling operations begin.

If the applicable mean stiffness module E_s and the mean permeability are known, c_v can also be calculated as:

$$c_v = \frac{k \cdot E_s}{\gamma_w} \text{ [m}^2\text{/year]}.$$

If this equation is used, it must be borne in mind that the permeability coefficient k may contain considerable errors, so that this procedure too can be recommended only with reservations. The consolidation coefficient c_v should therefore have been determined wherever possible in advance from a trial fill, during which the settlement and also, if possible, the pore pressures should be measured as a function of time. c_v is then calculated using fig. R 179-2 according to the following formula:

$$c_v = \frac{H^2 \cdot T_v}{t} \quad [\text{m}^2/\text{year}],$$

in which:

T_v = specific consolidation time [1],
t = time [years],
H = depth of soft soil stratum drained on one side [m].

For $U = 95\%$, i.e. virtually complete consolidation, $T_v = 1$ and thus:

$$c_v = \frac{H^2}{t_{100}} \quad [\text{m}^2/\text{year}].$$

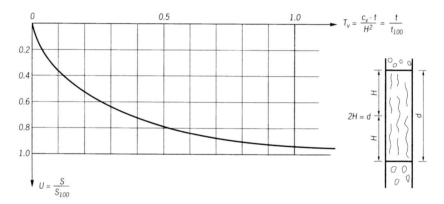

Fig. R 179-2. Relationship between time factor T_v and degree of consolidation U

7.13.5.2 Dimensioning of Preload Fill

For the dimensioning of the preload fill, the depth of the soft soil stratum and the c_v value must be known. Additionally, the consolidation time t_s (fig. R 179-1) must be specified (construction schedule).

$t_s/t_{100} = T_v$ is determined with $t_{100} = H^2/c_v$ and the required degree of consolidation $U = s/s_{100}$ under preloading is determined with the aid of fig. R 179-2. The 100 % settlement s_{100} of the permanent fill p_0 is determined with the aid of a settlement calculation in accordance with DIN V 4019-100.

The level of the preload p_v (fig. R 179-1) is then derived in accordance with [114] as:

$$p_v = p_0 \cdot \left(\frac{A}{U} - 1\right) \quad [\text{kN/m}^2].$$

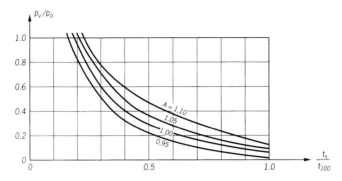

Fig. R 179-3. Determination of preload p_v as a function of time t_s

A is the relationship of the settlement s_s after removal of the preload to the settlement s_{100} of the permanent fill: $A = s_s/s_{100}$.
A must be equal to or greater than 1 if a complete removal of the settlement is to be achieved (fig. R 179-3).
Thus, for example, with a stratum drained on both sides, a depth $d = 2H = 6$ m, a c_v value $= 3$ m²/year and a specified consolidation time of $t_s = 1$ year:

$$t_{100} = \frac{3^2}{3} = 3 \text{ years},$$

$t_s/t_{100} = 0.33,$

$U = 0.66$ [1].

As shown in fig. R 179-3, with $A = 1.05$:

$$p_v = p_0 \cdot \left(\frac{1.05}{0.66} - 1\right) = 0.6 \cdot p_0 \ [\text{kN/m}^2].$$

7.13.6 Minimum Superficial Extent of Preload Fill

In order to save preload material, the soil stratum to be stabilised is generally preloaded in sections. The sequence of preloading is governed by the construction schedule. To achieve effective dissemination of stress in the soft strata which is as uniformly distributed as possible, the superficial extent of the fill must not be too small. An indication is that the smallest side dimension of the preload fill should equal two to three times the sum of the depths of the soft stratum and the permanent fill.

7.13.7 Check of Consolidation

The consolidation of the soft soil stratum can be checked by settlement measurements, pore pressure measurements and, at the edges, by inclinometer measurements.

For checks on site, it is recommended that a value be given for the settlement rate at which the preload fill can be removed again, e.g. in mm per day or in cm per month.

7.13.8 Secondary Settlement

It must be noted that secondary settlement, unrelated to pre-consolidation, can be predicted only to a very small extent by pre-loading (in highly plastic clays, for example). If secondary settlement is anticipated on a fairly large scale, special supplementary investigations are necessary.

7.14 Installation of Mineral Bottom Seals Under Water and Their Connection to Waterfront Structures (R 204)

7.14.1 Definition

A mineral underwater seal comprises natural, finely grained soil composed or prepared in such a way that it has very low permeability without the addition of additional substances to achieve the sealing effect.

7.14.2 Installation in the Dry

Mineral seals installed in the dry are dealt with in detail in [155].

7.14.3 Underwater Installation

7.14.3.1 General

When watertight harbour basins or waterways are to be deepened or expanded, it is often necessary to install bottom seals underwater, sometimes while vessel traffic is operating. Measures are to be taken against the prejudicial effects in respect of the stability of the structures involved and the effects on ground water quality and level of the water penetrating the subsoil for the event of resulting temporary seal failure. Special requirements are to be placed on the seal material to be installed, depending on the installation procedure.

7.14.3.2 Requirements

Mineral seals installed underwater can be mechanically sealed only to a limited extent, or not at all. They must therefore be prepared homogeneously and installed in a consistency such that a uniform sealing action is guaranteed from the outset, that the material installed evens out unevenness of the level plane without tearing, withstands the erosion forces from vessel traffic even during installation and is capable of guaranteeing the watertightness of the connections to the waterfront structures, even if deformation of these structures occurs.

If the seal is to be created on slopes, the installation strength must also be sufficiently great for the stability on the slope to be guaranteed.

On the basis of experience to date in the installation of mineral seals underwater, there is sufficient resistance to:

- the danger of disintegration underwater of the newly installed sealing material,
- erosion from the backflow of the vessel traffic adapted to the site conditions,
- the breaching of the seal in the form of thin tubes with coarsely grained subsoil (piping),
- slippage on slopes inclined up to 1 : 3, and
- the stresses when the seal is covered with filters and hydraulic structure blocks,

provided the sealing material fulfils the following conditions:

- proportion of fine sand
  (0.063 mm ≤ d ≤ 0.2 mm) $< 20\,\%$
- proportion of clay (d ≤ 0.002 mm) $> 30\,\%$
- permeability $k \leq 10^{-9}$ m/s
- undrained shear strength $15 \text{ kN/m}^2 \leq c_u \leq 25 \text{ kN/m}^2$
- depth (with 4 m water depth) $d \geq 0.20$ m

With artificial mixtures prepared with certain additives and a small proportion of cement, which are given certain stabilisation after installation, the installation strength can be reduced to $c_u = 5$ kN/m² even with subsoil predominantly comprising sand, if no significant removal takes place as a result of vessel movements in the period between installation of the seal and its being covered with erosion-proof protective layers, or if such removal is protracted through overdepth. The cement proportion must not impair the flexibility of the seal in the final state, this being checked with tests, e.g. in accordance with [156].

Special investigations are necessary if mineral seals are installed at greater water depth, or on gravely soil, or with large subsoil pore sizes or slopes steeper than 1 : 3, and also in the case of dimensioning the sealing material in respect of self-closure on gap formation and the sealing effect of butt joints (see [157] and [158]).

See [132] with regard to suitability and monitoring tests. Three patented procedures are currently available for the soft mineral seals which are today only single layer:

- the compressed clay sealing process,
- the soft sealing process,
- the vacuum bell jar process.

In all the processes, clay obtained from a formation as homogeneous as possible is taken to a processing plant and brought from there to the installation site.

In the *compressed clay sealing process*, the clay is taken to a box-type loader and thence to a worm extruder on the installation site. In the worm extruder, the final fine dosing with water takes place and the clay is further homogenised and brought to the inlet of the compressed clay laying unit.

To lay a continuous train of clay on the canal bottom, no failure points may be created as a result of separation of the clay train, and no enlargements or clay accumulations may be created by too high a dosage. A friction plate on the inlet compensates to a certain extent for the unevennesses of the level plane before the clay train is laid. Joint closure is achieved by close laying of the individual tracks with slightly inclined lateral edges laid one above the other.

In the *soft sealing process*, the prepared, pasty sealing material is installed with an undrained initial shear strength of $c_u = 5$ kN/m^2. To increase the erosion resistance against the backflow of passing vessels during and immediately after installation, an additive and a small amount of cement are added to the clay. As a result of this and of the consolidation, the necessary undrained shear strength c_u is achieved. To produce the sealing compound, the raw material is weighed in a positive mixer to which water is added under electronic control, and then brought to the pasty consistency necessary for further processing.

The mixture thus prepared is then fed to the inlet with a thick-matter pump. The inlet is always kept in the same position above the canal bottom by means of a parallel guide. The track is applied in a uniform, pasty consistency at a constant depth and width, with continuous monitoring of the flow rate and pressure, the consistency being selected so that positive contact between the individual tracks is guaranteed.

The *vacuum bell jar process* differs from the two processes already described in that the clay is removed in a relatively stiff consistency from a clay bed in 3/3 m to 4/4 m sections 0.2 to 0.3 m thick, using a vacuum bell jar, see fig. R 204-1. The clay is prepared in a similar way to the compressed clay sealing process. In comparison with the two processes already mentioned, the vacuum bell jar process has the advantage that fewer joints are created, so that the risk of the seal starting to leak is diminished. Because of the laying technique, every individual section is subjected to a watertightness test. For joint closure however, a careful laying and measuring technique is necessary.

With all laying processes, it is necessary to position the laying equipment rigidly with elevated piles and to protect the clay seal with a revetment immediately after laying.

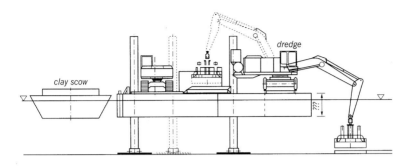

Fig. R 204-1. Vacuum bell jar procedure, laying principle

7.14.4 Connections

The connection of mineral bottom seals to structures is generally by means of a butt joint, the sealing material generally being pressed with suitable devices adapted to the shape of the joint (e.g., sheet piling profile). An amount of sealing agent corresponding to the sequence of the joints is applied in advance with suitable equipment. Since the sealing effect is generated by the direct contact stress between the sealing agent and the joint (see [157] and [158] in this respect), the pressing process must be undertaken with great care.

8 Sheet Piling Structures

8.1 Material and Construction

8.1.1 Design and Driving of Timber Sheeting (R 22)

8.1.1.1 Range of Application

Timber sheeting is advisable only when the existing soil is favourable for driving and when the required section moduli are not too large. In permanent structures, the pile tops must be at an elevation which is continuously wet, so that they will not decay. The risk of infestation from marine borers must be prevented. Timber sheeting is only advisable if no other materials can be used because of local conditions.

DIN 1052 – timber structures – is to be applied accordingly. Connecting elements of steel must be hot galvanised at least or have an equivalent corrosion protection.

8.1.1.2 Dimensions

Timber piles are usually made of highly resinous pine wood, but also of spruce and fir. Standard dimensions and design of the piles and the joint are indicated in fig. R 22-1. For straight or chamfer-tongued joints in general the tongue is made several millimetres longer than the depth of the groove, so that a snug fit develops as the pile is driven.

So-called "corner piles" are placed at the corners. These are thick squared timbers in which the grooves for the adjoining piles are cut corresponding to the corner angle.

8.1.1.3 Driving

Mostly double piles joined by pointed dogs are driven. Staggered driving or driving in panels is always required (see R 118; section 8.1.11) to protect the piles and increase the watertightness of the wall. The point of each sheet pile is bevelled on the tongue side so that the pile is pressed against the already standing wall. The cutting edge has a shoe which is reinforced by a 3 mm thick steel plate in soil which is difficult to drive into. The top of the steel pile is always protected from splitting by a conical ring of flat forced steel about 20 mm thick.

8.1.1.4 Watertightness

Timber sheet piling becomes somewhat tight as the result of swelling of the wood. As with other sheet piling, the watertightness of excavation enclosures in free water can be improved with ashes, fine slag, wood shavings and other environmentally compatible materials which are sprinkled in the water on the outside of the sheet piling while pumping out the excavation. Large leaky areas can be temporarily repaired by placing

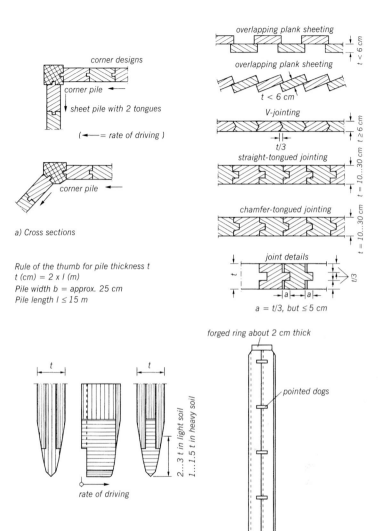

Fig. R 22-1. Timber sheet piling

Common name of wood	Botanical name	Mean density	Mois-ture	Abso-lute com-pressive strength	Elas-ticity module	Abso-lute bending strength	Shear strength	Durability as per TNO*		Teredo resistance
								in moist soils, in water or water chang-ing zone	exposed to the elements	
		kN/m³		MN/m²	MN/m²	MN/m²	MN/m²	years	years	
Demerara Greenheart	Ocotea rodiaei	10.5	dry wet	92 72	21 500 20 000	185 107	21 12	25	50	Yes, but some-what less than Basralocus
Opepe (Belinga)	Sarcoce-phalus	7.5	dry wet	63 50	13 400 12 900	103 92	14 12	25	50	Yes, but limited
Azobe (Ekki Bongossi)	Lophira procera	10.5	dry wet	94 60	19 000 15 000	178 119	21 11	25	50	Yes
Manbarklak (Kakoralli)	Eschweilera long pipes	11.0	dry wet	72 52	20 000 18 900	160 120	13 11	15–25	40–50	Yes
Basralocus Angelique	Dicorynia paraensis	8.0	dry wet	62 39	15 500 12 900	122 80	11.5 7	25	50	Yes, but limited
Jarrah	Eucalyptus marginata	10.0	dry wet	57 35	13 400 9 900	103 66	13 9	15–25	40–50	No
Yang	Dipterocarpus Afzelia	8.5	dry wet	54 39	14 600 12 300	109 80	11 10	10–15	25–40	No
Afzelia (Apa Doussie)	Afzelia africana	7.5	dry wet	66 30	13 000 9 900	106 66	13 9	15–25	40–50	

*) TNO = Nijverheisorganisatie voor Toegepast Natuurwetenschaappelijk Onderzoek

Table R 22-1. Characteristic values of tropical hardwoods

canvas over them, but must be permanently sealed by divers with wooden laths and caulking for permanency.

8.1.1.5 Protecting the Wood

Since native wood is usually well protected against rotting only under water, timber sheet piling subject to permanent loads must lie below the groundwater table and below low water in waterways. In tidal areas, they may extend to mean water level. In other cases, timber must be protected by means of an environment-friendly impregnating agent. This also applies to areas where there are marine borers, i.e. generally in water with a salt content exceeding 9 ‰. Tropical hardwoods are more resistant under such circumstances. The data for tropical hardwoods are listed in table R 22-1.

8.1.2 Design and Driving of Reinforced Concrete Sheet Piling (R 21)

8.1.2.1 Range of Application

Reinforced concrete sheet piling may be used only with the assurance that the sheet piles can be driven into the soil without damage and with tight joints. Their use should however be restricted to structures in which requirements in respect of watertightness are not high, e.g. groyns etc.

8.1.2.2 Concrete

Reinforced concrete sheet piling must be composed of concrete of a quality at least equal to B 35. The aggregate structure must lie in the favourable range between curves A and B (range 3) according to DIN 1045. A decided improvement of the mix can be achieved by the addition of high grade stone chips.

Cement of strength class CEM 42.5 (DIN 1164) should be used at the rate of about 375 kg/m^3 of concrete. The water-cement ratio must be between 0.45 and 0.48. The requirements of DIN 4030 are to be observed. However, it may not be too brittle either to prevent the concrete shattering during driving. The piles should therefore be driven soon after the required strength is reached.

8.1.2.3 Reinforcement

The cover over the main reinforcing bars shall be at least 5 cm in fresh water and in salt water. In other respect, the reinforced concrete sheet piling is rated according to DIN 1045, whereby DIN 4026 is to be observed for the loading cases lifting the piles out of the forms and lifting them before driving. Sheet piles generally have main longitudinal reinforcement of BSt 500 S. In addition, the sheet piles have a helical cross reinforcement of BSt 500 S or m or of wire rod $\varnothing$ 5 mm. The individual layout is shown in fig. R 21-1.

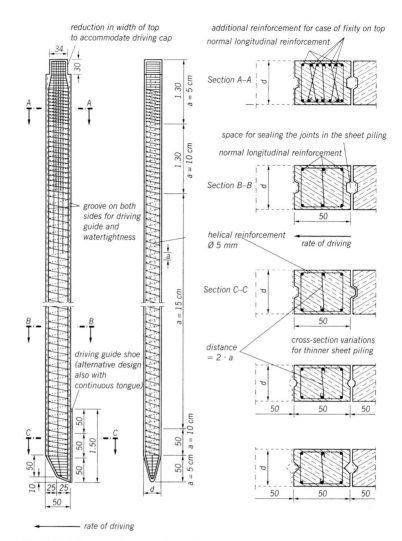

Fig. R 21-1. Reinforced concrete sheet piling

8.1.2.4 Dimensions

Piles to be driven must have a minimum thickness of 14 cm but should in general not be thicker than 40 cm for reasons of weight. Apart from the driving conditions, the thickness depends on construction details and structural requirements. The normal pile width is 50 cm, but the width of the pile butt is reduced to 34 cm if possible so that it fits into the normal driving cap. The piles are made up to 15 m long, or in exceptional cases 20 m long.

Normal groove shapes are shown in fig. R 21-1. The width of the grooves is up to 1/3 of the sheet pile thickness, but not greater than 10 cm. The groove runs continuously to the lower end of the pile on the leading side. On the opposite side, the point has a tongue about 1.50 m long which fits the groove. A groove runs upward from this tongue (fig. R 21-1). The tongue guides the points as the pile is driven. It may run from the point to the top of the pile and help create tightness, but only in non-cohesive soil if the soil is of such a nature, that a filter is automatically created behind every joint after minor leaching has taken place, to prevent any further leaching.

8.1.2.5 Driving

The point of the pile is bevelled to about 2 : 1 on the leading side, so that the pile presses against the already driven sheeting. This design is also retained for jetted piles. Driving is facilitated if the piles have a wedge-shaped point. The piles are always driven as individual piles, and in the tongue and groove design with the groove side forward. A driving cap should be used if the driving is done with a pile hammer. Piles should be driven with the heaviest hammers possible and with a small drop (0.50–1.00 m). Rapid stroke hammers are less suitable. In soils of fine sand and silt, better progress will be made if a jet is used.

8.1.2.6 Watertightness to Prevent Loss of Soil

If the sheet pile has only a short tongue, sufficient space for the insertion of joint leakproofing is available. Before this is placed, the grooves should always be cleaned with a jet of water. The space is then filled with a good concrete mixture using the contractor method. With large grooves, a jute sack filled with freshly mixed concrete can be lowered into place. A seal of bituminised sand and fine gravel can also be used. In any case, the leakproofing is to be so inserted that it fills the entire space without leaving any gaps.

The seal is best achieved with a C-shaped special section, with which a cavity of the appropriate size is obtained and which can also be lightly prestressed.

However, especially in surrounding non-cohesive, finely grained soil, above all with tidally-induced water level fluctuations, the achievable seal is restricted, so that the requirements relating to the seal cannot be too high, see section 8.1.2.1. Subsequent resealing of reinforced concrete sheet piling is possible only at great expense.

8.1.3 Steel Sheet Piling (R 34)

8.1.3.1 General

Steel sheet piling is frequently an ideal solution both as regards structural considerations and driving conditions (see R 106, R 119, R 200, R 201, sections 6.4, 6.5, 6.8 and 6.9), which is also capable of absorbing localised overloading without endangering overall stability. Damage caused by collisions and accidents often can be repaired simply. The sheet piling can be made watertight using measures according to recommendation R 117, section 8.1.20.

Corrosion and corrosion protection measures are dealt with in recommendation R 35, section 8.1.8.

8.1.3.2 Selection of the Profile and Steel Quality

In addition to the structural considerations and economic aspects, the choice of the type and the section depends on the driving conditions existing at the site, the stresses when being installed and in operation, the acceptable deflection, the watertightness of the interlocks and the allowable minimum wall thickness, whereby especially possible mechanical loads on the sheet piling from berthing manoeuvres of ships and ship groups and from sand abrasion effects are to be taken into account (R 23, section 8.1.9).

Satisfactory driving of the piling and, in permanent structures, adequate durability must be assured.

Combined steel sheet piling (R 7, section 8.1.4) is often economical for larger section moduli. Section strengthening with welded-on plates or locking bars can be taken into consideration, just so the selection of a higher-strength, weldable steel grade with minimum yield point above the values stated in table R 67-1, section 8.1.6.2. The choice of higher strength steels is to be agreed with the manufacturers on placing the order. Otherwise, recommendation R 67, section 8.1.6 is to be observed.

8.1.4 Combined Steel Sheet Piling (R 7)

8.1.4.1 General

Combined steel sheet piling is constructed by the alternate placing of different types of sections or driving elements. Long and heavy sections designated as bearing piles alternate with shorter and lighter intermediate piles. The most customary wall shapes and wall elements are described in detail in R 104, section 8.1.12.

8.1.4.2 Static System

From the static point of view, a distinction is made between walls without shear-resistant interlocks and walls with shear-resistant interlocks, so-called "piling with connected interlocks".

Usually, only the bearing piles can be assumed to absorb the vertical loads.

The horizontal loads which act directly on the intermediate piles must be transferred to the bearing piles.

Experience has shown that unwelded intermediate piles of Z-profile with a wall thickness of 10 mm are stable with a clear bearing-pile spacing of 1.2 m, and those of U-profile are stable with a clear bearing-pile spacing of 1.8 m up to a water pressure difference load of 40 kN/m^2. The prerequisite for this is an extensive relieving of the intermediate piles from earth pressure, for which a sufficiently compactly layered full backfill is necessary.

In the event of clearances and/or loads over and above this, the stresses must be checked. In such cases, horizontal intermediate wales can be used as supplementary support components.

Because of the interaction of the interconnected driving elements, connected sheet piling increase the virtual moment of inertia and section modulus. Particular advantages can be gained with the connected sheet piling in corrugated form described in R 103, section 8.1.5.1. The proportion in which the individual piles contribute to the cross-section depends on the degree of shearing force transmission by the interlocks. In calculations, the composite cross-section may be considered as a unit cross-section if full shearing force transmission can be demonstrated as per R 103, section 8.1.5.

In combined sheet piling, allowance for the out-of-centre intermediate piles as part of a composite cross-section is appropriate only if displacement of the axis through the centre of gravity is largely prevented by reinforcement of the bearing piles on the opposite side.

8.1.4.3 Construction Instructions

The material quality and manufacture of the bearing elements is to be certified according to DIN EN 10 204. Generally, commercially available general structural steels according to DIN 17 120 works test certificates according to previous DIN 50 049 are adequate. For special steel grades, certificates are to be provided according to DIN 50 049-3.1 B (acceptance test certificate B).

If required, slight deviations in shape and size from the standard are to be the subject of special agreements. Furthermore, existing contraction supervisory permits IfBt have to be taken into account with regard to the intended purpose when it comes to pipes of fine-grained structural steels to DIN 17 123 and DIN 17 124 and of thermo-mechanically treated steels. The load-bearing piles of tubes are welded with corresponding interlock sections for connection to the intermediate piles. For this purpose, the interlock connection must comply with the tolerances in R 67, section 8.1.6 and be able to transmit reliably the loads from the intermediate piles to the tubes. At the intersection points with girth and helical live welds, interlock sections must fit closely against the skin of the tube. Tube seams and interlock sections must be executed correspondingly at the intersection point.

Bearing tubes with interior interlocks are used particularly when low-noise, vibration free sinking is required using a rotary drill procedure, and hindrances have to be removed.

The steel grade of the tubes should comply with DIN 17 120 and fulfil all requirements otherwise made of sheet piling steels.

The load-bearing tubes of the tubular sheet piling are manufactured in the works in full length without girth welds, or of individual single lengths connected by girth welds. Differing piling thicknesses graded to the inside are usual for LN-welded tubes. Fully or semi-automatic machines can be used for joining on longitudinal, helical line, girth and transverse welds.

Longitudinal and helical line welds must be subjected to ultrasonic testing, in which colour-coded failure areas are repaired manually during the test procedure. Repaired weld seams can be documented under a new ultrasonic test by a connected printer log.

Girth welds between the individual pipe lengths and transverse welds between the coil ends of the SN pipes must be checked by X-ray [127, p. 132–134].

The intermediate piles consist generally of Z-shaped double piles or U-shaped triple piles. In the case of tubular bearing piles, they are usually arranged in the wall axis, and for box- or H-piles in the water-side interlock connection; here it must be noted that this does not produce a smooth berthing surface for shipping, particularly in the case of tubular bearing elements.

8.1.4.4 Installation

A combined sheet piling must be driven with particular care. The bearing elements (box piles, H-piles or tubular piles) are usually driven in. When driving the bearing elements, the recommendations E 104, section 8.1.12 and R 202, section 8.1.22 are to be observed. This is the only way to ensure that the bearing elements stand parallel to each other within the permissible tolerances at the planned spacing and without any distortion, so that the intermediate piles can be installed with the minimum possible risk of damage to the interlocks.

When driving in the pipes, any hindrances can be removed by excavating in the inside of the pipe. However, the prerequisite here is that there are no protruding structural elements in the inside of the pipe, such as interior lock chambers (see section 8.1.4.3). In addition, the inner diameter of the bearing pipe must be at least 1200 mm to allow operation of suitable dredging equipment. When there are many hindrances in the soil, it is advisable to drive the elements in using the vibration method to obtain early information about the hindrances.

8.1.4.5 Static Check

In the case of bearing pipes, the check of safety against bulges is not required, if the load-bearing tubes are concreted in or filled with non-cohesive material to the very top.

Reference is made to R 33, section 8.2.11 for the removal of the axial load from the tube into the subsoil. Where large axial loads and large pipe diameters are concerned, it may be necessary to weld a sheet metal cross or similar into the foot of the pipe to activate the necessary plug for the point resistance, which is to be verified by trial stress where applicable. The prerequisite for flawless formation of the plug is in any case adequate readiness for compaction of the soil in the area of the pile foot (see R 16, section 9.5). In these cases, any driving hindrances are to be removed in advance, by soil replacement, for example.

Furthermore it is also possible to use inside dog-bars adjusted to the pipe walls and supported on a synthetically inserted plug (concrete plug) to generate the required point resistance.

8.1.5 Shear-resistant Interlock Joining in Steel Sheet Piling (R 103)

8.1.5.1 General

When determining the cross-sectional values of connected sheet piling, all sheet piles are taken into consideration. The connected cross-section however may be calculated as a uniform cross-section only if full shear force absorption is certain.

Connected piles in wave-like form consist of "U" or "Z"-shaped sheet piles in which half the wave length consists of more than one individual pile. In this case, the uniform cross-section is achieved already by transmission of shear forces in every second interlock. In case of two individual piles per half wave length, the interlocks are alternately placed on the wall axis (neutral axis) and outside in the flanges. Here the uniform cross-section is only achieved when all interlocks on the wall axis are linked shear-resistant.

The interlocks in the flanges are the threading locks in construction, so that in this connected piling, all interlocks located on the wall axis can be drawn together in the workshop and prepared accordingly for the transmission of the shear forces, namely:

- by welding the interlock joints together in accordance with sections 8.1.5.2 and 8.1.5.3, or
- by pressure grouting the interlocks; here however only a partial connection can be achieved, as the interlocks at the pressure points displace by several millimetres on absorbing loads.

8.1.5.2 Bases of Calculation for Welding

The shear stresses in the welded interlocks originating from the main load-bearing system, and from the loading and supporting influences acting on the system, are determined by the formula:

$$\tau = \frac{V_d \cdot S}{I \cdot \Sigma a} \; [\text{MN/m}^2]$$

The symbols therein mean:

- V_d = design value of the shear force. For sheet piling where the moment component from active earth pressure may be reduced as per R 77, section 8.2.2, calculation may be simplified by using the shear diagram which results without active earth pressure redistribution. In case it is appropriate to assume the active earth pressure, the shear diagram resulting from that assumption is calculated with [MN],
- S = statical moment of the cross-section portion to be connected, referred to the centroidal axis of the connected sheet pile wall [m³],
- I = moment of inertia of the connected piling [m⁴],
- Σa = total of the welding seam thicknesses to be calculated with in each case at the position of the interlock under consideration. All seams mentioned in section 8.1.5.3 may be included proportionately in this total, insofar as they need not fulfil special additional functions [m].

For interrupted welds, the shear stress τ is to be taken correspondingly higher.

The verification of bearing capacity of the welds is to be provided according to DIN 18 800, part 1, issue 11/1990. Therein,

$$\frac{\sigma_{w,v}}{\sigma_{w,R,d}} = \frac{\text{design load}}{\text{design bearing resistance}} \leq 1$$

$$\sigma_{w,v} = \sqrt{\sigma_\perp^2 + \tau_\perp^2 + \tau_\parallel^2}$$

Verification of the interlock welds of sheet piling can normally neglect $\sigma_\perp$ and $\tau_\perp$. This results in $\sigma_{w,v} = \tau_\parallel = \tau$.

$$\tau = V_d \cdot S/I \cdot \Sigma a$$

- $\sigma_{w,R,d}$ = $\alpha_w \cdot f_{y,k}/\gamma_M$
- α_w = diminution factor (table 21, DIN 18 800)
- $f_{y,k}$ = yield point
- γ_M = 1.1 = partial safety factor for the resistance magnitude

8.1.5.3 Layout and Execution of the Welding Seams

The interlock welds shall be so laid out and executed as to achieve optimum continuous absorption of the shear forces. A continuous seam is best for this. If an interrupted seam is used, the minimum length is to be 200 mm, provided that the static shear stresses mentioned in section 8.1.5.2 do not require a longer weld. In order to keep the secondary stresses within bounds, the interruptions in the seam should be ≤ 800 mm.

Continuous seams are always to be used in areas where the sheet piling is subject to heavy concentration of stresses, especially near anchor connections and at the point where the equivalent force C is introduced at the foot of the wall (fig. R 103-1).

In addition to the requirements introduced by static forces, the effects of driving stresses and corrosion must be considered. In order to be able to cope with the driving stresses, the following measures are necessary:

(1) The interlocks are to be welded on both sides at the butt and point ends.
(2) The length of the weld is contingent on the length of the sheet pile and on the difficulty which may be encountered in driving.
(3) These seam lengths are to be ≥ 3000 mm in sheet piling walls for waterfront structures.
(4) Moreover, additional seams as per fig. R 103-1 are necessary for light driving, and seams as per fig. R 103-2 for difficult driving.

In areas where the harbour water is aggressively corrosive, a continuous welding seam of $a \geq 6$ mm (fig. R 103-2) is run on the outer side down to the sheet pile point.

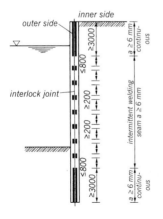

Fig. R 103-1. Diagram to show interlock welding at walls of killed, brittle-fracture resistant steel, easy driving and only slight corrosion in harbour and groundwater (6 mm = 15/64")

285

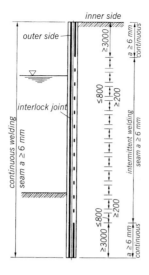

Fig. R 103-2. Diagram to show interlock welding at walls of killed, brittle-fracture resistant steel, difficult driving or stronger corrosion from the outside, in the harbour area (6 mm = 15/64")

If there is greater corrosion both in the harbour water and in the groundwater, a continuous seam of $a \geq 6$ mm must also be welded on the inner side of the wall.

8.1.5.4 Choice of the Steel Grade

As the amount of welding work on connected sheet piling is comparatively extensive, the sheet piles are to be manufactured of steel grades with full suitability for fusion welding. In view of the starting points of the welds not only at interrupted welding, killed, non-liable to brittle-fracture steels are to be used as per R 99, section 8.1.18.2.

8.1.5.5 Fundamentals of Calculation for Pressing of Interlocks

The design values of the shear forces on the pressing points arising from the main load-bearing system and the loading and supporting actions thereof are determined in accordance with the formula:

$$T_d = \frac{V_d \cdot S \cdot b_c}{I} \text{ [MN]}$$

in which V_d, S and I have the same meaning as in section 8.1.5.2, b_c = breadth of piling accounted for by one web [m].

The design value of the design bearing resistance of one pressing point under shear force is 100 kN, which can occur for displacement of up to 5 mm.

8.1.5.6 Layout and Execution of Pressing Points

The usual pressing point clearance is 400 mm. It can be reduced to 240 mm for dual piles and 330 mm for triple piles. It must be checked in each case whether the number of pressing points per breadth unit is sufficient for the magnitude of the overall shear force in a connected elevation range.

8.1.5.7 Welding On of Reinforcing Plates

In order to avoid rust forming on the inner contact surfaces of reinforcing plates, these must always be welded to the sheet piles around their full circumference. They should be tapered to reduce the leap in the moment of inertia. The seam thickness a should be at least 5 mm without corrosion and at least 6 mm where corrosion is severe. If a reinforcing plate spans an interlock located on the flange of the sheet pile, both sides of this interlock must be given a continuous weld in the area of the plate, plus an extension of at least 500 mm on each side of the plate. Even on the side opposite that on which the plate is located, the welds should be $a \geq 6$ mm thick, while the weld under the plate should be of a thickness which will permit seating the plate without subsequent machining. If this is not done, the welds attaching the plate may be seriously damaged during driving.

If unit welding of the interlocks is desired, the reinforcing plate, which otherwise would be welded on so as to cover the interlock, must be divided into halves longitudinally. Each half is then welded to the flange adjacent to but clear of the interlock.

8.1.6 Quality Requirements for Steels and Interlock Dimension Tolerances for Steel Sheet Piles (R 67)

This recommendation applies to steel sheet piles, trench sheeting and driven steel piles, which are all called steel sheet piles in the following. DIN EN 10 248-1 and -2 together with DIN EN 10 249-1 and -2 are applicable, replacing the previous Technical Conditions of Delivery for Steel Sheet Piles (TLS) – [106].

If steel sheets are stressed in the depth direction, e.g. in the case of special branch piles or branch piles for circular and flat cells (see R 100, section 8.3.1.2), steel grades with appropriate properties must be ordered from the sheet pile manufacturer in order to avoid terracing failure, see DASt Directive 0-14.

8.1.6.1 Designation of the Steel Grades

Grades of steel with the designation S 240 GP (formerly St Sp 37), S 270 GP (formerly St Sp 45) and S 355 GP (formerly St Sp S) are used for hot-rolled steel sheet piles in normal cases, as indicated in sections 8.1.6.2 and 8.1.6.3.

The qualities of steels with yield points up to 355 N/mm^2 should be verified with a works test certificate according to DIN EN 10 204, corresponding to DIN 50 049-2.2 – and higher quality steels with acceptance test certificate according to DIN 50 049-3.1 B stating the 14 alloy elements (as e.g. for S 355 J3 G3 as per DIN EN 10 025, formerly St 52-3). In special cases, for example for absorption of greater bending moments, steel grades with higher minimum yield points up to 500 N/mm^2 can be used, taking account of recommendation E 34, section 8.1.3. The higher steel grade is then to be verified by a works certificate, acceptance test certificate or acceptance test protocol according to DIN 50 049.

Steel grades S 235 JRC, S 275 JRC and S 355 JRC are possible for cold-rolled steel sheet piling.

In special cases, for example as stated in 8.1.6.4, fully killed steels to DIN EN 10 025 are used.

8.1.6.2 Characteristic Mechanical and Technological Properties

Steel grade	Minimum tensile strength Rm N/mm^2	Minimum yield point ReH N/mm^2	Minimum % elongation for measuring length of $L_0 = 5.56 \cdot \sqrt{S_0}$	Former name
S 240 GP	340	240	26	St Sp 37
S 270 GP	410	270	24	St Sp 45
S 320 GP	440	320	23	–
S 355 GP	480	355	22	St Sp S
S 390 GP	490	390	20	–
S 430 GP	510	430	19	

Table R 67-1. Characteristic mechanical properties for steel grades for hot-rolled steel sheet piling

The mechanical properties for steels S 235 JRC, S 275 JRC and S 355 JRC are described in DIN EN 10 025.

8.1.6.3 Chemical Composition

The ladle analysis is binding for verification of the chemical composition. A verification of the values for the bar analysis must be agreed specially for the acceptance test. The bar analysis should be used for subsequent checks in case of doubt.

Steel grade	Chemical composition % max. for ladle/bar					
	C	Mn	Si	P and S	N [*] [**]
S 240 GP	0.20/0.25	–/–	–/–	0.040/0.050	0.009/0.011
S 270 GP	0.24/0.27	–/–	–/–	0.040/0.050	0.009/0.011
S 320 GP	0.24/0.27	1.60/1.70	0.55/0.60	0.040/0.050	0.009/0.011
S 355 GP	0.24/0.27	1.60/1.70	0.55/0.60	0.040/0.050	0.009/0.011
S 390 GP	0.24/0.27	1.60/1.70	0.55/0.60	0.040/0.050	0.009/0.011
S 430 GP	0.24/0.27	1.60/1.70	0.55/0.60	0.040/0.050	0.009/0.011

[*] The stipulated values may be exceeded on condition that for every increase by 0.001 % N, the max. P level is decreased by 0.005 %; but the N-level of the ladle analysis may not be higher than 0.012 %.

[**] The maximum nitrogen value does not apply when the chemical composition has a minimum total level of aluminium of 0.020 %, or when there are sufficient N-binding elements. The N-binding elements are to be stated in the test certificate.

Table R 67-2. Chemical composition of the ladle/bar analysis for hot-rolled steel sheet piling

8.1.6.4 Weldability, Special Cases

An unlimited suitability of steels for welding cannot be presumed, since the behaviour of a steel during and after welding depends not only on the material but also on the dimensions and the shape, as well as on the manufacturing and service requirements of the structural member.

The suitability for arc-welding can be presumed for all sheet piling steel grades, if the general welding standards are observed. When selecting higher strength steels, the carbon equivalent CEV should not exceed the values of the steel grade S 355 as per DIN EN 10 025 table 4, in the interests of weldability.

In special cases with unfavourable conditions for welding because of external influences (e.g. plastic deformation due to difficult driving, at low temperatures) or the nature of the structure, spatial stresses and in the case of not predominantly static loads as per R 20, section 8.2.6.1 (2), fully killed steels are to be used as per DIN EN 10 025 in quality groups J 2 G 3, J 2 G 4, K 2 G 3 or K 2 G 4 with regard to the required brittle fracture resistance and ageing resistance. Killed steels are also to be given preference for thicknesses exceeding 16 mm (R 99, section 8.1.18.2 (1)).

The welding material is to be selected according to DIN EN 499, DIN 8557 and DIN EN 440, respectively according to the data provided by the manufacturer (R 99, section 8.1.18.2(2)).

8.1.6.5 Types of Interlocks and Coupling Connections

Examples of proven types of interlocks for steel sheet piling are shown in fig. R 67-1. Fig. R 67-2 corresponds to table 15 in DIN EN 10 248-2: 1995 but taking account of necessary corrections. The rated dimensions a and b marked in figs. R 67-1 and R 67-2, which can be queries from the manufacturers, are measured at right angles to the least favourable direction of displacement. The minimum interlock hook connection, calculated from a minus b, must correspond to the values in the pictures. In short partial sections, the values may not fall more than 1 mm below these minimum values. In forms 1, 3, 5 and 6, the required coupling must be present on both sides of the interlock.

8.1.6.6 Tolerable Dimension Deviations in Interlocks

Deviations from the design dimensions occur during rolling of sheet piles and locking bars. The tolerable deviations are summarised in table R 67-3.

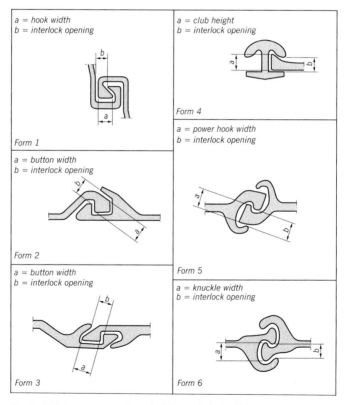

Fig. R 67-1. Proven types of interlock for steel sheet piling

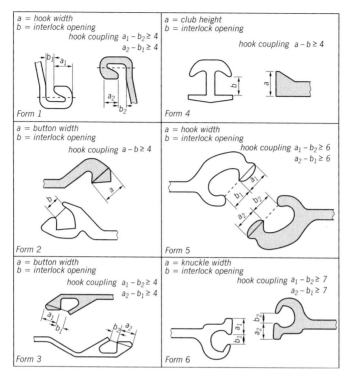

Fig. R 67-2. Types of interlocks and hook coupling

Form	Rated dimensions (to section drawings)	Tolerances of rated dimensions		
		Designation	plus [mm]	minus [mm]
1	Hook width a	Δa	2.5	2.5
	Interlock opening b	Δb	2	2
2	Button width a	Δa	1	3
	Interlock opening b	Δb	3	1
3	Button width a	Δa	1.5 … 2.5[1]	0.5
	Interlock opening b	Δb	4	0.5
4	Club height a	Δa	1	3
	Interlock opening b	Δb	2	1
5	Power hook width a	Δa	1.5	3.5
	Interlock opening b	Δb	3	1.5
6	Knuckle width a	Δa	2	3
	Interlock opening b	Δb	3	2

[1] depending on the section

Table R 67-3. Allowed interlock tolerances

8.1.7 Acceptance Conditions for Steel Sheet Piles and Steel Piles at the Site (R 98)

Although careful and workmanlike construction methods are of the greatest importance whenever steel sheet piles or steel piles are used as part of a structure, it is essential that the material is in satisfactory condition on delivery to the site. In order to achieve this, specific acceptance of the material at the site is necessary. As a supplement to the manufacturer's own shop inspection, a works acceptance can be agreed in each case. For shipments abroad, inspection is frequently carried out before the piles are loaded on board.

The acceptance procedure at the site should specify that every unsuitable pile will be rejected until it has been refinished in a usable condition, unless it is rejected out right. Acceptance on the building site is based on:

DIN EN 10 248-1 and -2 for hot-rolled sheet piling, or
DIN EN 10 249-1 and -2 for cold-formed sheet piling.

R 67, section 8.1.6.6 applies to the interlock tolerances.

8.1.8 Corrosion of Steel Sheet Piling and Counter-measures (R 35)

8.1.8.1 General Comments on Corrosion

When in contact with water and at the same time in the presence of oxygen, steel is subject to the natural corrosion process. Material abrasion from corrosion depends on the one hand on local (e.g. hydrological) conditions, on the other hand from the local (vertical) position regarding the water line. The latter means that different zones of corrosion are formed (fig. R 35-1). The degree of corrosion intensity is the decrease in thickness (rusting in mm), i.e. the decrease of the wall thickness, e.g. of a sheet pile. Referred to the time unit, one speaks of *rusting speed* (rusting rate in mm/a). Investigations of steel sheet piling with differing service lives indicate that the rusting speed decreases in time resulting from the formation of a cover layer, unless this cover layer is constantly destroyed by mechanical or chemical stress. Accordingly, when rating the decrease in thickness or rusting speed, the design period respectively service life of the member must also be stated.

8.1.8.2 Impairment of the Stability in the Light of Corrosion

As far as steel sheet pilings are concerned, the following influences as a result of corrosive action are to be taken into consideration:

(1) Decreases in thickness can result in reduced bearing capacity, contingent on the moment diagram and the location of the decrease in thickness.

Unprotected sheet piling should be designed in such a way that the prevailing bending moment (prev. m) in the area of the greatest corrosion is considerably smaller than the tolerable bending moment (tol. m). The greater the difference (tol. m – prev. m) and the initial thickness (t) are, the greater the reserve in wall thickness (Δt) which can be abraded by corrosion without exceeding the limits of load-bearing capacity. For bearing capacity verification, the section modulus and the sectional area of the sheet piling is to be reduced proportionally to the mean values of the decreases in thickness for the individual corrosion zones (fig. R 35-1).

(2) In zones of *maximum corrosion*, local rusting right through the piles can occur particularly in sea water. These are mainly observed on hill piles (from the water side); in U-piles in circular formations in the back, in Z-piles in slit formations at the transition from web to back (see fig. R 35-2). Similar generalising empirical values are not available for H-profiles. When the sheet piling rusts through, the backfilled soil can be flushed out causing settlement damage, which in turn results in safety risks and subsequent restrictions in use.

Damage of this kind requiring extensive repairs can be avoided in the initial planning stages by suitable corrosion protection measures (section 8.1.8.4).

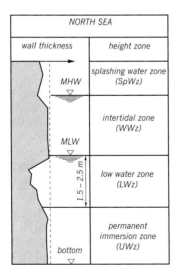
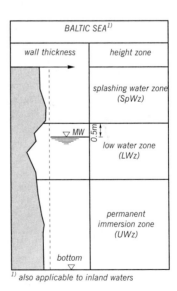

Fig. R 35-1. Qualitative diagram of the corrosion zones for steel sheet piling with examples of the North Sea and Baltic Sea

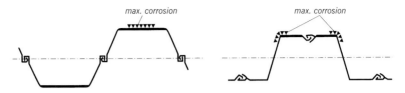

Fig. R 35-2. Zones of possible rusting through U- and Z-piles in the low water zone (sea-water)

Existing sheet piling should always be evaluated on the basis of wall thickness measurements by ultrasonic equipment, which provides the best possibility for registering local influences together with the mean and maximum rusting values. Information on carrying out, troubleshooting and analysing such measurements can be found in [176].

8.1.8.3 Decrease in Thickness and Abrasion Speed in Various Media

The following discourse states the average decreases in thickness of the individual corrosion zones for the various media. In view of the dependency on time, the calculations are based on a service life of 60 years. The derived abrasion speeds are suitable for pre-planning and can possibly be taken into account for execution planning. In view of locally varying corrosion stresses, it is advantageous to examine an existing adjacent structure and use it for corrosion forecasts.

(1) Atmospheric corrosion

Atmospheric corrosion, i.e. corrosion above the splashing water zone (fig. R 35-1) is generally low. In the case of water structures, it is practically negligible with a corrosion speed of approx. 0.01 mm/a compared to splashed water corrosion and underwater corrosion.

Higher abrasion speeds are found in the salty atmosphere of coastal regions, under permanently high relative humidity levels, under the action of defrosting salts and storage and handling of steel-corroding substances.

(2) Corrosion in fresh water

Normally in fresh water the corrosion load is low. The main zone of attack is frequently the area just underneath the water table (low water zone), where a mean corrosion rate of 0.03 mm/a can be expected.

Higher corrosion rates are found when the water contains aggressive substances and with strongly fluctuating water levels, and in some cases in the splash water zone. The decrease in thickness in fresh water is shown in fig. R 35-3.

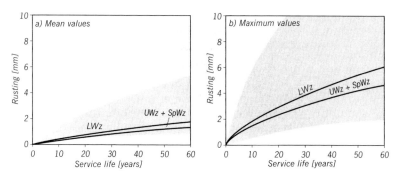

Fig. R 35-3. Decrease in thickness as a result of corrosion in the fresh water zone

(3) Corrosion in aggressive water and sea-water
The decrease in thickness is greatest in briny water and seawater in the low water zone (fig. R 35-1). The absolute values are comparable with internationally known abrasion speeds [151].

Fig. R 35-4 shows the decreases in thickness in relation to the service life, ascertained on the basis of extensive measuring data [151], [170] for the various corrosion zones. Various corrosion zones overlap extensively because of the widespread scatter of the values, so that in fig. R 35-4a), a joint balance curve is shown for the underwater zone (UWz), water fluctuation zone (WWz) and splashed water zone (SpWz) (thus the mean abrasion speed, referred to 60 years service life, is to be taken at 0.03 mm/*a*).

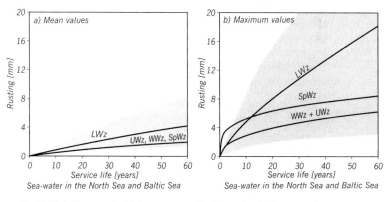

Fig. R 35-4. Decrease in thickness as a result of corrosion in the sea-water zone

Decreases in thickness can vary considerably depending on many kinds of influences (temperature, chemistry, microbes, mechanical stress, stray currents). Fig. R 35-4b) therefore also shows the maximum decreases in thickness. The curved lines represent the mean values of the individual corrosion zones.

(4) Corrosion in the soil
The anticipated decrease in thickness on both sides of steel sheet piling embedded in natural soils is negligible with 0.01 mm/*a*.
The same corrosion load applies in the case of necessary backfilling immediately behind the piling when sandy soil is placed in such a way that the troughs of the piling are also thoroughly embedded.
Highly aggressive soils should be kept away from the surface of the sheet piling as far as possible. Such soils include among others humus soil and carbonaceous soils, for example waste washings. An analysis according to DIN 50 929 can be helpful here. Similarly, contaminated surface water and leachate should be avoided on the backside of the sheet piling which promote the growth of bacteria aggressive to steel. Otherwise, increased abrasion speeds and uneven abrasion can be expected.

8.1.8.4 Corrosion Protection

The early planning stages should take account of the following factors when it comes to corrosion protection measures:

– planned use and total service life of the structure
– general and specific corrosion load at the location of the structure
– experience with corrosion phenomena in adjacent structures
– possibilities for including corrosion protection in the design
– profitability calculations for unprotected sheet piling should always take account of the costs for premature repairs (e.g. preplating).

Subsequent protection measures or complete renewal are extremely difficult so that particular care is required in the planning and execution of the protection system.
Specifically adjusted protection measures may be required, depending on the extent of the resulting corrosion load.

(1) Corrosion protection from coatings
According to available experience, coatings can delay the start of corrosion by more than 20 years.
The prerequisite is sandblasting to a standard degree of purity Sa 2½ and the selection of a suitable coating system, with execution and monitoring of the work according to DIN 55 928 and ZTV-W 218 [177].
With a view to health and environmental protection, bituminous systems will be replaced in future by coating systems with little or no solvents on

the basis of epoxy resins or polyurethane resins, or corresponding combinations of tar substitutes and hydrocarbon resins. The coating should be applied completely in the workshop so that only transport and installation damage needs to be touched up on the building site.

Planning of the coating system should make allowances for the possibility of subsequent installation of cathodic corrosion protection. Attention must be paid to corresponding compatibility of the coating materials.

If steel sheet piling is to be protected by coatings from sand grinding, the abrasion value (A_w) of the coating material must be taken into account [153].

(2) Cathodic corrosion protection
Corrosion under the water line can be substantially eliminated by electrolytic means by the installation of cathodic protection using impressed current or sacrificial anodes.

Additional coating or partial coating is an economical measure and usually indispensable in the interests of good current distribution and lower power requirements later on.

Cathodic protection systems are ideal for protecting those sections of sheet piling, e.g. the tidal low water zone, in which a renewal of protective coatings or the repair of corrosion damage in unprotected sheet piling is not possible or only at high financial and technical expense. Sheet piling structures with cathodic corrosion protection require special construction measures and must therefore be taken into account in the planning phase [178].

(3) Alloy additives
On the basis of available experience, the addition of copper to the steel does not increase its service life in the submerged zone. However, an alloy of copper in connection with nickel and chromium as well as with phosphorous and silicon increases the service life in and above the splash zone, especially in tropical regions with salt-laden moving air.

No differences in resistance to corrosion could be determined among the different types of sheet pile steels according to DIN EN 10 248 and the steels mentioned in DIN EN 10 025 (structural steels) and DIN EN 10 028 together with DIN EN 10 113 (higher strength fine grain structural steels). If higher strength is achieved by the addition of niobate, titanium and vanadium, this also has a positive effect on the corrosion behaviour.

(4) Corrosion protection by oversize
In order to extend the service life, with regard to exceeding the bearing capacity it is possible to select profiles with greater section modulus or higher steel grade according to E 67, section 8.1.6.1. Profiles with thicker backs and webs provide better protection from rusting through.

(5) Constructional measures
- Sheet piling structures with cathodic corrosion protection require special structural measures [59].
- As far as corrosion attack is concerned, constructions in which the back of the sheet piling is only partially backfilled or not at all are unsuitable.
- Clean sand should be used for backfilling the sheet piling as far as possible, and must be brought up to the piling head in order to protect the transitional zone air/soil which is particularly at risk from corrosion.
- Exposed open piles are exposed to corrosion on the whole periphery, closed piles e.g. box piles on the other hand essentially just on the outer surface. The inner surfaces of the pile are protected when the pile is filled with sand.
- In the case of free-standing sheet piling, e.g. flood protection walls, the coating of the sheet piling must be extended deep enough into the ground area; settlement must be taken into account.
- For placing tie rods a sand bed is recommended. The anchor heads should be closed carefully.
- At ships' berths, sheet piling should always be protected by friction timbers or similar from permanent scour from fenders, pontoons or ships, in order to avoid damage to the surface of the sheet piling. Otherwise, higher decreases in thickness must be expected which clearly exceed the data indicated in figs. R 35-3 and R 35-4.
- R 95, section 8.4.4 is to be taken into account for protecting capping profiles.

8.1.9 Danger of Sand Abrasion on Sheet Piling (R 23)

Heavy sand abrasion affects particularly walls of reinforced or prestressed concrete steel piles.

When steel sheet piling is used, it must be coated with a system which can permanently withstand the sand abrasion at the site of installation. According to DIN 55 928, the mechanical stress from sand abrasion can be divided into three groups, according to the quantity of transported sand and current.

Assessment of the necessary abrasion-resistance of the coating is performed in accordance with a procedure described in the "Directive for testing coating agents for corrosion protection in steel hydraulic engineering" [153].

8.1.10 Driving Assistance for Steel Sheet Piling by Means of Loosening Blasting (R 183)

8.1.10.1 General

If difficult driving is anticipated, it is necessary to check which driving aids can be used in order to prepare the subsoil in such a way that driving progress is economic, whilst as the same time avoiding overloading the driving equipment or overstressing the driving elements, and reducing power requirements. The latter will also result in less driving noise and vibration. It should also be ensured that the necessary embedment is reached.

In rocky soils, loosening blasting is frequently used.

Blasting of the rock with the aid of an individually tamped borehole charge generates a wedge-shaped ejection cone, the tip of which is located at the base of the borehole.

If a dosing with several such individual blastings is placed in the sheet piling line, a V-shaped disturbed area is created in which the rock is broken up to a greater or lesser degree, depending on the charge and tamping. After blasting, the subsoil corresponds to a loose soil (crushed stone), depending on the degree of fragmentation. Correspondingly large deformations should be anticipated at the base of the sheet piling during embedment in soils loosened in this way. Such blasting is called loosening blasting. The embedment of piles in zones disrupted in this way is possible provided the stone fragments are not too large and can still be shattered to the necessary extent during driving without the base of the pile being damaged to any appreciable degree. Special importance is attached to the selection of the section, the steel grade and possible base reinforcement here.

8.1.10.2 Blasting Methods

Suitable methods have been publicised in [121] and [112]. Their aim is to create a loosened, roughly vertically delimited hearth-like area along the line of the sheet piling, into which the piles can be driven as far as they will go until they encounter the untouched rock. The spacings between the boreholes and their charging with explosive charges must be selected in such a way that the rock in the loosened area is deliberately crushed and made amenable to driving. Since the loosened zone is restricted no appreciable lateral displacement of the base of the sheet piling occurs. The original strength of the rock is virtually unreduced and can be fully used for dimensioning.

In the method according to [121], the so-called "shock blasting" method, boreholes arranged vertically in the sheet piling axis and extending down to the planned base of the sheet piling are driven with axial clearance 0.6 to 1.2 m, if possible without directional deviations. Plastic pipes fitting as closely as possible are inserted into these. They are sealed at the base in such a way that no groundwater can penetrate the pipes. The pipes with extended explosive charges (fuses with torpedoes) are introduced into those areas in which the hard subsoil is to be shattered or loosened. In order to retain sufficient expansion space for the gas discharge following detonation of the explosive, the percent by volume of the explosive charges is selected to be significantly smaller than the pipe volume. The borehole clearance and the size of the charges are governed by the strength of the subsoil. This means that in the event of changes in the subsoil, both these parameters must be amended in line with the strength of the rock in order for the optimum driving progress to be achieved and so that the stratum will be disturbed as little as possible.

When work is carried out in the dry, at least the explosive charges of two adjacent boreholes are detonated, but not more than those in 6 to 8 boreholes simultaneously. The pressure waves generated by the explosions coincide effectively and shatter or loosen the rock subsoil without blasting the rock away. A roughly vertical delimited, loosened zone is created between the boreholes at the level of elevation of the explosive charges, and this can be approx. 0.4 to 0.8 m wide depending on the rock.

8.1.10.3 General Experience

Experience of fundamental importance has been gained with various sheet piling structures erected with blasting, e.g.

(1) The axis of the boreholes and the set piling must lie on one plane, even with increasing depth, so that the sheet piles are always driven into the zone loosened by blasting.

(2) The driving of the piles must follow blasting rapidly, since its effect in the loosened area will decline.

(3) For the driving of sheet piles into the loosened area, stringent requirements are laid down for loosening blasting in respect of the sheet piling section with any reinforcements, of the installation method and the driving equipment so that even in critical situations, uninterrupted, undamaged sheet piling is ensured. Vibration of the piles in the loosened subsoil is not recommended, since the loosened material can become compressed again thereby.

(4) Stringent co-ordination of the boring and driving work and a continuous exchange of information about the work performed are necessary and will improve the success of the work. Under certain circumstances, partially-driven piles must subsequently be under-blasted.

(5) It is necessary to test the subsoil with trial bores and trial explosions before work is started, in order to obtain information about the optimum borehole clearance and the charging of the blast holes. In doing so, operating time graphs from ultrasonic measurements before and after the blasting for example, can be used for estimation of the loosened area.

(6) In populated areas, evidential measurements must be taken in respect of environmental stresses before the actual blasting work is carried out.

(7) Blasting has been used to prepare the following soil types, for example for driving: marl clay, marl lime, marly till, stoney clay with geode interstratification, consolidated sands, sandstone, new red sandstone, shell marl, limestone. The effectiveness of blasting is limited in stratified and in tenacious rock as well as in the case of broad crevices with soft rock veins.

8.1.11 Driving Corrugated Steel Sheet Piles (R 118)

8.1.11.1 General

Driving of sheet steel piles represents a widely used and proven construction method. The method which seems to be relatively simple may, however, mean that driving work is carried out without sufficient special knowledge and in a negligent manner. The resident engineer and the driving team should always adhere to the target of the assigned task, that steel sheet piles are so driven as not to adversely affect the intended purpose of the wall, and to achieve a closed sheet piling wall of maximum safety.

The more difficult the soil conditions, the longer the piles, the greater the embedment depth and the deeper the later dredging in front of the sheet piling, the greater will be the need to insist on a high order of competence in the construction work. Poor results can be expected if long driving elements are driven one after the other to final penetration, because the exposed interlock is then too short and therefore proper initial guidance to the next pile is not given.

A working basis for judging the behaviour of the ground with respect to pile driving is obtained from borings and soil mechanics investigations, as well as from load and penetration tests, see R 154, section 1.9. In critical cases however, test drivings are required. If it is necessary to adhere closely to the designed position of the sheet piles, the behaviour of these test piles at selected locations should be observed with the aid of measurement facilities to determine the amount of deflection which may occur.

The success and quality of the sheet piling installation depend largely on the correct manner of driving. This demands as a prerequisite that in addition to suitable reliable construction equipment, the contractor himself must possess broad experience and therefore be capable of making the best use of the skills of qualified technical and supervisory personnel.

Echeloned driving as described in section 8.1.11.4 produces the best results.

8.1.11.2 Driving Elements

Corrugated steel sheet piling of U- or Z-type sections is generally driven in pairs. Triple or quadruple piles may also have technical and economical advantages in specific cases.

Sheet piles joined in pairs should, as far as possible, be securely connected so as to form a unit by pressing or welding the middle interlocks. This facilitates handling of the driving elements, as well as the driving so that elements already in place mostly do not need to be dragged along. The driving of single piles should be avoided if possible.

Due to technical driving requirements, it can be necessary from the very first to use sheet piles with heavier wall thickness when difficult subsoil and/or deeper embedment is concerned, or to use a higher steel grade than required for purely structural reasons. The pile tip and, if necessary, also the pile butt may have to be reinforced occasionally, which is particularly required for example for driving in soils with stone interstratification and rocky soils.

Reference is made to R 98, section 8.1.7 for the acceptance conditions for steel sheet piles at the site.

8.1.11.3 Driving Equipment

The size and efficiency of the required driving equipment are determined by the driving elements, their steel grade, dimensions and weights, the embedment depth, the subsoil conditions and the driving method. The equipment must be so constructed that the driving elements can be driven with the necessary safety and careful handling, and at the same time be guided adequately, which is above all essential in the case of long piles and deep embedment.

Slow stroke drop hammers, diesel hammers, hydraulic hammers and rapid stroke hammers may be used for pile driving. The efficiency of a driving operation generally improves when the ratio of hammer weight to weight of driving element, including pile cap, is increased. In the case of drop hammers (free drop hammer, hydraulic hammer, single acting steam hammer), a ratio of hammer weight to weight of driving element plus cap of 1 : 1 is especially advantageous. A ratio of 1 : 4 should be preferred for rapid stroke hammers.

Rapid stroke hammers tend not to damage the driving element and are especially well-suited for driving in non-cohesive soils. Slow-stroke, heavy hammers are universally used particularly in cohesive soils.

Driving in rock is best carried out with heavy hammers and corresponding small drop heights. A similar effect can also be achieved using hydraulic hammers whose hammer energies can be adjusted and controlled to the corresponding driving energy requirements. Vibration hammers can also place the driving elements with great care. This is dealt with in greater detail in R 154, section 1.9.3.2 and R 202, section 8.1.22.

When driving with drop hammers, driving caps are urgently required. They must fit well to avoid the caps from bursting due to reinforced compression deformation on the pile top.

The severity of driving vibrations is often exaggerated and the effects on neighbouring structures are often overestimated (see [7]). The anticipated vibrations in the buildings can be estimated by forecasting procedures or trial driving.

During driving work in the vicinity of buildings, equipment should be used which avoids extreme stress for persons and damage to buildings. indicative values are contained in DIN 4150, parts 2 and 3.

Reference is made to R 149, section 8.1.14 as regards noise protection.

8.1.11.4 Driving Sheet Piles

The following must be observed when driving:

The driving blow should generally be introduced centrally in the axial direction of the driving element. The effect of the interlock friction which acts only on one side, can be countered if required, by a suitable adjustment of the point of impact.

The driving elements must be so guided with regard to their stiffness and driving stresses that their design position is achieved in the final state. For this, the pile driver itself must be adequately stable, firmly positioned and the leads must always be parallel to the inclination of the driving element. The driving elements should be guided at least at two points which are spaced as far apart as possible. A strong lower guide and spacer blocks for the driving element in this guide are specially important. The leading interlock of the pile being driven must also be well guided. When driving without leads, care must be taken to ensure tight contact between the hammer and the driving element by the use of well-fitting leg grips. When a floating pile driver is employed, it must be securely moored to restrain its movements as fully as possible.

The first driving elements must be positioned with special care at the intended inclination of the wall. In this way, good interlock engagement is ensured when driving the remaining elements in deep water. In some cases, the interlock engagement can be improved by dredging a trench as deep as possible before piles are driven. This will also result in a reduced embedment depth. If this is done, possible deterioration of the soil conditions must be taken into consideration if backfilling must be carried out under water.

In case of difficult subsoil conditions and for deeper embedment, a driving method with two-sided interlock guidance of the driving elements is recommended, if not already required for other reasons. This refers to a situation when normal continuous driving does not produce the desired result because of increased driving resistance, and because it becomes impossible to hold the pile in the design position. In case of difficult subsoil conditions and for deeper embedment, a driving method with two-sided interlock guidance of the driving elements is recommended, if it is not required for other reasons as well. This refers to a situation when normal continuous driving does not produce the desired result because of increased driving resistance and because it becomes impossible to hold the pile in the design position.

In such cases, driving should be echeloned (e.g., initial driving with a light hammer and redriving with a heavy one) or in panels whereby several driving elements are pitched and then driven in the following sequence: $1 - 3 - 5 - 4 - 2$. This method of driving, particularly for long and deeply embedded piles, should be required by specification and always insisted on. The staggering should be based on the results of the test piles, but should not exceed 5 m at extreme depths.

When constructing sheet piling enclosures, driving is also to be carried out in panels.

U-shaped sheet piles tend to lean forward in driving direction with the pile head, whereas Z-shaped piles tend to lean backward.

This can be prevented by driving echeloned or in panels. For normal continuous driving of U-shaped sheet piles, the web extending in driving direction can be bent out by several millimetres so that the theoretical pile width is somewhat enlarged. In the case of Z-shaped sheet piles, the leading web can be slightly pressed in toward the trough. If leaning over cannot be prevented by these methods, taper piles must be used. Care must be taken with these that they are properly designed for driving, so that the flanges do not plough up the soil. For this, the corrugation of the taper pile must be similar at both ends, and the connecting flange with a welded-in wedge must lie in the driving direction (see fig. R 118-1a).

The chamfering of the points of either U- or Z-shaped sheet piles can lead to damage to the interlocks and is therefore to be avoided.

If the unit spacing of certain stretches of sheeting must be maintained with great accuracy, the width tolerance must be observed. If necessary, compensating piles (fig. R 118-1b) must be inserted.

Driving can be facilitated by loosening blasting to R 183, section 8.1.10, loosening bores, soil replacement or jetting to R 203, section 8.1.23.

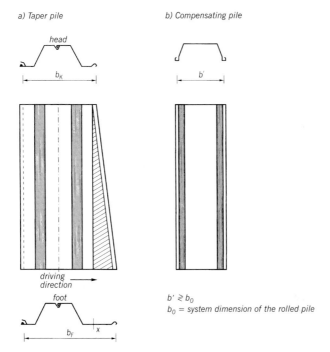

$b_K < b_F$ taper pile when pile leans forwards
$b_K > b_F$ taper pile when pile leans backwards

Fig. R 118-1. Diagram to show taper and compensating piles

Rocky substructures can be perforated by bores sunk at intervals corresponding to the pile width and thus loosened to such an extent that the sheet pilings can be driven in.

The energy required for driving is less and the driving progress greater, the greater the care used to position and guide the driving elements, and the better the hammer and driving procedure are in line with local conditions.

When driving piles with slow stroke hammers, the penetration rate should not fall below 2 cm per 10 blows (1 heat); with rapid stroke hammers, the penetration should not or only very briefly fall below 15 cm per minute, so that the equipment is not subject to undue stress.

8.1.12 Driving of Combined Steel Sheet Piling (R 104)

8.1.12.1 General

In view of the long lengths, particularly for bearing piles, the greatest possible care must be used when driving. This is the only way to ensure satisfactory success with the desired penetration and undamaged interlock connections.

8.1.12.2 Piling Shapes

Combined steel sheet piling (R 7, section 8.1.4) consists of bearing piles and intermediate piles.

Rolled or welded I-beams or I-shaped steel sheet piles are especially well suited for the bearing piles. In order to increase the section modulus, additional plates or locking bars can be welded on. I-beams welded together to form a box pile or double piles can also be used. Special constructions, such as bearing piles as welded box piles of U- or Z-shaped sections, which are connected to each other with web plates, may also be considered.

LN or SN-welded pipes are suitable bearing pipes with welded corner sections or individual piles (see R 7, section 8.1.4). In special cases, interlock chambers are welded flush to the outer edge of the pipes in the pipe wall.

Intermediate piles are generally corrugated steel sheet piles in the form of double or triple piles. Triple piles may require adequate stiffenings for structural and statical reasons. Other suitable constructions may also be considered if they can properly transmit the acting forces to the bearing piles and can be installed undamaged.

8.1.12.3 Types of Piling Elements

If intermediate piles with interlock types 1, 2, 3, 5 or 6 as per R 67, section 8.1.6 or DIN EN 10 248-2 are used, matching interlocks or sheet pile cuts are to be attached to the bearing piles by shear resistant welded joints. The outer and inner welding seams of these joints should be at least 6 mm thick. The individual units of the intermediate piles are to be secured against displacement by welding or pressing their interlocks together.

If intermediate piles with interlock type 4 as per R 67 section 8.1.6 are used, bearing piles with conforming interlock design are to be chosen. Type 4 locking bars are as a rule mounted on the intermediate piles, but occasionally also on the bearing piles.

When piles are driven to greater embedment depth, the locking bars, which are laterally mounted on the intermediate piles, are welded to the latter only on their upper end to maintain the rotation flexibility of the connection and reduce interlock friction during driving. The length of these welds depends on the length of the piles, embedment depth, soil conditions and such driving difficulties as will probably be encountered. Generally, the length of the weld will be between 200 and 500 mm. With especially long piles and/or difficult driving, an additional safety weld at the butt of the pile is recommended. At lesser embedment depth, a shorter safety weld at the top of the piles is generally sufficient. Care must be taken that the driving cap covers the outer locking bars, but not completely, so that enough play is left between bearing piles in case it should be necessary to drive the intermediate piles so that the elevation of their tops is below that of the bearing piles.

If the locking bars are mounted to the bearing piles, they are welded-on with shear resistant joints ($a \geq 6$ mm), if a higher moment of inertia and section modulus are desired. However, reduced rotational flexibility of the connection must be accepted.

If the bearing piles consist of U- or Z-shaped sections connected to each other with web plates, the plates are to be welded continuously to the U- or Z-shaped sections on the outside, and inside at the ends of the bearing piles for at least 1000 mm. The welding seams must be at least 8 mm thick. Furthermore, the bearing piles must be stiffened at tip and butt with wide plates between the web plates so the driving energy can be transmitted without damaging the bearing piles.

When deep penetration and/or especially long bearing piles are required, these should consist of box piles or double piles having broad-flange or box sheet pile sections, because they provide the desired greater rigidity referred to the z-axis and greater torsional rigidity. The resulting increased driving costs must be accepted.

8.1.12.4 General Requirements of Piling Elements

In addition to the otherwise customary requirements in accordance with R 98, section 8.1.7, the bearing piles must be straight with the gage as a rule no more than $\leq 1\ ‰$ of the pile length. They must be free of warp and, in the case of long piles or deep penetration, must have adequate flexural and torsional rigidity.

The top of the bearing pile must be finished plane and at right angles to the pile axis, and must be so shaped that the hammer blow is introduced and transmitted over the entire cross-section by means of a well-fitting driving cap. If reinforcement is placed at the point of the pile, e.g. vanes for increasing axial bearing capacity, these must have a symmetrical layout so that the resultant force of the resistance to penetration will act along the centre of gravity of the pile, eliminating the tendency of the pile to creep out of alignment. Furthermore, the vanes should terminate far enough above the pile point in order to provide some guidance of the pile during driving.

The intermediate piles are to be so designed that they have the greatest adaptability to changes in position, and can thus follow acceptable deviations of the bearing piles from the design position. In intermediate piles with exterior interlocks (Z-shaped), better adaptation to greater positional changes of the bearing piles from the design position is possible with free rotation. In intermediate piles with interlocks on the axis (U-shaped), adaptation is possible only by deformation of the section. They must at the same time be so designed that horizontal sagging under later loading also remains within reasonable limits.

The interlocks must engage in a snug fit and be capable of absorbing such lateral tensile stresses as may occur (see R 97, section 8.1.6). Special care must be taken that matching interlocks fit properly and are not distorted.

8.1.12.5 Driving Procedure

Driving must be so carried out that the bearing piles are embedded straight, vertically or at the prescribed batter, parallel to each other and at the designed spacing. Prerequisites for this are good guidance of the piles during positioning and driving, as well as maintenance of the correct driving sequence. Furthermore, a suitable, heavy, adequately rigid and straight pile driver frame, adapted to the length and weight of the piles is necessary.

Floating pile drivers are basically suitable when the machine is large enough with an experienced crew working with all due care and attention.

On building sites particularly exposed to the elements and sea action (including waves from passing ships), mobile platforms have proven ideal. These can be used flexibly and have the required stability for the driving precision required for combined piling. The position must be constantly checked because the mobile platform can shift as a result of driving vibrations.

When driving bearing pipes, torsional movements, for example from guiding the interlocks, must be counteracted.

Moreover, a satisfactory result will be more likely if a fixed guide wale is placed at the lowest practicable elevation. The spacing of the bearing piles is indicated on the wale by means of welded-on frames, taking any width tolerances into account. Furthermore, the pile butt shall be guided by means of the driving cap by the leads during driving to ensure that the top of the pile is always held in the position required by the design. Care must be taken to assure that the play between the pile and the cap and between the cap and the leads is always as small as possible.

The driving sequence of the bearing piles must be determined so that the pile point encounters compacted soil uniformly on its total circumference, and never on only one side. This is achieved by driving in the following sequence:

1 – 7 – 5 – 3 – 2 – 4 – 6 (large driving step).

At least the following sequence should be observed:

1 – 3 – 2 – 5 – 4 – 7 – 6 (small driving step).

In general, all of the bearing piles should be driven to full penetration without interruption. Subsequently, the intermediate piles can be successively placed and driven.

In unusually deep water, or if the free portion of the sheet pile wall is very high, vertical guidance may be employed in driving the bearing piles. For this purpose, use may be made of structural steel guide frames, which can be adjusted as necessary to fit exactly to the heights and widths concerned. The sides of these frames are fitted with locking bars which match the bearing piles. In addition, in order that the completed wall is straight, an arrangement for horizontal guidance must be provided above the water level.

An effective driving procedure such as the following should be used. A guide frame is threaded into the first, fully embedded bearing pile, so that it lies in the line of the driving direction. This frame is then lowered to the bottom, or in some cases suspended from the bearing pile. The next bearing pile is then threaded into the interlocks of this guide frame, but only driven to partial penetration. Subsequently a second guide frame is installed, into which the next bearing pile is threaded and driven to full penetration. Only then should the next to the last bearing pile be driven to design depth. The guide frames may now be removed and the intermediate piles driven into the space formerly occupied by the guide frames.

In soil which is suitable for jetting, and which is free of rocks, the bearing piles, and if necessary, also the intermediate piles, can be driven with the aid of a water jet. In this case, the jetting devices are to be installed symmetrically and properly guided. Drifting of the piles from the design position can be prevented by careful handling.

The certainty of constructing a faultless, tight wall is increased, if a trench is excavated, as deep as practicable, before driving begins, thus increasing the guided depth of the wall, while at the same time decreasing the driving depth.

If rubble is encountered, hard strata excavation is recommended. For building sites on land, the slotted piling method with adjusted sheet piling is feasible.

8.1.13 Observations During the Installation of Steel Sheet Piles, Tolerances (R 105)

8.1.13.1 General

During the installation of steel sheet piling, the position, setting and condition of the elements must be constantly controlled and suitable measurements carried out to check when the design position is reached. Together with the correct starting position, intermediate phases are also to be checked, particularly after the first few metres of driving. This should make it possible to detect even the slightest deviations from the design position (slant, out-of-line, distortion) or deformations of the pile butt so that early corrections can be made and, if necessary, suitable counter-measures initiated.

The penetration, line and setting of the elements are to be observed frequently and with particular care when the subsoil is heavy with hindrances. If a driving element no longer moves, i.e. unusually slight penetration, the driving should be stopped straight away. In the case of continuous driving, the subsequent piles can be inserted first. Later a second attempt can be made to drive the protruding pile in deeper. In the case of unusual penetration of bearing piles, they have to be pulled and driven again under special precautions. Driving elements which move only with great difficulty just before reaching the design depth so that there is a risk of damages to the point area, should not be driven further. Individual, shorter but undamaged elements are to be given preference to sheet piling which is true to specification but possibly damaged. Lower reached final depths should not impair the total concept (e.g. all-roundness) and stability.

If observed peculiarities such as extreme distortion or skewed driving elements give rise to the assumption that they have been damaged, an attempt should be made to inspect the piles by dredging them partially free and pulling them to investigate the subsoil for driving hindrances.

8.1.13.2 Interlock Damage, Signal Transmitters

When a driving element runs out of the interlock connection (burst lock), it is usually not possible for this to be detected by observations, particularly in the case of increasing penetration resistance. Interlock damage cannot be completely ruled out even when driving is handled with care, so that care should be taken before driving sheet piling to ensure that in the case of Z-profiles, the threading locks of the double piles and quadruple piles are on the landside so that in the case of repairs to any damaged points according to R 167, section 8.1.16, there will be more space behind the line of the sheet piling. If the harbour bottom is to be dredged clear, burst locks can be checked by divers.

In some cases, the interlock connections can be monitored by signal transmitters (fig. R 105-1). Method a) is the only method for continuous

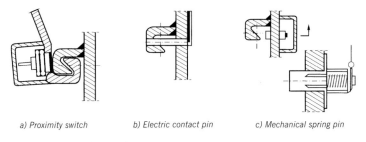

a) Proximity switch   b) Electric contact pin   c) Mechanical spring pin

Fig. R 105-1. Signal transmitters

measurement of whether the interlock connection is still intact during driving. The signal transmitter (proximity switch) is located at the foot of the threading pile. A burst lock is indicated straight away so that measures to rectify the situation can be taken early before the design depth is reached. In methods b) and c), the signal transmitter is installed in the element that is driven. When the interlock of the threading pile reaches the signal transmitter, a contact is registered, thus indicating that the interlock connection is intact. Otherwise there is a burst lock whose starting point cannot be defined. In this case, measures should not be considered until dredging down to the harbour bottom.

8.1.13.3 Driving Deviations and Tolerances

With increasing driving depth, the deviations from the vertical increase. The following tolerances should be included in the calculations at the planning stage for deviations of the piles in compliance with DIN EN 12 063:

± 1.0 % for normal soil conditions and driving on land,
± 1.5 % of the driving depth for driving in water,
± 2.0 % of the driving depth with difficult subsoil.

Verticality is to be measured at the top meter of the driving element. The deviation of the sheet piling butt vertical to the pile may not exceed 75 mm for driving on land and 100 mm for driving in water.
For combined piling, the precise position of the bearing elements has to fulfill strict requirements, therefore the tolerances are always to be fixed in any special case.

8.1.13.4 Measurements, Equipment

The correct starting position and also intermediate positions of the driving elements can be checked easily by using two theodolites each checking the position in the y- or z-axis. This method should generally be prescribed for driving bearing piles in combined piling walls. At the end of the driving procedure and after removal of the guides, each bearing pile is to be surveyed in the installed position, to draw any necessary conclusions for driving the filling piles.

When working with spirit levels, these must be long enough (at least 2.0 m), if necessary with ruler. The use of a theodolite is however preferable. The checks are to be repeated at different points to compensate for local irregularities.

8.1.13.5 Records

Small driving reports are to be kept according to the sample forms in DIN 4026 as records of the driving observations. Under difficult driving conditions, the driving curve for the whole driving procedure should also be recorded for the first 3 elements and then for every 20th element according to the sample form in DIN 4026.

Modern pile drivers record the driving results on data carriers so that the information about the driving results can be made available by computers and suitable software in next to no time. This is recommended particularly under difficult driving conditions in changing soils.

8.1.14 Noise Protection, Low-Noise Driving (R 149)

8.1.14.1 General Comments on Sound Level and Sound Propagation

Various different levels are used in acoustic and measuring engineering to quantify, compare and evaluate the noise produced by a source. Together with the frequently used sound pressure level, which is the characteristic value to describe a noise related to a place or person, recently the sound power level is increasingly also used as a distance-independent characteristic value for a sound source.

The sound level is evaluated in dB(A), which is the ten-fold logarithm of the measured sound power, evaluated to the so-called A-curve. As far as measuring technique is concerned, this results in an increase in the sound level of 3 dB(A) when the noise is doubled, an increase of 6 dB(A) when the noise is quadrupled, etc.

However, from an audio-physiological point of view, the human ear registers 10 dB(A) as a doubling of the noise. For individual levels of different loudness, the level of the loudest individual noise source corresponds approximately to the total level of all noise sources, which also results from the logarithmic addition. The noise level from the driving procedure is composed of various individual sound noises produced both by the pile driver and the impact surface and by the driving element itself. When several sound sources combine together, it should be noted that equally loud levels increase the total level by 3 to 10 dB(A), depending on whether 2 to 10 individual levels of the same loudness are present (see fig. R 149-1a). It should also be noted that for single levels of different loudness, the total level is only just above the level of the loudest individual source (see fig. R 149-1b).

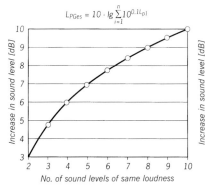
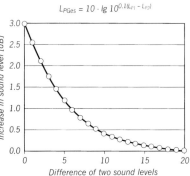

Fig. R 149-1a. Increase in sound level when several sound levels of the same loudness come together

Fig. R 149-1b. Increase in sound level for two levels of different loudness

Consequently, measures to counteract noise can only be effective when initially the loudest individual noise levels are reduced. The elimination of weaker individual noise levels only makes a slight contribution to noise reduction.

Given ideal free-field propagation in the infinite semi-spherical room, the sound pressure of a specific sound source is reduced with every doubling of the distance by 6 dB(A) on account of the geometric propagation of sound energy on the four-fold surface.

$$\Delta L_P = -20 \cdot \lg \frac{s}{s_0}$$

s	= distance 1 to sound source	[m]
s_0	= distance 2 to sound source	[m]
ΔL_p	= change in sound pressure	[dB]

In addition, the sound is attenuated by up to 5 dB(A) in larger distances over grown uneven ground on account of air and ground absorption and, if existing, vegetation or buildings.

On the other hand, it must be taken into account that the simple sound reflection on a structure in the vicinity of the sound source or at concrete or asphalt surfaces can cause an increase of up to 3 dB(A) in the sound level, depending on the absorption and propagation degree of the surface. When there are several reflecting surfaces, each can be substituted by a theoretical mirror sound source with the same loudness as the original sound source and the resulting increase in level calculated taking account of the calculation rules for the interaction of several sound sources (see fig. R 149-1a).

With sound propagation over larger distances, it must also be taken into account that the decrease in sound level from meteorological influences, such as wind currents and temperature stratification, can have both a positive effect, i.e. in the sense of a larger decrease in sound level, and a negative effect. For example, a positive temperature gradient (increase in air temperature at altitude = ground inversion) results in an amplification of the sound level because of the deflection of the sound rays back to the ground at locations approx. 200 m away from the sound source. This effect is to be found in particular over water surfaces, which are generally colder than the ambient air which heats up more quickly, thus resulting in a positive temperature gradient, similar to the earth surface cooling down rapidly after sundown.

In interaction with the ground reflection, the curvature of the sound rays also means that the propagation of the sound remains limited to a corridor between the ground and the inversion layer, so that the geometric propagation attenuation is reduced to half.

The influence of the wind can be compared to that of temperature. Here again, the lower reduction in sound level in the wind direction is caused by a change in the horizontal wind speed with increasing height and the resulting deflection of the sound rays downwards. This effect is particularly noticeable on cloudy or foggy days, when the wind can be observed with a speed of up to 5 m/s as laminar air current. On the other hand, turbulence and vertical air circulation caused particularly during the day by sunshine can result in a higher reduction in sound level from scatter and refraction of the sound rays.

8.1.14.2 Regulations and Directives

Special attention must be paid to the following regulations:

- General Administrative Regulation for Protection against Construction Noise – Emission of Noise. The Federal and State Regulations for Reduction of Construction Noise, CARL HEYMANNS Verlag KG, Köln, 1971.
- General Administrative Regulation for Protection against Construction Noise – Emission Measuring Methods. CARL HEYMANNS Verlag KG, Köln, 1971.
- Directive 79/113EEC from the European Council dated December 19, 1978, for Harmonisation of the Legal Regulations of the Member States Referring to Ascertaining the Noise Emission Level of Construction Machinery and Equipment (Abl. EU 1979 No. L 33 p. 15).
- 15th Regulation for Enforcement of the Emission Protection Law dated November 10, 1986 (Noise Protection in case of Construction Machines).
  – VDI Directive 2714 [91/88] Sound propagation outside – calculation method.

Furthermore, the higher regulations from the state legislation regarding the quiet periods to be observed at night and on public holidays must be complied with, together with the laws and regulations regarding occupational safety and protection (accident prevention regulations UVV, GDG i.V.m. 3., GSGV).

The tolerable noise imissions in the influence area of a noise source are stipulated graduated according to the need to protect the surrounding areas from construction noise. The need for protection results from the structural use of the areas as stipulated in the local development plan. If no local development plan exist or the actual use deviates considerably from the use intended in the local development plan, the need for protection results from the actual structural use of the areas.

The effective level generated by the construction machine at the emission site may be reduced by 5 or 10 dB(A) when the average daily operating period is less than 8 respectively 2 ½ hours. On the other hand, a nuisance surcharge of 5 dB(A) is to be added when clearly audible sounds such as whistling, singing, whining or screeching can be heard from the noise.

If the evaluation level of the noise caused by the construction machine evaluated in this way exceeds the tolerable emission indicative value by more than 5 dB(A), measures should be initiated to reduce the noise. This however is not necessary when the operation of the construction machines does not cause any additional danger, disadvantages or nuisances as a result of not merely occasionally acting outside noises.

The emission measuring procedure serves to register and compare the noises of construction machines. For this purpose, the construction machines are subjected to a minutely described measuring procedure during various operating procedures under defined overall conditions. As part of the standardisation of EU regulations, the noise emissions of a construction machine is now stated as sound power level L_{WA} referred to a semi-spherical surface of 1 m^2. The sound pressure level L_{PA} is still frequently used, referred to a radius of 10 m around the centre of the sound noise or at the operator's workstation in combination with occupational safety regulations.

Emission levels have been defined for the certification or use of various construction machinery which can be observed easily using state-of-the-art technology. Up to now, no binding indicative values have been stipulated for pile drivers.

8.1.14.3 Passive Noise Protection Measures

In the case of passive noise protection, suitable measures prevent the sound waves from spreading unhindered. Baffles prevent the spreading of sound in certain directions. The baffle may not have any leaks or open joints.

In addition, the baffle must be lined with sound-proofing material on the side facing the noise source, otherwise reflections and so-called standing waves will reduce the effect of the baffle. The effectiveness of a sound baffle depends on the effective height and width of the sound baffle and the distance from the sound source being baffled. Basically the baffle should be erected as close as possible to the sound source.

So-called encapsulated solutions with sound jackets, sound aprons or sound chimneys surround the sound source completely with sound-proofing material. A sound-proofing jacket enclosing the pile driver and pile can reduce the noise level by up to 25 dB(A). Unfortunately, encapsulated solutions to reduce the driving noise make working conditions much more difficult and can only be used when driving free-standing piles.

Up to now, sound chimneys are only used for smaller driving units.

8.1.14.4 Active Noise Protection Measures

The most effective and, as a rule, cheapest measure to reduce noise levels both at the workstation and in the neighbourhood is to use construction machines with low noise emissions. For example, compared to the striking hammer, the use of vibration hammers can reduce the sound level by up to 40 dB(A). Hydraulic jacking-in of sheet piles and the insertion of bearing piles using the pile rotating method can be classified as low-noise levels in any case. But the scope for using these low-noise driving methods depends essentially on the quality of the existing subsoil.

Active measures also include the construction methods which make it easier to drive the sheet piling or piles into the subsoil, thus reducing the energy required for the driving procedure. Together with loosening borings or loosening blasting and jets, these also include limited soil replacement in the area of the driving elements, as well as setting the sheet piling in suspension-supported slots. However, these measures can only be used if they are permitted by the soil and construction conditions.

8.1.14.5 Planning a Driving Site

In planning the construction procedure, an attempt should be made to keep the anticipated environmental nuisance of the alter building site to an absolute minimum. Necessary baffling measures are cost-intensive and frequently not taken into account in the calculation. The assurance and observance of only short construction periods with severe disturbances followed by adequately long periods of low or no noise, for example over the midday period – should be striven for. If a certain reduction in capacity has to be taken into account in the calculations from the very beginning, then this must be mentioned separately and clearly in the tender.

8.1.15 Driving of Steel Sheet Piles and Steel Piles at Low Temperatures (R 90)

At temperatures above 0°C and with normal driving conditions, steel sheet piles of all steel grades can be driven without hesitation.

If driving has to be done at lower temperatures, special care is required in the handling of the driving elements as well as during driving.

In soils favourable for driving, driving is still possible down to temperatures of about −10°C, particularly when using S 355 GP (formerly St Sp S) and higher steel grades.

Fully killed steels as per DIN EN 10 025 are however to be used when difficult driving with high energy expenditure is expected, and when working with thick-walled sections or welded driving elements.

If even lower temperatures prevail, steel grades with special cold workability are to be used.

8.1.16 Repair of Interlock Damage to Driven Steel Sheet Piling (R 167)

8.1.16.1 General

Interlock damage can occur during driving of steel sheet piles or from other external actions. However, the risk is the less, the better the recommendations dealing with the design and construction of sheet piling structures are observed with care and diligence. In this connection, special reference is made to the following recommendations; R 34, section 8.1.3, R 73, section 7.4, R 67, section 8.1.6, R 98, section 8.1.7, R 104, section 8.1.12 and R 118, section 8.1.11.

These recommendations also contain numerous possibilities of precautions to limit the risk.

If damages still occur, one advantage of structural steel is that it can be repaired with comparatively simple means, and the repair possibilities are extremely adaptable.

8.1.16.2 Repair of Interlock Damages

If the driving behaviour shows that interlock damages may be expected over an extended area and the sheet piling cannot be extracted again, for example due to pressure of time, repairs are considered consisting chiefly of large-area grouting of the soil behind the wall. Particular success has been had here when repairing combined steel sheet piling with the high-pressure injection method (HDI) (see fig. R 167-5). For permanent structures however, the interlock damage should be additionally secured in the following manner.

Individual interlock damages are subsequently repaired on the spot, in which case the work can only centre on the sealing of the interlock damage. Restoration of the statical effectiveness of the locks is practically impossible.

The type of sealing of interlock breaks is contingent above all on the size of the interlock opening and on the sheet piling section. Various methods have proven themselves here. Repair work is generally carried out on the outboard side. Smaller interlock openings can be closed with wooden wedges. More extensive damage can be temporarily sealed for example with a rapid-setting material, such as lightning cement or two-component mortar, which will be placed in sacks. For permanent safety however, the opening must be fully bridged with steel parts. This applies especially to the parts of a damaged sheet piling which protrude beyond the quay face. Thereby an adequate and secure fastening of the forward placed steel parts at the sheet piling must be provided. In addition, the damage area must be concreted in order to prevent the soil from washing out later. The installed concrete must possibly be reinforced in order to make it less vulnerable to the impact of ships. The measures to be taken must be well thought over, since they are also quite costly. It is recommended that an additional protective layer, e.g. of ballast, be installed in the bottom area to prevent scouring.

The work must be executed mostly below water and therefore always with diver assistance. Very high demands are therefore to be placed on the technical ability and reliability of the divers.

The installed sealing elements must extend at least 0.5 m, but better 1.0 m below the design bottom (R 36, section 6.7) in front of the waterfront structure. Expected scourings from currents in the harbour water or screw action must be taken into account. The smoothest possible sheet piling surface on the water side is to be striven for. Protruding bolts for example are to be burnt off after the damage area has been concreted, once they have fulfilled their function as formwork element. The steel plates and the sheet piling are to be joined to the concrete with anchor elements, so-called rock claws. When repairing sheet piling, the wale frequently located on the land side has to be taken into account, as well as possibly existing sheet piling drainage, including gravel filter.

The solutions shown in figs. R 167-1 to R 167-5 for the repair of underwater areas have proven successful in practice, but lay no claim to completeness. Interlock damage to combined sheet piling is repaired if still possible by rear driving of intermediate steel piles (fig. R 167-4) and otherwise from the water side according to fig. R 167-3.

Repair in accordance with figs. R 167-1 to R 167-3 presumes that there are no obstructions to the repair work from leaking soil, either because there is no prevailing internal pressure difference or the temporary sealing measures described above have a sufficiently stabilising effect.

If for reasons of soil mechanics or for technical, economic or other reasons, such temporary sealing measures are not possible, driving elements combining all the necessary functions can be inserted as shown in fig. R 167-4.

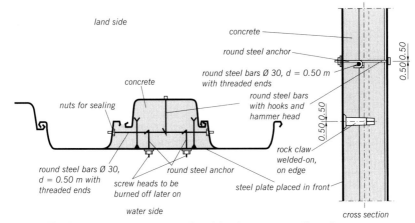

Fig. R 167-1. Executed example of repairing damage at a small opening

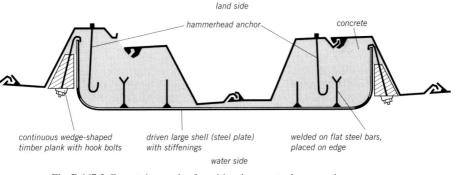

Fig. R 167-2. Executed example of repairing damage at a large opening

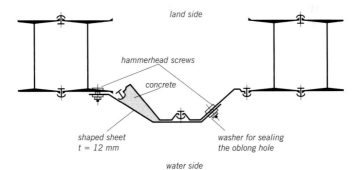

Fig. R 167-3. Executed example of repairing damage at an opening in combined sheet piling

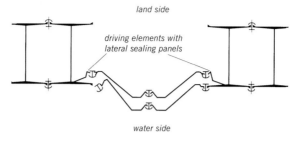

Fig. R 167-4. Executed example of repairing damage by rear driving with a driving element at an opening in combined sheet piling

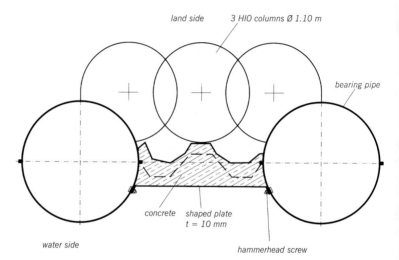

Fig. R 167-5. Executed example of repairing damage to a burst interlock in combined tubular sheet piling

In *rear driving*, e.g. according to fig. R 167-4, the driving element is pressed by the internal pressure difference against the intact parts of the sheet piling during dredging on the water side, this being either combined or corrugated sheet piling.

In *forward driving*, suitable measures must be taken to ensure that the base of the driving element is always pressed against the bearing pile, in other words, it has to be located as near to the bearing pile as possible. Before dredging takes place, the head of the driving element must be secured to the bearing pile, using hammerhead screws, for example. The elevation of the first dredging cut must be coordinated with the bearing capacity of the driving element between this upper fixed support and the lower, more flexible earth support. After this has been done, the driving element must be secured to the bearing pile in the exposed area, etc. During dredging operations, constant investigations of the repaired area by divers are necessary. Any local leaks can be sealed by injections over an appropriate area.

8.1.17 Design of Pile Driving Trestles (R 140)

8.1.17.1 General

Insofar as driving operations are not possible from existing or hydraulically filled land, the following possibilities may be considered:

(1) Driving from a pile driving trestle.
(2) Driving from a mobile working platform.

8.1.17.2 Design of Pile Driving Trestle

The pile driving trestle can be built of either steel, timber or reinforced concrete piles. The design of the trestle must make allowance for the following, especially for reasons of economy:

(1) The trestle piles can be driven from a pontoon or a barge. The inaccuracies resulting from this method must be taken into account in the construction. Occasionally, it may be possible to drive some of the piles from existing ground or from an existing trestle.
(2) The length of the trestle depends on the progress made by the pile driver, and on whether the trestle will be needed for work later on. Parts which are no longer needed are reclaimed and used to extend the trestle. Trestle piles which are no longer required, are extracted. If this cannot be done, even with jetting and vibrating, they should be cut off below the elevation of the existing and planned harbour bottom. In this connection, dredging tolerances and possible future deepening of the harbour should be taken into consideration. When extracting the trestle piles, any side effects on the structure must be taken into consideration.
(3) The design and construction of the trestle should be simple, keeping in mind the need to salvage and reuse the structural elements, with minimum waste.
(4) The danger of scour should be given special attention in the case of uniform fine sand on the harbour bottom. To this end, the piles should be of a generous length as a matter of course. Moreover, the area in the vicinity of the trestle should be constantly observed during the construction period, especially if strong currents and sand fill are present. An example of the latter is a foundation which has been improved by soil replacement. If scour occurs, the affected area should be immediately filled with graded gravel.

(5) The trestle piles should be driven well clear of the permanent structure, so that they will not be dragged down by the driving of the structural piles. The separation required depends on the type of soil. Open holes left in the bottom after extraction of piles from cohesive soils should, if necessary, be filled with suitable material.

(6) Pile driving trestles built close to the shore can consist of a row of piles in the water and a bearing on shore for the rail beam. Where such a shore connection does not exist, the trestle is supported by two or more rows of piles. Structure piles can also be used for this, if they are in the right place on planning.

8.1.17.3 Load Application

The design of the foundation of the driving trestle should take account of the inherent loads and the operating conditions etc. corresponding to DIN EN 996, section 4.1.

Together with the loads from the pile driver and the undercarriage, and, when used, the tower crane, the current pressure load, wave action and ice pressure load are to be taken into account for the driving trestle (trestle piles and tie braces). Insofar as the driving trestle is not safeguarded by additional measures (safety dolphins) from ship contact – e.g. pontoons or similar equipment delivering the driving elements to the site, vessel impact and, where applicable, also line pull forces of up to 100 kN each are to be taken into account in rating the driving trestle and its piles, even in the most unfavourable possible position.

8.1.17.4 Safety Factors

DIN 15 018 applies to calculation of the travelling undercarriage. Since the undercarriage is used only during construction, the allowable stress may be increased by 10 %, and the safety factor for stability may be reduced by 6 %, in loading case HZ, when considering the most unfavourable loads. No other relaxation of stress requirements may be made unless the building authorities approve.

The tie beams are to be designed for the γ-fold loads caused by the travelling undercarriage, taking account of the combination factor Ψ of the basic combination according to DIN 18 800 T1 point 7.2. Using the elastic-elastic verification method, the stresses resulting from these loads are to be compared to the bearing ability of the used steel grades.

For verification that the loads are carried off safely into the subsoil is to take account of the characteristic values of the soil concerned in the same way as for the water structure being built.

The piles are to be designed with the due foundation loads. The actions on the bearing piles resulting from the tie beams and travelling undercarriage are calculated with the partial safety factors of the load combinations in DIN 18 800 T1 and taken as design value of action on the piles. Additional loads from further actions are to be increased to the design value by the partial safety factors in the limit states LS1B and LS1C for load case 1 of table R 0-3 in section 0.2.

The resistances of the soil here are to be decreased accordingly with the partial safety factors as per table R 0-4 in section 0.2.

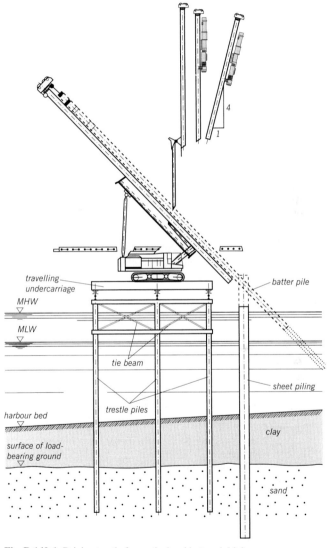

Fig. R 140-1. Driving trestle for vertical and battered driving

In the elastic-elastic verification method, the resulting tensions in the pile are to be compared as per DIN 18 800 T1 to the bearing ability of the corresponding steel grades.
Reference is made to DIN 24 096.

8.1.18 Design of Welded Joints in Steel Sheet Piles and Driven Steel Piles (R 99)

This recommendation applies to welded joints in steel sheet piles and steel piles of every type.

8.1.18.1 General

(1) General requirements

The dimensioning, design and workmanship must conform to DIN 18 800 part 1 and part 7, and to DIN 8563, part 1 and part 2. Wind and weather protection measures are to be provided for welding on the building site. The joints are to be dried and cleaned.

(2) Technical documents

The welding of joints may not be started until the following technical documents are on hand:

Verified strength calculation stating the occurring and permissible welding seam stresses, taking into special account the working conditions when making the joint in the workshop or under the pile driver.

Working drawings with data on base material, material thickness, seam shapes, seam dimensions with evaluation and, if necessary, welding drawings with data on welding methods, welding additives and welding sequences, as well as about tests on the welding connections. A possibly intended procedure test must be completed prior to the welding work.

Reference is made to R 67 section 8.1.6.1 as regards necessary certification.

(3) Proof of competence to weld

Proof of competence of the welding firm is to be submitted regarding welding work in the workshop and on site. In general, the "major proof of competence" is required.

8.1.18.2 Materials

(1) Base material

Sheet piling steel grades according to R 67 and steels according to DIN EN 10 025 can be used with welding suitability described in section 8.1.6.4.

The steel grades used must be supported at least by a works test certificate according to DIN 50 049 fig. 2.2, indicating both mechanical and technical properties as well as the chemical composition (R 67, section 8.1.6.1).

(2) Welding material
The welding materials are to be selected by the welding engineer of the executing firm licensed for the work, taking into account the proposals of the sheet piling and steel pile manufacturer.
With unkilled steels, basic electrodes or welding material with a high degree of basicity may generally be used (filler wire, powder).

8.1.18.3 Classification of the Welded Joints

(1) Fundamental comments
The butt joint is intended to replace as fully as possible, i.e. to 100 %, the steel cross-section of the sheet piles and driven piles.
The percentage of the effective butt weld sectional area in relation to the full steel cross-section, called effective butt coverage in the following, is however contingent on the construction type of the elements, the offset of edges at the joint ends and on the prevailing conditions on site, (table R 99-2).
If the steel cross-section of the sheet piles or piles is not achieved by the butt weld cross-section, and if a full joint is required for statical reasons, splice plates or additional sections are to be used.

(2) Stresses in butt welds
Verification for butt and fillet welds is to be provided as per DIN 18 800, part 1, Nov. 1990.

Verification is required

$$\frac{\sigma_{w,v}}{\sigma_{w,R,d}} \leq 1 \quad \text{(equation 71, DIN 18 800-1)}$$

with

$$\sigma_{w,v} = \sqrt{\sigma_\perp^2 + \tau_\perp^2 + \tau_\parallel^2}$$

and

$\sigma_{w,R,d}$ = $\alpha_W \cdot f_{y,k} / \gamma_M$
$\sigma_{w,v}$ = comparative value of the existing weld stresses, to be ascertained with the partial safety factor γ of the actions (see e.g. DIN V 1054-100)
$\sigma_{w,R,d}$ = limit weld stress
α_w = coefficient as per table 21 of DIN 18 800-1
$f_{y,k}$ = characteristic value of the elongation limit
γ_M = partial safety factor for the resistance variable $\gamma_M = 1.1$ in case no other value is prescribed in the specialist standards

Steel grade		Limit weld stress [N/mm^2]
S 240 GP S 235 JR S 235 JR G2 S 235 JR G3	(DIN EN 10 248) (DIN EN 10 025)	218[1]
S 270 GP S 320 GP	(DIN EN 10 248)	245[1] 291[1]
S 355 GP S 355 J2 G3	(DIN EN 10 248) (DIN EN 10 025)	323[1]
S 380 GP	(DIN EN 10 248)	291[1]
S 390 GP	(DIN EN 10 248)	355[1]
S 430 GP	(DIN EN 10 248)	391[1]

[1] The stated limit weld stresses for all weld qualities are to be reduced by 20 % for shear load ($\alpha_w = 0.80$)

Table R 99-1. Limit weld stresses for tested welds ($\alpha_w = 1.0$)

Construction type of sheet piles or piles		Effective butt coverage as %	
		in the workshop	under the pile driver
a) Pipe sections, calibrated joint ends, root welded through		100	100
b) Piles of I-shaped sections, box sheet piles Cross section reduction with material removed from the throats		80–90	80–90
c) Sheet piles	Single piles	100	100

	Double piles
	  Interlock area only with one-sided welding
	  U-shaped sheet piles
	  Z-shaped sheet piles
		90
		80
			~ 80
			~ 70

d) Box piles consisting of individual sections
  Individual sections jointed then assembled
  Jointed box pile
		100
		70–80
			50–70

Table R 99-2. Effective butt coverage as %

The following limit weld stresses of table R 99-1 for pressure and tensile load can be used for killed steels in groups JR, JQ, JRG2, J2G3 and for unkilled steels with wall thicknesses ≤ 12.5 mm, when radiation and ultrasonic tests verify that there are no cracks, lack of fusion or root defects. If the weld quality is not verified by the stated tests, the limit weld stresses are to be ascertained according to table 21 of DIN 18 800-1. The table values are to be halved for unkilled steels with wall thicknesses > 12.5 mm.

(3) Effective butt coverage
The effective butt coverage is expressed as a percentage and is the ratio between the butt weld cross-section and the steel cross-section of sheet piles or piles.

8.1.18.4 Making Weld Joints

(1) Preparation of the joint ends
The cut of the section to be welded should lie in a plane at right-angles to the axis; an offset in the joint is to be avoided.

Special attention is to be paid to ensuring a good fit between the cross-sections and, in the case of steel sheet piling, to preservation of a fair passage in the interlocks as well. Differences in width and height between pieces to be welded should not exceed ± 2 mm so that offsets in the welding edges will not exceed 4 mm.

It is recommended that hollow piles composed of several sections first be manufactured in the required full length, and then cut into working lengths after having been suitably match-marked (e.g. for transportation, driving, etc.).

The ends intended for the butt joint are to be checked for doublings for a length of about 500 mm.

(2) Edge preparation for welding
In the shop, butt welding is generally performed with V- or Y-groove welds. Both edges of the butt joint are to be suitably prepared.

If a butt joint must be made in the field on driven steel sheet piles or steel piles, the top of the driven element must first be trimmed as required by R 91, section 8.1.19. The extension piece is to be prepared for a butt weld, with or without a root pass.

(3) Welding procedure
All accessible surfaces of the butted section are to be fully connected. Where possible, the roots are gouged and sealed with root passes.

Root positions which are not accessible require a high degree of accuracy in fitting the sections to be joined and careful edge preparation.

The proper welding sequence depends on various factors. Special care must be taken that residual tensile stresses from the welding process will not be superimposed on tensile stresses which will occur when the structure is in service.

8.1.18.5 Special Details

(1) Joints are to be placed as far as possible into a cross-section where stresses are low, and staggered by at least 1 m.

(2) In preparing H-shaped sections for welding, the throat areas of the web are to be drilled out in such a way that the drilled openings form a semi-circle with a diameter of 35 to 40 mm open to the flanges and sufficient to ensure full penetration of the flanges, using a root pass. The surface of these openings must be machined to remove any notches after completion of welding.

(3) If flange splice plates are required for the butt coverage for structural reasons, the following rules are to be observed:
   a) The thickness of splice plates shall not exceed the thickness of the spliced section members by more than 20 %. The maximum thickness of the splice plates is 25 mm.
   b) The width of the plates shall be dimensioned in a way that they can be welded all-round on the flanges without end craters
   c) The ends of the plates shall be tapered to 1/3 b with each side converging at a rate of 1 : 3.
   d) Before the plate is placed, the butt weld is to be ground down flush.
   e) Non-destructive tests must be completed before the splice plates are placed.

(4) If butt joints in service are not subjected to predominantly static load within the meaning of R 20, section 8.2.6, splice plates over the joints are to be avoided wherever possible.

(5) If butt joints are scheduled e.g. as a result of transportation or driving practice, only killed steels should be used.

(6) Butt joints under the pile driver are to be avoided as far as possible for economic reasons and because unfavourable weather conditions could have a negative effect on the welding.

(7) If welded joints have unavoidable openings through which the soil could leach out, such openings must be sealed in a suitable manner (see section R 117, section 8.1.20).

8.1.19 Burning off the Tops of Driven Steel Sections for Load-bearing Welded Connections (R 91)

If the tops of driven steel sheet piling or steel piles are to be fitted with welded load-bearing connections (e.g. welded joints, bearing fittings or the like), these may not be placed in areas with driving deformation. In such cases, the top ends of the piles must be cut off to a point below the limit of the deformation, or all welding seams must be run beyond the deformed area.

This measure is to remove embrittlement which would have a detrimental effect on the load-bearing welded connections.

8.1.20 Watertightness of Steel Sheet Piling (R 117)

8.1.20.1 General

Because of the required play in the interlocks, steel sheet piling walls are not absolutely watertight, which is however not necessary in general. The degree of watertightness of interlocks joined at the shop (W interlocks) is mostly less than for the driven threaded interlocks (B interlocks) which fill partially with soil in the area below ground surface. Progressive self-sealing (natural sealing), due to corrosion with incrustation as well as the accumulation of fine particles in sediment-laden water, can usually be expected in the course of time.

8.1.20.2 Assisting in the Natural Sealing Process

With water pressure difference acting on one side and at walls standing free in water, for example at an excavation pit sheeting, the natural sealing process can be assisted if necessary by pouring in sealing material, as for example boiler ash etc., provided the interlocks are constantly under water and the sealing materials are poured close to the interlocks into the water.

When dewatering excavations, especially high pumping rates are necessary at the beginning, so that the greatest possible differential head is quickly built up between inner and outer water.

Thereby, the interlocks join together well. Furthermore, this causes an adequately strong flow of water to the excavation interior and thus effectively flushes the sealing materials into the interlocks. The costs for sealing and pumping are to be adjusted to each other optimally. However, this sealing method produces no permanent success under fluctuating water pressure difference from the sides and with movements of the sheet piling in free water through wave impact or swells, etc.

8.1.20.3 Artificial Sealing

Sheet piling interlocks may be artificially sealed both before as well as after installation.

(1) Sealing method before the sheet piles are driven:
   a) Filling the interlocks with a durable, environmentally compatible, sufficiently plastic compound, namely the W-interlocks at the shop and the B-interlocks at the plant or at the site.

b) A noticeable improvement in watertightness is achieved by providing the B-interlocks at the plant with a durable elastic, profiled compound, which adheres firmly to the surface.

When employing this method, also those B interlocks can be sealed which are no longer accessible after driving, for example the areas below the bottom of excavation or waters. Reference is made to a) concerning the position of the sealed joint.

With both sealing methods, the achievable watertightness of the interlocks depends on the water pressure difference and the embedment method. Driving places little stress on the seal, since the movement of the pile in the interlock takes place in one direction only. In the case of vibration, the stress is greater, depending on the frequency selected. It cannot be ruled out here that the seal will be damaged as a result of friction and temperature increases.

c) Interlock joints of the W interlocks are welded tight, either at the plant or at the site. In order to avoid cracks in the watertight seam during the installation procedure, additional seams are required, for example on both sides at the head and foot of the driven element, as well as counterseams in the area of the tight seam. The tight seam must lie on the correct side of the sheet piling, for example for dry docks and navigation locks on the air/water side.

(2) Sealing methods after installation of the sheet piling:
    a) Caulking the interlock joints with wooden wedges (swelling effect), with rubber or plastic cords, round or profiled, with a caulking compound capable of swelling and setting, e.g. fibrous material mixed with cement.

The cords are tamped in with a blunt chisel. Light pneumatic hammers have also proven themselves.

The caulking work can also be carried out for water-bearing interlocks. B interlocks can generally be made watertight better than squeezed W interlocks.

The interlock joint must be cleaned of clinging particles of soil before caulking is begun.

    b) The interlock joints are welded tight. As a rule, these are only the B interlocks, as the W interlocks were already welded tight before the piles were driven, see section 8.1.20.3 (1) c).

Direct welding of the joint is possible for dry and properly cleaned joints. Water-bearing joints should be covered with steel plates or sections, which are welded to the sheet piling by two fillet welds. With this method, a fully watertight sheet piling may be achieved.

c) At the completed structure, plastic sealing compounds may be placed into the accessible joints above the water level at any time, or PU foam injected into the interlock chamber by impact or screw nipples. Care must be taken in such cases to ensure that the flanks of the plastic compound are applied on a dry surface. This can be achieved following previous temporary sealing of the joints.

In the case of box sheet piling with double locking bars, sealing can also be achieved by filling the emptied cells with a suitable sealing material, e.g. with underwater concrete.

Special emphasis is drawn to the fact that interlocks which are not sealed behave like vertical drains in strata of low permeability.

At greater water table differences and especially where wave action may possibly prevail, the interlocks must be sealed with extra care if fine sand or coarse silt is present behind the sheet piling, as this non-cohesive material can be easily washed out through the voids of the interlocks.

8.1.20.4 Sealing of Penetration Points

Aside from the watertightness of the interlocks, special attention must be paid to adequate sealing of the points of penetration of anchors, wale bolts and the like.

Lead or rubber washers are to be placed between sheet piling and base plates, as well as between base plates and nuts. In order not to damage the sealing washers, the anchor must be tensioned by means of a turnbuckle, and the wale bolts with the nuts on the wale side.

The holes in the sheet piling for the wale bolts and, if necessary also for the anchor, are to have ridges cleanly removed so that the base plate rests fully.

8.1.21 Waterfront Structures in Regions Subject to Mining Subsidence (R 121)

8.1.21.1 General

Predictable ground movements and their changes in the course of time are to be taken into account in the planning work. A distinction must be made here between:

(1) movements in the vertical direction, subsidence, and
(2) movements in the horizontal direction.

Since the movements as a rule follow on at differing time intervals, subsidence, tilting, torsion, tearing and squeezing may occur in changing sequence.

Where local subsidence occurs, the elevation of the groundwater table usually remains unchanged. This is also true of the water level in navigation channels.

Insofar as any information might be available on planned or current underground mining operations, the company carrying out this work should be informed of proposed waterfront construction as early as possible. The plans must be submitted to these companies, and it must be left to their discretion whether to propose safety measures and have them carried out, or assume responsibility for making good any damages caused by the mining operations.

It is not usually possible to predict the locations or extent of failures. If at the outset the responsible mining company is not willing to undertake measures to prevent mining damages, or does not consider such measures necessary, the owner cannot be advised to undertake any extra expenditure for safety measures in advance. However, it would be wrong to adopt types of construction which are especially susceptible to damage or failure due to mining operations, or which are especially difficult to repair, if it appears likely that harmful effects from mining operations may occur later. It must be mentioned in this connection that solid waterfront structures have frequently been heavily damaged by being torn or squeezed, as well as by torsion. In contrast, no appreciable damage to structures of U- or Z-shaped steel sheet piles has been confirmed up to now. Such structures can therefore generally be recommended at waterfront regions where mining subsidence may occur. The following suggestions must be given particularly careful consideration in planning, design, calculation and construction.

8.1.21.2 Suggestions for Planning

The magnitude of the earth movements to be expected must be ascertained from the responsible mining company. The ground elevations and load assumptions are determined on the basis of this data.

Vertical movements can make it necessary to establish the elevation of the top of the waterfront structure higher to allow for the probable magnitude of subsidence; this is generally more economical than increasing the height of the wall following subsidence. If subsidence of varying depth is expected along the length of the waterfront structure, about which the mining surveyor can make quite reliable and correct predictions, and if it is not planned to increase the height of the structure later on, the elevation of the top of the structure, when placed, must be increased by varying amounts to compensate for the anticipated local subsidence. Thus in the final state, the top of the wall will reach a horizontal position throughout most of its length. It is frequently more practicable however to increase the height of the waterfront structure only in later years. However, the resulting load increases on sheet piling and anchoring should be taken into account in the original design, in order to avoid the need for later reinforcing, which is as a rule quite expensive. The effect of the water pressure difference is also to be determined exactly for all stages of the height increase and taken into account in the calculations.

Longitudinal pulling and squeezing of a waterfront structure, built with U-or Z-shaped sheet piling generally do not suffer from damages because the accordion effect enables the structure to accommodate ground movements.

A transverse push against the waterfront structure creates a negligibly slight displacement of the wall toward the water. A pull transverse to the water structure may severely overstress the anchor rods only if unnecessarily good stability exists in the lower failure plane due to overlength anchor rods. This can also be the case with firmly embedded overlength anchor piles. If dredging should become necessary in front of the waterfront structure to meet a temporary requirement, the harbour bottom shall not be dredged deeper than is absolutely necessary to expose as little as possible of the sheet piling.

8.1.21.3 Suggestions for Design, Calculation and Construction

Beyond taking into consideration the interim stages and the final stage, the waterfront sheet piling generally does not require any overdesigning, unless this is requested by and paid for by the mining company. This also applies to a reinforced concrete capping beam and its reinforcement, provided that it steel remains above water after the mining subsidence takes place. Any damaged places can easily be demolished and repaired afterwards.

In order to minimise the susceptibility of the structure to harmful mining effects, the portion of the sheet piling above the anchoring should be as little as possible, so that the anchor rods and wale should be placed as close as possible below the top of the sheet piling; this is why the sheet piling moment should not be reduced for active earth pressure redistribution due above all to expected pulling.

The steel grades for the sheet piling can be chosen according to R 67, section 8.1.6. For the capping beam and the wale, the steel grades S 235 J2G3, S 235 J2G4 and S 355 J2G3 should be selected in accordance with DIN EN 10 025.

The latter also applies to anchor rods. If round steel anchor rods are used, upsetting in the thread area of the anchor rod is permissible, if requirements as per R 20, section 8.2.6.3 are fulfilled. Upset round steel anchors offer the advantage of a grater elongation path and higher flexibility than round steel anchors without upsetting in the thread area; besides this, they are easier to install and cheaper.

When the sheet piles are delivered, special attention should be paid to inspecting the interlock tolerances for compliance with R 67, section 8.1.6.

The motion possibilities of the sheet piling are not impaired in horizontal direction when the interlock joints are welded together, for example because of watertightness. However, the interlock welding does hinder the vertical possibilities of motion of the sheet piling. This can still be maintained to an adequate degree however, if the interlock joints are only welded together at certain intervals. In order to also ensure the watertightness in these joints, they are filled with an elastic, profile sealing compound before being joined at the shop, which does not hinder the vertical motions.

Accessible interlock joints can also be sealed at the site, for example by installing an elastic sealing compound in front of the joint. This compound is supported by a plate construction, which will not deter vertical movements. For the other structural members, welded constructions are to be avoided if possible, if they impair the motion possibilities of the sheet piling.

The foregoing remarks apply similarly to the interaction of reinforced concrete structural members and the sheet piling. It is especially important that the flexibility of the sheet piling is not limited by the presence of solid structural members. The waterfront structure and the craneway are to be designed on separate foundations to give opportunities for independent settlement and realignment. The same applies to a reinforced concrete capping beam, whose expansion joints should be spaced at about 8 to 12 m, depending on the magnitude of the expected varying subsidence. If the craneway is not laid on ties, see R 120, section 6.16.2.1 (2), but rather on reinforced concrete, the reinforced concrete beams should be connected to each other by sturdy tie beams to maintain a true gauge. It is advisable to omit electric contact line channels and depend on trailing cables instead.

Wales consisting of two U-channels are preferable to other arrangements because the wale bolts can more easily adapt themselves to deformations. They are to be generously designed and to be so manufactured that wale reinforcing will not become necessary later on.

Oblong holes should be provided in the joints of the wales and the steel capping beams to allow longitudinal movement of the wall. As a substitute, circular holes may be enlarged carefully with circle oxy-cutters. They are to be finished as far as necessary in order to avoid or eliminate notches which can produce cracks in the steel.

If a sheet piling wall must later be increased in height, this should be anticipated and taken into account in the original design of the capping beam (simple dismantling).

Anchor connections in a capping wale are to be avoided.

Horizontal or gently sloped anchor rods are recommended for the lowest possible additional stresses produced by differential settlement between anchorage and wall. The anchor connections are to be made completely flexible. The end hinges are to be placed if possible in the waterside trough of the sheet pile section, so that they are accessible and can be easily observed.

8.1.21.4 Remarks on Structures
Waterfront structures in regions subject to mining subsidence require periodic inspection and checks on measurements. Even if the mining company is liable for any damage, the owner of the facility still remains responsible for its safety.

8.1.22 Vibration of U- and Z-shaped Steel Sheet Piles (R 202)

8.1.22.1 General
Vibration hammers are used for vibration. They generate vertically directed vibrations through unbalances rotating synchronously in opposition. The vibrator must be connected rigidly with clamping jaws to the material being driven, whereby the soil is induced to covibrate when the vibrations and frequencies have been correctly selected. This can considerably reduce skin friction and peak resistance in non-cohesive soils. Good knowledge of the interactions of the vibrating hammer, the material being driven and the soil is an important prerequisite for planning an operation.

In R 118, section 8.1.11.3 and R 154, section 1.9.3.2, reference is made to the influences of soil and driving material.

As a result of the effects of vibration, the recommendations for percussive embedment when using a vibration hammer can be applied to a limited extent. This applies in particular if vibrated elements are subject to tensile and/or pressure load loss in operation.

8.1.22.2 Terms and Ratings for Vibration Hammers
Important terms and ratings are as follows:

(1) Type of drive
- electrical
- hydraulic
- electro-hydraulic

(2) Required power P [kW]
It ultimately determines the efficiency of the hammer. At least 2 kW should be available per 10 kN centrifugal force.

(3) Effective moment M [kg m]
This is the product of the total mass m of the unbalances, multiplied by the spacing r of the centre of gravity of the individual unbalance from tits axis of rotation.

$$M = m \cdot r \, [\text{kg m}]$$

The effective moment also plays a part in determining the vibration width or amplitude.

(4) Revolutions per minute n (U min^{-1})
The revolutions per minute of the unbalanced shafts have a quadratic effect on the centrifugal force. Electric vibration hammers operate with constant speed and hydraulic hammers with fully variable speed adjustment.

(5) Centrifugal force (exciting force) F [kN]
This is the product of the effective moment and the square of the angular velocity

$$F = M \cdot 10^{-3} \cdot \omega^2 \, [\text{kN}] \quad \text{with} \quad \omega = \frac{2 \cdot \pi \cdot n}{60} \, [\text{sec}^{-1}]$$

In practice, the centrifugal force is a comparative variable of different machines. However, the revolutions per minute and the effective moment at which the optimum centrifugal force is reached must also be taken into account here. The new generation of vibrators allows for fully variable centrifugal force at constant frequency.

(6) Vibration width S, amplitude $\bar{x}$ [m]
The vibration width S is the total vertical shift of the vibrating unit in the course of one revolution of the unbalances. The amplitude $\bar{x}$ is half the vibration width. In the manufacturer's equipment lists, the quoted amplitude – the value of S is frequently mistakenly quoted for $\bar{x}$ – is the quotient of the effective moment (kg m) and the mass (kg) of the oscillating vibrator.

$$\bar{x} = \frac{M}{m_{\text{Ham,dyn}}} \, [\text{m}]$$

On the other hand, the "working amplitude" $\bar{x}_A$ necessary in practice is an unknown variable. The divisor hereby is the total covibrating mass.

$$\bar{x}_A = \frac{M}{m_{\text{dyn}}} \, [\text{m}]$$

where $m_{\text{dyn}} = m_{\text{Ham,dyn}} + m_{\text{Driving material}} + m_{\text{Soil}}$. In prognoses, m_{Soil} should be assumed to be: $\geq 0.7 \cdot (m_{\text{Ham,dyn}} + m_{\text{Driving material}})$.
A theoretical "working amplitude" of $\bar{x}_A \geq 0.003$ m should be aimed for.

(7) Acceleration a [m/s^2]

The acceleration of the driving material acts on the grain structure of the encountered soil. The grain structure stratification should be constantly moved during vibration in order to approach the "pseudo-liquid" state in the ideal case.

The product of the "working amplitude" and the square of the angular velocity yields the acceleration of the driving material.

$$a = \bar{x}_A \cdot \omega^2 \ [\text{m/s}^2] \quad \text{with} \quad \omega = \frac{2 \cdot \pi \cdot n}{60}$$

Experience shows that a should be ≥ 100 m/s^2.

8.1.22.3 Connection Between Equipment and Driving Element

An extensively rigid connection must be made between the hammer and the driving material with the clamping jaws. Hydraulic clamping jaws are mostly used for this. Since the hammer should be located in the centroidal axis of the driving material during vibration as during driving, a dual clamping jaw is helpful for dual piles. Only thus can the vibration energy be introduced optimally into the driving material (see R 154, section 1.9.3.2 "wobbling or shimmying effect" in this respect). In selecting the hammer, therefore, the possibility of using various clamping jaws must also be borne in mind.

8.1.22.4 Criteria for the Selection of the Vibrator

For uniform, rearrangeable and water-saturated soils, a hammer should be selected with 15 kN centrifugal force as the minimum for each m driving depth and with 30 kN centrifugal force as the minimum for each 100 kg driving material mass.

$$F \stackrel{\wedge}{=} 15 \left(t + \frac{2\, m_{\text{Driving material}}}{100} \right) [\text{kN}]$$

where:
t = embedment depth [m]
$m_{\text{Driving material}}$ = driving material mass [kg]

In other cases – which are the most frequent in practice – a compromise should be aimed at with the aid of the above formula and use of the theoretical standard values stated under section 8.1.22.2. In the case of larger construction operations, a calibration test with theoretically suitable vibration hammers and an adequate number of different driving elements is recommended.

8.1.22.5 General Information on Experience Gained

(1) The action and effects of the vibration are virtually impossible to predict.

If the vibration is effective and penetration speeds of ≥ 1 m/min occur, no damaging effects need to be anticipated as a result of shock propagation.

In the event of penetration speeds ≤ 0.5 m/min, vibration should be terminated. Short-term lower penetration speeds, which can occur for example when passing through consolidated strata, should be accompanied by vibration measurements.

When driving work is to be carried out in the vicinity of buildings, a prediction should be drawn up in advance to select a machine which guarantees that the vibrations do not exceed the indicative values of DIN 4150, part 2/A1 and part 3. It must be noted that low penetration speeds can lead to heating and bonding together of the interlocks. Constant water cooling can prevent overheating.

(2) In soils which are not very amenable to rearrangement or in dry soils, jetting can be used (R 203, section 8.1.23).
Loosening bores with small advance intervals or soil exchange should also be considered as aids.

(3) The compaction effect mentioned in R 154, section 1.9.3.2, is more likely to occur at high revolutions per minute. It can be expedient in such cases to continue the work with an equivalent vibrator, but one which operates at a lower speed. Determination of a in accordance with section 8.1.22.2 (7) is useful here for orientation.

(4) Reference is made particularly to R 117, section 8.1.20.3 (1) b) regarding the achievable watertightness of artificially pre-sealed interlocks.

(5) R 118, section 18.1.11 applies accordingly for the execution. Vibration records should contain as a minimum the time of each 0.5 m penetration.

(6) Vibration is generally a low-noise method of embedment. High noise levels can occur with defective vibratory action as a result of covibration of the piling and driving jaw, through hitting against one another. Covibration can be intensive in the case of high piling and staggered or compartmental embedment. The use of an embedment aid in accordance with section 8.1.22.5 (2) or padded driving jaws can provide a remedy.

(7) The instructions in DIN 4150 for evaluating the resulting vibrations are to be taken into account.

(8) Even when using modern high-frequency vibrating machines with variable unbalances, the risk of settlement must be taken into account in the vicinity of buildings.

8.1.23 Jetting when Driving Steel Sheet Piles (R 203)

8.1.23.1 General

Reference is made to the embedment aid of a "water jet" in recommendations R 104, section 8.1.12.5, R 118, section 8.1.11.4, R 149, section 8.1.14.4, R 154, section 1.9.3.4 and R 202, section 8.1.22. To summarise, jetting can be used in the embedment methods of driving, vibrating and pressing, in order to:

a) generally facilitate the embedment,
b) prevent overloading of the equipment and overstressing of the driving elements,
c) achieve the statically required embedment depth,
d) reduce soil shocks,
e) reduce costs through shortening embedment times and reducing power requirements and/or enabling lighter equipment to be used.

8.1.23.2 Low-pressure Jetting Procedure

A water jet is directed at the base of the driving elements through flushing pipes. The subsoil is loosened by the water injected under pressure and the loosened material is carried away in the jetting flow. Essentially, the peak resistance is reduced hereby. Depending on the soil structure, the skin friction and the interlock friction are also reduced by the flowing, rising water. The method is limited by the strength of the subsoil, the number of jetting lances, the magnitude of the water pressure and the volume of water introduced. In order to establish the necessary parameters, it is recommended that trial driving operations are carried out before the method is used.

(1) Ratings:
- Jetting lances dia. 45–60 mm,
- Jetting pressure at pump 7 to 21 bar (50 to 70 bar in special cases),
- Required nozzle action can be obtained by constricting the tip of the lance or with special jetting heads,
- Water required: approx. 1000 l/min, delivered by centrifugal pumps.

The low-pressure jetting method is used for loose, closely-stratified soils, especially dry mono-grained soils or in sandy soils mixed with gravel. Depending on the difficulty of embedment, the jetting lances are flushed in next to the driving material, or secured directly to the driving material. Reductions in the soil properties and settlement may occur as a result of the introduction of relatively large volumes of water.

(2) More recent experience:
A special low-pressure jetting method, combining the jetting process with placing the driving material by vibration, has been used with success for some time. It enables sheet piles to be driven into very compact

soils, which would be very difficult to drive without this aid. Because of its environmental compatibility, the low-pressure jetting method is also used in residential areas and inner cities.

Success depends on the correct adjustment of the jetting, and selection of the vibrator for the soil encountered.

The vibrator should be selected in accordance with R 202, section 8.1.22, and equipped with adjustable effective moment and stepless speed regulations.

Usually, two to four lances of 45–60 mm diameter are secured to the driving material (dual pile). The lance tip ends flush with the base of the pile. The optimum arrangement is the use of one pump per lance.

Jetting begins simultaneously with vibration in order to prevent the opening of the lance from becoming clogged by the ingress of soil material. If a penetration speed of ≥ 1 m/min are achieved, jetting can be retained until the statically required embedment is reached. The soil properties previously determined for the sheet piling calculation generally then also applies. The transmission of high vertical forces requires trial loading.

8.1.23.3 High-pressure Jetting Method (HVT)

Ratings:
- Jetting lances (precision tubes), e.g. dia . 30 x 5 mm,
- Jetting pressure 250–500 bar (at pump),
- Special nozzles in screw-in nozzle container (generally circular-section jet nozzles, dia. 1.5–3 mm; occasionally, fan-jet nozzles may also be appropriate),
- Water required: 60 to 120 l/min per nozzle, supplied by reciprocating pump.

The high-pressure jetting method can enable sheet piles to be driven in rock of varying solidity. The relatively low volumes of water mean that high-pressure jetting is also an expedient driving aid under unfavourable circumstances, e.g. in areas subject to settlement risk. It is ideal for firmly stratified, highly pre-loaded cohesive soils, e.g. in silt and clay rocks and mellow sandstone.

Economic use is achieved only if construction operations are geared toward reclamation of the jetting lances. The lances are fed into pipe clamps welded to the pile and the lance jetting heads are secured to the sheet pile in such a way that the nozzle lies approx. 5 to 10 mm above the bottom elevation of the pile.

The HVT method can be adapted particularly well to local conditions. During the course of construction operations, intensive observations are required for fine adjustments to be made. For example, it may be necessary to change the nozzles frequently because of unusual wear, or to readjust the lance arrangement or the nozzle diameter to prevailing conditions resulting from changes in the soil.

8.2 Calculation and Design of Sheet Piling

8.2.0 General

The relevant actions (active earth pressure, passive earth pressure and water pressure) used for calculation of the internal forces (bending moments, axial forces, shear forces), support reactions and embedment depths in the subsoil which form the basis for designing the sheet piling, are determined according to section 2.0 as per DIN V ENV 1997-1. In the normal case, the following limit states are to be examined:

– limit state 1B, failure of the structure or a component due to exhaustion of material strength,
– limit state 1C, failure in the ground.

The most unfavourable calculated results of the internal forces, support reactions and embedment depth are to be used as basis for designing the sheet piling.
Verification of serviceability limit state, corresponding to LS 2 as per DIN V 1054-100, is characterised by actions which make the structure unserviceable without loss of the bearing capacity. Normally it does not need to be investigated for waterfront structures.
Verification of bearing capacity for steel sheet piling and anchoring elements of steel (verification of internal bearing capacity) is to be provided according to DIN 18 800, 11/1990 or DIN V ENV 1993-5 respectively.
Verification of bearing capacity for sheet piling of reinforced concrete or timber is to be provided fundamentally according to DIN V ENV 1992-1 or DIN V ENV 1995-1. In so far as the verification is provided according to the prevailing safety concept, i.e. the standards DIN 1045 for reinforced concrete structures and DIN 1052 for timber structures, respectively still in force, attention is drawn to the fact that the calculated internal forces S_d of the sheet piling calculation contain the partial safety factors of the new safety concept, so that the internal forces S_d, relevant for the verification of the bearing capacity, have to be reduced accordingly.

8.2.0.1 Partial Safety Factors γ for Determining the Internal Forces (Bending Moments, Axial Forces and Shear Forces) S_d Relevant for the Design

When calculating sheet piling structures, anchor walls and anchor plates (deadmen), verification of the limit states 1B and 1C is fundamentally based on the partial safety factors γ for actions as per table R 0-2. However, the partial safety factors for the soil properties in LS 1C may be selected as follows:

$$\gamma_\varphi = 1.1; \quad \gamma_c = 1.3; \quad \gamma_{cu} = 1.3$$

According to DIN V ENV 1997-1 section 2.4.1 (17), active earth pressure and passive earth pressure in limit state 1B are to be considered as a joint action (resulting load) to be multiplied by the same partial safety factor.

Section 0.2.3 applies to verification of serviceability limit state.

8.2.0.2 Determining the Relevant Internal Forces S_d for Unfavourable Load Combinations

Since the unfavourable load actions rarely coincide and an earth pressure redistribution takes place under certain circumstances as a result of deflection of the sheet piling wall, but on the other hand irregularities in the soil and also in the materials can have a negative influence, the internal forces S_d relevant for bearing capacity verification for the loading cases stated in R 18, section 5.4, may be determined by assuming the following reduction factors.

Loading case to R 18, section 5.4	LC 1	LC 2	LC 3
Sheet piling, anchor walls and wales	1.00	1.15	1.30
Round steel anchors and wale bolts	1.00	1.00	1.15

The prerequisite here are meticulous static calculations and structural design, perfect deliveries and proper execution of the construction work.

8.2.1 Sheet Piling Structures, Fully Fixed in the Ground Without Anchorage (R 161)

8.2.1.1 General

Contingent on the bending resistance of the wall, sheet piling without an anchorage can be economical, especially if there is comparatively little difference in ground surface elevation or when the installation of an anchoring or of another head bracing would be very costly and if relatively large displacements of the pile head are harmless.

8.2.1.2 Design, Calculation and Execution of Work

In order to attain the required stability, the following demands are to be particularly observed in the design, calculation and execution of work for sheet piling without anchorage. If in doubt, sheet piling without anchorage should be eliminated.

(1) The soil properties are to be carefully determined and must lie on the safe side along the entire wall, especially to ensure the fixity of the sheet piling in the ground.

(2) Also, other loads are to be ascertained as accurately as possible, such as compaction pressures when backfilling in accordance with DIN V 4085-100. This applies particularly to loads acting in the

upper section of the sheet piling, as these can substantially affect the fixing moment and the embedment depth.

(3) An exact ascertainment of the loading cases, especially of LC 2 and 3, as well as of unusual scour formation and special water pressure differences, must be possible without doubt.

(4) The bottom level may in no case fall below the designed bottom level in the sheet piling area. The bottom elevation (design depth) used for calculation is therefore to be fixed with adequate certainty from the very beginning.

(5) The statical calculation may be carried out according to BLUM [61], whereby active and passive earth pressures are applied in "classic distribution", taking R 4, section 8.2.4 into consideration.

(6) The respective theoretical driving depth, determined according to R 56, section 8.2.9, must be absolutely attained in the construction work.

(7) Verification of the equilibrium of the vertical forces must always be provided.

(8) If the load assumptions, soil properties and other important facts have been clarified, the translation and deflection of the wall are to be investigated and checked for their compatibility with the structure itself, with the subsoil (e.g. in respect of the formation of cleavages in cohesive soils on the active earth pressure side which could fill up with water) and with the other project requirements. This is especially applicable when overcoming larger differences in ground levels. Backfilled sheet piling behaves more favourably, as displacement and deflection of the wall occur in the construction state and are therefore generally harmless. Should the loads in the construction state become larger than in the final state, e.g. due to compaction pressures according to section 8.2.1.2 (2) and thus become a determining factor for translation and deflection of the wall, only insignificant changes need be expected in the final state.

(9) The translation of the sheet piling depends on the subsoil and on the extent of the utilisation of the passive earth pressure. The deflection of the wall coping is contingent on the rotation and translation of the earth support and on the elastic deflection of the wall. In general, the first named actions prevail.

(10) The driving inclination of the sheet piling is generally selected so that overhanging of the wall coping is avoided with certainty under maximum load and thus maximum deflection.

(11) If a concrete slab is placed during the construction stage and if this also serves to brace the sheet piling at the same time, this considerably shortens the effective cantilever of the wall for all loads occurring after the placing of the concrete slab.

(12) The coping of the sheet piling without anchorage shall be provided with a sturdy load-distributing beam of steel or reinforced concrete, in order to prevent non-uniform translations and coping deflections as far as possible.

8.2.2 Calculation of Single-anchored Sheet Piling Structures (R 77)

The sheet piling calculation method of BLUM [7] and [6] has proven successful. When the prerequisites are satisfied, bearing capacity and embedment depth are to be verified with redistribution (readjustment) of the classical earth pressure as per COLOUMB according to figs. R 77-1 and R 77-2.

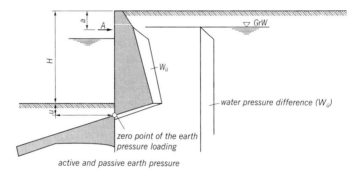

Fig. R 77-1. Design values of the earth pressure and water pressure difference

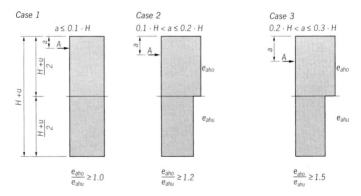

Fig. R 77-2. Redistribution of the earth pressure

As a rule, the illustrated redistribution of the earth pressure takes place above the zero point of the earth pressure loading.
Redistribution of the earth pressure is only permissible when the anchoring

(1) consists of driven, bored or grouted piles which are proven to undergo only slight head movement and which are tightly (i.e. without slippage) connected to the structure,
(2) consists of anchors which are slightly or not at all pre-tensioned, but tightly connected to the structure, with the anchors undergoing only slight deformation under the design load,
(3) consists of grouted anchors according to DIN 4125 which have been tested to section 11 of that DIN, pre-tensioned and connected to the structure with at least 80 % of the design load,
(4) consists of anchors whose head movement is indeed not negligibly small but where an earth pressure redistribution can still take place following a comparatively large deflection of the sheet piling as well as translation of its embedded portion.

In this context, reference is made to EAB-100, issue 1996 [186].
Earth pressure redistribution is however not permissible

(1) when the sheet piling between bottom and anchoring is mainly backfilled and subsequently not dredged sufficiently in front of the sheet piling to allow for additional deflection,
(2) when there is cohesive soil behind the sheet piling which is not adequately consolidated,
(3) when the soil surface behind the sheet piling does not nearly reach the elevation of the anchoring.

Reduced earth pressure redistribution must also be presumed with increasing flexural rigidity, i.e. with thick reinforced concrete diaphragm walls. In the case of waterfront structures with particularly large flexural rigidity, a deformation study is to be carried out as per DIN V 4085-100, section 6 with characteristic values of actions and resistances to see whether earth pressure redistribution may take place or not.
Sheet pilings can also be calculated using horizontal moduli of subgrade reaction [67], [68], [69], [70], and [71]. Here it should be noted that the soil reaction stresses cannot be larger than the passive earth pressure stresses calculated with the partial safety factors.
Once sheet piling walls have undergone deflection, the slippage of soil will allow only a partial retrogression of this deflection. The effects of the construction states on the final state are therefore to be taken into account if they have a decisive influence on the stresses and/or on the deformation of the completed structure.

For high prestressing of high-strength steel anchors, R 151, section 8.4.13 gives information on the earth pressure redistribution and anchor force increase then to be applied.

8.2.3 Calculation of Double-anchored Sheet Piling (R 134)

In contrast to R 133, section 8.4.7 which deals with questions about auxiliary anchoring, true double-anchored sheet piling are discussed here (figs. R 134-1 and -2). For such structures, both anchors are valid components of the total structure and therefore exercise a constantly effective bearing function. The total load on the sheet piling wall from active earth pressure and water pressure difference, apart from earth support, is carried off by both anchors together. Due to the difference in load and the nature of the statical system, the preponderant portion of the load to be anchored is absorbed by the lower anchor B.

Both anchors are run to a common anchor wall and connected to it at the same elevation (figs. R 134-1 and R 134-2). In verification for stability, the anchor direction is taken to be that resulting from anchor forces A and B. With separate anchors (for instance, with grouted anchors according to DIN 4125), both anchors are to be included independently of each other in the stability verification. Verification of stability of the anchoring is to be provided according to DIN V 4084-100.

The calculation of double-anchored sheet piling walls may be carried out graphically or analytically according to LACKNER [73]. Displacements of anchor points A and B and the end supports at the point of the sheet piling can thereby be taken into account without difficulty.

The resultant load areas from active earth pressure and water pressure difference are applied in different ways in this calculation. It is recommended to compute active earth pressure and passive earth pressure ini-

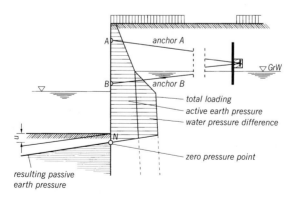

Fig. R 134-1. Load assumptions in sheet piling computation without active earth pressure redistribution upward

tially according to the classical theory, taking account of R 4, section 8.2.4, and water pressure difference according to R 19, R 113 and R 114, sections 4.2, 4.7 and 2.9 (fig. R 134-1). When calculating the internal forces (bending moments, axial forces and shear forces) including anchor force B, the earth pressure redistribution indicated in R 77, section 8.2.2 applies accordingly. In this case, the earth pressure redistribution applies to the earth pressure between the middle of the span $A - B$ and the zero pressure point. This results in the relevant stresses for support in the soil, the anchor force B and the sheet pile wall. The internal forces and support reactions above anchor layer B are to be calculated without earth pressure redistribution.

Frequently however, certain conditions prevail in which a decisive upward redistribution of the active earth pressure occurs. This is especially the case for waterfront structures, when the sheet piling structure is erected in high-lying ground and the anchors are installed as prestressed grouted anchors. When such a redistribution is to be expected, an additional calculation is carried out with loads according to fig. R 134-2. The active earth pressure above the zero pressure point N according to fig. R 134-1 is taken into account here in the form of an equivalent rectangle, and the calculation is also carried out according to LACKNER [73], for the conditions in fig. R 134-2.

In each case, the least favourable results are used as a basis for the design.

The calculations can also be carried out using horizontal moduli of subgrade reaction [67], [68], [69], [70] and [71]. Here it should be noted that the soil reaction stresses cannot be larger than the mobilisable passive earth pressure stresses, calculated with the partial safety factors.

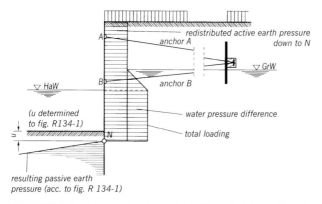

Fig. R 134-2. Load assumptions in the sheet piling calculation with active earth pressure redistribution upward

However, a calculation according to R 133, section 8.4.8 must nevertheless be undertaken by way of comparison, only for the design of the wall coping and of anchor A. The result must be used as a basis for the design of these parts, as the case may be.

Otherwise, verification of bearing stability, calculation and design of the anchors, wales, capping beams etc. and other design elements are subject to the pertinent recommendations of the EAU.

8.2.4 Assumed Angle of Wall Friction and of Adhesion (R 4)

8.2.4.1 Reference is made to DIN V 4085-100 as regards assumptions of the direction of earth pressure (angle of wall friction).

8.2.4.2 If the sheet piling is subject to a large water pressure difference, to inclined anchor tensile forces directed upward or the like, the direction or sign of the wall friction angle δ_p must be checked and adjusted where necessary, in the course of verification of the equilibrium of vertical forces. Any resulting reduction in passive earth pressure must be taken into account.

8.2.5 Assumed Angle of Wall Friction of Passive Earth Pressure for an Inclined Embankment of Non-cohesive Soil in Front of Sheet Piling (R 199)

The magnitude of the passive earth pressure for an inclined embankment (β negative) depends greatly on the angle of inclination β of the surface of the embankment. Even the smallest inclinations produce a perceptible reduction in passive earth pressure.

In this case, the passive earth pressure coefficients may not be determined with plane but with curved failure surfaces, e.g. as per DIN V 4085-100.

8.2.6 Bearing Stability Verification of Sheet Piling Structures (R 20)

8.2.6.1 Quay Wall

(1) Predominantly static stress
Bearing capacity verification is to be provided here as per DIN 18 800, 11/1990, and/or DIN V ENV 1993-5. respectively.

(2) Predominantly alternating stress (direction of stress reversed)
If, in special cases, the sheet piling is not backfilled, i.e. not statically stressed by active earth pressure but rather dynamically stressed as by wave action, so that a large number of stress cycles occur in the course of time, fatigue assessment procedure is to be provided according to DIN 19 704-1. In addition, reference is made to DIN 18 800, 11/1990, part 1, El. (741).

In order to prevent adverse action from the notch effect, such as from structural welding seams, tack welds, unavoidable irregularities on the surface due to the rolling process, pitting and the like, killed steels as per DIN EN 10 025 should be used, see R 67, section 8.1.6.4

8.2.6.2 Anchor Wall, Wales, Capping Beams and Base Plates

(1) Predominantly static stresses
Section 8.2.6.1 (1) applies to verification of bearing capacity.
In the case of wales and capping beams, where applicable the partial safety factors may have to be increased according to R 30, section 8.4.2.3.

(2) Predominantly alternating stresses
Section 8.2.6.1 (2) applies to verification of bearing capacity. Turned structural bolts of at least strength class 4.6 are to be used for bolted connections in wales and capping beams. Fatigue assessment procedure is to be provided according to DIN 18 800, 11/1990, part 1, section 8.2.1.5, El. (811).

8.2.6.3 Round Steel Anchors and Wale Bolts

(1) Predominantly static loads
The materials for round steel anchors and wale bolts are stated in R 67, section 8.1.6.4.
The design is based generally on the anchor forces arising from the loads as per loading case 2 (R 18, section 5.4.2). Verification of bearing capacity is to be provided according to DIN 18 800, 11/1990, part 1, section 8.2.1.3. The relevant reduction factors for loading cases 1 to 3 as per section 8.2.0.2. are to be taken into consideration. DIN 4125 applies to the design and execution of sheet piling anchoring with grouted anchors.
Cut, rolled or hot rolled threads can be used as per R 184, section 8.4.8. Prerequisite for this is that the anchors are equipped with proper flexible connections and perfectly installed to take the best possible consideration of any settlements or sagging by sloping them upwards. Upsetting in the thread area of the anchor rod as well as round steel anchors with hinged eyes is only permissible:

- if using steel grades J2G3, J2G4, K2G3 and K2G4, but no thermo-mechanically rolled steels of grades J2G4 and K2G4 (R 67, section 8.1.6.1 must be taken into account),
- if other steel grades such as e.g. S 355 J0 are used, accompanying tests must ensure that the strength values after the normalisation procedure of the forging process do not fall below those required in DIN EN 10 025,
- if the mechanical and technological values according to the selected steel grade prevail in all areas of the anchor,

- if this does not affect the course of the fibre,
- if detrimental structural disturbances are thereby safely avoided.

(2) Predominantly fluctuating stresses (direction of stress not reversed)
Anchors are generally subjected to static loads. Highly fluctuating stresses only occur in anchors in exceptional cases (see section 8.2.6.1 (2)) but occur more frequently in wale bolts.

Only specially killed steels to DIN EN 10 025 may be used where fluctuating stresses occur.

Verification of the bearing capacity is to be provided according to DIN 18 800, 11/1990, part 1, section 8.2.1.5.

If the basic static load is equal to or less than the stress amplitude, it is recommended that the anchors or wale bolts are permanently prestressed in a controlled way beyond a value which prevents the anchors or wale bolts from becoming stress-free (loose) and subsequently failing at load increase under a suddenly applied, severe stress.

A certain prestressing is applied to all anchors and wale bolts for installation reasons, even if the exact amount required cannot be determined. In such cases without controlled prestressing, only a stress $\sigma_{Rd} = 80$ N/mm^2 may be assumed for the thread of the anchors or wale bolts, regardless of loading class and steel grade, neglecting the prestressing.

Care must be taken in any case that the nuts of the wale bolts cannot loosen at the changes in stress.

8.2.6.4 Steel Cable Anchors

Steel cable anchors are only used for predominantly static loads. They are to be designed in a way that at:

loading case 2: the condition $N_d = N_{failure} / 1.5$ and at
loading case 3: the condition $N_d = N_{failure} / 1.3$ are fulfilled
(N = tensile force of the anchor).

The mean modulus of elasticity of patented steel cable anchors should not be below 150 000 MN/m^2 and must be guaranteed by the manufacturer to within ± 5 %.

8.2.7 Consideration of Axial Loads in Sheet Piling (R 44)

In general, sheet piling is stressed chiefly by bending. If an additional compressive force acts in the axis of the wall, verification of bearing capacity is to be provided according to DIN 18 800, 11/1990, or DIN V ENV 1993-5, respectively.

The bending stress of the sheet piling can be positively influenced by eccentric introduction of the vertical load P.

In verification of the bearing capacity, the effective length for consideration of buckling can generally be taken with adequate accuracy as the spacing between the zero points delimiting the bending moment of the span due to the horizontal loading.

Reduced cross-section values are to be considered if extreme corrosion is to be expected in the area of the largest bending moment.

8.2.8 Selection of Embedment Depth of Sheet Piling (R 55)

Construction considerations, operational and economic requirements are relevant to the selection of the embedment depth for sheet piling, along with theoretical structural calculation. Any foreseeable future deepening of the harbour and the danger of scour should be considered along with possible increases in vertical sheet piling loads. Furthermore, the need for safety from slope and foundation failure, failure by heave and erosion influences the required embedment depth.

These requirements usually result in such a large minimum embedment depth that, apart from foundations in rock, full or partial fixity in the earth will exist in permanent structures. However, if in exceptional circumstances with a non-rocky subsoil, free earth support should be sufficient, it is still advisable to enlarge the embedment depth as a precaution. Since in any case full use should be made of the pile section under consideration, partial fixity in the earth is frequently the most practical solution. It can be calculated using BLUM's equivalent force method [61] without any difficulties.

If the sheet piling is also subject to large axial loads, a sufficient number of driving units are driven to such a depth as bearing piles so that the loads can be safely transmitted into the load-bearing subsoil. When the piles themselves are used to transfer the axial loads, it should be noted that with an increase in the vertical load for balance reasons, the wall friction angle can change its sign. This must then also be taken into account when calculating the earth pressure.

8.2.9 Determination of Embedment Depth with Partial or Full Fixity of the Point of the Sheet Piling (R 56)

The limit state 1C is applicable to the embedment depth of the sheet piling wall. Here the soil properties and all resulting values are to be assumed with the design values.

8.2.9.1 If sheet piling is calculated according to BLUM's equivalent force method for full or partial fixity in the earth, the extra length Δx required for absorption of the equivalent force C_h can be approximated in the calculation with $\Delta x = 0.2\ x$. But this extra length can increase up to $\Delta x = 0.5\ x$ in walls subject primarily to water pressure difference and in unanchored (cantilever) walls.

The extra length for non-cohesive soils can be more accurately calculated according to the further development of the formula of LACKNER [73] with the following equation:

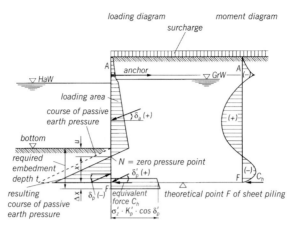

Fig. 56-1. Equivalent load at full fixity in the earth

$$\Delta x = \frac{C_h}{2 \cdot \sigma'_z \cdot K'_p \cdot \cos \delta'_p}$$

With reference to fig. R 56-1, the symbols mean:

- t = $u + x + \Delta x$ = required embedment depth [m],
- u = depth of zero pressure point N below the bottom in front of the sheet piling [m],
- x = depth of theoretical point F of sheet piling below N [m],
- Δx = extra length for assumed equivalent force C_h [m],
- σ'_z = design value of the vertical soil stresses in the depth F taking account of the surcharge [kN/m²],
- δ_p = angle of wall friction of the passive earth pressure in front of the point of the sheet piling according to R 4, section 8.2.4 [degree],
- K'_p = K_p value for the soil behind the sheet piling in area F for angle of wall friction δ'_p,
- δ'_p = angle of wall friction in area F behind the sheet piling [degree],
- C_h = horizontal component of the equivalent force according to BLUM [61] [kN/m].

Deviating from R 4, section 8.2.4, the angle of wall friction is assumed in general $\delta'_p = +1/3\ \varphi'$, unless verification of stability requires $\delta'_p > +1/3\ \varphi'$, or conversely if a large axial sheet piling force causes $\delta'_p < +1/3\ \varphi'$. δ'_p can be taken up to a value of $-2/3\ \varphi'$. For larger negative angles δ'_p, verification of bearing capacity must also take into account that calculation according to BLUM [61] results in too large a C-force, because the resulting passive earth pressure is calculated as being fully active down to the theoretical sheet piling point F. To compensate for this error, an equivalent additional force must be applied on the C-force side (in order to reinstate equilibrium). Actually, C acts therefore in half the arithmetically determined magnitude. In verification of the vertical bearing capacity, the total vertical force acting upward is therefore to be reduced by the vertical component of $2 \cdot C_h / 2 = C_h$ at full application of the passive earth pressure. On the other hand, a certain amount of point resistance at the sheet pile point may be considered, insofar as the conditions for assuming point resistance prevail according to R 33, section 8.2.11.

8.2.9.2 For cohesive soils, the same procedure is used accordingly, taking account of the corresponding consolidation state (c_u or φ' and c' respectively).

8.2.10 Staggered Embedment Depth of Steel Sheet Piling (R 41)

8.2.10.1 Application

Alternate driving units (generally double piles) are frequently driven to different depths for technical and economical reasons. The extent of this staggering (difference in embedment) depends on the stresses in the region of the point of the longer piles and on structural considerations. For technical driving reasons, staggering within a driving unit is not recommended for sheet piles having U- or Z-shaped sections.

A uniform continuous passive earth pressure sliding zone forms along the point of staggered sheet piling, just as for close lying anchor plates. Therefore, the passive earth pressure can be assumed to be effective to the very tip of the deeper sheet piles. However, the bending moment which is present at the end of the shorter piles must then be absorbed by the longer piles alone. In corrugated steel sheet piling therefore, only adjacent driving units (at least double piles) are staggered (figs. R 41-1 and R 41-2). A length of 1 m is usual. In practice, it has been found that a statical check in this case is unnecessary. For greater staggering, verification of stress absorption is required from bending moment, shear force and axial force.

8.2.10.2 Wall Fixed in the Earth

Staggering can be fully utilised for saving steel in walls which are fixed in the earth (calculation according to BLUM [61]). Only the long sheet piles need to reach down to the assumed lower edge of the wall (fig. R 41-1).

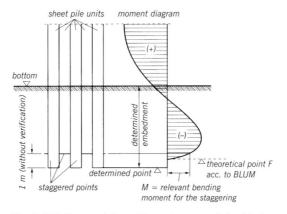

Fig. R 41-1. Staggered sheet piling points at a wall fixed in the earth

8.2.10.3 Free Earth Support Wall

When the sheet piling is freely supported in the earth, staggering may be used only to increase the safety of the earth support (fig. R 41-2).

In this case, the longer sheet piles must be driven deeper by that length by which the shorter piles end above the determined point. If the staggering length is more than 1 m, verification of bearing capacity must be provided according to fig. R 41-2.

The same applies to reinforced concrete or timber sheet piling, providing the pile joints have sufficient strength to guarantee that the shorter and the longer sheet piles will act together.

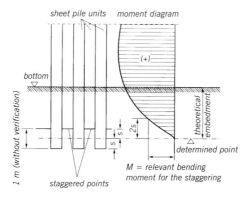

Fig. R 41-2. Staggered sheet piling points at a free earth support wall

8.2.10.4 Combined Sheet Piling

The conditions are different for sheet piling composed of bearing and intermediate piles (see R 7, section 8.1.4). The zero point of the load diagram indicates the approximate location of lower edge of the intermediate piles, but water pressure difference, safety from failure by heave (R 115, section 3.2) and scour danger must also be considered. The embedment depth of the intermediate piles for high harbour walls in good bearing soil should be at least 2.50 m, for low walls with slight water pressure difference, it should be at least 1.50 m.

If soft or very soft layers prevail in the harbour bottom area, the embedment depth of the short or of the intermediate piles respectively should be determined by special investigations.

8.2.11 Vertical Load Bearing Capacity of Sheet Piling (R 33)

8.2.11.1 General

Sheet piling can support axial loads just like other piles, if they are well embedded in firm soil and become adequately firm when driven in. In steel sheet piling in general, the vertical load bearing capacity increases in time as a result of progressive incrustations on the steel surface.

8.2.11.2 Vertical Equilibrium

All unfavourable influences have to be taken into account when evaluating the vertical equilibrium, such as the vertical component of active earth pressure, the vertical anchor force component, vertical surcharges and the vertical components of the passive earth pressure etc. Under great compressive load and fixity of the sheet piling in the soil, the wall friction angle both of the active earth pressure and the equivalent force C calculated as per BLUM [61] can become negative.

8.2.11.3 Verification of Bearing Capacity

Verification of bearing capacity $S_d \leq R_d$ is to be provided as per DIN V 1054-100.

S_d = design value of the sheet piling axial force including the vertical earth pressure component,

R_d = totality of the design values of the axial resistances including the vertical components of the passive earth pressure and the C-force at the point of the sheet piling.

The comments about the magnitude of C according to R 56, section 8.2.9.1, must be considered in the determination of the axial forces.

8.2.11.4 Axial Compressive Resistance

The axial resistances R of the sheet piling, or the sheet piling bearing elements respectively, for absorbing compressive loads are to be determined as per DIN V 1054-100. On mobilising the axial resistances, it should be noted that skin friction forces are already mobilised after only slight relative slip, whereas point resistance requires large movements, unless the driving elements are adequately firm after being driven on the basis of local experience. If the formation of plugs cannot be expected within the bearing elements, the otherwise allowable point resistance may be considered at only half the magnitude. This is not necessary when the driving depth is increased by at least 1.5 m, or when flat or sectional steels are welded into the piling point for plug formation. The latter are to be welded into the point in a manner that the driving process is affected as little as possible, but the welded-in elements survive undamaged and can transfer the occurring pile loads with adequate safety. In the case of stout corrugated or box-shaped profiles, the point resistance may be applied to the area delimited by the enveloping sides of the wall/pile cross-section.

In the case of combined sheet piling with I-shaped bearing piles, the mentioned flat or sectional steels can be welded into the point to promote plug formation when the clear width between flanges of the bearing piles exceeds 400 mm. The area is to be reduced in sheet piling of corrugated profiles with a mean web spacing of 400 mm. The skin friction resistance taken into account, must be compatible with the assumptions in the earth pressure calculation. This means that the sign of the wall friction angle of the active earth pressure can change.

8.2.11.5 Axial Tensile Resistance

The same prerequisites as for compressive loads also apply here accordingly. Following unfavourable effects according to section 8.2.11.6, tensile loads on waterfront sheet piling should be avoided as far as possible.

8.2.11.6 Influence on the Sheet Piling Calculation

An axial compressive load has a favourable effect on the passive earth pressure in the sheet piling calculation. An increase in the active earth pressure occurs if the vertical movement of the sheet piling is the same as or larger than that of the earth pressure slip wedge.

The active earth pressure may decrease under axial tensile load, but the resisting passive earth pressure decreases relatively much more.

Therefore verification of the vertical equilibrium is required in any case for tensile loads in the sheet piling calculation, see also R 4, section 8.2.4.2.

8.2.12 Capacity of Steel Sheet Piling to Absorb Horizontal Longitudinal Forces Acting Parallel to the Shore (R 132)

8.2.12.1 General

Combined and corrugated steel sheet piling is comparatively yielding to horizontal forces acting parallel to the shore. When such loads occur, a check must be made as to how these forces are absorbed by the sheet piling and to determine whether additional measures are required.

In many cases, longitudinal stressing of sheet piling structures deriving from active earth pressure and from water pressure difference can be avoided, if an appropriate design is chosen, for example, by crossed anchoring of quay wall corners according to R 31, section 8.4.11, or in the case of circular quay wall corners or pier heads, by radial anchoring to a slab laying at the centre of the circle. This slab is in turn held fast in the direction of the bisectors of the sector angles, by further anchors to a transverse anchor wall further to the rear.

8.2.12.2 Transmission of Horizontal Longitudinal Forces into the Sheet Piling

The transmission can take place by the available construction members such as capping beam and wale, if these are suitably designed, or by additional measures such as the installation of diagonal braces behind the wall. Under certain circumstances, welding the interlocks in the upper section will suffice.

The longitudinal forces from line pulls act at the mooring devices, the major longitudinal forces from wind at the wheel arresting points of cranes and those from ship friction at the fenders. Furthermore, frictional forces may act at any point of the wall. This also applies to the wall coping with regard to the longitudinal forces from crane braking. The construction elements which transmit longitudinal forces to the sheet piling can be designed so that this transmission will be distributed over a considerable length of the wall instead of concentrating it in a short section.

For this purpose, the flanges of steel wales should be bolted or welded to the soil-sided sheet piling flanges (fig. R 132-1).

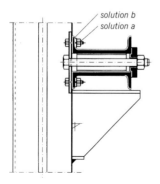

Fig. R 132-1. Transmission of longitudinal forces by turned structural bolts in the wale flanges (solution a) or by welding seams (solution b)

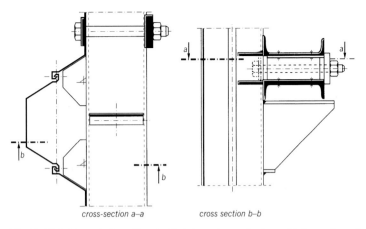

Fig. R 132-2. Transmission of longitudinal forces by steel strips welded onto the wale

The transmission of longitudinal forces can also be effected by steel strips which are welded onto the wale and braced against the sheet piling webs (fig. R 132-2).

When the wale consists of 2 U-channels, the wale bolts can be used to transmit longitudinal forces only when both wale channels are joined by a vertically welded plate with holes where they rest on the soil-sided sheet piling flanges. The force from the wale bolts is transmitted into this vertically welded-in plate with the holes drilled in advance, causing bearing stresses at the plate, whereas the bolts are subject to shearing stresses (fig. R 132-3).

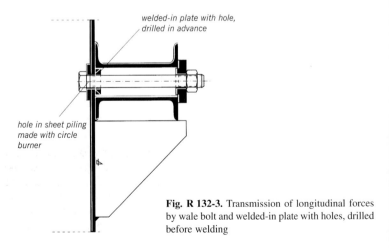

Fig. R 132-3. Transmission of longitudinal forces by wale bolt and welded-in plate with holes, drilled before welding

When longitudinal forces occur, the capping beam and the wale, inclusive of their joints, should be dimensioned for both direct and shear stress. The sheet piling wall must be embedded adequately in the capping beam in order to absorb the longitudinal forces from a reinforced concrete capping beam. The concrete in this embedment zone must be suitably reinforced for force transmission.

8.2.12.3 Transmission of the Horizontal Forces into the Ground Through the Sheet Piling

The horizontal longitudinal forces are transmitted into the ground by friction on the land-sided sheet piling flanges and by resistance in front of the sheet piling webs. The latter however may not be of greater magnitude than the friction in the ground for the length of the sheet piling trough.

In non-cohesive soils, the force absorption can therefore be calculated entirely with reference to friction, for which a reasonable arithmetical mean value of the friction coefficient between earth and steel as well as between earth and earth is assumed to be the friction coefficient. This force transmission in non-cohesive soils is all the more effective, the larger the angle of internal friction and the larger the degree of density of the backfill are, or in cohesive soils, the higher its shear strength and its consistency are.

The additional bending moments of a sheet piling wall from longitudinal forces, introduced through the capping beam or the wale and transmitted into the ground in the described manner, can be calculated in the same way as are the bending moments in a fixed or freely supported anchor wall. Instead of the passive earth pressure however, the aforementioned average wall friction or a corresponding shear resistance are assumed in this case.

As a rule, only shear resistant welded double piles should be considered as bearing elements for the absorption of these additional stresses. Unwelded piles may be considered only as single piles.

When absorbing horizontal forces parallel to the shore, the sheet piles are stressed in two planes by bending. The occurring superimposed stresses may exceed the allowed limit for axial stress $\sigma_{R,d}$ by 10 % according to DIN 18 800, 11/1990.

By taking into account wall friction in the horizontal direction, in the determination of the active earth pressure for the sheet piling calculation only a reduced wall friction may be applied. The vectorial combination of the two components may not exceed the maximum possible wall friction.

8.2.13 Calculation of an Anchor Wall Fixed in the Earth (R 152)

Obstacles in the subsoil, such as conduits, supply lines and the like, sometimes mean that it is not possible to erect the anchor walls with a central anchor connection. The high-lying anchor must then be connected in the upper area of the anchor wall, which leads to an increase in the cost of the anchorage. Above all, the anchor wall becomes substantially longer and heavier.

Verification of stability as well as calculation and design of the anchor wall are to be carried out according to the following assumed loads:

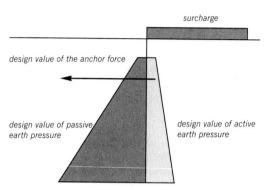

Fig. R 152-1. Load assumption for stability verification of an anchor wall fixed in the earth

The calculation is to be executed for limit states 1B and 1C. The partial safety factors are stated in section 8.2.0.1 and the reduction factors in section 8.2.0.2.

The required extra embedment depth Δx is calculated according to R 56, section 8.2.9, the passive earth pressure distribution in front of the wall under consideration of the vertical equilibrium.

Staggering of the anchor wall according to R 42, section 8.2.14 is allowable only at the lower end, but can be carried out here up to 1.0 m, without any special verification, at deep-reaching anchor walls.

In order to avoid any appreciable water pressure difference at predominantly horizontal groundwater flow, an adequately large number of weepholes must be foreseen in the anchor wall. Otherwise, the water pressure difference must be taken into account in the determination of the resulting anchor resistance.

8.2.14 Staggered Design of Anchor Walls (R 42)

In order to save on materials, anchor wall piling may be staggered in the same manner as the waterfront sheet piling. Either the bottom or the top end, or both ends may be staggered. In general, the stagger should not be more than 0.50 m. When the wall is staggered at both ends, double piles may be made 0.50 m shorter than the designed height of the anchor wall.

They are driven alternately, so that every second double pile lies at the designed level of the upper or lower edge of the anchor wall with its higher upper or deeper lower end, respectively. A greater staggering than 0.50 m is permissible only for deep lying anchor walls, if the load absorption in the ground and the stress absorption from moment, shear force and axial force in the sheet piling wall can be proven. Verification of the bearing capacity is also required for a stagger of 0.50 m if the height of the anchor wall is less than 2.50 m. It must also be proven here that the upper and the lower anchor wall moments are distributed among the piles.

A similar procedure may be used for reinforced concrete and timber piles, if the pile joints have sufficient strength to ensure that all the piles act together.

8.2.15 Steel Sheet Piling Driven into Bedrock or Rock-like Soils (R 57)

8.2.15.1 When bedrock shows a fairly thick decomposed transition zone, with solidity increasing with depth, or when the rock is soft, experience has shown that steel sheet piles can be driven so far into the rock to achieve at least sufficient free support.

8.2.15.2 In order to make it possible to drive sheet piles into bedrock, the piles must be modified and strengthened at the point, and if need be, also at the top, depending on the pile section and the type of rock. It is recommended that the sheet piling be of special steel S 355 GP (formerly St Sp S) (R 67, section 8.1.6), in consideration of the high driving energy required. Heavy hammers and a correspondingly smaller drop height are effective for this work. A similar effect can be achieved by use of hydraulic hammers, the impact energies of which can be regulated in a controlled manner in line with the particular driving energy requirement (R 118, section 8.1.11.3).

If there is strong, hard rock up to the upper surface, test piles and rock investigations are indispensable. If necessary, special measures must be taken for protection of the pile point and for ensuring proper alignment. Borings of diameters 105 mm to 300 mm are sunk at a spacing corresponding to the pile width of the sheet piling wall to perforate and destress the subsoil in such a manner as to facilitate driving of the sheet piles.

The same effect can be achieved by high-pressure jetting in the case of rock with changing hardness and the like. (See R 203, section 8.1.23.3).

8.2.15.3 If deeper driving embedment into bedrock is required, precision blasting can be used to loosen the bedrock in the area of the sheet piling and facilitate driving. In selecting the section and the steel grade, possible irregularities in the foundation soil and resulting driving stresses must be taken into account. For loosening blastings, reference is made to R 183, section 8.1.10.

Additionally, specific pre-boring on its own is possible, offering the advantage of retaining the rock properties in an undisturbed state. Compared to blasting, this has positive effects on the lower support reaction of the sheet piling. The pre-boring depth is also less than the depth requiring blasting.

8.2.16 Waterfront Sheet Piling in Unconsolidated, Soft Cohesive Soils, Especially in Connection with Undisplaceable Structures (R 43)

For various reasons, harbours and industrial plant with waterfront structures today must sometimes be constructed in areas with poor foundation soil. Existing alluvial cohesive soils, sometimes with layers of peat, are surcharged by a layer of fill and thus altered into an unconsolidated state.

The resulting settlement and horizontal displacement require special construction features and a structural design treatment that is best suited to the particular site.

In unconsolidated, soft cohesive soils, sheet piling structures may be built as "floating" only when neither the serviceability nor the stability of the entire structure and its parts will be endangered by the resulting settlement and horizontal displacement and/or their differences. In order to evaluate the conditions and to adopt the required measures, the expected settlement and displacement must be calculated.

If a quay wall is constructed in unconsolidated, soft cohesive soils, in connection with a structure on a foundation with practically no settlement to be expected, such as a rigidly founded structure on piles, the following solutions may be used:

8.2.16.1 The sheet piling may be anchored or supported so that it is free to move in a vertical direction so that the stability or operability of the connection to the structure is not impaired. The solution is quite straightforward apart from the settlement and displacement calculations. For operational reasons however, it can generally be used at structures on piles only for sheet piling lying behind them. The vertical friction force occurring at the support must be taken into consideration in the design of the piles supporting the structure. Slotted holes are not sufficient at the anchor connections of front sheet piling. In fact, freely sliding anchoring is then required.

8.2.16.2 The sheet piling is supported against vertical movements by driving a sufficient number of driving units deep enough to penetrate into the bearing, deep-lying foundation soil. In this case all vertical loads of the sheet piling must be safely absorbed by the deeper-driven piles, namely:

(1) the dead load of the wall,
(2) the clinging of the soil on the sheet piling due to negative wall friction and adhesion, and,
(3) a possible axial load on the wall.

The solution is technically and operationally practicable in case of forward-positioned piling. Since the soil which is settling clings to the sheet piling, the active earth pressure decreases. If the supporting soil in front of the sheet pile point also settles, the potential passive earth pressure also decreases, due to negative wall friction. This must be taken into consideration in the sheet piling calculation.

In calculating the vertical load on the sheet piling arising from soil settlement, negative wall friction and adhesion for the initial and final state are taken into consideration.

8.2.16.3 Apart from the anchoring or support against horizontal forces, the sheet piling should be so suspended from the structure that the loads mentioned in section 8.2.16.2 are transmitted to the structure and then into the load-bearing soil.

In this solution, the sheet piling and its structural suspension elements are calculated in accordance with the information in section 8.2.16.2.

8.2.16.4 If the load bearing soil is not at an excessive depth, the entire wall is driven down into the load-bearing soil. The passive earth pressure in compact soil is calculated with the usual wall friction assumptions and the partial safety factors as per table R 0-2 and the reduction factors as per section 8.2.0.2 and taken into full consideration. Only partial mobilisation of the passive earth pressure may be presumed for the passive earth pressure in the soft yielding soil above, see DIN V 4084-100, section 10.

8.2.17 Effects of Earthquakes on the Design and Dimensioning of Waterfront Structures (R 124)

Dealt with in section 2.13.

8.2.18 Design and Dimensioning of Single-anchored Sheet Piling Structures in Earthquake Areas (R 125)

8.2.18.1 General

Careful checks must first be made on the basis of the soil findings and the soil mechanics investigations, as to the effect which the vibrations occurring during a probable earthquake may have on the shear strength of the subsoil.

The results of these investigations can be the decisive criteria for the design of the structure. For example, when soil conditions are such that liquefaction must be expected as per R 124, section 2.13, no high-lying anchor wall or anchor plates may be used unless the mass of earth supporting the anchoring is adequately compacted in the course of the construction work, and the danger of liquefaction thus eliminated.

Reference is made to R 124 section 2.13 regarding the magnitude of the seismic coefficient k_h and other effects, as well as the allowable effects of actions and the required safety factors.

8.2.18.2 Sheet Piling Calculation

Taking into consideration the sheet piling loads and supports determined according to R 124, sections 2.13.3, 2.13.4 and 2.13.5, the calculation can be carried out as per R 77, section 8.2.2, but without redistribution of the active earth pressure.

The active and passive earth pressures determined with the imaginary angles of inclination are generally used as a basis for the calculations, although tests have shown that the active earth pressure from an earthquake does not increase linearly with the depth, but rather is comparatively higher near the ground surface. The anchoring is therefore to be generously dimensioned.

8.2.18.3 Sheet Piling Anchoring

Verification of stability of anchoring is required according to DIN V 4084-100, taking account of the additional horizontal forces occurring at reduced live loads due to acceleration of the supporting mass of earth and the pore water contained therein.

8.3 Calculation and Design of Cofferdams

8.3.1 Cellular Cofferdams as Excavation Enclosures and as Permanent Waterfront Structures (R 100)

8.3.1.1 General

Cellular cofferdams are constructed of flat sections with high interlock tensile strength ranging from 2000 to 5000 kN/m, depending on the type of steel and section. Cellular cofferdams offer the advantage that they can be designed as stable gravity walls without wale and anchoring, even if embedment of the steel walls is not possible because of a rocky bottom or substratum. However, the nature of the cell fill material must be carefully specified and controlled.

Cellular cofferdams become economical for greater water depths, high retaining heights and longer structures. They are especially advantageous when bracing or anchoring is impossible or uneconomical. The additional requirement in sheet piling area is thus compensated for by savings in weight as compared to an otherwise required heavier and longer sheet pile section, and by the omission of wales an anchors.

8.3.1.2 Construction of Cofferdams

A distinction is made between cofferdams with circular cells (fig. R 100-1a) and with flat cells (fig. R 100-1b).

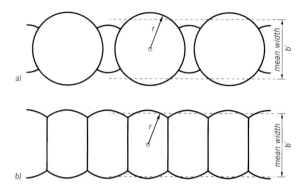

Fig. R 100-1. Schematic diagrams of plan views of cellular cofferdams
a) with circular cells, b) with flat cells

(1) Circular cells, which are connected by small, arched diaphragms, have the advantage that each cell can be individually constructed and filled, and is therefore independently stable. The diaphragms required for sealing can be installed later.

In order to keep the unavoidable extra stresses at the branches low, the clearance of the circular cells and the radius of the diaphragms should be kept as small as possible. If need be, bent piles may be used in the diaphragms. Information on the calculation is to be found in [188].

(2) Flat cells. In certain circumstances, the use of circular cells becomes unfeasible because the cell diameter would be so large that the tensile stresses in the interlocks and the pile webs themselves would become excessive. In these cases, flat cells may be used. These cells are placed successively in a straight line, adjacent cells having a common transverse partition. Since the individual flat cells are not independently stable, a cofferdam of this type must be filled in stages unless other measures are taken to ensure stability during construction. It follows that the ends of a flat cell cofferdam wall must be constructed as individually stable units. In the design of long structures it is recommended to provide intermediate fix points, especially if there is danger by ramming or by storm damage. In this case local failures can lead to destruction of large portions of the structure.

Under otherwise equal prerequisites, flat cell cofferdams required more steel per linear meter than circular cell cofferdams.

For welded branches, steel grades with appropriate properties must be ordered from the sheet piling manufacturer in order to prevent terracing failure (see R 67, section 8.1.6.1).

8.3.1.3 Calculation

(1) Verification of stability

The verification of stability of cellular cofferdams used as excavation enclosures is based on LS 1C and depicted in figs. R 100-2 to R 100-4. Mean width b' according to fig. R 100-1 is to be taken as the design width of the cofferdam. It results from the conversion of the actual ground plan into a rectangle of equivalent area.

If a cofferdam rests directly on rock (fig. R 100-2), a convex slip line occurs in a cross sectional plane when there is failure between the toes of the cofferdam walls. This slip line is approximated for practical purposes by a logarithmic spiral for the design value of the friction angle φ' depending on the limit case being investigated according to DIN V 1054-100 or DIN V ENV 1997-1.

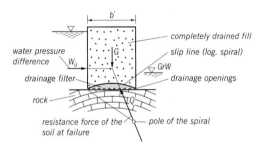

Fig. R 100-2. Cofferdam resting on rock, with drainage

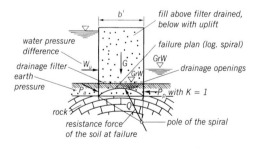

Fig. R 100-3. Cofferdam resting on rock overlain with other soil strata, with drainage

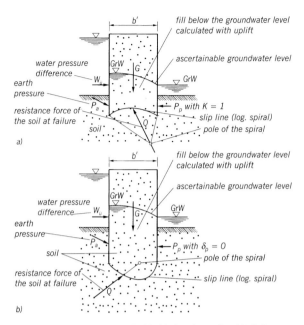

Fig. R 100-4. Cofferdam embedded in bearing soil, with drainage
a) In case of flat embedment
b) Additional investigation in case of deep embedment

Note: G, $W_ü$ and Q are design values which are to be determined depending on the limit case with the partial safety factors according to DIN V 1054-100.

Stability is given when all forces with their design values (moment of $W_ü$ and P_a) intersect in one point and equilibrium of all design values of forces is given with the least favourable slip line.

If the cofferdam rests on rock which is overlain by other soil strata (fig. R 100-3), or if the cofferdam is embedded in bearing soil (fig. R 100-4), the actions have to be added by the active earth pressure P_a on the driving side and by the passive earth pressure support P_p on the opposite side. Considering that the deformations shall be kept small, the latter is to be applied at only a reduced magnitude, as a rule with $K = 1$ and at deeper embedment in the soil with K_p for $\delta'_p = 0$.

Water pressure difference $W_ü$ is above all to be taken into account here as an action. It is the difference between the exterior water pressures W which acts on the cofferdam down to the elevation of its bottom, but the most unfavourable water level in the excavation may also lie above the excavation level.

367

The required stability when the foundation is in soil can be achieved not only by widening the cofferdam, the choice of a better fill material with larger γ and φ', as well as a precision cell drainage, but can also be achieved by staggered deeper driving in small steps of the piles erected as a wall. In this case, stability verification is also to be carried out with concave slip lines (fig. R 100-4b)). The spiral is then to be so placed that its centre point lies in no case above the line of action of P_p for $\delta_p = 0$ (fig. R 100-4). Both overturning safety and sliding safety are confirmed by this stability calculation.

(2) Calculation of the sheet piling

When calculating the circumferential tensile forces, it may be assumed that the water and active earth pressures, acting from the load side, are directly absorbed by the cofferdam fill. As a rule, the investigation of the cross-section at the level of excavation bottom or river bed is sufficient, as the governing circumferential tensile force generally occurs there. It is determined by the formula $Z = p_i \cdot r$. The active earth pressure at rest, calculated with $K_0 = 1 - \sin \varphi'$, and, as far as existing, the water pressure difference acting on the exposed side of the wall are to be applied as inner pressure p_i. Verification is provided as per LS 1B. In addition, reference is made to section 8.3.1.3 (4).

(3) Bearing capacity of subsoil

For cofferdams not resting on rock, verification of bearing capacity of the subsoil is to be made in accordance with DIN 4017-100 in combination with DIN V 1054-100, in which the mean width b' is taken as the cofferdam width. Section 8.3.1.3 (4) is also referred to in this connection. If necessary, safety against slope failure and overall stability must also be verified (DIN V 4084-100).

(4) Effect of water flow

In the above mentioned calculations, any existing water flow pressure is to be taken into account. Furthermore, safety from hydraulic ground failure and erosion is also to be checked. In the case of cofferdams on fissured or softening rock special sealing measures are required at the sheet piling tips to rule out the risk of such failure.

8.3.1.4 Construction Measures

Cellular cofferdams may be constructed only on good bearing soil. As soft strata, especially if they occur near the bottom of the cofferdam, decidedly reduces its stability, they are to be removed from the inside of the cofferdam. Fine-grained soil according to DIN 18 196 may not be used for the fill. For cellular cofferdams used as excavation enclosures the fill must be particularly permeable.

The stability of the cofferdam depends, among other factors, on the weight density (taking buoyancy into account), and on the angle of internal friction φ' of the fill. Therefore a soil of high weight density and with a large angle of internal friction is to be used as fill. Both of these characteristics can be increased by vibrating the soil in the cofferdam.

(1) Excavation enclosures

At excavation enclosures with foundation on rock, the uplift in the cofferdam shall, as far as possible, be eliminated by means of effective drainage which can be constantly controlled by observation wells. Drainage openings at the bottom of the exposed wall, filters at the level of the excavation floor and good permeability of the entire fill are essential.

Experience has shown that the permeability of sheet piling interlocks under tension is low.

The load-side portion of the excavation enclosure must ensure adequate watertightness. It can be practical in some cases to plan additional sealings on the outboard side, for example underwater concrete or similar.

(2) Waterfront structures

At waterfront structures, especially in deep water, much of the cell fill is constantly submerged. Deep-lying drainage is therefore superfluous.

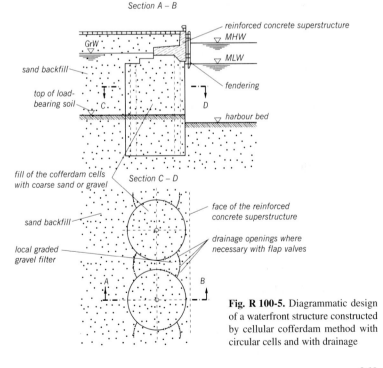

Fig. R 100-5. Diagrammatic design of a waterfront structure constructed by cellular cofferdam method with circular cells and with drainage

When there are large and rapid fluctuations in the level of the water table, drainage of the cell fill and of the structure backfill can be of advantage by reducing the water pressure difference (fig. R 100-5).

The superstructure is to be designed and constructed, with fenders, if necessary to keep dangerous ship impacts clear of the cofferdam cells (fig. R 100-5).

8.3.2 Double-wall Cofferdams as Excavation Enclosures and as Permanent Waterfront Structures (R 101)

8.3.2.1 General

In double-wall cofferdams, the parallel steel sheet piling walls are driven into or otherwise placed on the bottom depending on the prevailing subsoil and hydraulic conditions, as well as on the structural requirements. The two walls must be connected by tie rods, and if the double-wall cofferdam rests on rock, at least two horizontal rows of tie rods must be provided.

Transverse walls and anchor cells as shown in fig. R 101-1 may also assist in the planning and executing of the construction work. They are also recommended for use in permanent structures of considerable length because they will confine any damage to the section in which the damage occurs. The length of the individual sections in which the cofferdam is constructed, including tie rods and fill, depends on the spacing of the transverse walls and anchor cells.

The remarks made in R 100, section 8.2.1. regarding the fill also apply here.

For a cofferdam subject to high hydrostatic pressure, such as might be encountered at an excavation enclosure, an effective and highly reliable drainage arrangement is vital to the stability of the cofferdam. Even in waterfront structures, a drainage system may be useful. For excavation enclosures the fill is drained into the pit, and for waterfront structures

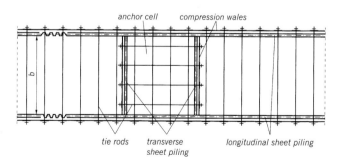

Fig. R 101-1. Plan elevation of a double-wall cofferdam with anchor cells

into the harbour. In the former case, weepholes as per R 51, section 4.4 are sufficient, whereas drainage openings into the harbour for waterfront structures must be fitted with flap valves as per R 32, section 4.5 if there is any risk of clogging by dirty harbour water.

8.3.2.2 Calculation

(1) Verification of stability

For calculation purposes, the width of the double-wall cofferdam is taken as the centre-to-centre distance b between the two sheet piling walls. For verification of the stability of double-wale cofferdams, essentially the same principles apply as for investigating the stability of cellular cofferdams – see R 100, section 8.3.1.3 (1). In contrast to figs. R 100-4a) and b), the passive earth pressure P_p in front of the front sheet pile wall is fully applied at the angle specified in R 4, section 8.2.4, because of its greater bending possibility than in normally anchored sheet piling.

If the front sheet pile wall has free earth support, the logarithmic spiral can be brought to the tip of the wall or, if the wall has fixed earth support, to the shear force zero point. The starting point of the spiral is generally at the same level at the load-side wall. However, if this is shorter than the front sheet pile wall, the spiral must be run to the existing point. If the outer stability cannot be verified, the equilibrium can be achieved by widening the cofferdam, selecting a better fill material with larger φ' and γ', through compacting the cofferdam fill inclusive of subsoil and from case to case, also by deeper embedding of the cofferdam sheet piling. Additional tie rods are also possible, but it must be decided on a case to case basis whether the possibly difficult installation (under water etc.) is practical.

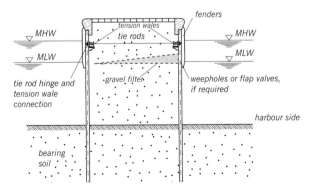

Fig. R 101-2. Diagram to show a mole structure in double-wall cofferdam structure

(2) Calculation of the sheet piling and anchoring (LS 1B)
It is taken for granted that the cofferdam fill has been so thoroughly drained that the water pressure and possibly additional active earth pressure acting from the load side are transmitted into the load-bearing subsoil directly through the cofferdam fill. A pressure higher than the active pressure acts on the front sheet pile wall due to the non-uniform distribution of the vertical stresses in the cofferdam (moment action from the water pressure difference). The increase in active earth pressure can generally be taken into account with adequate accuracy by increasing the active earth pressure calculated for $\delta_a = +\,2/3\,\varphi'$ by a quarter. To this must be added, insofar as it still exists, the residual water pressure difference acting on this sheeting.

If the front sheet pile wall is embedded in bearing soil, the supporting passive earth pressure can be calculated as usual, with the angle of wall friction as per R 4, section 8.2.4 The design of this sheet piling for LS 1B can be executed for single anchoring as per R 77, section 8.2.2 and for double anchoring as per R 134, section 8.2.3.

The load-sided sheet piling is generally installed with the same section and same length as the front sheet pile wall. Deviations from this rule are possible if the load-sided wall is to be checked for its cross-section and its embedment or if the requirements on the watertightness and underflow have to be fulfilled.

Bearing stability verification for the load-sided wall shall take account of the following actions:

- the anchor force of the front sheet pile wall,
- the water pressure of the load-sided wall,
- the active and where applicable also passive earth pressure of the outer surface of the load-sided wall,
- the earth support acting on the load-sided wall from the cofferdam side,
- ship's impact and other horizontal actions.

The earth support must be so distributed over the wall height that at least the equilibrium conditions of all actions on the wall are fulfilled.

(3) Bearing capacity of subsoil
See R 100, section 8.3.1.3 (3).

(4) Effect of water flow
See R 100, section 8.3.1.3 (4).

8.3.2.3 Construction Measures

See the general remarks in R 100, section 8.3.1.4.
Furthermore, special reference is made to the relevant recommendations of the EAU for the individual structural members. This applies above all to the tie rods and their proper installation.

(1) Excavation enclosures

The remarks in R 100, section 8.3.1.4 (1) are valid here except for the last paragraph.

The weepholes in the base area of the front sheet pile wall are burnt into the web of the sheet pile section for expediency.

The wales for the transmission of the tie rod forces are mounted on the outer side of the sheet piling as compression wales, if this does not create an objectionable hazard to ships. With this method, wale bolts are not necessary and installation of the tie rods is simplified. However, the anchor duct must be made watertight on the water side.

(2) Waterfront structures, breakwaters and moles

The remarks in R 100, section 8.3.1.4 (2) are valid here accordingly (fig. R 101-2).

8.3.3 Narrow Partition Moles in Sheet Piling Construction (R 162)

8.3.3.1 General

Narrow partition moles in sheet piling construction are cofferdams at which the spacing of the sheet piling is only a few metres and thus considerably less than a customary cofferdam (R 101, section 8.3.2). These partition moles are subject to load chiefly from water pressure difference, vessel impact, ice impact, line pull and the like.

The sheet piling walls are anchored against each other at or near the coping and stiffened for transmission of the external loads.

The space between the sheet piles is filled with sand or gravelly sand of at least medium degree of density.

8.3.3.2 Calculation Assumption for the Partition Mole as Cantilever Beam

For the absorption of the external loads acting normal to the partition mole centre line and for their transmission into the subsoil, the partition mole is viewed as a free-standing sheet piling structure, fully fixed in the ground. The effect of the soil fill between both sheet piling walls is neglected. It is furthermore generally assumed that both walls are tied against each other by a hinged connection in the coping area. In exceptional cases, a flexural rigid connection may also be planned which however leads to large bending moments at the head of the sheet piling and thus to expensive structural measures to realise the junction. In addition, it produces axial forces in the sheet piling walls, which reduce the passive earth pressure on the tension side.

As both sheet piling walls deflect in an extensively uniform manner from external loads, the total bending moment can be distributed over both walls in the ratio of their bending stiffness, that is to say as a rule in the ratio of the moments of inertia (2nd moment of area). Afterwards the stress proof is to be carried out separately for each of both walls according to LS 1B with the moment percentages in consideration of the respective moments of resistance.

The passive earth pressure mobilised on the passive side in the soil cannot be applied in full magnitude however, as a part of it has already been utilised for absorption of the active earth pressure and an eventual water pressure difference from the cofferdam fill. This percentage is determined in advance and subtracted from the possible total resistance.

8.3.3.3 Calculation Assumption for Sheet Piling Walls Anchored Against Each Other

The individual sheet piling walls are loaded by active earth pressure from the fill and the surcharge on the partition mole, as well as by external loads. In addition, water pressure differences have to be considered if the water table is higher in the partition mole than in front of the sheet piling. The anchoring is to be designed taking these assumptions into account. The bending moment relevant to the design of the sheet piling walls generally corresponds to the moment for fixed support in the earth mentioned in section 8.3.3.2 of the sheet piling structure fully fixed in the ground.

8.3.3.4 Design

The wale, anchoring and stiffening must be calculated, designed and installed in accordance with the pertinent recommendations, according to the occurring loads.

Special significance is attached to the absorption of the external loads and also those which act parallel to the centre line of the mole on the sheet piling. The check is to be made according to R 132, section 8.2.12. Transverse walls or anchor cells are to be foreseen as per R 101, section 8.3.2.

8.4 Anchorings, Stiffeners

All anchoring elements are to be so designed, that their bearing capacity corresponds with the full internal bearing capacity of the anchors.

8.4.1 Design of Steel Wales for Sheet Piling (R 29)

8.4.1.1 Arrangement

The wales must transmit the anchor forces (support reactions) from the sheet piling into the anchors, and the resistance forces of the anchor walls into the anchors. Further more, they stiffen the sheet piling and facilitate alignment of the piling.

As a rule, these wales are installed as tension wales on the inboard side of the waterfront sheet piling. At anchor walls they are generally placed as compression wales behind the wall.

8.4.1.2 Design

Wales should be of heavy construction and ample design. Heavier wales of S 235 JR (formerly St 37-2) are preferable to lighter ones of S 355 JO (formerly St 52-3). Splices, stiffeners, bolts and connections must be properly made in accordance with the best steel construction and welding practice. Stressed welds must be made at least 2 mm thicker than is structurally required because of the danger of corrosion. The wales are conveniently constructed of two closely-spaced U-shaped steel channels, whose webs are at right angles to the plane of the sheet piling (see R 132, section 8.2.12.2, figs. R 132-1 to R 132-3). Where possible, the U-shaped steel channels are placed symmetrically about the point of application of the anchor rod, so that the anchor rod can rotate freely by the anticipated amount. Proper spacing of the two U-shaped steel channels is maintained by U-shaped steel channel stiffeners or by web plates. Additional stiffening of the U-shaped wale channels is necessary in the area of the anchor force transmission in heavy anchor systems, or if there is a direct connection between the anchor rod and the wale.

Splices of the wale should be placed where stress is at a minimum. A full cross-section splice is not required, but it must be adequate to carry the calculated stresses at the section.

8.4.1.3 Attachment

The wales are either supported on welded brackets or, especially with limited working space beneath the wales, suspended from the sheet piling. The design and attachment must be so that any vertical loads on the wales are satisfactorily transmitted into the sheet piling. Brackets facilitate the installation of the wales. Suspensions should not weaken the wales, and should therefore be welded to the wales or attached to the base plate of the wale bolts.

If the anchor force is transmitted directly (through hinges) into the tension wale at the inboard side, the wale must be attached to the piling with special care. The anchor force is transferred from the sheet piling into the wales through heavy bolts. They are placed in the centre between the two U-shaped wale channels and transmit their load through base plates, which are attached by tack welding to the wales. The wale bolts are made extra long, so that they can be used to align the sheet piling against the wales.

8.4.1.4 Inclined Anchors

The connection of the inclined anchors must also allow for the transmission of vertical forces.

8.4.1.5 Extra Wale

Sheet piling which has become severely misaligned due to improper driving is to be realigned by means of an extra wale, which remains in the structure.

8.4.2 Verification of Bearing Capacity of Steel Wales (R 30)

Wales and wale bolts should be designed for at least the force which corresponds to the bearing capacity of the selected anchorage. Additionally, they must be so designed that all horizontal and vertical loads which would otherwise be applied are absorbed and transmitted into the anchors or into the sheet piling and anchor wall. The following loads are to be taken into consideration:

8.4.2.1 Horizontal Loads

(1) The horizontal component of the anchor tensile forces, whose magnitude can be taken from the sheet piling calculation. Taking into consideration future deepening in front of the quay wall, it is recommended that the wales be of generous size, and capable of absorbing the permissible tensile force derived from the chosen anchor rod diameter.
(2) Design values of direct acting hawser pulls on mooring devices.
(3) Design value of vessel impact, depending on the size of the vessel, the berthing manoeuvre, current and wind conditions. Ice impact may be neglected.
(4) Compulsive forces which are introduced when aligning the sheet piling.

8.4.2.2 Vertical Loads

(1) The dead load of the wale channels including stiffeners, bolts and base plates.
(2) That portion of the soil surcharge between the rear surface of the sheet piling to a vertical plane through the rear edge of the wale.
(3) The portion of the live load on the quay wall between the rear edge of the sheet piling capping and a vertical plane through the rear edge of the wale.
(4) The vertical component of the active earth pressure which acts on the vertical plane through the rear edge of the wale from the bottom edge of the wale to the surface of the ground. The active earth pressure is in this case calculated for plane sliding surfaces with $\delta_a = + \varphi'$.
(5) The vertical component of an inclined anchor tensile force with tension and compression wales, according to section 8.4.2.1 (1).

The loads stated under (1) to (5) are to be considered with their design values for LS 1B.

8.4.2.3 Loads for the Calculation

In the statical calculation of the wales, the only horizontal loads generally included are the component of the anchor tensile force (section 8.4.2.1 (1)) and the hawser pull (section 8.4.2.1 (2)). On the other hand, the vertical loads (section 8.4.2.2) are all included. In order to make some allowance for the stresses from vessel impact and aligning the piling, it is recommended that the partial safety factors for the resistance be increased by 15 %. When several wales lie over each other, the vertical loads are divided among the wales. In order to ensure the safe design of the attachment of the wale brackets, the loads are assumed to act at the rear edge of the wale.

8.4.2.4 Method of Calculation

The loads included are resolved into component forces vertical and parallel to the sheet piling plane (main axes of the wale). It is to be assumed in the calculation that for the absorption of the forces, acting at right angles to the plane of the sheet piling, the wales are supported by the anchors, and for the parallel forces by brackets or suspensions. If the anchors are connected to the sheet piling, the pressure of the piling on the wale in the areas around the anchor connection has an adequate supporting effect so that it is sufficient here to suspend the wale on the inner side, as is normal for compression wales. The moment at support and the span moment resulting from the design value of the sheet piling support reaction force are generally calculated according to the formula $q \cdot l^2 / 10$.

8.4.2.5 Wale Bolts

The wale bolts should be designed using the same principles as for the anchoring of the sheet piling, see R 20, section 8.2.6.1, but generously so in order to allow for corrosion and the stresses introduced when aligning the piling. With double anchoring, the bolts of the upper wales, statically only slightly loaded, should be at least 32 mm (1¼"), preferably 38 mm (1½") in diameter, to allow for vessel impact. The base plates of the wale bolts are to be designed in a manner that their bearing capacity corresponds with that of the wale bolts.

8.4.3 Wales of Reinforced Concrete for Sheet Piling with Driven Steel Anchor Piles (R 59)

8.4.3.1 General

In quay walls, it has been found that anchors consisting of steel anchor piles driven with a 1 : 1 batter are practicable and very economical.
This is especially valid when there are high lying layers of poor soil which make other anchorings difficult or impossible, and circumstances where extensive earthwork would otherwise be necessary.

If the anchor piles are driven first and if the sheet piles cannot be placed with precision, the anchor piles are not always in the designed position relative to the sheet piling.

However, inaccuracies of this type are of no consequence when the sheet piling wale is constructed of reinforced concrete, and the local structural dimensions have already been considered in the reinforcement plans (fig. R 59-1).

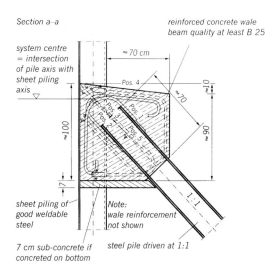

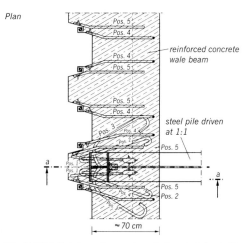

Fig. R 59-1. Reinforced concrete wale for steel sheet piling

If the reinforced concrete wale is constructed at a greater distance above the existing terrain, it would be advisable to equip the sheet piling wall with an auxiliary steel wale and to leave this in place until the piles are connected and the reinforced concrete wale has achieved its bearing capacity.

8.4.3.2 Construction of Sheet Piling Wales

Reinforced concrete wales are anchored to the sheet piling by round or square steel bars which are welded to the sheet pile webs, evenly spaced except that additional bars are required at expansion joints (fig. R 59-1, pos. 4 and 5). The anchor force is transferred into the anchor piles in a similar manner (fig. R 59-1, pos. 1 to 3).

The steel connectors welded to the sheet piles and the anchor piles are generally made of S 235 JO (formerly St 37-3). They are forged flat at the connection points. Similarly, round steel bars of BSt 500 S may also be used. Welding may be performed only by qualified welders under the supervision of a welding engineer. Only material whose welding properties are known, which are of uniformly good quality and are compatible with each other, may be used (see also R 99, section 8.1.18). The concrete shall be at least of strength class B 25 with an aggregate grading in the favourable range between curves A and B. BSt 500 S is generally chosen for the reinforcing bars.

8.4.3.3 Construction of the Connection Between Piles and Wale

If there are layers of considerable thickness, composed of soils which are very susceptible to settlement, or if non-compacted backfill is to be placed to a great depth behind the wall, the pile connections should be constructed as hinges. With more favourable soil conditions when considerable settlement or subsidence is not to be expected, the steel piles should be fixed into the reinforced concrete wale. Even with settlement-sensitive, thin-layered soils or with well compacted backfill of non-cohesive soil, such a connection can also be an economical solution. In order to allow for residual settlement or subsidence of the soil, and for the fixed-end influences, including those resulting from the deflection of the sheet piling, the fixed-end moment of the anchor pile along with the other horizontal and vertical forces acting on the wale must be taken into account in the design of the pile connection. In extremely yielding subsoil, this is to be determined for the yield strength $f_{y,k}$ taking account of the characteristic value of the axial force N_k acting in the pile. It is to be unfavourably introduced in the verification of the bearing capacity of the pile connection.

If the anchor piles are driven through relatively thin layers of soils susceptible to settlement or if the backfill is shallow, a correspondingly smaller additional connection moment may be applied.

The introduction of the internal forces in the anchor pile into the reinforced concrete wale at its point of connection with the wale is to be verified at this point. Thereby, the combined stressing of the pile top by axial force, shear force and bending moment is to be observed. If necessary, reinforcing plates may be welded laterally on the anchor pile, in order to improve the absorption of the forces. The reinforcing bars, which are otherwise formed as loops, may be welded to these plates. The voids which tend to form along the web of the anchor pile as a result of this construction must be carefully filled with concrete.

Only specially killed steels, resistant to brittle fracture, such as S 235 JO (formerly St 37-3) or S 355 JO (formerly St 52-3) may be used for piles and their connections in all quay walls with pile anchorings exposed to comparatively large uncontrollable bending stresses.

8.4.3.4 Calculation

The wale loads are to be applied according to R 30, section 8.4.2. The horizontal component of the anchor force according to the sheet piling calculation is considered as a horizontal load action to be applied at the system centre (= intersection of the axis of the sheet piling with the axis of the pile). The wale, including its connections to the sheet pilling, is calculated as uniformly supported. Dead load, vertical surcharges, pile forces, bending moment and shear force of the anchor piles are considered to be actions and introduced with their design values.

The internal forces at the pile connection, which result from the soil surcharges on the pile by the backfill or the layers susceptible to settlement, are calculated on an assumed equivalent beam fixed in the wale and in the load-bearing foundation soil. If shielding of the sheet piling loading by the anchor piles has not been taken into account, the fixed-end moment and the shear force acting on the pile connection have to be considered only with reference to the connection between the wale and the sheet piling, and need not be calculated as acting on the sheet piling itself.

A reduction of the anchor pile cross-section at the point where the pile is fixed into the wale in order to reduce the connection moment and the accompanying shear force is not allowable, because such a reduction can lead to pile failure especially if the workmanship has been faulty.

If instead of the fixed connection a flexible one is chosen, the additional internal forces at the pile connection arising from settlement or subsidence of the soil are also in this case to be checked and safely absorbed.

Verification of bearing capacity is to be provided for the design values of the internal forces S_d, which may be reduced depending on the loading case, as stated in section 8.2.0.2.

When the connection moment and the accompanying shear force in the anchor pile are based on the yield strength $f_{y,k}$, then the connection elements themselves may also be designed on the basis of the yield strength $f_{y,k}$.

For practical construction reasons, the dimensions of the concrete wale should not be smaller than those shown in fig. R 59-1. In order to allow for variations in the acting forces and in the pile anchorings, the cross-section of the reinforcing steel should be increased by at least 20 % beyond calculated requirements.

8.4.3.5 Spacing of Expansion Joints

Reinforced concrete wales generally have expansion joints with horizontally acting keying at about 15 m spacing (fig. R 59-2). Larger joint spacing even corresponding to a normal section length of approx. 30 m, is only allowable when the wale is constructed so high above the natural surrounding ground, that it can shrink without any appreciable hindrance from the sheet piling and if furthermore, adequate additional reinforcement is placed for absorbing the increased tensile forces in longitudinal direction of the wale.

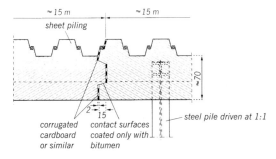

Fig. R 59-2. Joint keying of a reinforced concrete wale

8.4.3.6 Equipping of Top of Steel Anchor Piles to Transmit Forces into a Reinforced Concrete Superstructure

The equipping of the top of steel anchor piles must be arranged, designed and dimensioned in such a way that the anchor forces of the connected structure can be absorbed in the framework of the allowable effects of actions. Additional stresses from deflection and shear of the anchor pile should be kept to a minimum in the connection zone. For this purpose, the pile should be embedded to about twice its height in the reinforced concrete (fig. R 59-3). It is then sufficient to dimension the connection steels and their welds in such a way that about the whole cross-sectional area of the anchor pile is connected.

With yielding subsoil under the anchor piles, stresses in the reinforced concrete superstructure are to be verified within the framework of the

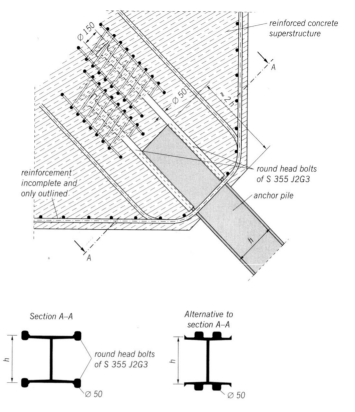

Fig. R 59-3. Example of an anchor pile connection to a reinforced concrete superstructure using so-called round head bolts

allowable effects of actions according to loading case 3. This is valid not only for the full anchor pile force but also for the loads from shear force and the bending moment at the anchor pile connection when the pile is stressed up to the yield strength.

Fig. R 59-3 shows a favourable connection solution with so-called "round head bolts" – as already used for bollard anchoring. Here one end of the round steel bar is upset to form a plate of up to three times the diameter of the round steel bar at the head. The end of the round steel bar to be welded to the tension pile is flattened to allow for good welding.

End anchoring in the concrete can also be achieved by welding cross bars or plates of a corresponding size to the round and square anchor bars.

8.4.4 Steel Capping Beams for Waterfront Structures (R 95)

8.4.4.1 General

Steel capping beams are designed with a view to meeting structural, operational and construction requirements. Furthermore, R 94, section 8.4.6.1 applies accordingly.

8.4.4.2 Structural and Construction Requirements

Steel capping beams serve as upper closure of sheet piling (fig. R 95-1). With adequate flexural rigidity (fig. R 95-2), capping beams can also be used for absorbing forces arising during the alignment of sheet piling tops, as well as for meeting certain operational requirements (fig. R 95-2). The sheet piling top can be aligned only if the sheet piling is sufficiently flexible and free during alignment to permit the necessary corrective deflections. With close spacing between capping beam and wale, the alignment of the sheet piling will be accomplished mostly with the wale. In service, the capping beam distributes non-uniform loads to the sheet pile tops and prevents non-uniform deflections of the sheet pile heads.

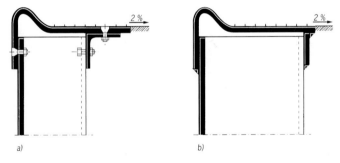

Fig. R 95-1. Rolled or pressed steel capping beam with bulb plate as nosing, bolted or welded onto steel sheet piling

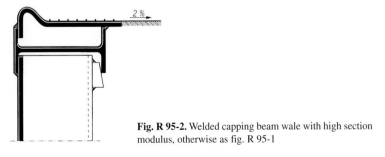

Fig. R 95-2. Welded capping beam wale with high section modulus, otherwise as fig. R 95-1

Figs. R 95-1 a) and b) show standard types of capping beams. With sheet pile sections of larger depth, the angle section in fig. R 95-1a) is reversed.

The greater the distance to the wale, the more important is an adequately high moment of inertia of the capping beam. Fig. R 95-2 shows a strengthened capping beam or a capping beam wale respectively.

Vessel impact is to be taken into account in dimensioning the capping beam. In order to avoid deflection or buckling, the capping beams shown in figs. R 95-1 a) and b) will be provided with stiffeners welded to the capping beam and sheet piling, if the troughs in the sheet piling are so wide as to make this necessary.

If the capping beam also serves as wale, this capping beam wale is to be designed and dimensioned in accordance with R 29, section 8.4.1 and R 30, section 8.4.2.

8.4.4.3 Operational Requirements

The top surface of the capping beam must be such that hawsers run over it will not be damaged. To protect the personnel working on the quay against slipping off, a portion of the capping beam should project slightly above the surface of the quay.

Horizontal capping beam surfaces should if possible be studded, chequered or similarly treated (figs. R 95-1 and R 95-2). For heavy vehicle traffic, a guard rail as shown in fig. R 95-3 is recommended.

If there is an outboard crane rail (R 74, section 6.3.4, fig. R 74-3), this is made part of the guide rail.

The berthing side of the capping beam must be smooth. Unavoidable edges are to be chamfered if possible. The design must be such that ships cannot catch under it and that the danger of capping beam sections being turned out by crane hooks is minimised (fig. R 95-4).

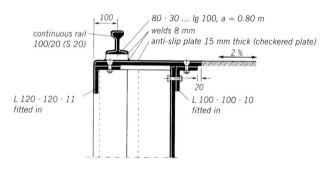

Fig. R 95-3. Bolted steel capping beam with welded-on rail as nosing

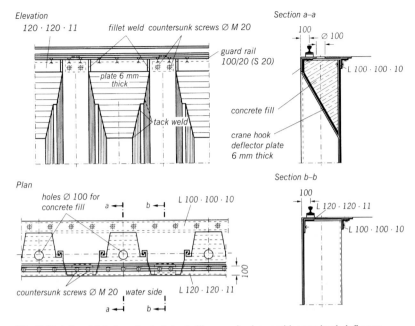

Fig. R 95-4. Special design of a steel sheet piling capping beam with crane hook deflector plates

8.4.4.4 Delivery and Installation

The capping beam members are to be delivered unwarped and true to size. During manufacture in the workshop, the tolerances for the width and depth of the sheet pile sections and deviations during driving are to be taken into account. Where necessary, the capping beams are to be adjusted and aligned at the construction site. Capping beam joints are to be designed as full joints, bolted or welded.

After installation of the capping beam, when it is located at an adequate height above HHW and is safe from wave action, sand is to be densely deposited in the area of the top of the sheet piling and replaced where necessary, in order to reduce settlement and protect the ground side of the sheet piling and the capping beam from corrosion.

If the capping beam can be flooded or inundated, lies in the wave action zone or is designed to lie below the water level and the water level can be lowered by passing ships, the risk exists that sandy backfill material will be washed out. The reason for this is that there is not a tight junction between the steel capping beam and the top of the sheet piling, as a rule. In order to prevent the washing out of sandy material, a tight connection is to be made in such cases between the steel capping beam and the sheet piling, for example by backfilling the top of the sheet piling with concrete. In so doing, the vertical flange of the capping beam or the capping beam angle should embed adequately deep into the concrete and the concrete should be adequately secured in its position by welded-on claws or bolts. Furthermore, the backfill material in the area of the paving is to be covered with an adequately thick graded gravel filter of well-balanced composition.

8.4.5 Reinforced Concrete Capping Beams for Waterfront Structures (R 129)

8.4.5.1 General

Reinforced concrete capping beams can be used as upper closure of waterfront structures. Statical, structural, operational and construction aspects determine their design.

8.4.5.2 Statical Requirements

In many cases, the capping beam serves not only to cover the sheet piling but also as stiffener and thus also to absorb horizontal and vertical loads. If in addition, it acts as a capping beam wale for transmitting anchor forces, it must be designed for ample strength, especially when it must carry a crane rail, resting directly on its upper surface.

Reference is made to the applicable portions of R 30, section 8.4.2 concerning horizontal and vertical loads. In case bollards or other mooring facilities occur, the forces acting on these members must be added (R 153, section 5.11, R 12, section 5.12, R 102, section 5.13) insofar as line pull is not absorbed by special structural members. In addition, if a crane rail rests directly on a reinforced concrete capping beam (fig. R 129-2), the horizontal and vertical crane wheel loads are also to be absorbed (R 84, section 5.14).

In the structural calculations, it is advisable to treat the reinforced concrete capping beam as a flexible beam, elastically supported on the sheet piling both horizontally and vertically. In doing so, for heavy capping beams on quay walls for seagoing vessels, a modulus of horizontal subgrade reaction $k_{s\,bh} = 25$ MN/m^3 can generally serve as criterion for the horizontal direction. The modulus for the vertical direction $k_{s\,bv}$ depends largely on the section and length of the sheet piling, as well as on the width of the capping beam. $k_{s\,bv}$ must therefore be specially determined for each structure, whereas $k_{s\,bv} = 250$ MN/m^3 can be used in preliminary calculations. Limit considerations are required for final dimensioning with the design rated on the least favourable case.

Anchorings connected to the sheet piling structure or to bollard foundations are to be taken into account separately. Special attention should be paid to the absorption of shrinkage and temperature stresses, because expansion and/or contraction of the capping beam can be greatly restricted by the connected sheet piling and by the soil backfill.

In order to allow for irregularities in the support by the sheet piling and possibly by anchorings, the reinforcing steel should be increased by at least 20 % over calculated requirements, in accordance with R 59, section 8.4.3.

Reference is made to R 72, section 10.2, regarding concrete grade, reinforcement and design.

The vertical loads to be absorbed in the plane of the sheet piling are generally introduced centrically into the sheet pile top. On this account, to prevent cleavage, sufficient lateral tension reinforcement is placed in the reinforced concrete capping beam directly above the sheet piling. For corrugated steel sheet piling, the concrete capping beam can be designed in a way, that the vertical forces will be transmitted directly into the sheet piling. Hereby, the cross-sectional area of the piling is acting as a very narrow bearing. This so-called "edge bearing" must be designed in accordance with a construction supervisory permit issued by the IfBt. In case there are large concentrated loads, e.g. from a crane rail, a plate effect of the sheet piling should be ensured by properly welding the interlocks. Geometric conditions (see R 74, section 6.3) may make it necessary to support the craneway out of centre of the capping beam.

Verification is required of the safe transfer of all internal forces (e.g. bending moments, shear forces, etc.) in the transition range piling/capping beam.

8.4.5.3 Construction and Operational Requirements

The sheet pile top must be aligned as required before concrete is placed. For this, a steel wale included in the design or a temporary steel wale can be used. However, the sheet piling top can only be aligned by these means if the wale protrudes far enough out of a more or less yielding soil at the completion of driving. It is then possible to give the sheet pile structure good alignment at the top with the aid of the reinforced concrete capping beam. If it is necessary to ensure a concrete cover of adequate thickness over the water side of the sheet piling top, the width of the capping beam should be appropriately increased. Contingent on the design, the pile top should have a concrete cover of at least 15 cm on both water and land sides, and the depth of the capping beam should be at least 50 cm (figs. R 129-1 and R 129-2). Furthermore, the length of the sheet pile top embedded in the capping beam should be about 10 to 15 cm.

In order to prevent a ship's hull from catching under it, the capping beam may be designed as shown in fig. R 129-2, where it is provided with a wide universal bent plate at 2 : 1 or steeper at the water side. The lower edge of this steel plate is welded to the sheet piling.

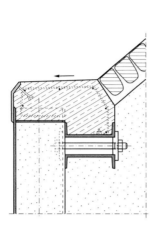
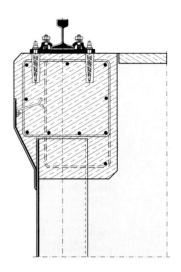

Fig. R 129-1. Reinforced concrete capping beam for corrugated sheet piling without water-sided concrete cover at a partially sloped bank

Fig. R 129-2. Reinforced concrete capping beam for corrugated sheet piling with concrete cover on both sides and craneway resting directly on the surface

If the concrete cover over the water side is omitted, a universal steel plate is generally installed on the water side of the sheeting (fig. R 129-1). It is welded to the flange of the sheet piling, as this method is more economical than a bolted connection. Anchor claws should be used over the sheet pile troughs in order to produce a solid connection between the steel plate and the concrete. The steel plate is to be chamfered on top (fig. R 129-1). Irregularities in the alignment of the sheet pile top up to about 3 cm can be corrected by the use of fillers.

The capping beam is provided with nosing and skid protection as per R 94, section 8.4.6, or DIN 19 703, at least at facilities serving sea-going traffic. The other provisions of the aforementioned section are also to be observed as applicable.

The stirrup reinforcement must be so designed that the portions of the capping beam which are separated by the embedded sheet piling are securely connected. To this end, the stirrups should either be welded to the webs of the sheet piles or inserted through holes or placed in slots, burnt in the sheeting. Such measures are not required with the certified "edge bearing" as per section 8.4.5.2. If transverse tension reinforcement to prevent cleavage is placed above the top of the sheet piling to carry off vertical loads, and if the stirrups associated with this reinforcement lie immediately above the upper surface of the sheeting, additional

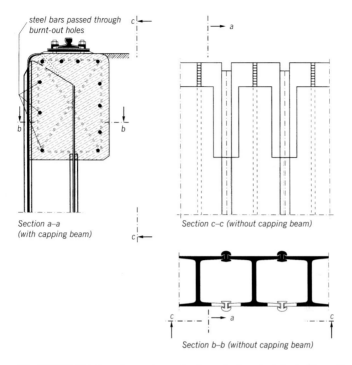

Fig. R 129-3. Reinforced concrete capping beam for box steel sheet piling without concrete cover on water side, with craneway resting directly on surface

stirrups should be provided to ensure the structural integrity of the capping beam, including its lower portions.

Bulkheads composed of box steel sheet piling may be also built with a capping beam which requires no concrete cover over the water side of the pile tops (fig. R 129-3). In this case, the reinforcing steel must be placed in the sheet piling cells. To make this possible, the webs and flanges are cut away as necessary and holes are burnt in them for inserting the longitudinal rods.

The same procedure can be followed for combined steel sheet piling.

Reinforced concrete capping beams can also be used as foundations for bollards, if the necessary local reinforcement is provided. Fig. R 129-4 shows an example of this for a heavy quay wall for seagoing vessels. In such cases, hawser pulls of large magnitude are best absorbed by heavy round steel anchor rods, in order to keep the elongation of the anchor connection and thus the bending moments in the capping beam as slight as possible.

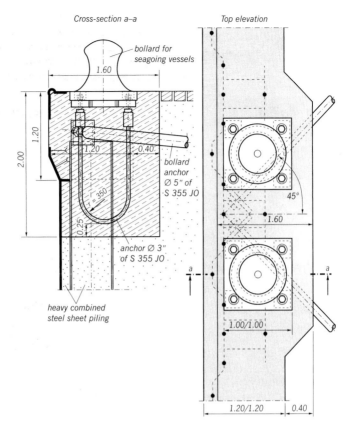

Fig. R 129-4. Heavy reinforced concrete capping beam of a quay wall for seagoing vessels, design at an anchored bollard foundation

Prestressed steel cable anchors may prove to be a disadvantage if later excavation work behind the capping beam becomes necessary.

8.4.5.4 Expansion Joints

In general, expansion joints are placed in reinforced concrete capping beams at approx. 15 m intervals. larger joint spacing, even corresponding to a normal section length of approx. 30 m, is only allowable if the capping beam is so constructed high above the natural surrounding ground, at the same time with an equally long reinforced concrete wale as per R 59, section 8.4.3, that it can shrink without appreciable hindrance by the sheet piling or by the reinforced concrete wale. Furthermore, adequate additional reinforcement must then be placed for the absorption of the increased tensile forces in the longitudinal direction of the capping beam. The joints themselves must also be so constructed that the changes in length of the reinforced concrete capping beam at these points are not hampered by the sheet piling. To accomplish this for corrugated steel sheet piling, the following methods are suggested:

(1) The expansion joint is placed directly above a sheet pile web, which is coated with elastic material so that the required movements are possible.

(2) The expansion joint is placed above a sheet pile trough. The pile or piles of this trough should then be only slightly embedded in the reinforced concrete capping beam and must be covered with a substantial plastic coating, which at the same time ensures watertightness in the vicinity of the joint.

An example for a joint in cases without the requirement of transferring shear forces is shown in fig. R 129-5.

Expansion joints in heavy reinforced concrete capping beams are keyed for the transmission of horizontal forces. A certain key effect can be achieved in not so heavily stressed capping beams by the use of a steel dowel.

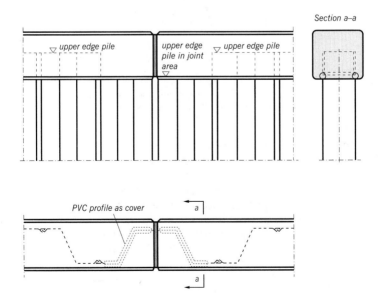

Fig. R 129-5. Expansion joint of a reinforced concrete capping beam

8.4.6 Top Steel Nosing for Reinforced Concrete Walls and Capping Beams at Waterfront Structures (R 94)

8.4.6.1 General

For practical purposes, edges of reinforced concrete waterfront structures are provided with a carefully designed steel nosing. This is to protect both the edge and the lines running over it from damage caused by ship operations, and serves as a safety measure to prevent line handlers and other personnel working in this area from slipping off. The nosing must be so constructed that ships cannot catch under it. The same also applies to crane hooks (R 17, section 10.1.3).

If waterfront structures in inland harbours are frequently flooded so that there is a danger of ships grounding thereon, the nosing may not have any bulges or moulding.

8.4.6.2 Examples

Fig. R 94-1 shows the standard design for navigation locks which can also be used for waterfront structures in ports.

It is possible to build such waterfront structures, especially where cargo is handled, so that surface run-off will flow to the land-side, making weepholes as per fig. R 94-1 unnecessary.

The steel nosing shown in fig. R 94-1 can also be supplied with aperture angles $\neq 90°$, so that it can be fitted to a sloping upper surface or front face of the waterfront structure. It is supplied in lengths of about 2500 mm, which are bolted or welded before installation.

The design in fig. R 94-2 depicts a special section developed in the Netherlands and frequently used there with success. It is made of steel sheet of augmented thickness and has reinforced steel anchors, so that the upper hollow space created during concreting need not be grouted. The upper ventilation openings must be closed after concreting however, in order to minimise corrosion attacks on the inner surface.

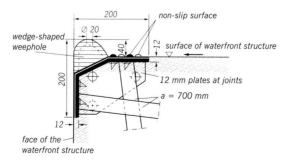

Fig. R 94-1. Nosing with weephole

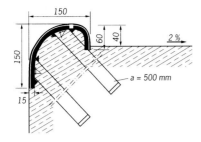

Fig. R 94-2. Special section of nosing frequently used in the Netherlands

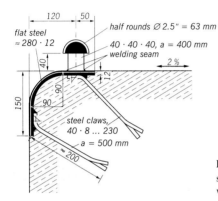

Fig. R 94-3. Nosing made of rounded sheet with foot railing in seaports and without in inland harbours

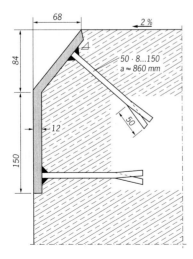

Fig. R 94-4. Nosing made of angular sheet without foot railing for non flood-free embankments in inland harbours

The designs shown in figs. R 94-3 and -4 have proven themselves at numerous German waterfront structures.

All types in figs. R 94-1 to -4 must be carefully aligned, placed in the shuttering and securely fastened. The types in figs. R 94-3 and -4 must be concreted flush against all contact surfaces in the course of concreting the quay wall. The inner surface of the nosing is to be cleaned of adhering rust with a wire brush before installation.

8.4.7 Auxiliary Anchoring at the Top of Steel Sheet Piling Structures (R 133)

8.4.7.1 General

For structural and economic reasons, the anchoring of a waterfront sheet piling wall is in general not connected at the top of the wall, but rather at some distance below the top. This applies especially to walls, when there is a large difference in elevation between harbour bottom and top of the wall. In this manner, the span in case of a single-anchored wall is decreased and thus also the moment in the span and the fixed end moment. Furthermore, increased redistribution of earth pressure takes place.

In such cases, the section above the anchor is frequently given auxiliary anchoring at the top, even if the customary sheet piling statics (R 77, section 8.2.2) show that it sustains no load. Its function is to prevent the flexible upper sheet piling end from too large deflection during the final stage of construction and later at the occurrence of large, local unexpected operational loads. The auxiliary anchoring however is not taken into consideration in calculating the structural main system of the sheet piling structure.

8.4.7.2 Aspects for the Installation

The height of the section above the anchor, for which the installation of an auxiliary anchor is useful, depends on various factors, such as the flexural rigidity of the sheet piling, the magnitude of the horizontal and vertical live loads, on operational requirements for the alignment of the top off the sheet piling wall and the like.

When a waterfront sheet piling wall is designed for crane loading, an auxiliary anchor should be added as near as possible to the top, unless circumstances make it advisable to place the main anchors quite near the top of the wall.

As a rule, loads on the section above the anchor by mooring hooks or posts also call for auxiliary anchoring. Although the anchoring for major line pull forces is likewise connected near the top of the wall, it is generally run to the main anchor wall and incorporated in the main anchoring system.

8.4.7.3 Design, Calculation and Dimensioning of the Auxiliary Anchoring

Round steel anchor bars with flexible connections at both ends are generally used for the auxiliary anchoring. For the calculation of the auxiliary anchoring an equivalent system is used as a basis in which the section above the anchor is considered to be fixed at the level of the main anchor. The load acts on this system similar to the load in the statics for the main system. It is to be observed that the load applied to the section above the anchor must be fully absorbed by both the auxiliary anchoring as well as by the main anchor.

In some cases, the auxiliary anchoring must also be calculated with the load assumptions according to R 5, section 5.5.5.

R 5, section 5.5, especially section 5.5.5 and R 20, section 8.2.6, as well as R 29, section 8.4.1 and R 30, section 8.4.2, also apply to the design, calculation and dimensioning of the auxiliary anchor wale. With regard to future need to straighten the top of the wale, and to absorb the force of moderate vessel impact, the auxiliary anchor wale should be dimensioned stronger than theoretically required. In fact, it is customary to make it identical with the main anchor wale.

If the sheet piling wall capping beam is also used for the connection of the auxiliary anchor, R 95, section 8.4.4 and/or R 129, section 8.4.5 are also to be observed.

The stability of the auxiliary anchoring is to be checked against both heave of the anchoring soil as well as for the lower failure plane, which extends to the connection point of the main anchor (fig. R 133-1). R 10, section 8.4.9, applies accordingly here.

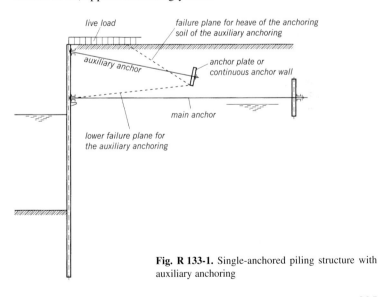

Fig. R 133-1. Single-anchored piling structure with auxiliary anchoring

8.4.7.4 Execution of Construction Work

It is advisable to dredge the harbour bottom in front of the quay wall sheet piling after the auxiliary anchoring has been installed. If the reverse procedure is followed, the top of the sheet piling wall may move uncontrollably, so that later adjustment with only the auxiliary anchoring may not always be successful.

8.4.8 Threads of Sheet Piling Anchors (R 184)

8.4.8.1 Types of Threads

The following thread types are used:

(1) Cut thread (cutting thread) (fig. R 184-1)
The outside thread diameter is equal to the diameter of the round bar steel or the upsetting.

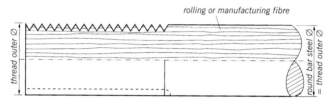

Fig. R 184-1. Cut thread

(2) Rolled thread (non-cutting thread produced in the cold state) (fig. R 184-2)
When using steel grades S 235 JR (formerly St 37-2) and S 355 JO (formerly St 52-3), the round bar steel or any upsetting must be machined or skimmed to the necessary extent before rolling the thread, in order to obtain a thread conforming to standards. For anchors with a rolled thread, the diameter of the round bar steel or upsetting may be somewhat smaller

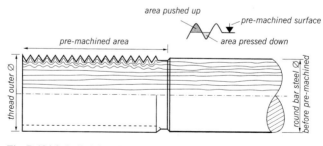

Fig. R 184-2. Rolled thread

than for anchors with cut thread, without the load-bearing capacity declining.

With the methods shown, an outside thread diameter is produced, depending on the pre-machining, which is greater than the diameter of the starting material.

Drawn steels (up to ⌀ 36 mm) do not need to be pre-machined.

(3) Hot-rolled thread (non-cutting thread) (fig. R 184-3)

During hot rolling, the thread shaft receives two rows of thread flanks which lie opposite and supplement one another to form a continuous thread.

With the hot-rolled thread, the additional processing stage of rolling or cutting the thread is not required. The nominal diameter is applicable to the thread shaft. The actual cross-sectional dimensions differ slightly from this. Associated elements must be used for end anchorages and butt joint designs.

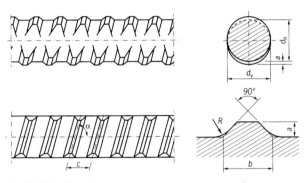

Fig. R 184-3. Hot-rolled thread

8.4.8.2 Required Safety

Reference is made to R 20, section 8.2.6.3 in respect of the verification of bearing capacity and special production information.

8.4.8.3 Further Information on Thread Types

- Rolled threads have a high profile accuracy.
- When the thread is rolled, cold forming occurs. This increases the strength and the yield point of the thread root and flanks, which has a favourable effect in respect of central loads.
- The thread root and flanks are particularly smooth in rolled threads and therefore have higher fatigue strength under dynamic loads.

- The production times for rolled threads are shorter than those for cut ones, but this advantage is more than offset by the necessary machining or skimming, provided drawn grades are not used.
- The course of the steel fibres is not interrupted in rolled or hot-rolled threads.
- Rolled threads with larger diameters are used primarily in centrally loaded anchors with dynamic stresses.
- Compared with the cut thread, the rolled thread achieves a weight saving of e.g. 14 % in anchor rods with ∅ 2"and 8 % with ∅ 5".
- In round bar steel anchors with rolled thread, no nuts, couplers or turnbuckles with rolled internal thread are necessary, especially since the stress on an internal thread is always smaller than on an external thread. When the internal thread is loaded, ring tensile forces are generated, and these provide support. Therefore, a combination of rolled external thread and cut internal thread can be selected without hesitation.

8.4.9 Verification of Stability of Anchoring at Lower Failure Plane (R 10)

8.4.9.1 Stability in Non-cohesive Soils

Stability for the lower failure plane is verified for LS 1 C according to DIN V 4084-100. It may be simplified compared to DIN V 4084-100 following the suggestions of KRANZ [74], working on the basis of an external cut (cut on the passive side of the sheet piling). The length of the anchor or the depth of the anchor wall is sufficient when a driving additional force $\Delta T \geq 0$ in the direction of the lower failure plane can be absorbed with the design values of the actions and resistances as per DIN V ENV 1997-1 for LS 1C.

Symbols used in fig. R 10-1:

G_1 Total weight force of the sliding body *FDBA*,
H Horizontal action,
U_a Water pressure, harbour-side,
P_p Passive earth pressure as per LS 1C,
U Water pressure on the lower failure plane *FD*,
U_1 Water pressure on the anchor wall *DB*,
P_1 Active earth pressure with live load on the anchor wall *DB* as per LS 1C,
Q_1 Failure surface force under φ' to the normal line on the lower failure plane *FD* acting against the direction of movement. Its magnitude results from the force polygon.

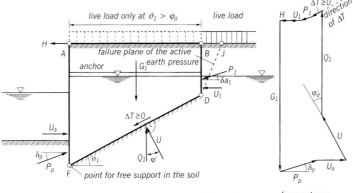

Fig. R 10-1. Verification of stability for the lower failure plane in non-cohesive soils

force polygon for fig. R 10-1

The graphic investigation of stability is carried out by drawing a force polygon (fig. R 10-1). A polygon is formed from the design values of the known variables H, U_1, P_1, G_1, P_p, U_a, U and the direction of Q_1. This results in the additional absorbable force $\Delta T \geq 0$ in the driving direction when there is adequate safety for the mechanism being examined. As in all other earth pressure assumptions, the equilibrium condition for the acting moments is not taken into account, so that no statement is made about the type of force distribution. KRANZ has performed comparative calculations to ascertain that the straight connection DF can be assumed with adequate accuracy to be the major failure plane. Stability can also be examined purely by computation [72]. In the case of several tie rods, the procedure stated in DIN V 4084-100 is to be used.

A refined calculation with curved failure surfaces (e.g. circular or logarithmic spiral) is generally not required. When the groundwater table falls off toward the sheet piling, the water pressure difference is referred to the earth pressure failure plane belonging to the anchor wall (R 65, section 4.3 and R 114, section 2.9).

If the influence of the flowing groundwater on stability in the lower failure plane is to be taken into account more precisely for a groundwater table falling off toward the sheet piling, a flow net according to R 113, section 4.7 is required. This can be used to determine the water pressure in the lower failure plane FD and also for the anchor wall plane DB referred to the active failure plane DJ. They are then integrated in the force polygon (fig. R 10-1) with the directions at right angles to the planes on which they act.

8.4.9.2 Stability in Cohesive Soils

The investigation is undertaken as in non-cohesive soils (section 8.4.9.1), except that the design value of the cohesion force C' or C_u respectively also acts in the lower failure plane. The angle of internal friction is to be taken as $\varphi_u = 0$ for non-consolidated, water-saturated, first-loaded, cohesive soils.

Fig. R 10-2 shows the application to the force polygon.

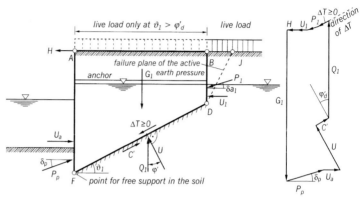

Fig. R 10-2. Verification of stability for the lower failure plane in cohesive soils

force polygon for fig. 10-2

The symbols for fig. R 10-2 correspond to those for fig. R 10-1, but in addition:

C' cohesion force in the lower failure plane FD as per LS 1C.

8.4.9.3 Stability in Differing Layers

The calculation is made (fig. R 10-3) by first splitting up the soil mass between the sheet piling and anchor wall into as many sections as there are layers cut by the lower failure plane. This is done by passing imaginary vertical planes through the intersections of the lower failure plane with the boundaries of the layers. Now a graphical method is applied to all component sections in turn. Verification of stability is provided when the force polygon results in an additional absorbable force $\Delta T \geq 0$. If cohesion exists in individual layers, it is taken into consideration in the corresponding component sections as per fig. R 10-3.

Symbols for fig. R 10-3:

G_1 Total weight force of the sliding body F_1DBB_1 plus live load for the case $\varphi'_{1d} < \vartheta_1$,

G_2 Total weight force of the sliding body FF_1B_1A,
H Horizontal action,
U_a Water pressure, harbour side,
P_p Passive earth pressure as per LS 1C,
U_1 Water pressure on the lower failure plane in section F_1D,
U_2 Water pressure on the lower failure plane in section FF_1,
P_1 Active earth pressure with live load on the anchor wall DB as per LS 1C,
U Water pressure on the anchor wall DB,
C'_1 cohesion force in the lower failure plane F_1D as per LS 1C,
Q_1 Failure surface force under φ'_1 to the normal line on the lower failure plane F_1D acting against the direction of movement. Its magnitude results from the force polygon,
Q_2 Failure surface force under φ'_2 to the normal line on the lower failure plane FF_1 acting against the direction of movement. Its magnitude results from the forces polygon,
U_{21} Water pressure on the vertical separation plane F_1B_1,
P_{21} Earth pressure in the vertical separation plane F_1B_1 assumed to be horizontally acting.

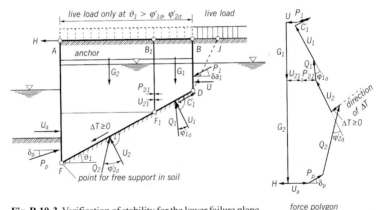

Fig. R 10-3. Verification of stability for the lower failure plane with differing layers of soil

force polygon to fig. R 10-3

8.4.9.4   Stability for Fixed Earth Support of Sheet Piling

The method can be used with sufficient accuracy also in the case of fixed earth support, if the shear force zero point in the fixity area is taken as the theoretical point of the sheet piling to which the lower failure plan runs. This point coincides with the position of the greatest fixed end moment. Its position can therefore be taken from the sheet piling calculation. The bending moment at the section in the sheet piling is analytically without influence on the stability of the anchoring, as long as uniform soil exists in the lower failure plane. Otherwise, a more detailed calculation can be made, taking the effect of the moments into consideration.

8.4.9.5 Stability of a Fixed Earth Support Anchor Wall

If the anchor wall is fixed on the bottom, the lower failure plane should run to the theoretical point at the level of the shear force zero point in the fixed-end area of the anchor wall, according to section 8.4.9.4.

8.4.9.6 Concluding Remarks

By making use of the lower failure plane concept as described above, the stability of the anchoring can be determined without undue effort, even under difficult conditions. Since experience has shown that in certain cases, the stability at the lower failure plane requires anchor lengths greater than those calculated by the previous method using the failure surfaces of the active and passive earth pressures (high-lying failure surface method), the stability for the lower failure plane must be confirmed in each design of a sheet piling structure. As an aid in the investigation, the rule of thumb can be used that the required anchor length for a free earth support wall is about equal to the length of the sheet piles.

It must be observed in the investigation that high ground water levels frequently lead to a decrease in stability.

Due to the importance of sufficiently stable anchoring for the lower failure plane, and the comparatively large sensitivity of ΔT to small changes in the friction angle in the lower failure plane, all loading cases must be expected to observe the partial safety factors of LS 1C according to section 8.2.0.1.

8.4.9.7 Safety Against Failure of Anchoring Soil

When the stability of the lower failure plane has been confirmed, the determination of the anchor length by means of the high-lying failure surface method may be omitted.

However, in order to avoid failure of the anchoring soil mass, and the consequent upward displacement of the anchor plate or anchor wall, it must be proven that the design values of the resisting horizontal forces from the bottom edge of the anchor plate or anchor wall to the surface of the ground are greater than or equal to the sum of the horizontal component of the anchor force, the horizontal component of the active earth pressure on the anchor wall and possibly a water pressure difference on the anchor wall.

The active and passive earth pressures on the anchor wall can be determined for example, from the tables of KREY, JUMIKIS or CAQUOT-KERISEL/ ABSI [10]. A live load may be considered in the design only in an adverse position behind the anchor wall or anchor plate. Unfavourable, high ground water levels, where they can occur, are also to be taken into consideration. In calculating the passive earth pressure on the anchor wall, the angle of wall friction shall not be assumed to be greater than required to counteract the total of all acting vertical forces, including dead load and soil surcharge (condition $\Sigma V = 0$ at the anchor wall). When the anchor tension is inclined upwards, the vertical component should be multiplied by a factor of 1.5 for purposes of computing the stability of the anchoring.

This investigation and those under R 10, section 8.4.9 replace the generally required slope failure investigation according to DIN V 4084-100, unless special conditions prevail, for example poor soil strata behind the structure, high surcharges behind the anchor wall or particularly long anchors.

In the case of simply supported anchor plates and walls, the anchor connection is generally located at half the height of the plate or wall. For further details, see R 152, section 8.2.13 and R 50, section 8.4.10.

8.4.10 Sheet Piling Anchorings in Unconsolidated, Soft Cohesive Soils (R 50)

8.4.10.1 General

If subsoil conditions exist which affect the design of sheet piling bulkheads as treated in R 43, section 8.2.16, special measures are also required in the anchorings of these walls in order to avoid the harmful effects of differential settlement.

Even a sheet pile bulkhead, designed as floating, generally has its point in a layer of soil which is more firm than the upper layers. Therefore, in such cases, a movement of the soil in the region of the anchor relative to the sheet piling should be taken into consideration. This condition is aggravated, the more the soil settles and the less the sheet piling is displaced downwards. This can cause a considerable rotation of the anchor connection of the sheet piling.

Inclinations of anchor rods of 1 : 3 have already been measured at quay walls of medium height, where the anchor rods were originally installed horizontally.

If the land end of the anchor rods is connected to a firmly founded structure, the situation is similar.

Differential settlement relative to the anchors is generally slight for anchor walls on floating foundation.

As observations of completed structures have shown, the anchor rod is taken downward by settlement of the soil, even in soft soil. It hardly indents into the downward pressing soil, so that it must bend considerably at its connection to a firmly founded structure.

Under the conditions described above, settlement of the subsoil in the entire anchoring area may vary greatly, so that greater or lesser settlement differences may occur even along the anchor length. Therefore, the anchor rod must be able to bend without being damaged.

Round steel anchors with upset thread sections are to be recommended here, since they always have a greater elongation and a higher flexibility than anchors without upsetting.

8.4.10.2 Steel Cable Anchors

As an example, this requirement can be met with steel cable anchors. They are sufficiently flexible in all practical cases, without it being necessary to lower the allowable effects of actions. However, because of corrosion danger, only patently keyed and sealed anchors, or prestressed anchors approved as permanent anchors should be used. It is important that the ends of the steel cable are perfectly insulated at the transition to the cable head or to a concrete structure.

Steel cable anchors must have provisions at one end for prestressing because of the great elongation which occurs in the anchor cable, and for restressing when anchoring is accomplished with floating anchor walls. The waterside end is usually used for this purpose. The steel cable anchor shaft ends there in a cable socket in which the cable wire ends are held by poured babbit metal. A short round steel bar with a continuous thread is attached to the cable head. It projects through the sheet piling, and is connected to the sheet piling on the water side by a hinged joint. The tensioning device is placed on the protruding threaded end. The surplus threaded end is burnt off after pre- or restressing has been completed.

For hinged anchor connection see R 20, section 8.2.6.3. In order to give the anchor end sufficient freedom of movement for the expected large rotation about the hinge, the two U-shaped wale channels must be spread wide apart. The required spacing however often exceeds that which is structurally allowable, so that the anchor must be attached below the wale. Care must then be taken that there is a satisfactory transmission of forces into the anchor by reinforcement on the sheet piling, or by supplementary attachments on the wale.

8.4.10.3 Floating Anchor Wall

In a floating anchor wall, the usual spacing of both U-shaped wale channels is generally sufficient; the anchor rod passes between them and is attached to the compression wale behind the anchor wall by means of a hinged joint (fig. R 50-1).

If the steel cable anchor terminates on the land side in the reinforced concrete wale of a floating anchor wall or the like, the end should be unravelled and broomed before it is embedded in the concrete. If the

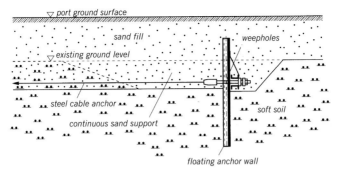

Fig. R 50-1. Floating anchor wall with eccentric anchor connection

anchor cable is attached to a structure which has a rigid foundation, the connection must also be a flexible one, because even a steel cable anchor must not be bent more than 5°.

8.4.10.4 Wale

Especially soft and weak zones in the subsoil must be expected in these cases, even when they are not apparent from the test borings. In order to bridge these intruding layers, the wales of the sheet piling and the anchor wall must be liberally overdesigned. In general, U-400 channels of S 235 JR (formerly St 37-2) should be used for wales, in larger structures S 355 JO (formerly St 52-3). Reinforced concrete wales must have at least the same bearing capacity. They are placed in sections of 6.00 to 8.00 m length, with joints keyed against horizontal movement (R 59, section 8.4.3).

8.4.10.5 Anchoring at Battered Pile Bents or Anchor Walls

The design of the rear anchoring depends on whether or not considerable displacement of the quay wall coping can be accepted. If no displacement is permissible, anchoring must be by means of pile bents or the like. If some displacement is not objectionable, anchoring onto a floating anchor wall is possible. In this case, the horizontal pressures in front of the anchor wall must remain so slight that undesirable displacements will not occur. If local experience is not available, soil investigations and calculations, are required regarding the partial mobilisation of the passive earth pressure. If necessary, test loading is to be performed.

8.4.10.6 Design

If the harbour land area is filled to the top surface with sand, the design shown in fig. R 50-1 is recommended. The soft soil in front of the anchor wall is excavated to just below the anchor connection, and replaced by a compacted sand buffer of sufficient width. The anchors can be laid in trenches, which are either filled with sand or carefully tamped with suitable excavation soil. The anchor wall is then eccentrically connected, so that the allowable effects of horizontal actions are not exceeded in either the sand fill or in the area of the soft soil. In this case, it is sufficiently accurate if uniformly distributed pressures in both areas are used in design calculations. The magnitude of the pressure follows from the requirement that equilibrium must exist with reference to the anchor connection.

In this solution, irregularities in the anchor wall bedding are equalised. In order to prevent the build-up of water pressure difference, weepholes must be placed in the anchor wall for the percolation of water.

8.4.10.7 Stability

The stability at the lower failure plane must be checked here with special care. The usual investigations according to R 10, section 8.4.9. for the final state are not sufficient in this case. The shear strength must be determined for the unconsolidated state, and used as a basis for the verification (initial strength). If the shear strain of the soft cohesive soils in the triaxial test as per DIN 18 137-2 exceeds 10 %, the degree of shear strength utilisation is to be reduced according to DIN V 4084-100.

8.4.11 Design and Calculation of Protruding Corner Structures with Round Steel Anchoring (R 31)

8.4.11.1 Impractical Design

The quay wall sheet piling at corner structures should not be held by anchors running diagonally from quay wall to quay wall and thus are not connected at right angles to the wall axis as is customary. Otherwise damage can occur because the anchor forces create high, additional tensile forces in the wales whose highest stress occurs at the last diagonal anchor connection. Since corrugated steel sheet piling yields comparatively easily under horizontal loads parallel to the shore, a considerable length of the quay wall is required for the transfer of tensile forces from the wale, through the sheet piles and into the foundation soil. The wale joints are especially endangered thereby. special reference here is made to R 132, section 8.2.12.

8.4.11.2 Recommended Crossed Anchoring

The tensile forces mentioned in section 8.4.11.1 do not occur in the wale if crossed anchoring is used as shown in fig. R 31-1. In order to prevent the anchors from interfering with each other, the rows of anchors and the wales must be offset in height. Clearance at the turnbuckles should be somewhat more than the diameter of the anchor rod.

Edge bollards should have independent, additional anchoring.

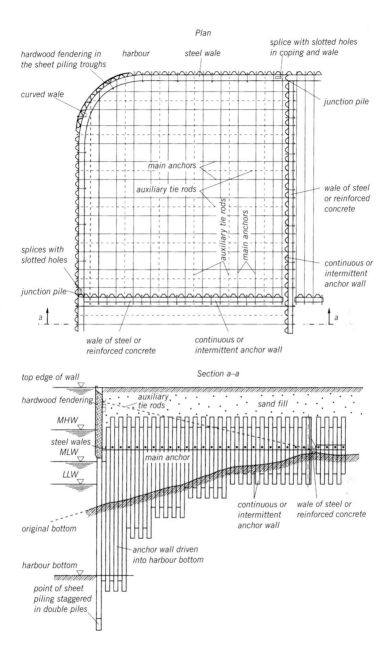

Fig. R 31-1. Anchoring of a protruding quay wall corner in a sheet piling structure at seaports

8.4.11.3 Wales
The wales at the sheet piling are steel tension wales, shaped to fit the quay wall. The wales at the anchor wall are pressure wales of steel or reinforced concrete. The transition from the wales of the corner section to the wales of the quay wall and anchor wall, as well as the intersection of the anchor wall wales, are so arranged that the wales can move independently of each other. The wales and copings have spliced joints with slotted holes where the anchor walls abut the quay wall.

8.4.11.4 Anchor Walls
The position and construction of the anchor walls in the corner section depend on the design of the quay walls. The anchor walls at the corner are carried through to the quay wall (fig. R 31-1) but are driven staggered and down to the harbour bottom in the end stretch so that in case of damage, the loss of backfill at the especially critical corner area will stop at the anchor wall. This precaution is also recommended when individual anchor plates are used, such as reinforced concrete slabs, instead of anchor walls.

8.4.11.5 Timber Fendering
Ships and the wall itself are protected if the sheet pile troughs at wall corners are fitted with suitable wharf timbers. This fendering should project about 5 cm beyond the outer face of the sheet piling (fig. R 31-1).

8.4.11.6 Rounding Off and Concrete Reinforcing of Wall Corners
Since protruding quay wall corners are especially exposed to damage by ship traffic, they should be rounded off if possible, and if need be, also strengthened by strong reinforced concrete wall.

8.4.11.7 Protection by a Dolphin in Front of the Corner
Each quay wall corner should be protected by a resilient dolphin if ship traffic permits.

8.4.11.8 Verification of Stability
Verification of stability is carried out individually for each quay wall, according to R 10, section 8.4.9. A special check for the corner section is not necessary if the anchorings of the quay walls are carried through to the other wall as shown in fig. R 31-1.
Self-supporting pier heads are designed and calculated according to other principles.

8.4.12 Design and Calculation of Protruding Quay Wall Corners with Batter Pile Anchoring (R 146)

8.4.12.1 General

Protruding corners of quay walls are especially exposed to damage from passing ship traffic. In many cases, quay wall corners are at the end position of a berth for a large ship and hence must accommodate a heavily loaded bollard, as described in R 12, section 5.12.2. In seaports, they are also equipped with the required fendering, which must possess higher energy absorption capacity than that on the adjacent quay wall sections. Corners should be of sturdy construction and be as rigid as possible.

Batter pile anchoring can also be a beneficial design for such quay wall corners, so that the solution with pile anchoring elements is dealt with here as a supplement to the solution with round steel anchors according to R 31, section 8.4.11.

8.4.12.2 Design of the Corner Structure

The most appropriate design of quay wall corners with batter pile anchoring can differ greatly due to local conditions and future utilisation as a port facility. It depends largely on the structural design of the adjoining quay walls, on the difference in ground surface elevation and on the angle enclosed by the wall sections forming the corner. As far as construction is concerned, the final design will be decisively influenced by the existing water depth and the nature of the subsoil.

In order to ensure proper positioning of the overlapping inclined piles occurring at the corners, definite requirements must be adhered to regarding the clearance between these piles at all points where they cross. Whereas the clearance between crossing piles which overlap above the existing bottom can be kept comparatively small (about 25 to 50 cm), the clearance between long piles below the bottom at all crossing points should be at least 1.0 m, or preferably 1.5 m, especially in compact soils where driving is difficult. In stony soils, where driving can still be executed, longer piles will probably drift out of line. Under these conditions, the clearance at greater depths should be at least 2.5 m. When calculating the clearance between piles, existing steel vanes are always to be taken into account.

In order to be able to meet these requirements, the spacing and inclination of the piles must be properly varied, but the inclination should be kept fairly uniform, because of the varied bearing behaviour of individual piles of an interdependent pile group.

Should a deep foundation also be necessary at the quay wall corner for highly loaded bollards or other items of equipment such as anchoring constructions of conveyor belts and the like, the construction of a special reinforced concrete corner section with a deep-founded relieving plat-

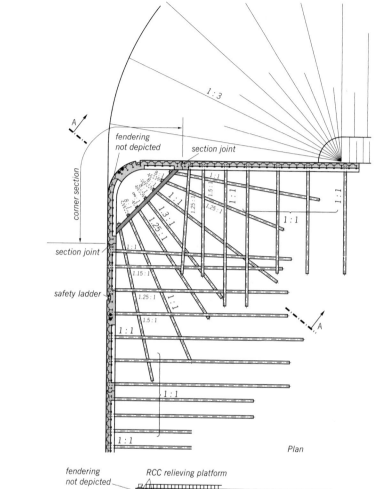
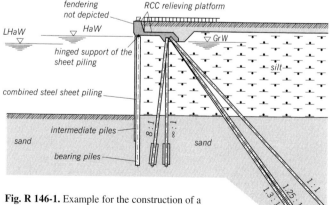

Fig. R 146-1. Example for the construction of a protruding quay wall corner with steel pile anchoring

form is recommended in most cases. The latter has a hinged support on the sheet piling. This also applies in general to quay wall corners at which proper positioning of the piles cannot be achieved by altering pile inclinations and clearances. In such corner constructions, the tension piles required in the corner area are appropriately placed at the rear portion of the relieving platform. As a result, they lie within a plane which differs from that of the tension piles of the bordering quay wall sections, so that interference between the piles can be more readily avoided. Due to the additional compression piles required at the rear slab edge, as well as to the required relieving platform, such designs are much more costly. However, they ensure sound and proper construction. Fig. R 146-1 shows a typical example of this.

Sections 8.4.11.5 and 8.4.11.7 in R 31 also apply to quay wall corners with batter pile anchoring.

8.4.12.3 Use of Scale Models

In order to avoid future driving difficulties, a small but adequately exact scale model should be built and be available for checking during the project planning of difficult corner constructions. A larger model scaled at about 1 : 10 should be used later on the building site. In this model, each pile must be placed in the position in which it was actually installed, to allow for necessary corrections in the position or inclination of the remaining piles.

8.4.12.4 Verification of Stability of the Corner Sections

In all designs of corner constructions with batter pile anchoring, verification is required of the stability of all piles in the entire corner area. Reference is made to section 9. In so doing, each wall in the corner is to be considered individually. At corners with additional loads, from corner stations of a conveyor facility, from bollards, fenders and other items of equipment, it must be verified that the piles are also able to absorb these additional forces satisfactorily.

Should extensive changes take place in the piling during construction, their effects are to be verified in a supplementary calculation.

8.4.13 High Prestressing of Anchors of High-strength Steels for Waterfront Structures (R 151)

8.4.13.1 Anchors for waterfront structures, particularly for sheet piling bulkheads and also for subsequent securing of other structures, such as walls on pile foundations and the like, are usually made of steel grades S 235 JR (formerly St 37-2), S 235 JO (formerly St 37-3) or S 355 JO (formerly St 52-3). However, in certain cases high pre-stressing of the anchors can be beneficial, but this is only advisable when they consist of high-strength steels.

The high prestressing of anchors of high-strength steels can be practical or necessary for the following, among others:

- for limiting displacements, especially at structures with long anchors, making allowance for existing sensitive structures or when joining to subsequently driven front sheet piling (R 45, section 11.3.6) and

- for achieving load transmission with pronounced active earth pressure redistribution (strongly reduced moment of span at increased anchor force), with the prerequisite that the structure be situated in at least medium densely deposited non-cohesive or stiff, cohesive soil.

For permanent anchors of high-strength steels, special significance is to be ascribed to technically sound corrosion protection in all cases. Any existing licences for example for grouted anchors according to DIN 4125 are to be observed.

8.4.13.2 Effects of High Anchor Prestressing on the Active Earth Pressure

Anchor prestressing always lessens the displacement of the waterfront structure toward the water side, especially in its upper part. High prestressing can favour an increased redistribution of the active earth pressure toward the top. In this case, the resultant of the active earth pressure can move from the bottom third point of the wall height h above the harbour bottom to about $0.55\ h$ upward, whereby the anchor force to be absorbed increases correspondingly. This active earth pressure redistribution is especially pronounced at quay walls extending above the anchor (cantilever).

In case an active earth pressure diagram deviating from the classic distribution according to COULOMB is to be achieved using fully-loaded, high-strength anchoring steels, the anchors must be locked off at about 80 % of the anchor force determined for loading case 1 by prestressing.

8.4.13.3 Time for Prestressing

The prestressing of the anchors may not begin until the respective prestressing forces can be absorbed without appreciable undesirable movements of the structure or its members. This requires corresponding backfill conditions and is to be considered in the planning of the construction stages and in the assumption regarding the absorption/transmission in the structure.

Experience shows that the anchors must be prestressed briefly beyond the planned value, since part of the prestressing force is lost again through yielding of the soil and of the structure when the adjacent anchors are stressed. This can be extensively avoided if the anchors are prestressed in several steps, which however can create difficulties in the execution of the work.

The prestressing forces are already to be subjected to spot checks during the construction period, in order to be able to make a correction in the scheduled prestressing, if necessary.

DIN 4125 is chiefly applicable to pressure-grouted anchors.

8.4.13.4 Further Suggestions

For high anchor prestressing in limited waterfront sectors, the produced locally differing displacement possibility of the waterfront structure is to be taken into account. The prestressed zones act as rigid points, which are acted on by increased spatial active earth pressure and are to be adequately designed fore this. The anchor forces in such areas are always to be rechecked.

Where at all possible, one end of the prestressing anchor should be fabricated so as to always be accessible and be so designed that if necessary, the prestressing force or related displacement can also be checked subsequently and corrected. Otherwise, the anchor ends should have hinged connections.

Since bollards are loaded only from time to time, their anchors should not be made prestressed of high-strength steel but rather of strong round steel bars of S 235 JR (formerly St 37-2), S 235 JO (formerly St 37-3) or S 355 JO (formerly St 52-3). The latter show only slight extension when subject to loads. Difficulties would crop up during earthwork behind bollard heads, if highly prestressed anchors of high-strength steels were used. However, these are frequently required for operational reasons, for example for laying lines of various types.

8.4.14 Hinged Support of Quay Wall Superstructures on Steel Sheet Piling (R 64)

8.4.14.1 General

The support of a quay wall superstructure on steel sheet piling may be either hinged or fixed. The hinged support is preferable in fully backfilled, high sheet piling. Fixed support results in a large fixed-end moment, which frequently requires plate reinforcement at the top of the sheet piling. The fixed-end moment requires heavier anchoring, increases the load on the pile foundation and introduces strong additional bending stresses into the superstructure, which must absorb the fixed-end moment of the sheet piling. Structural movements and possible later deepening of the harbour would have an adverse effect on the entire structure.

8.4.14.2 Advantages of Hinged Support

With a hinged connection, the sheet piling is in a large measure separated from the more rigid superstructure from which it differs widely in behaviour and deformation. The hinge force is transmitted to the superstructure at a favourable location.

Unavoidable structural movements and future increases in harbour depth have an insignificant effect on the superstructure. Moreover, the hinge connection results in the least possible anchor force, and thus in an especially economical foundation for the superstructure.

The greatest sheet piling moment occurs in the span below the zone of the strongest corrosion attack. In backfilled sheet piling, it is smaller than the moment at the top of a pile having a fixed connection.

These advantages of the hinged support can only be fully realised however if the hinge connection complies in all respects with the rules for steel construction. With unsatisfactorily hinged support, damage may occur in the support area, particularly in high structures with a large horizontal load on the sheet piling.

8.4.14.3 Structural Design

The bearing must be adapted to the sheet piling system and the superstructure.

(1) Corrugated sheet piling

Fig. R 64-1 shows characteristic design examples for corrugated sheet piling with welded or bolted anchors and individual bearing, or a continuous, eccentrically placed, welded bearing plate. The latter guarantees a tight connection at the joint, and is therefore especially recommended in cases in accordance with section 8.4.14.5.

Fig. R 64-2 shows a design in which the outboard flanges of the sheet piling are tied into the superstructure in the form of a flexible joint.

This achieves particularly great eccentricity of the axial force. For the satisfactory transmission of forces, the superstructure must extend about 30 cm beyond the face of the sheet piling. In order to prevent ships from catching under the edge, a steel deflector plate should be placed on every second double pile in tidal areas. In harbours without tide, intervals of up to five double pile widths are permissible.

The slight eccentricity of the horizontal anchoring with respect to the vertical bearing which appears in the designs shown in figs. R 64-1 and R 64-2 is of no consequence, since the flexibility of the sheet pile with respect to the cross-section and the elasticity of the anchors guarantee adequate flexibility.

(2) Combined sheet piling

Figs. R 64-3 and R 64-4 show practical examples for combined sheet piling. In cases with pure hinge connection as in figs. R 64-3 and R 64-4, the superstructures must be firmly secured, also against displacement in the axial direction of the sheet piling, so that lifting from the bearing is avoided, even with greater forces due to percussion or pressure wave loads.

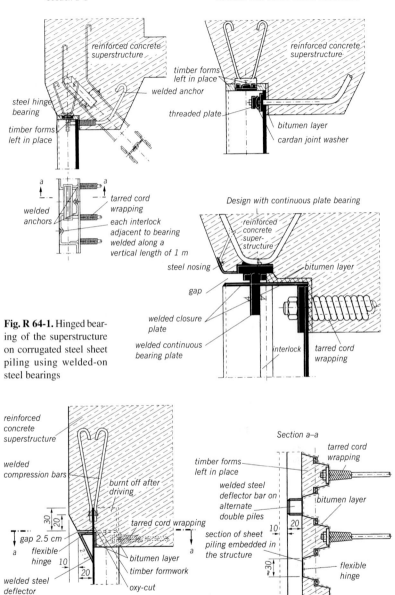

Fig. R 64-1. Hinged bearing of the superstructure on corrugated steel sheet piling using welded-on steel bearings

Fig. R 64-2. Hinged bearing of the superstructure on corrugated steel sheet piling, using a flexible bearing, designed to be stressed to the yield point

415

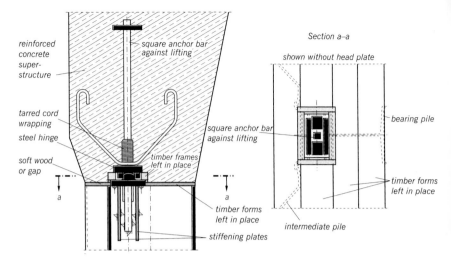

Fig. R 64-3. Hinged bearing of the superstructure on combined steel sheet piling, using welded-on single hinges

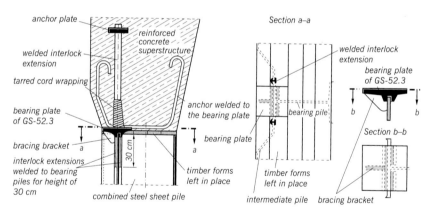

Fig. R 64-4. Hinged bearing of the superstructure on combined steel sheet piling using cast steel hinge parts

8.4.14.4 Material

The structural parts of the hinged bearing are generally made of S 235 JR (formerly St 37-2) and overdesigned to allow for corrosion.

Other satisfactory support designs are also technically possible and practicable.

8.4.14.5 Sealing the Hinged Bearing Joint

If the elevation of the ground behind the sheet piling is at or above that of the hinged connection, so that wave action or water pressure difference may cause wash-outs, the hinged bearing joint must be thoroughly sealed to prevent this. In order to prevent scouring, granular filters and/or geotextiles are also suitable from case to case. Satisfactory sealing of the hinge joint is also required to prevent penetration of fine particles into a designed hollow space behind the piling.

8.4.15 Hinged Connection of Driven Steel Anchor Piles to Steel Sheet Piling Structures (R 145)

8.4.15.1 General

A hinged connection between driven steel anchor piles and sheet piling structure allows the desired and largely independent reciprocal rotation of the structural members to take place freely, thus creating an uncomplicated statical condition which helps to keep the costs low.

Rotation in the proximity of the connection between the anchor pile and the sheet piling occurs inevitably as a result of deflection of the wall. It can however also occur at the head of the steel anchor pile, especially when a downward movement of the active earth pressure sliding wedge is accompanied by severe settlement and/or subsidence of the natural or filled ground behind the sheet piling. In such cases, the hinged connection is preferable to the fixed connection described in R 59, section 8.4.3. In accordance with the fundamentals of steel construction, all parts of the connection must be designed for safety and effectiveness.

8.4.15.2 Pointers on Designing the Hinge Connection Parts

The capability to rotate can be achieved by the use of single or double hinge pins or by the plastic deformation of a structural member (designed to be stressed to the yield point) suitable for this purpose. A combination of pins and flexible joint is also possible.

(1) Hinges, designed to be stressed to the yield point, should be located at a sufficient distance from butt and fillet welds, to prevent the yielding of welded connections as far as possible.
Fillet welds parallel to acting forces should lie in the plane of forces or in the plane of the tension transmitting member, in order to guard against the welds loosening or sealing off. If this is not possible, other means must be employed to accomplish this.

(2) Every weld transverse to the planned tensile force of the anchor pile can become effective as a metallurgical notch.

(3) Construction welds in difficult positions, which are not made true to stress and welding requirements, increase the probability of failure.

(4) In difficult connection designs, even with hinged connection, it is recommended that a calculation be made of the effect on the cross-sectional area probably forming a yielding hinge by the action of the planned axial forces, in connection with possible supplemental stresses and the like (see R 59, section 8.4.3). When verifying hinges, designed to be stressed to the yield point, DIN 18 800, 11/1990 is to be taken into consideration.

(5) Notches caused by sudden discontinuity of stiffness, for example when there are oxy-burned notches in a pile and/or metallurgical notches from cross welds, as well as abrupt increases of steel cross-sections, for example due to welded-on, very thick straps, are to be avoided, especially in possible yield areas of tensile anchor piles, because they can produce sudden failures without deformation (brittle fracture).

Several characteristic examples of hinged connections of steel anchor piles are shown in figs. R 145-1 to R 145-7.

8.4.15.3 Execution of Work

Contingent on local conditions and on the design, the steel anchor piles may be driven either before or after the sheet piling. If the location of the connection with respect to certain members of the sheet piling is critical, as would be the case if the connection must be made in the trough of a corrugated sheet pile, or on the bearing pile of combined sheet piling, it is important that the upper end of anchor piles be as close as possible to their designed position. This is best accomplished if the anchor piles are driven after the sheet piling. However, the construction of the connection must always be such that certain deviations and rotations can be compensated for and absorbed.

If the steel anchor pile is driven directly above the top of the sheet piling, or through a "window" in the sheet piling, the sheet piling can provide effective guidance for driving the anchor pile. If an anchor pile must be driven behind a double pile, a driving "window" can be provided by burning off the upper end of the double pile, hoisting it clear and later returning it to its original position and welding it in place.

Steel piles which are not embedded in the soil to the upper end leave a certain scope for aligning the head in the connection.

In determining the lengths of the anchor piles, an allowance of extra length should be made. This extra length depends on the type of the connection and is for the purpose of allowing the butt of the pile to be burned off in case it is damaged in driving or for the driving itself.

Slots for connection plates shall if possible be cut into the sheet piling and into the anchor piles only after the piles have been driven.

8.4.15.4 Structural Design of the Connection

The hinged connection to corrugated sheet piling is generally placed in the trough, especially if the interlock connection is on the centroidal axis, or, for combined sheet piling, on the web of the bearing piles.

Where a wall is only lightly loaded, especially in an open canal section, the steel anchor pile may be connected to the capping beam, which is mounted on the top of the sheet piling (fig. R 145-1), or to a wale behind the sheet piling by means of a splice plate (anchor plate) and a flexible connection. Special attention must be paid to the danger of corrosion in such cases. Reference is made to R 95, section 8.4.4 for waterfront structures with cargo handling and for berthing places.

Tension elements of round steel rods (fig. R 145-3), flat or universal flat steel straps (tension straps) are frequently installed between the connec-

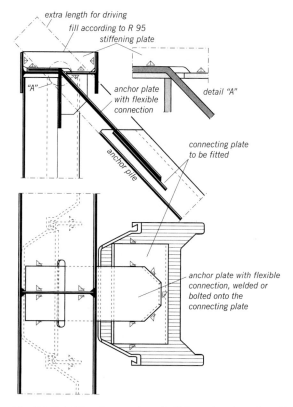

Fig. R 145-1. Hinged connection of a light steel anchor pile to light steel sheet piling, using anchor plate and flexible connection, designed to be partly stressed to the yield point

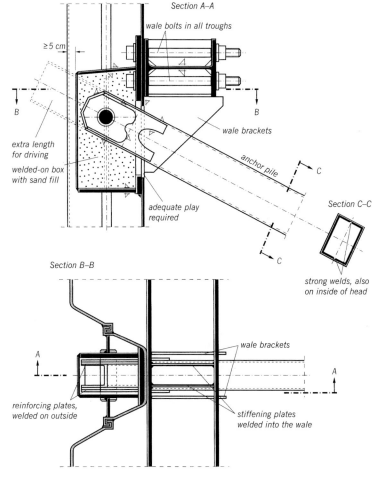

Fig. R 145-2. Hinged connection of a steel anchor pile to heavy steel sheet piling, using a hinge pin

tion in the trough or at the web and the upper pile end (figs. R 145-4 and R 145-5). Connections consisting of a threaded steel rod, nut, base plate and joint plate have the advantage that they may be tensioned.

In addition to the hinged connection in the sheet piling trough, in the capping beam or at the web of the bearing pile, an additional hinge may, in special cases, be installed near the top end of the anchor pile. This solution, depicted in fig. R 145-5 for the case with double bearing piles, can also be employed for single bearing piles, somewhat varied. The

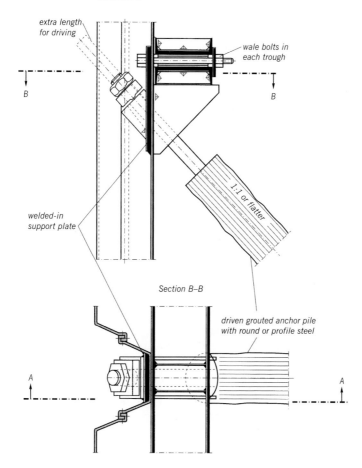

Fig. R 145-3. Hinged connection of a driven grouted anchor pile to heavy steel piling

slots (burned openings) in the flanges of the bearing piles are to be taken deep enough under the connection plates to create sufficient freedom of movement for pile deformations and to rule out any compulsive forces which can arise from unwanted fixity, should contact between plates and pile occur. Care is also required to ensure that the intended hinge effect is not impaired by incrustations, sintering and corrosion at the connection elements. This is to be checked from the individual case and taken into account in the structural design.

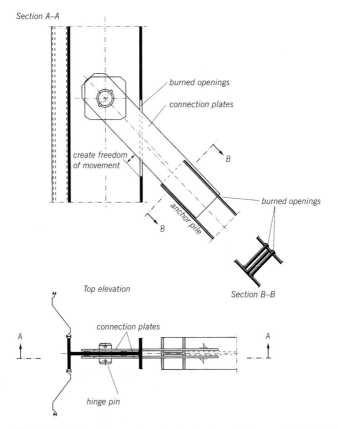

Fig. R 145-4. Hinged connection of a steel anchor pile to a combined steel sheet piling with single bearing piles, using a hinge pin

The anchor pile can also be driven through an opening in a sheet piling trough and the hinge connected there by means of a supporting construction welded to the sheet piling (fig. R 145-2).

If the connection is made at the water-side trough of a sheet piling wall, all construction members must terminate at least 5 cm behind the sheet piling alignment. Furthermore, the point of penetration should be carefully protected against soil running out and/or washing out (for example by means of a protective box as in fig. R 145-2).

Depending on the design selected, preference should be given to those connections which can be manufactured largely in the workshop and which provide adequate tolerances. Extensive fitting work at the site is very expensive and is therefore to be avoided if at all possible.

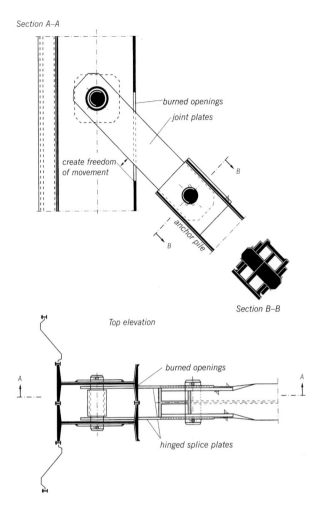

Fig. R 145-5. Hinged connection of a steel anchor pile to a combined steel sheet piling with double bearing piles, using hinged splice plates

The connection shown in fig. R 145-6, for example, fulfils these conditions. All bearing seams on the bearing pile are welded in the workshop in the box (flat) position. However, this solution should only be used if the jaw bearing plates are installed after driving the anchor piles.

In the solution shown in fig. R 145-7, the connection between anchor pile and sheet piling is created by loops which enclose a tubular wale welded into the sheet piling. Care is required here because the free rota-

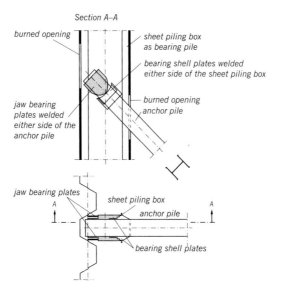

Fig. R 145-6. Hinged connection of a steel anchor pile to combined steel sheet piling using jaw bearings/bearing shells

tion of the connection is prevented by friction between loops and tube. The loops are therefore to be dimensioned to allow for the adequate bearing reserves required for the resulting unequal load distribution. As a rule, deflections of the anchor piles and consequently displacement of the pile heads are to be expected, so that the connection has to be designed for transfer of shear forces besides the anchor tensile forces. This can be achieved with a bracket-type extension of the piles through to the tubular wale, or other suitable measures.

8.4.15.5 Verification of Bearing Capacity of the Connection

The anchor force determined by the sheet piling calculations is the primary factor in the design of the anchor connection. It is recommended however to design all anchor connection parts for the internal forces, which can be transmitted by the anchor system used. Loads from the water side, such as vessel impact, ice pressure or the subsidence of mines etc., can at times reduce the tensile force present in the steel pile or even change it into a compressive force. If necessary, checks should be made of the stresses in the connection and of the buckling stress on the free-standing portion of a partially embedded pile or on the pile connection. In some cases, ice impact must also be taken into account.

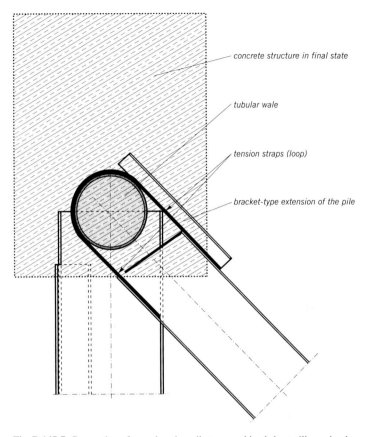

Fig. R 145-7. Connection of a steel anchor pile to a combined sheet piling using loops

If possible, the connection should lie at the intersection of the sheet piling and the pile axis (figs. R 145-1, R 145-2, R 145-4 to 145-7). If there is a substantial deviation from this, additional moments in the sheet piling are to be taken into account.

The vertical and horizontal components of the anchor pile force are also to be taken into account in its connection to the sheet piling and, if every stressed wall element is not anchored, in the wale and its connections. If a vertical load due to soil effects must be expected, it must also be allowed for in the reaction forces and in the verification of the bearing capacity of connections. This is always the case when deflection of the anchor piles is to be expected. When the opening angle between anchor pile and sheet piling changes, the bearing capacity verification must also take account of the resulting changes in the tensile forces in the area of the connection structure, as well as resulting shear forces.

In case of connections in the trough, the horizontal force component is to be transmitted into the sheet piling web (fig. R 145-3) by a support plate of adequate width. The weakening of the sheet piling cross-section should be considered because reinforcing of the sheet piling may be necessary in this area.

A constant flow of tensile forces and safe transfer of the shear forces is to be observed in the connection elements, particularly in the area of the upper anchor pile end. If the flow of forces cannot be satisfactorily checked and confirmed at difficult, highly loaded connection structures, the calculated dimensions and stresses should be checked by at least two test loadings to failure on full sized working parts.

8.4.16 Armoured Steel Sheet Piling (R 176)

8.4.16.1 Necessity

The increasing size of ships' vessels, the traffic of ship formations and the changed nature of locomotion have led to increased operational requirements being placed on waterfronts in inland harbours and waterways. In order to prevent damage to sheet piling structures also, these should have a surface which is as even as possible (R 158, section 6.6). With dual piles of large widths, the susceptibility to damage increases as a result of diminution of the approach angle and the increased clearance between the backs of the piles. The requirement for a largely even surface is achieved with an armouring for the sheet pilling, in which steel plates are welded into or over the sheet piling troughs (fig. R 176-1).

8.4.16.2 Scope

However, owing to the technical and economic resources involved, armouring is recommended only for stretches of waterfront subjected to special loads.

In inland harbours, these are waterfronts with very heavy traffic, especially with push tows and large motor vessels, as well as waterfronts in areas of high risk, such as changes of direction in plan and guide walls in lock entrances.

8.4.16.3 Elevation

The sheet piling armouring is required for the elevation area of the waterfront in which contact between a ship and the sheet piling is possible from the lowest to the highest water level (fig. R 176-1).

8.4.16.4 Design

The design of the armouring and its dimensions depend primarily on the opening width B of the sheet piling trough. This is determined by the

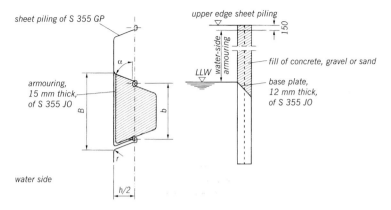

Fig. R 176-1. Armouring of a U-shaped sheet piling section

system dimension b of the sheet piling, the inclination α of the pile leg, the section height h of the piling wall, and the radius r between the leg and the back of the sheet piling (fig. R 176-1).

8.4.16.5 Pile Shape and Production Method for Armouring

In designing the armouring, it must also be made clear whether the sheet piles are Z-shaped or U-shaped. It is also of significance whether the armouring is to be fitted by the workshop, and therefore before driving, or on site, and therefore after driving.

In the case of Z-shaped piles and armouring produced on site, the steel plates are supported over their full width as far as the interlocks. Because they project beyond the alignment, they also protect the interlocks (fig. R 176-2).

For Z-shaped piles, production in the workshop in accordance with fig. R 176-2 cannot be recommended, because driving may then be impaired due to a lack of flexibility of the dual piles.

The subsequent installation of armouring is also possible with U-shaped piles, a completely even waterfront surface being achievable (fig. R 176-2).

During installation on site, pressure and adjustment measures must not be neglected with either section type. This will produce a waterfront structure of uniform rigidity, no longer comprising individual elements and having lost flexibility.

The best solution for installation and operation is offered by the piling wall of U-shaped piles with armouring produced in the workshop.

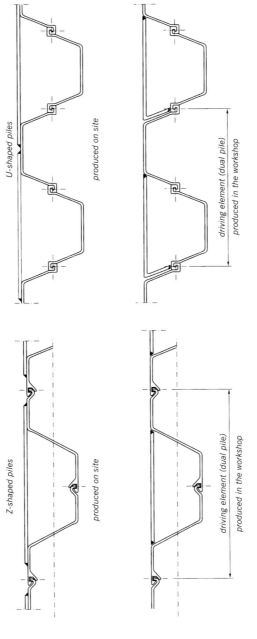

Fig. R 176-2. Ways of producing armouring for Z- and U-shaped piles

8.4.16.6 Information on Installation by the Workshop of Armouring for U-shaped Piles

In the case of installation by the workshop, the armouring is fixed to the interlock of the trough pile and the back of the peak pile. For technical reasons relating to driving, the interlock of the dual pile must also be welded, so that a rigid connection is created. Only then can the weld seams of the armouring withstand the driving process without damage. The pile back and the free leg of the peak pile are still available for elastic deformation (fig. R 176-2).

The armour plates can be welded together or bent (fig. R 176-3). The welded version produces the smaller width of the remaining cleavage, because limits are set for the minimum radius for cold forming in the bent version. The cleavage width in the welded version is approx. 20 mm. With a system dimension for the dual pile of 1.0 m, an even piling wall is then achieved to approx. 98 %.

8.4.16.7 Dimensioning

The armouring is generally not taken into account in the static calculations for the sheet piling. The thickness of the plate is derived from the support width corresponding to the opening width of the sheet piling troughs.

However, since the support width of the impact armouring is always greater than the width of the back of the peak pile, the armouring plates

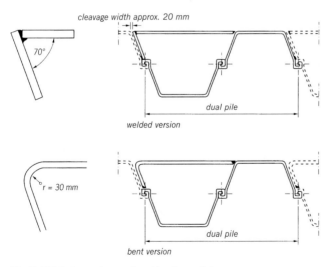

Fig. R 176-3. Armouring produced by the workshop

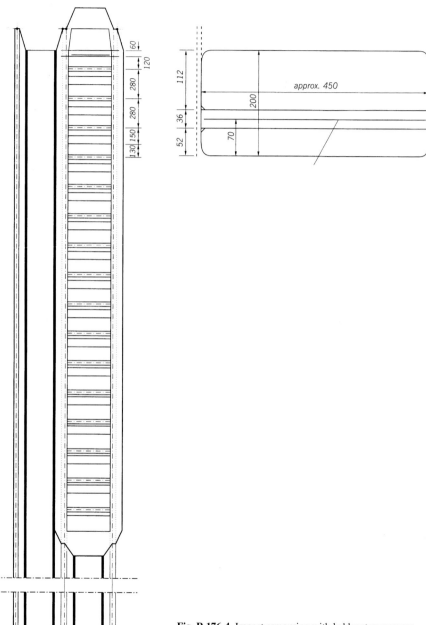

Fig. R 176-4. Impact armouring with ladder step recesses

must be thicker than the backs of the piles. To avoid plate thicknesses ≥ 15 mm, it is recommended that the remaining space between armouring and the dual pile be filled up.

8.4.16.8 Filling

A base plate is welded in to close off the bottom (fig. R 176-1) so that the space between armouring and the dual pile can be filled in. Sand, gravel or concrete can be used for filling purposes.

8.4.16.9 Access Ladders and Mooring Facilities

The impact armouring is usually interrupted in sheet piling where there are access ladders and recess bollards. In order to keep the surface as smooth as possible here, armouring with step recesses (fig. R 176-4) can be used at the ladder pile, or a continuous step recess box (fig. R 176-5) can be used, leaving a space which can be filled according to section 8.4.16.8, insofar as this solution appears technically feasible and economically acceptable. Fig. R 176-6 shows the arrangement of recess bollards in an armoured wall. The requirements of R 14, section 6.11 and R 102, section 5.13, are observed.

8.4.16.10 Costs

The extra costs for armoured sheet piling as compared to unarmoured solutions depend primarily on the ratio of the length of the armouring to the overall length of the sheet pile, and on the section. The delivery charges for the piling wall material are increased by 25 to 40 %. In the case of U-shaped piles, it is cheaper and technically better to include armouring in the design from the outset and have it installed by the workshop.

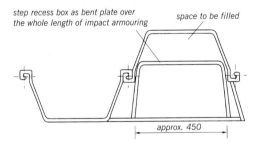

Fig. R 176-5. Impact armouring with integrated step recess box

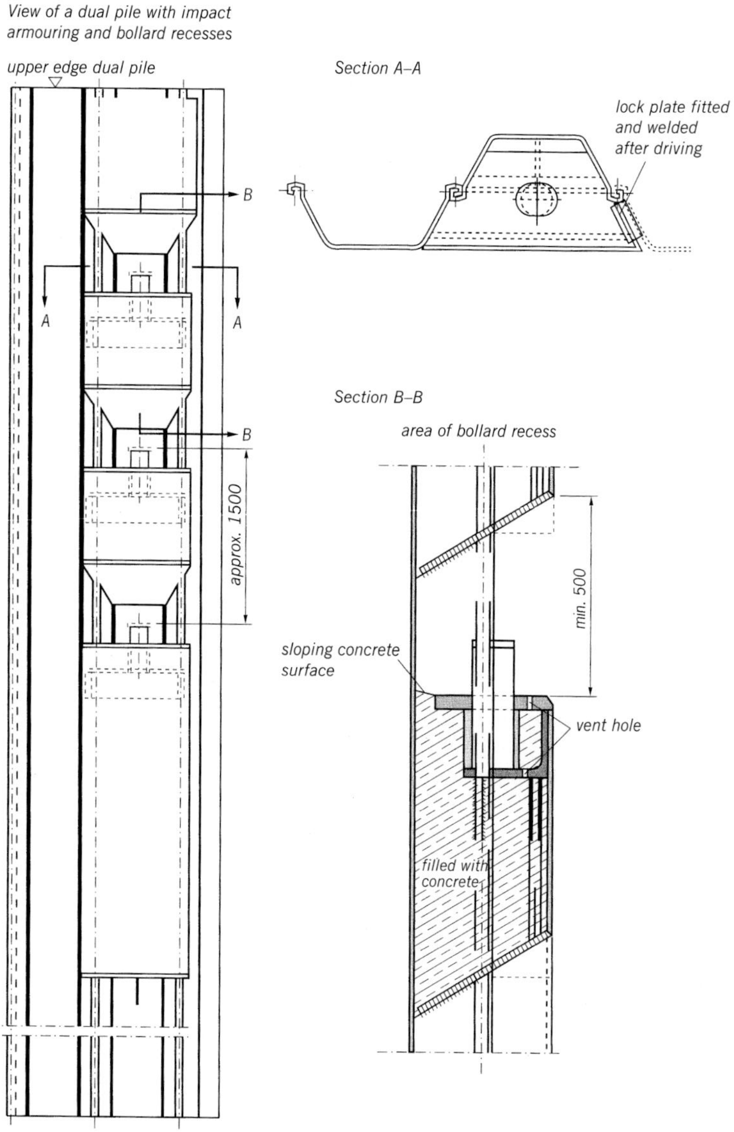

Fig. R 176-6. Impact armouring with recess bollards

9 Anchor Piles

9.1 General

Vertical constructions as waterfront structures, particularly quay walls, usually have to be equipped with corresponding anchoring elements for tilting and sliding safety, absorption of the horizontal forces from earth and water pressure, forces from the superstructure such as bollard tension and vessel impact. In the case of slight differences in terrain elevation, these forces can also be carried away by means of corresponding design of the pile-founded structure over pile bents. Larger differences in terrain elevation, as those regularly encountered in modern seaports and inland ports, require special anchoring elements.

The recommendations featured in this section for the structural calculations are based on the safety concept of verification of theoretical bearing capacity in limit cases using partial safety factors (see section 0.1).

9.2 Anchoring Elements, General Explanations

The anchoring of waterfront structures consists essentially in the structural elements described below.

Quay walls are preferably anchored with *anchor piles*. These can be designed as

driven steel piles, or
driven piles with grouted skin (VM-piles).

The latter resemble the steel piles as far as the bearing element is concerned (these are displacement piles because of the design of the point/pile shoe); they are grouted around the skin periphery with an additional cement suspension injected through a specially fitted grouting line during the driving process.

Steel piles as anchor piles are rolled products which are supplied as individual sections and more rarely even as assembled sections. Their outstanding features include good adjustment to the prevailing structural, geo-technical and driving-technical conditions. They can be driven in wherever necessary for the local conditions. They are relatively insensitive to driving hindrances (even a pile which has been slightly deformed by a hindrance can still bear its load without any problems) and heavy driving. Steel piles can be easily connected with other structural components.

Normally, steel piles are made from S 235 or S 355 as per DIN EN 10 025, but also from S 240 GP, S 270 GP, S 355 GP, S 390 GP or S 430 GP to DIN EN 10 248. As normal hot-rolled broad flange girders they correspond to EN 53-62 respectively DIN EN 10 025.

VM piles are driven as a rule at a batter of 3 : 1 to 1 : 1 and more rarely also 1 : 2, and grouted with cement suspension in a predefined length along their skin surface – usually in non-cohesive layers. In addition to the normal displacement (compression) of the soil during driving, grouting with cement suspension also produces a good connection between the steel and the surrounding soil. This can produce considerable improvements in the bearing capacity in non-cohesive soils. In addition, the cement suspension can also act as driving aid during the driving process, as long as it has not started to bond by then.

An alternative to VM piles is *tubular grouted piles*. In contrast to the steel driving piles, which are either vibrated or driven in, these are usually drilled in. They offer the advantage of low-vibration and low-noise placement. In this system, the drilling tube is also the bearing element. Alternatively, a solid steel element can also be used as bearing element. *Grouted piles (in-situ concrete and composite piles) with a small diameter* to DIN 4128 can also be used as further anchoring elements. These are not-prestressed grouted piles with shaft diameters ≤ 300 mm, in which the shaft diameter must amount to

at least 150 mm in in-situ concrete piles
at least 100 mm in composite piles

according to DIN 4128.

The *grouted in-situ concrete piles* with small diameter are placed by drilling, and marketed by some manufactures under the name root piles. They have a continuous longitudinal reinforcement of concrete steel and are usually filled with concrete or cement mortar.

The *composite piles* have a continuous pre-fabricated bearing element consisting of steel or more rarely reinforced concrete. They can also be placed by driving, e.g. as driven grouted piles with small cross section. For both types of pile, grouting can be carried out with compressed air (for root piles) or liquid pressure on the grouting element (cement suspension). Subsequent grouting is always possible, but all necessary precautions must be taken here (re-grouting tubes, sleeve pipe).

In order to avoid the zipper effect, the characteristic value of the actions on the anchor piles should be limited to max. 1500 kN for each pile. When this is exceeded, verification of the failure of an anchor pile must be provided.

Anchoring elements for waterfront structures can also consist of grouted anchors as per DIN 4125. Since these are steel tension elements of high-strength prestressing steel, corresponding corrosion protection to DIN 4125 is to be guaranteed. However, in each individual case the endurance and verifiability of functioning capacity must be examined separately. They may only be used as permanent anchors when there are no media present which corrode the concrete of the grouted element. In view of the special conditions prevailing in seaports, these anchoring elements are generally not suitable here. The following prestressing steels are certified for the steel tension elements: St 835/1030, St 1080/1230, St 1325/1470, St 1420/1570 and St 1570/1770.

The corrosion protection for these steels in the grouting section consists of:

(1) a cement suspension filled between the steel tension element and the corrugated or pressure pipe,
(2) the corrugated or pressure pipe,
(3) the outer grouting body of cement suspension.

During manufacture, a permanently elastic mass (usually Vaseline) is pressed into the free length between tension element and surrounding tube. On passing the anchor through the wall being anchored, care must be taken to ensure that the sealing is applied after prestressing and fixing the anchor in the head area, in such a way that any penetration of moisture to the prestressing steels or wedging devices is ruled out.

Normal tension forces for injection anchors are 300–600 kN, or up to 1200 kN for the larger anchor types.

In addition, the titanium anchor by Ischebeck is possible when certified accordingly. The anchor body consists of a corrugated steel tube through which the drilling and grouting process takes place. These anchors are manufactured in diameters from 30 mm (100 kN) to 103 mm (900 kN).

9.3 Safety Factors for Anchoring (R 26)

9.3.1 General

According to NAD to DIN V ENV 1997-1, it is recommended to provide safety verification according to DIN V 1054-100.

9.3.2 Battered Piles

In the case of battered piles with an incline steeper than 1 : 2, the design values of the pile resistances are to be reduced contingent on the incline compared to the design values as per DIN V 1054-100. The diminution factor amounts to 0.75 for piles battered to 2 : 1 and steeper; linear interpolations between 1.0 and 0.75 are allowed for inclines between 1 : 2 and 2 : 1 (see fig. R 26-1). Otherwise DIN V ENV 1997-1 applies with the NAD.

9.3.3 Grouted Anchors

DIN 4125 is applicable for grouted anchors.

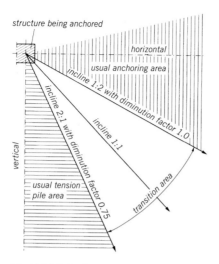

Fig. R 26-1. Resistance diminution factors of battered anchor piles compared to the design resistance to DIN V 1054-100

9.4 Limit Tension Load of Anchor Piles (R 27)

9.4.1 The limit tension load is taken to be the load which starts on pulling out the pile. If this is not clearly revealed in the load lifting line, the limit tension load is taken to be the load which does not yet endanger the presence and use of the structure on lifting the pile (in the pile axis). In the case of waterfront structures, the remaining lift can generally amount to around 2 cm. The limit tension load can be determined as "limit load of displacement" over the creeping dimension k_s according to DIN 4125. The creeping dimension should not exceed 2 mm.

9.4.2 The limit tension load for grouted anchors – both for temporary purposes and for lasting anchoring – is stipulated according to DIN 4125. The limit tension load for driven piles and grouted piles is stipulated as per DIN V 1054-100.

9.4.3 In preliminary designs the limit load can also be approximately determined by static penetrometer tests, if both the end bearing and the local skin friction are measured with suitable penetrometers. In so doing, the results of borings sunk in the vicinity must also be taken into account, in order to ascertain to which type of soil each of the sounding results refer. Indicative values for the characteristic pile resistances are stated in table R 4 in DIN V 1054-100, appendix E.

In the case of VM piles, preliminary designs may increase the stated skin friction tensions by factors varying between 2 and 4.

9.4.4 When determining the calculative limit tension load of the individual pile within groups of piles, the group effect must also be taken into consideration.

9.4.5 If the limit tension load is not achieved during trial loading, the greatest tensile force reached during the test is taken as theoretical limit tension load.

Under certain circumstances, the so-called "hyperbola method" [5] and [6] can be used to estimate the fracture loads – as asymptotic lines of the load-displacement curve. It must be noted that the hyperbola method can also lead to grossly incorrect estimations, particularly in compactly stratified, non-cohesive soils and in cohesive soils with a semi-firm to firm consistency [189].

9.5 Design and Embedment of Driven Steel Piles (R 16)

9.5.1 Design

In the choice of the piles, the loss of energy occurring during driving in the inclined position must be considered. With suitable subsoil, steel piles can be equipped with welded-on steel vanes, thus improving their bearing capacity. Vane piles should only be used in soils without any hindrances, and preferably in non-cohesive soils, and must reach far enough into the bearing soil. In the case of cohesive layers, the vanes should be positioned underneath these layers, and open driving channels should be closed, for example by grouting. Such vanes must be designed and arranged so as not to impede the driving process too much, without damage to the vanes themselves. The design of the vanes and the elevation at which they are arranged must be carefully adapted to the respective soil conditions. Care must be taken that saturated, cohesive soils are displaced but not compacted during driving. In non-cohesive soils, a highly compacted solid packing of soil can develop mainly in the vane area through driving vibrations, which correspondingly increases the bearing capacity, but at the same time makes the driving process difficult. Therefore, the subsoil must be thoroughly tested by borings followed by very careful exploration with penetrometer tests and soil mechanics laboratory tests before using vane piles in the execution of works. Pile cross-sections suitable for driving and resistant to bending should be used in heavy soil or with great pile lengths. In non-cohesive soil, the vanes of the vane piles should be at least 2.00 m long so that the required soil arching (packing together) is achieved in the cells. The clear intervals between cell walls should be less than 30 to 40 cm to safeguard packing together (see also [7], part 3, page 186).

The vanes are placed symmetrically around the pile axis and begin generally just above the end of the pile point, so that at least an 8 mm weld can be placed between vanes and pile point. The upper end of the vanes must also be given a correspondingly strong transverse weld. The welds are subsequently extended for a length of 500 mm on both sides of the vanes in the longitudinal direction of the pile. Intermittent weld beads are sufficient for the rest (interrupted weld seam).

The connecting surface of the vanes shall be sufficiently wide in consideration of the compulsive forces (in general at least 100 mm). Cross-section and positions of the vanes should promote cell formation. Depending on the soil conditions, the vanes can also be mounted higher on the pile shaft.

9.5.2 Driving

Secure guidance is required when driving piles with a flat batter. Slow striking hammers are basically preferable to rapid striking hammers because of their longer force action but also for environmental aspects (noise, vibration). However, rapid striking hammers can result in an increase in the bearing capacity in non-cohesive soils because of their "vibration" effect. Energy loss resulting from the inclined position must be taken into account when rating the hammer weight. The free length of the pile shaft projecting beyond the driving guide should be such that the tolerable bending stresses in the pile will not be exceeded during installation. The consequences of a possible jetting aid must be taken into account. For further details see [78].

9.5.3 Embedment Length

The resistance of flat battered anchor piles to being pulled out (limit tensile force), i.e. determination of the embedment length of the piles in the bearing subsoil, can be estimated using the indicative values for limit skin friction stated in table E 4 of DIN V 1054-100, appendix E. The skin friction is to be referred to the outer, unwound pile skin surface. These values must be checked by adequate subsoil investigations, e.g. cone penetration tests, with the prerequisite that the piles are not subject to any notable vibrations.

Settlement of the ground creates shear forces on the side or skin surfaces of the piles resulting from negative skin friction; in battered piles, the additionally occurring deformations and loads in the piles are to be taken into account, together with their effect on the pile connection structure. The design value of the anchor pile tension force is to be calculated on the basis of the limit tension force, depending on the load case, using the partial safety factors γ_M for resistances stated in table R 0-4, section 0.2.1. The ultimate resistance to being pulled out has to be stipulated in every case by an adequate number of trial loadings.

9.6 Design and Loading of Driven Piles with Grouted Skin (VM Piles) (R 66)

9.6.1 General

A VM pile is a driven steel pile that is driven into the ground under simultaneous grouting with mortar and is suitable for absorbing large tensile and compression forces. VM piles differ regarding production and bearing capacity in many things from anchors in accordance with DIN 4125, but have much in common with piles as per DIN 4128.

Use and type presume accurate knowledge of the soil conditions and soil properties, above all in the bearing foundation area. Non-cohesive soils with comparatively large pore volumes are particularly suitable.

Suitability of the subsoil is to be verified carefully, particularly for absorbing a tensile load as regards skin friction and movement of the piles under permanent load, particularly in cohesive soils.

VM piles can be made above and below the groundwater table. VM piles are specially recommended for anchoring waterfront walls, because the inner indentation of the hardened grouting mass with the subsoil produces good utilisation of the high inner bearing capacity of the steel pile. Permanent environmental compatibility of the grouting material with groundwater/soil must be guaranteed. This is the case as a rule with cement compounds. Any risk to the grouting compound from aggressive substances in the groundwater or soil must be checked with examinations as per DIN 4030.

9.6.2 Calculation of Piles

The bearing capacity of VM piles depends essentially on the following factors:

statically effective grouting length, circumference of the pile shoe, type of soil and depth of cover.

The circumference of the pile shoe is to be taken as the circumference of the grouting body. The skin friction surface is thus defined as a product of the pile shoe circumference and statically effective grouting length.

The driving grouting length and anchor batter should be selected so that the statically effective anchoring length l_w lies in a uniform, bearing strata as far as possible. This makes it easier to determine the values to be included in the design (in strata with differing bearing capacity, the values of the strata with lower bearing capacity have to be taken to avoid the so-called zipper effect).

Determination of the theoretical limit load in the preliminary design stage can be based for skin friction on the values in table E 4 of DIN V 1054-100, appendix E.

These values still have to be checked by trial loadings. The design value of pulling-out resistance results from the division of these values by the corresponding partial safety factor to DIN V 1054-100. The required length of a VM pile can be determined according to fig. R 66-1 for limit state 1C. Verification of stability of the anchoring should always be carried out according to DIN V 4084-100. This verification can be simplified in accordance with R 10, section 8.4.9; such simplification is generally on the safe side when the pulling-out resistance which can still be mobilised behind the foot point of the substitute anchor wall is not taken into consideration. Both anchoring force and embedment depth of the wall should have been determined previously with the partial safety factors for limit state 1C. In each individual case it must be checked to what extent the earth pressure force P_1 on the substitute anchor wall has to be taken battered or horizontal. The presumption of a wall friction angle $\delta_a = 0$ for the earth pressure force on the substitute anchor wall is normally on the safe side.

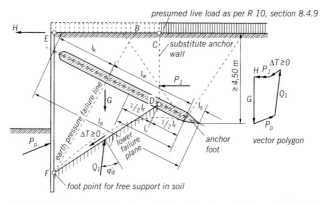

Fig. R 66-1. Stability for the lower failure plane for anchoring with VM piles

Note: the actions and internal forces shown in fig. R 66-1 are to be understood as design values.

The symbols therein mean, in [m]:

l_a = length of the pile
l_s = length of the pile foot
l_r = required minimum anchoring length determined from the design value of anchor force, skin surface, design value of skin friction
l_k = upper pile length, statically ineffective. Begins at the anchor pile head and ends on reaching the active earth pressure failure plane or in the upper surface of the bearing soil, if this is deeper

l_w = statically effective anchoring length. Begins at the active earth pressure failure plane or lower in the upper surface of the bearing soil respectively and ends at a depth at which the following three conditions are fulfilled:
(1) $l_w \geq l_r$
(2) stability for the lower failure plane as per R 10, section 8.4.9
(3) embedment depth l_w in the bearing soil min. 5.00 m

The length of the pile is thus: $l_a = l_k + l_w + l_s$

Verification of the stability of a waterfront structure anchored with VM piles can be simplified compared to the procedure in DIN V 4084-100. To do so, it must be ensured that the anchor force required for stability verification of the waterfront structure in limit state 1C behind the active earth pressure wedge is fully available. If this is the case, the following procedure is possible based on verification as per KRANZ [74] respectively R 10, section 8.4.9 (fig. R 66-1): the foot point D of a substitute anchor wall DC lies in the middle of the force introduction line starting from the anchor foot. The lower failure plane goes from this point D to point F of the waterfront structure. The position of F is determined according to R 10, section 8.4.9. The active earth pressure force P_1 determined with the design values of the shear parameters acts on the substitute anchor wall DC under the presumption of a wall friction angle $\delta_a = 0$. The passive earth pressure on the waterfront structure, also determined with the design values of the shear parameter according to DIN V 4085-100, is taken to be resistive. The stability for the lower failure plane is guaranteed when a driving additional force $\Delta T \geq 0$ in the direction of the lower failure plane can be absorbed, taking the design values of the shear parameters and neglecting the pulling-out force of the anchor which can be mobilised under the foot point of the substitute anchor wall. When the anchor spacing is greater than ½ l_r, the soil body $CDFE$ including the earth pressures P_p and P_1 may only be taken with a width of ½ l_r vertical to the sketched plane. Alternatively, the substitute anchor wall DC may be displaced by the dimension ½ l_r to the air side so that the width of the considered soil body is reduced accordingly. This procedure is equally applicable to other piles and also to grouted piles as per DIN 4125. Regardless of the stability for the lower failure plane, safety from foundation failure is to be verified to DIN V 4084-100 when there are unfavourable soil layers (soft layers underneath the anchoring zone) or high loads behind the active earth pressure wedge on the substitute anchor wall.

The limit tension force as per R 27, section 9.4. and tolerable load as per R 26, section 9.3, taken as basic criteria for a project, must be checked in any case by adequate trial loadings. A minimum of 2 trial loadings is specified in DIN V 1054-100 for grouted piles, but at least 3 % of the piles.

If the grouting length is the same as the pile length, i.e. grouting takes place along the length l_k, the characteristic pulling-out resistance for use in the stability verification is:

$$Q'_k = Q_k \cdot \frac{l_w}{l_k + l_w},$$

whereby Q_K is the limit tension load determined in the test.

9.6.3 Execution of Construction

The cross-section of the pile shoe is adapted to the shape of the pile shaft. In general, the dimensions for the cross-section vary between 450 and 2000 cm^2, for the circumference between 0.80 and 1.60 m.

The spacing of the piles can be selected according to requirements. It must be ensured however that this produces a technically sound installation. Therefore the centre-to-centre spacing should be at least 1.60 m. When anchoring sheet piling in which the anchor piles have been placed directly in the trough, the spacing is many times the double sheet pile width. The space and time sequence of the pile construction is to be coordinated in such a way that the setting of grouting mortar in neighbouring piles is not disturbed.

The bearing capacity of the VM piles is chiefly contingent on proper and expert execution of construction. Installation may therefore only be awarded to firms with experience and which provide a guarantee for diligent work. Particular significance is attached to the grouting work. The capacity of the mixing and grouting equipment is to be adjusted to the capacity of the driving gear. At the lower end of the shaft, the pile has a projecting, closed, wedge-shaped point of l_s length (fig. R 66-1). Depending on the pile type, this pile shoe is firmly welded or detachably screwed onto the shaft. It produces a void in the ground while being driven which is constantly filled with grout compound under pressure. A steel pipe or plastic hose is to be fastened to the pile shaft for conveying the grout compound to the pile shoe. Interruptions may not take place during driving so that the grout compound does not set before the pile has been completely manufactured.

The grout compound consists of cement, fine sand, water, trass and usually a swelling agent. The choice of the grout compound components depends on the respective soil type and on the degree of density of the foundation soil. Grouting compounds consisting only of cement stones produce greater adhesion and friction forces between girders and grouting bodies, and between cast body and soil; they should however only be used for scarcely permeable strata.

The consumption of grout compound per m pile depends on the following factors: theoretical hollow cross section, degree of density and volume of pores in the subsoil, as well as grouting pressure. The ratio of the consumption of grout compound to theoretical hollow space is designated as consumption factor. It can be generally expected to amount to at least 1.2, but can be substantially higher. Values of up to 2.0 are not unusual. If the consumption factor is under 1.2, the piles should be checked.

Trial loadings as per DIN V ENV 1997-1 are to be carried out as per DIN V 1054-100 section 5.4.2. The trial loading may not be carried out until the grout compound has hardened.

9.7 Design and Loading of Tubular Grouted Piles (R 207)

9.7.1 General

The tubular grouted pile is a drilled, thick-wall steel tubular pile with shaft and foot grouted from cement suspension. This pile is suitable for bearing high tensile loads of up to 2500 kN (characteristic value), is produced in a low-vibration and low-noise process and is thus a sensible alternative to driven and structurally connected anchoring elements in emission-risk areas and in cases requiring subsequent verification of the load on the tension pile. The tubular grouted pile as high-load anchoring element has not yet been covered by any DIN standard. Consultation is required in each individual case before certification is obtained from the building authorities. Good experience has been obtained with these elements for example in the port of Hamburg. This type of pile cannot be used at all or only with special additional measures in subsoil where large drilling jetting losses can occur.

9.7.2 Execution of Construction and Loading

The tubular grouted pile consists of individual tubular steel sections of approx. 3 m in length which are screwed together during installation. Tubes of steel grade S 355 JO (formerly St 52) are used for the diagonally placed tension piles. Tube dimensions from 73 to 102 mm outer diameter with wall thicknesses of 12.5 to 20 mm are used to absorb tension loads from 750 to 1500 kN (characteristic value), tube dimensions of 114×28 mm are used for 2500 kN piles. The tube has a rolled-on thread for better adhesion of the grouting mortar to the tube.

The piles are drilled section by section in position using the rotation jetting method with outer jetting. A drilling shoe 30–50 cm long at the point of the pile centres the tube during placement and produces a circular space around the pile shaft according to its larger outer diameter.

The jetting agent consisting of a water/cement mixture (cement suspension) comes out of the drilling point, loosens the prevailing soil and conveys it to the surface outside the tube. Depending on the geological conditions and jetting pressure, a column diameter of up to 70 cm is achieved in the foot area of the anchoring by the jetting stream emerging radially at the drilling point. This extension of the pile foot is grouted with a viscous cement suspension under pressure of up to 80 bar. The outer hollow around the steel tube remains filled with hardened cement suspension up to the pile head.

The production method selected according to the technical regulations of DIN 4128 results in possible live loads of up to 1500 kN depending on soil conditions and grouting element length. Following adequate hardening of the cement stone, the bearing capacity should be tested by trial loadings according to DIN 4128, section 9.

9.8 Anchoring with Piles of Small Diameter (R 208)

As explained in section 9.2, the production of grouted piles with small diameter is stipulated in DIN 4128.

The in-situ grouted piles with continuous longitudinal reinforcement are usually made of concrete, sometimes with cement mortar. Their bearing capacity (characteristic value) depends as a rule on the outer bearing capacity, i.e. that of the subsoil supporting the pile, and is around 300 kN for piles with a diameter of 150 mm and approx. 900 kN for piles with a diameter of 300 mm. It may be possible to verify higher bearing capacity from case to case by means of trial loadings, but this is to be limited to the inner bearing capacity, rated to DIN 1045.

Composite piles usually have higher outer bearing capacities than the in-situ concrete grouted piles, particularly when driven. In both cases, trial loads are recommended before stipulating the outer bearing capacities.

The so-called *single-rod pile* is an example of the composite piles with small diameter laid down in DIN 4128. This consists of ribbed concrete steel St 500 S-Gewi with ribs rolled on both sides. The steel used here means that this pile is also sometimes called GEWI pile. The rods are available in diameters from 16 to 60 mm. When processed in the pile, they are subject to certification by the building authorities.

The individual rods can also be extended by a sleeve joint using the rolled on thread. A sleeve joint subject to tension must always be locked with a lock nut.

9.9 Connections of Anchor Piles to Reinforced Concrete and Steel Structures

See R 59, section 8.4.3.

9.10 Transmission of Horizontal Loads via Pile Bents, Slotted Wall Plates, Frames and Large Bored Piles (R 209)

9.10.1 Preliminary Remarks

The complete solution for anchoring horizontal loads to land-side pile bents is contained in section 11, pile foundation structures, which can also be used for all intermediate stages through to single pile bents, both with and also without upper screening plate, and with intermediate tension element, e.g. round steel anchor.

In special cases however, simpler solutions are possible, as explained in section 9.10.2.

In addition, under special conditions, anchoring to vertical structures can also be sensible and economical. As examples of this, slotted wall plates and large bored piles are described in section 9.10.3.

9.10.2 Pile Bent Anchorage in Special Cases

The relatively large pressure pile loads are frequently transmitted by means of in-situ driven piles for economic reasons. If these consist of typical point pressure piles under corresponding conditions, there will be no resulting action on the waterfront structure, when the settlement depth reaches to a straight rising under 1 : 2 which starts at the shear force zero point of the waterfront structure. The influence on the waterfront structure in this case is negligibly small. This principle applies both to single pile bents as per fig. R 209-1 and to several rows of pressure piles as per fig. R 209-2.

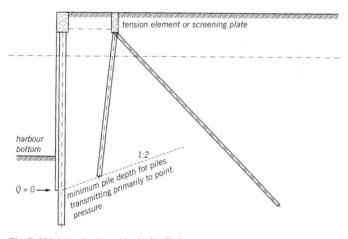

Fig. R 209-1. Anchoring with single pile bents

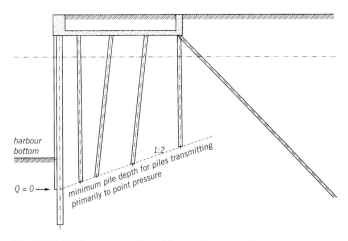

Fig. R 209-2. Pile bent anchoring with several pressure piles

If the corresponding conditions are not present, the pile forces create active earth pressures, i.e. actions on the waterfront structure which are then to be multiplied as usual with the partial safety factors. The magnitude of the actions depends particularly on the properties of the subsoil and the pile batter. It is to be determined according to DIN V 4085-100. If necessary, the effects of the peak pressure on active earth pressure as per DIN V 4085-100 are to be taken into account.

As regards designing the pressure piles, it should be noted that – depending on construction method and point in time of manufacturing the pressure piles with respect to manufacturing the waterfront structure –, deflection of the waterfront structure creates moments in the pressure piles depending on their distance to the waterfront structure. On the other hand, an improvement in the soil friction angle behind the waterfront structure resulting from the compaction effect of driving the pressure piles can also have a positive effect on the whole waterfront structure.

9.10.3 Special Anchoring

In general, the most favourable way of absorbing anchor forces in a waterfront structure will be when the anchor forces are transferred to the anchoring elements by the shortest possible route. With favourable soil conditions, this is horizontal anchoring to anchor plates, on pile bents in lower-lying bearing soil layers, or direct anchoring to the waterfront structure by means of battered piles.

But when the soil conditions for driving the waterfront structure and pile bents are unfavourable, in the form of hindrances or rocky formations in the subsoil, special structures may be necessary.

The slotted wall technique is representative for such structures.

In principle, transverse slotted wall plates can be used to obtain stability in a waterfront structure in slotted design without additional anchoring. Examples of such structures are known. Attention should be drawn to the fact that there are not yet any generally accepted calculation procedures for such solutions. In some cases, the land-side slotted wall plate has been replaced by large bored piles.

10 Waterfront Structures, Quays and Superstructures of Concrete and Reinforced Concrete

The recommendations for static calculations featured in this section are based on the safety concept of the verification of theoretical bearing capacity in limit states with the aid of partial safety factors (see section 0.1). The recommendations formulated in the context of designing concrete and reinforced concrete structures are currently still subject to the previous safety concept to DIN 1045 (July 1988). This means that the actions are to be taken as characteristic values, because DIN 1045 still works on the basis of the global safety concept which includes the safety factors for actions.

10.1 Design of Quays and Superstructures, Structures on Pile Foundations (R 17)

10.1.1 Concrete

In walls with large cross-sections, it must not be overlooked that the hydration heat inside the concrete causes self-stresses, which can lead to surface cracks, especially when there is simultaneous cooling of the outer portions, and can thus shorten the durability of the wall, particularly in sea water. All necessary data concerning the measures to be undertaken and considered for concrete to be used in sea water is contained in DIN 4030 and DIN 1045. The most important requirement is therefore the production of impervious concrete with high resistance to the corresponding form of attack.

Exposed surfaces of beams and slabs have to be protected against the influence of anti-icing salt and harmful substances.

Special reference is made to R 72, section 10.2.3.

10.1.2 Attack by Waters and Soils Harmful to Concrete

DIN 4030 deals with action, source, evaluation and investigation of matter aggressive to concrete and must be carefully observed. The concrete technology measures to be undertaken are to be found in DIN 1045, section 6.5.7.5. The use of liquefying additives such as concrete liquefiers or solvents to reduce the water/cement ratio is generally better than an increase in the cement content to over 350 kg/m^3. Fresh concrete is especially susceptible to washing out, and should therefore not come in contact with aggressive water during curing and for some time thereafter. This requires good possibility of consistency of the concrete and impervious formwork.

Care must be taken in the selection of the concrete mix to consider whether the structural parts are exposed to fresh or sea water, keeping in mind that groundwater is also frequently aggressive to concrete (see DIN 4030). Depending on the composition of water and soil, special standard Portland or blast-furnace cements are used, such as Portland cement low in tricalcium aluminate, blast-furnace cement low in clinker, or equivalent cements. It may also be advantageous to partially replace cement by fly ash.

10.1.3 Front Face of the Wall, Nosing

Concrete walls are constructed vertically and have a 5 to 5 cm chamfer at the upper edge of the wall, or are correspondingly rounded off, or when cargo is handled above them, are stabilised on the water and land side with steel brackets, whereby R 94, section 8.4.6 and R 64, section 8.4.14 must be observed where applicable. A specially installed nosing for the protection of the wall and to safeguard line handlers from slipping, must be constructed in such a way that the water can easily drain off. At quay walls with a sheet pile front and reinforced concrete superstructure, the reinforced concrete cross-section is extended about 15 cm beyond the front face of the sheet piling. The waterside lower edge of the reinforced concrete superstructure should however be positioned at least 1 m above mean tidal high water or the mean water level, to avoid any corrosion occurring here to a specially high degree. The bevel underneath is constructed at about 2 : 1 so that vessels and load handling equipment cannot be trapped underneath. The bevel is fitted with a folded steel plate protection which is connected flush to the face of both the sheet piling and the upward concrete wall.

10.1.4 Facing

Facing of the concrete can be omitted if satisfactory concrete is used. If a facing is required as protection against unusual mechanical or chemical wear, or for artistic purposes, the use of basalt, granite or hard brick is recommendable. Ashlar or slabs at the front upper edge of a wall must be secured on their rear side against displacement and against lifting, by a sufficiently thick reinforced concrete moulding or similar device. Concrete with high wear resistance (DIN 1045, section 6.5.7.6) is to be recommended on the coping of the wall, but is only required if there is very heavy traffic and if vessels are moored with steel hawsers.

10.1.5 Expansion and Construction Joints

All quay walls are provided with expansion joints in order to enable the absorption of movements deriving from shrinkage, temperature and differential settlements.

The length of the sections between the expansion joints is generally about 30 m. However, spliced constructions have been completed with far larger block sections without any problems for the structure. But the section length must be considerably shortened, if the shrinkage and temperature movements are hampered for example by embedding in solid subsoil (rock) or by connection to slabs which have already been concreted.

Construction joints without continuous reinforcing can also be arranged. The expansion joints are keyed for reciprocal horizontal support of the sections. The indentations should be designed so as not to hinder any changes in section length. In special cases, dowelling can also be of advantage. In the case of pile-founded walls, horizontal keying is accommodated in the relieving platform.
Joints should be protected against washing out of the backfill.
As far as the construction joints are concerned, reference is made above all to R 72, section 10.2.4.
Construction joints are to be arranged in the front wall of L-shaped concrete walls in such a way that only the finest possible shrinkage and temperature cracks occur. Vertical construction joints with continuous horizontal reinforcement are to be avoided. The reinforcement is to be designed in such a way that it also forms a safety net to prevent shrinkage and temperature cracks taking into account the crack width restriction as a result of environmental conditions (see also R 72, section 10.2.3), so that special crack prevention meshes are not required.

10.2 Construction of Reinforced Concrete Waterfront Structures (R 72)

10.2.1 Preliminary Remarks

In general, the following standards are to be taken into account:
DIN 488, DIN EN 499, DIN 1045, DIN 1048, DIN 1084, DIN 1164, DIN 4030, DIN 4099, DIN 4226 and DIN 4227, in future also DIN EN 196, DIN V ENV 206, DIN EN 450.

In the design of the reinforced concrete members of quay walls, the future behaviour of the quay wall and its service life in comparison to presumed savings should be considered by means of special statical and structural measures. The special stresses in marine construction and the resulting increased demands on the construction and maintenance of structural members in contact with water are to be given special consideration with the following rules.

10.2.2 Zone Division for Reinforced Concrete Structural Members in Contact with Water

Within the meaning of the preliminary remarks, the following zones are to be differentiated on the water side according to the elevation of the structural members:

- Zone A
  above MHW or above MHWS
- Zone B
  between MWH and MLW or MHWS and MLWS
- Zone C
  below MLW or below MLWS.

For these three zones, varying requirements are to be established as to the quality of the concrete, dimension and design of the reinforcing steel with regard to the possibility of cracking, as to the covering of reinforcing steels and as to the design of the construction joints, etc., while being mindful of special local conditions.

10.2.3 General Principles

Waterfront structures are subject to attacks by changing water levels, waters and soils harmful to concrete, ice and ship impact, chemical actions of transhipped and stored goods, etc. It therefore does not suffice to design the reinforced concrete members of waterfront structures only in accordance with statical demands.

Of utmost importance are a highly impervious concrete with high abrasion resistance, intensive post-treatment and sufficient covering of the reinforcing steel. The latter should be greater than required by DIN 1045, table 10 and should be in general min. 5 cm, better 6 cm. DIN 1045, section 17.6 must be taken into consideration with regard to the limitation of the cracking width under the live load.

The strength class of the concrete shall be at least B 25. In consideration of the danger of leaching out, the cement content should not be less than 325 kg/m^3, consequently the volume of fine aggregates (≤ 0.250 mm) shall be about 450 kg/m^3 including cement.

The water-cement ratio shall not exceed 0.50. If fines are lacking in the concrete aggregate, coal fly ash, trass, suitable stone flour or the like can be added to improve the workability and density of the concrete.

10.2.4 Construction Joints

The usual rules apply to the design of the construction joints in zone A. Construction joints should be avoided in zones B and C, if the other conditions so permit, and if it cannot be ensured that the adjoining surface will be clean when the concreting of the next section is begun.

10.2.5 Precast Structural Elements

These can be used. However, provisions must be made for joining them securely with in-situ concrete, and for the proper transmission of forces. They are useful in the solution of construction problems in zones B and C respectively in avoiding reductions in quality in these zones. The unavoidable construction joints in this case, especially at points in the structure where the stresses are high, must therefore be carefully designed and their construction must be under constant supervision.

Prestressed precast structural members can be used, in which case however, the minimum dimensions indicated in DIN 1045 shall not be applied. As a rule, the use of precast units is limited to zone A because of the special requirements which must be met when a unit is prestressed in place.

10.2.6 Weldability of the Reinforcing Steels

According to DIN 488, part 1, the reinforcing steel grades BSt 420 S and BSt 500 S are generally suited for welding by manual metal-arc welding (E), metal-active gas welding (MAG), gas pressure welding (GP), resistance flash welding (RA) and resistance projection welding (RP).

10.2.7 Hair Cracks and their Allowable Width

Owing to the danger of severe corrosion, all waterfront structures shall be so constructed that no appreciable cracks occur. Inasmuch as good concrete has a fairly high tensile strength which is not considered in the design calculations, a design according to state II is also permissible. It is important to follow a well-planned concreting sequence in which the inherent deformation of freshly placed concrete sections from dissipating hydration warmth and shrinkage is not hindered too much by previous elements which have already essentially cooled down. This can be accomplished in large structures such as piers by casting the beams and slabs as a unit. In order to avoid the resulting additional costs for formwork by concreting the beams in advance for use as a support for the slab formwork, it is possible, for example, for the slabs to be prestressed in one direction only to avoid the formation of hair cracks; the slabs are in this case cast in strips up to a width of 5 m with sealed construction joints between adjacent strips without continuous reinforcing.

If these measures are not used, extensive hair cracks cannot be avoided in larger structures, even when increasing the amount of reinforcing steel used. But these are in general harmless and will disappear again if the mean crack width of $w_{cal} = 0.25$ mm (e.g. in accordance with table 2 of the leaflet of the German Concrete Association: Limitation of Crack Formation in Reinforced Concrete and Prestressed Concrete Construction [117]) is not exceeded. However, when there is severe danger of corrosion, particularly in the tropics, all hair cracks which do not close of their own accord must be permanently sealed by injecting polyurethane resin or other similar material.

Restriction of crack widths with large cross-sectional dimensions can be achieved with a minimum reinforcement according to the Supplementary Technical Contractual Conditions – Hydraulic Engineering (ZTV-W) for hydraulic engineering works in concrete and reinforced concrete (work area 215) section 3.4 [118].

10.2.8 Special Remarks

Since DIN 1045 deals chiefly with the design and dimensions of reinforced concrete structures without extreme external influences, difficulties occur in some cases in its application on foundation and marine/hydraulic engineering structures. The dimensions of these structures tend to conform less to statical requirements. Special importance is given in general for these structures to demands for simpler construction work without difficult formwork, for technically sound embedment of sheet piling, piles and the like, and for adequate safety against uplift, sliding and impact of vessel.

Therefore requirements of DIN 1045, insofar as not of a fundamental nature, but which in their present working are a hindrance to practical dimensioning of civil engineering structures, should be modified in co-ordination between designer, examining engineer for structural design and the responsible building inspection authority, so that an adequately safe, structurally satisfactory and economic design and dimensioning will be possible.

10.3 Formwork in Marine Environment (R 169)

10.3.1 General

Besides the loads from the lateral pressure of the fresh concrete and a precautionarily applied live load of approx. 0.25 kN/m^2, formwork in marine environment is also subjected to the stresses from wind, wave action, high water levels, impact of drifting objects and rigs, impact of barges or tugs and the like.

10.3.2 Principles for the Design of the Formwork

(1) In the zone influenced by the tide and/or wave action, formwork should be avoided if possible, for example by using prefabricated elements (fig. R 169-1, water side), by choosing a higher bottom level of the concrete construction, or similar measures.

(2) Concreting work in the zone influenced by the tide and/or waves should take place as far as possible in calm weather periods.

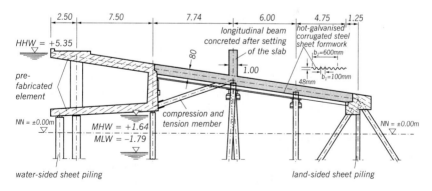

Fig. R 169-1. Execution example of a quay wall in marine environment, with prefabricated RCC element and rear corrugated steel sheet formwork

(3) Areas difficult to reach, such as the underside of pier slabs, should, if possible, be placed in forms with the use of sacrificed formwork, such as concrete slabs, corrugated steel sheets, or similar.

10.3.3 Construction of the Formwork

(1) Formwork which is to be used several times should be sturdy, easy to repair and to shift around.

Prefabricated wooden formwork or large-surface steel formwork elements have proven themselves; these can be installed and moved rapidly and in large units, so that their employment in an endangered area is limited to comparatively brief periods. Reference is also made here to portable formwork.

(2) Formwork in marine environment should largely be able to react elastically if struck by wave action, which is the case for example with a corrugated steel sheet bottom-formwork at proper design and elevation (fig. R 169-1).

The corrugated steel sheet panels must be secured from lifting and equipped for the connection to the concrete, as sacrificed formwork, for example with galvanised wires embedded in the concrete, or with other anchorages.

The corrugated steel sheet formwork shall not form a part of the corrosion protection for the slab reinforcement, because of the joints between the individual panels.

10.4 Design of Reinforced Concrete Roadway Slabs on Piers (R 76)

10.4.1 Roadway Slabs for Regular Traffic
Refer to R 5, section 5.5 for the load actions.

10.4.2 Slabs for Occasional Traffic
For only occasional use by individual vehicles, e.g. rescue or fire brigade vehicles of up to 12 t, it is possible to revert to DIN 1055, part 3 with special attention to section 6.3.2 and DIN 1045, section 20, after obtaining approval from the responsible building inspection authority. Contrary to DIN 1045, section 20.1.3 however, a minimum thickness of 20 cm is recommended for pier slabs.

10.4.3 Special Cases
If the design must provide for cargo stacked on the pier slabs, it is recommended that a uniformly distributed live load of 20 kN/m^2 be considered in the most adverse positions.

If pier slabs also support a railway, a slow speed may be assumed, so that a vibration factor according to R 5, section 5.5 can be presumed.

10.5 Box Caissons as Waterfront Structures in Seaports (R 79)

10.5.1 General

Box caissons often offer economical solutions for retaining heavily loaded, high vertical banks in areas with good load-bearing soil, and especially for advancing construction in the open harbour water.

Box caissons consist of rows of floating reinforced concrete sections which are designed for floating stability as a rule with additional ballast. After being launched and positioned over a properly prepared bearing area, they are then filled and backfilled with sand, stones or other suitable material. The water-side cells are often not filled to reduce the edge pressures. When they are in place, the top is barely above the lowest working water level. A reinforced concrete superstructure is placed on top of the caisson to give additional rigidity to the structure and to form the coping of the front wall. Any irregular settlement and horizontal displacement which occur when the caissons are set in place and during backfilling can be equalised by a suitable shaping of the reinforced concrete top.

The front wall of the caissons must be resistant to mechanical and chemical wear.

10.5.2 Calculation

Apart from the verification of stability in the final state, structural conditions such as floating stability of the caissons are to be investigated during construction, during launching, when they are set in place and during backfilling. The following are to be verified for the final state in addition to the requirements of DIN V 1054-100:

- Limit state 2 (serviceability)
- Safety against bottom erosion

In contrast to DIN V 1054-100, the bottom joint may not gape under any action combination of the characteristic loads.

Verification of stability for a box caisson must also be provided for the longitudinal direction. This must take account of centre support as well as support at the ends only. In these limit case investigations, the partial safety factor γ which applies to reinforced concrete in the normal loading class, may be divided by 1.3.

10.5.3 Sliding Danger and Factor of Safety Against Sliding

An especially careful investigation is required to determine if silt will be deposited on the foundation surface in the period between the completion of the founding and setting the box caissons in place. If this is possible, it must be verified that a sufficient safety factor remains against sliding of the caissons on the dirty foundation bottom, even under the existing circumstances. The need for this precaution is similarly valid for the surface between the existing subsoil and the fill placed over a dredged area.

Sliding danger can be reduced at reasonable costs by a roughened bottom surface of the concrete bottom. The degree of roughness must be matched by the mean grain size of the foundation joint material. If the bottom surface of the concrete is correspondingly roughened, the angle of friction between the concrete and the foundation surface is to be assumed to be equal to the angle of internal friction φ'_r of the foundation material, at smooth bottom surface, only with $^2/_3\, \varphi'_r$ of that of the foundation material. The danger of sliding can also be decreased by a greater foundation depth. The least favourable combination of water pressures is to be presumed at the bottom and sides of the caissons. These can be caused by hydraulic backfilling or by tidal changes, precipitation, etc. Hawser pull is also to be taken into consideration. Under simultaneous acting of all most unfavourable loads, the verification of safety from sliding must still be provided with the characteristic values, when there is full clarity about the characteristic values of the acting loads and shear resistances in the foundation bottom. Verification is also required of compliance with sliding safety as per DIN V 1054-100.

10.5.4 Structural Design

In order to avoid excessive stresses in the longitudinal direction, the box caissons should generally be constructed approx. 30 m long, but no longer than approx. 45 m, even at high structures.

The joint between two adjoining box caissons must be so designed that the unexpected uneven settlement of the caissons when being set in place, during filling and backfilling, will be absorbed without damage. On the other hand, in the final state it must be sufficiently tight to prevent leaching out of the backfill.

A design with tongue and groove joint extending from top to bottom may only be used when the relative movement of adjoining caissons is expected to be slight, even when there is a satisfactory solution to the tightness problem.

A solution according to fig. R 79-1 has proven practical. Here four vertical reinforced concrete ridges are so arranged on each of the side walls of the caisson, that they face each other on both sides of the joint and form three chambers when the caissons are in place. As soon as the adjoining caisson is installed, the two outer chambers are filled with gravel of suitable gradation as a seal. The centre chamber is flushed clean after backfilling of the caissons when settlements have subsided for the greater part, and is carefully filled with underwater concrete or concrete in sacks. The danger of washout under the foundation slab exists when there is a great difference in water level between the front and rear of the caissons.

In such cases, the layers must be mutually filter-stable and with regard to the subsoil. To reduce high water pressure differences, backwater drains can be successfully used as per R 32, section 4.4.

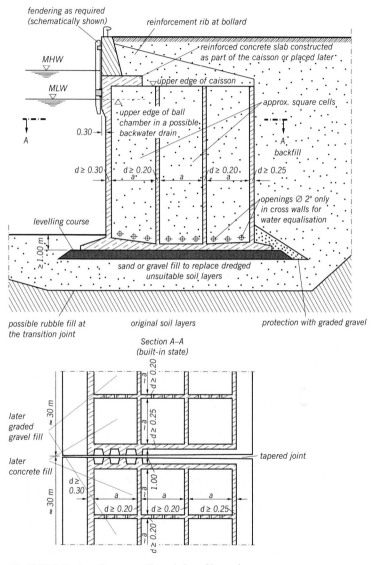

Fig. R 79-1. Design of a quay wall consisting of box caissons

When there is a risk of scour from water current and wave action, adequate scour protection is to be provided as per R 83, section 7.6.

10.5.5 Construction

The box caissons must be set on a well levelled foundation of stones, gravel or sand. When weak soil layers exist in the foundation area, these must be removed and replaced with sand or gravel before setting the caisson (R 109, section 7.9).

10.6 Pneumatic Caissons as Waterfront Structures in Seaports (R 87)

10.6.1 General

Pneumatic caissons may offer an advantageous solution for retaining high banks when their installation can be accomplished from the shore. The pneumatic caissons for the quay wall are then first sunk making use of accessibility from the existing terrain, and subsequently, the dredging work which is limited to the harbour basin, is carried out.

Pneumatic caissons can also be constructed as floating caissons, when an adequate bearing foundation does not exist or cannot be provided in the sinking area, or when the levelling of the foundation bed creates special difficulties, as with a rocky bottom. The construction fundamentals given in R 79, section 10.5.1 are then equally applicable to pneumatic caissons.

10.6.2 Calculation

R 79, section 10.5.2 remains valid. This is joined for the sinking situations of the caisson into the ground by the usual calculation for bending and shear forces in the vertical direction as a result of uneven surcharge on the caisson cutting edges, and for the horizontal bending and shear forces due to unequal active earth pressure.

In view of the location and preparation for the foundation and the firm keying of the pneumatic caissons edges and the concrete plug in the working chamber with the foundation layer, the caisson may be assumed to have a normal level foundation. Hence in contrast to R 79, section 10.5.2 paragraph 2, the bottom joint may gape, but the minimum distance of the resultant from the front face of the caisson from actions of characteristic loads shall not be less than $b/4$.

At high water pressure difference, the danger of washouts in front of and under the foundation bed is to be investigated. If necessary, special safeguards are to be provided to prevent washouts, such as soil stabilisation from the working chamber or similar. However, it may be economically preferable to deepen or widen the foundation bottom.

In the final state, a special verification of stresses resulting from uneven bearing for the longitudinal direction is not necessary for pneumatic caissons.

With especially large dimensions however, it is recommended that a verification of the stresses be provided for a bottom pressure distribution according to BOUSSINESQ.

10.6.3 Sliding Danger and Factor of Safety Against Sliding
Sliding safety is to be guaranteed according to DIN V 1054-100. Otherwise, R 79 section 10.5.3 is valid unchanged.

10.6.4 Structural Design
R 79, section 10.5.4, para. 1 to 3 remains valid. Fig. R 87-1 shows a method which has been used successfully in practice for closing the joint between sections of pneumatic caissons. Sheet pile interlocks are em-

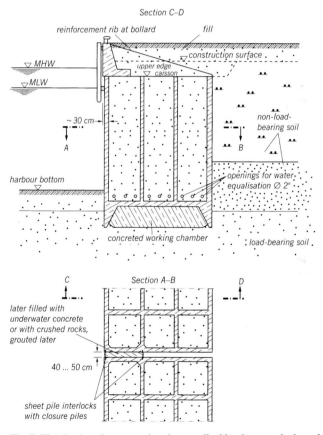

Fig. R 87-1. Design of a pneumatic caisson wall with subsequent harbour dredging

bedded in the side walls during construction of the caissons. After the caissons are in their final position, with the customary 40 to 50 cm space between them, flexible closure piles are driven into the two pairs of interlocks, thus closing off the space between the caissons. The enclosure formed by the two closure piles is cleaned out and filled with underwater concrete if the subsoil is firm. When poor subsoil, the space is filled with crushed rocks and can be grouted later. The front closure pile may lie flush with the face of the caisson. However, it may also be set back in order to form a shallow recess for installing an access ladder or the like. If large fluctuations in water level occur, the elevation of the bottom joint must be such that an adequate safety factor against underscouring is ensured, or the differences in water pressure can be compensated by using suitable backfill and drainage devices.

10.6.5 Construction

Pneumatic caissons installed from the shore are sunk from the ground on which they were previously constructed. As a rule, the soil in the working chamber is excavated almost exclusively under compressed air, or jetted and pumped out. If the soil at the predicted foundation depth proves to be too weak, the caisson is sunk to a corresponding deeper level.

When the required foundation depth has been reached, the bed is adequately levelled and the working chamber concreted under compressed air.

Floated-in pneumatic caissons must first be set down on the existing or deepened bed. In general, a rough levelling of the bed suffices, as the caisson cutting edges penetrate easily into the soil with their narrow cutting width, so that smaller irregularities of the foundation surface are of no importance. Subsequently, the caissons are sunk and concreted in the manner already described.

10.6.6 Frictional Resistance During Sinking

The frictional resistance depends on various characteristics of the subsoil and on the construction. It is influenced by:

(1) Type of soil, density and strength of the existing strata (non-cohesive and cohesive soils).
(2) Groundwater level.
(3) Embedded depth of the caisson.
(4) Plan and size of the caisson.
(5) Batter of the outer wall surface.
(6) Use of lubricants.

The determination of the necessary "sinking overweight" for any given sinking condition is less a matter of exact calculation than of experience.

It is sufficient in general if the "overweight" (total of all vertical forces without taking friction into account) is adequate to overcome a skin friction of 20 kN/m² on the embedded caisson surface. At lower overweight (modern reinforced concrete caissons), a special investigation is recommended with the use of additional measures such as lubrication and the like.

10.7 Design and Dimensioning of Quay Walls in Blockwork Construction (R 123)

10.7.1 Basic Remarks on Design and Construction

10.7.1.1 Waterfront structures in blockwork construction can be built successfully only if load-bearing soil is present below the base of the foundation, its bearing capacity can be improved (for example by compacting) or the non-load-bearing soil can be replaced.

10.7.1.2 The dimensions and the weight of the individual blocks must be determined in consideration of the available construction materials, the fabrication and transportation potentials, the capacity of the apparatus for placing the blocks, the conditions to be expected with regard to the site location, including wind, weather and wave action during construction and under operating conditions. If it is possible to transport the blocks to the site in a submerged condition, the decrease in effective weight resulting from the buoyant force may facilitate delivery of the blocks.

However, the chief advantage gained by this decrease in effective dead load is the corresponding increase in the working radius of the crane used to place the blocks. In any case however, the blocks must be sufficiently large or heavy to withstand wave action. This requirement will in turn obviously lead to the need for construction equipment of greater capacity. Blocks which appear to be too light for the expected stresses can, if the occasion so demands, be provided with adequately large pockets and interconnected with sturdy anchors concreted into the pockets. Since the wall will be subjected to frequent and severe stress reversal, these block anchorings must be designed for a service life of ample length. In this connection, consideration of corrosive action is especially important.

Especially when using a floating crane, frequently blocks of 600 to 800 t effective dead load are chosen for placing.

10.7.1.3 The concrete for the blocks must be watertight. The strength class of the concrete must be at least B 25. The same also applies to the reinforced concrete wall coping constructed at the site. Greater resistance from seawater attack is achieved by using cement with low aluminate levels.

In addition to blast furnace cement, Portland cement can also be used if part of the cement (up to 25 M-%) is replaced by fly ash. A tight concrete structure and thus high resistance to chemical attack can also be achieved by adding silica dust (up to 10 M-%).

A sucking formwork section can be used to improve the edge zone of the concrete. Surplus water from the edge zone is drained to lower the water/cement ratio here and make the concrete structure tighter.

10.7.1.4 The blocks have to be so shaped and placed that they will not be damaged during installation and produce good keying transverse to the wall. They should extend along the entire width of the wall if possible. In weakly deformable subsoil, smooth continuous vertical joints should be avoided in any case. This is achieved, for example, if the blocks are bedded at an inclination of 10 to 20° from the vertical. To accomplish this, a bearing surface must be provided, consisting of horizontally placed blocks, a sunk floating caisson or the like. Wedge-shaped blocks form the transition. The latter can also be utilised if it becomes necessary to correct the inclination. The sloped position of the blocks facilitates their placement and also promotes a minimum joint width between the individual blocks. However, this does increase the number of block types. All blocks for this type of construction are made with tongue-groove type interlocks to key their adjacent sides together. The tongue projection lies on the exposed side of the blocks already in place, and the succeeding blocks are guided so that their groove engages this tongue, along which they then slide downward into place.

If the blocks are simply stacked and not placed on a slope, which is recommendable in settlement-sensitive subsoil, large joint widths can be avoided only at great expense. These joint widths can however be accepted if suitable backfill material is used. Generally the optimum cost solution should be sought regarding allowable joint width and backfill material. The walls can be erected of blocks with tongue and groove joints or with I-shaped blocks.

10.7.1.5 A foundation bed of hard rubble at least 1.00 m thick is placed between the load-bearing subsoil and the block wall. As a rule, the bed must be carefully graded and levelled with special gear and with the aid of divers. In sediment-laden waters, this bed must be thoroughly cleaned before the blocks are placed, so that the foundation joint does not become a failure plane. This is especially important where the blocks are simply stacked instead of being place on a slope.

10.7.1.6 In order to prevent the rubble bed from settling under the surcharge, especially in fine-grained, non-cohesive subsoil, its voids must be filled with suitably graded gravel. Otherwise, a filter layer of graded gravel must be placed between the foundation bed and the bearing subsoil. If the foundation soil is very fine-grained but not cohesive, a plastic screen cloth should also be placed under the graded gravel filter as a stabilisation measure between the foundation soil and the graded gravel filter.

10.7.1.7 Depending on the equipment used, the blockwork construction may be particularly successful in areas with severe wave action and in countries in which there is a scarcity of skilled workers. Besides the use of suitable heavy gear however, this method chiefly requires an unusually large amount of diving work in order to ensure and check the required careful execution of both the foundation bed and the block placing and backfill work.

10.7.2 Nature of Acting Forces

10.7.2.1 Active and Passive Earth Pressures

The assumption of the active earth pressure is adequate, as the wall movements which actuate it may be taken for granted. With the generally very slight foundation depth of the block walls, the passive earth pressure is not to be taken into account.

10.7.2.2 Water Pressure Difference

If the joints between the individual blocks have good permeability and if rapid water table equalisation is ensured by the proper choice of backfill material (fig. R 123-1), the water pressure difference on the quay wall

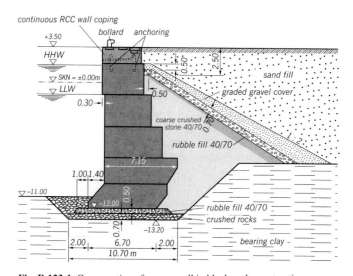

Fig. R 123-1. Cross-section of a quay wall in blockwork construction

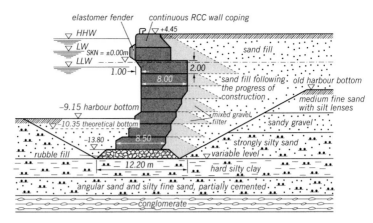

Fig. R 123-2. Design of a quay wall in blockwork construction in an earthquake region

need be assumed to be only half the height of the highest waves to be expected in the harbour basin, at the most unfavourable water level as per R 19, section 4.2. Otherwise, the water pressure difference as per R 19 is to be added to half the wave height. In cases of doubt, reliably operating backwater drainages can be installed, even if wave action is present. On the other hand, experience has demonstrated that there is no method for properly sealing the block joints.

An effective filter layer, which will without question prevent washouts, is to be installed behind the quay wall or between a backfill of coarse material and a subsequent fill of sand and the like (figs. R 123-1 and R 123-2).

10.7.2.3 Stress Due to Waves

When waterfront structures of blockwork construction must be built in areas in which high waves can occur, special stability investigations are required. In doubtful cases, a special investigation must be made by means of model tests, to determine whether breaking waves can occur. If this is the case, the risks which must be anticipated with regard to the stability and service life of a block wall are so great that this method of construction can no longer be recommended. The relation between the water depth d at the wall to the wave height H can be used as a criterion for evaluating whether breaking or only reflected waves occur. If $d \geq 1.5\,H$, it can generally be assumed that only reflected waves will occur, see also R 135, section 5.7.2 and R 136, section 5.6.

Wave pressures act not only on the face of the wall but are also transmitted to the joints between the individual blocks. The resulting increased water pressure in the joints is momentarily greater than the buoyant force, with the result that the effective block weight, and consequently the friction between blocks, is reduced to the point where the stability of the wall is endangered. At the moment the wave recedes, the pressure drop in the narrow joints, which is also influenced by the groundwater, takes place more slowly than along the outer surface of the quay wall, so that higher water pressure occurs in the joints than that attributable to the water level in front of the wall. At the same time however, the active earth pressure and the water pressure difference from behind remain fully effective.

10.7.2.4 Hawser Pull, Vessel Impact and Crane Loads

The relevant recommendations such as R 12, section 5.12, R 38, section 5.2, R 84, section 5.14 and R 128, section 13.3, are applicable here.

10.7.3 Calculation, Dimensioning and Further Remarks on Design

10.7.3.1 Base of Wall, Ground Pressure, Stability

The block wall cross-section must be so designed that the ground pressure at the foundation joint due to dead load is distributed as uniformly as possible. This can be achieved as a rule without any difficulty, by a suitable design of the base with a spur on the water side projecting beyond the face of the wall, and by placing a "knapsack" protruding to the land side (figs. R 123-1 and R 123-2).

In order to prevent the formation of voids under blocks which project beyond the back of the wall, such blocks must be trimmed at an angle steeper than the angle of internal friction of the backfill material (figs. R 123-1 and R 123-2).

The eccentricity of the resultant and thus the concentration of the stresses at the heel of the bottom joint should also be kept as low as possible even when all unfavourable loads are applied simultaneously. According to DIN V 1054-100, no gaping bottom joint is allowed to occur in verification of serviceability (LS 2) as a result of constant characteristic actions. The ground pressures are also to be checked for all important stages of construction. The quay wall must be backfilled at about the same time as the blocks are placed to counteract landward tilting movements or excessive pressures at the heel of the foundation joint (fig. R 123-2). Besides the allowable ground pressures, the safety from sliding, foundation failure and slope failure must be verified.

Reference is made above all to R 79, section 10.5.3 with respect to the danger of sliding and sliding safety.

Possible changes in the harbour bottom due to scouring, but above all due to foreseeable deepening, are to be taken into account. In the course of later port operation, checks of the condition of the harbour bottom in front of the wall must be made at regular intervals, and should such prove necessary, suitable protective measures are to be undertaken immediately.

In anticipation of a known tendency of the wall to tilt toward the water, the quay wall must be initially designed with a slight batter. The crane track gauge can change later due to unavoidable wall movements and should always be designed in an adjustable way.

10.7.3.2 Horizontal Joints in the Block Wall

Safety from sliding and the position of the resultant of the acting forces must also be checked in the horizontal joints of the block wall for all essential construction stages and for the final state. In contrast to the foundation joint, theoretical tension in the joints may be allowed here up to the centroidal axis, if all unfavourably acting forces are applied simultaneously.

10.7.3.3 Reinforced Concrete Wall Coping

The reinforced concrete beam, which is to be fabricated at the site and installed on the top of every block wall, serves to equalise inaccuracies in block placement, to distribute concentrated horizontal and vertical loads and to equalise local variations in earth pressures and support conditions in the base, as well as construction inaccuracies. Since unequal settlement of the block wall will occur, concrete should not be placed before settlement has essentially ceased. In order to accelerate the settling process, a temporary surcharge of the wall, e.g. of additional layers of concrete blocks, is practicable. In doing so, the settlement behaviour must be measured regularly. However, as even this method cannot rule out subsequent settlement differences and in view of shrinkage and temperature stresses, the wall coping blocks should not be longer than 15.00 m. They should be fabricated in at least three longitudinal sections with continuous reinforcement.

In calculating the internal forces of the wall coping from vessel impact, hawser pull and side thrust of crane wheels, it can be generally assumed that the coping beam is rigid in comparison to the block wall supporting it. This assumption lies generally on the safe side.

In the calculation for the vertical forces on the wall coping, above all for the crane wheel loads, the bedding module method can generally be applied. In case major non-uniform settlement or subsidence of the block wall is to be expected, the internal forces of the wall coping however are to be determined by means of comparison investigations with various support conditions, riding on the supported middle section or in the end areas. In this case, the dead load of the wall coping is also to be taken into account. If required, the joint intervals are then to be reduced.

The coping beams are keyed at the block joints only for transmitting horizontal forces. Because the manner in which the block wall will eventually settle is not clearly predictable, keying for the vertical forces is not recommended.

Swelling, which occurs in cohesive soil after excavation, need not generally be considered in the statical calculations, because it will soon recede under the increasing load of the wall.

Structural protection against abrupt changes in alignment due to settlement must be provided for rail supports at block joints. For this purpose, short bridges are inserted across the block joints, permitting the crane rail to be laid across the bridge without a rail joint.

For the transmission of horizontal forces between wall coping and block wall, they should be reciprocally and effectively keyed.

10.7.3.4 Limit States

Insofar as they do not conflict with the pertinent recommendations of the EAU, the requirements of DIN 1045, DIN V 1054-100, DIN V 4017-100 and DIN V 4084-100 are to be observed.

10.8 Design and Dimensioning of Quay Walls in Blockwork Construction in Earthquake Areas (R 126)

10.8.1 General

In addition to the general conditions contained in R 123, section 10.7, R 124, section 2.13 must also be taken into consideration.

In determining the horizontal inertia forces of the block wall, it must be observed that they are derived from the mass of the blocks and the earth wedge in the backfill. In this connection, the mass of the pore water in the soil is also to be taken into account.

10.8.2 Active and Passive Earth Pressures, Water Pressure Difference, Live Loads

The remarks in the sections 2.13.3, 2.13.4 and 2.13.5 of R 124 are also applicable here.

10.8.3 Safety Aspects

Reference is made above all to R 124, section 2.13.6.

Even when considering the seismic forces, the eccentricity of the resultant in the horizontal joints between the individual blocks may only be so large that no theoretical gaping occurs over and beyond the centroidal axis under characteristic loads. This also applies to the foundation joint, in which no gaping is allowed in the case without earthquake (R 123, section 10.7.3.1, para. 3).

10.8.4 Other Solid Waterfront Structures

The foregoing remarks are also similarly valid for other solid waterfront structures such as box caissons, as per R 79, section 10.5, pneumatic caissons as per R 87, section 10.6., etc.

10.9 Construction and Design of Quay Walls Using the Open Caisson Method (R 147)

10.9.1 General

Open caissons – previously called open wells – are used in seaports for waterfront structures and berthing heads, but also as foundation elements for other types of structure, although this is very rare. Similar to the pneumatic caissons as per R 87, section 10.6, they can be constructed on terrain lying on the site of the sinking area over the water table, or in a driven or floating spindle frame, or be floated into position and then sunk as finished boxes using a jack-up platform or float elements. Open sinking requires lower wage and installation costs than pneumatic caissons, and can be used to far greater depths. However, the same degree of positioning accuracy cannot be achieved. In addition, this solution does not produce equally reliable load- bearing conditions in the foundation bottom. Any hindrances met during the sinking process can only be by-passed or removed with considerable difficulties. Placing on sloping rock surfaces always requires additional measures.

The construction principles for waterfront structures stated in R 79, section 10.5.1 for box caissons are accordingly applicable to open caissons. In other respects, reference is made to section 3.5 "Caissons" in the Grundbau-Taschenbuch [7].

10.9.2 Calculation

R 79, section 10.5.2 and section 10.5.3, and R 87, section 10.6.2 and 10.6.3 are to be taken into account.

10.9.3 Structural Design

Open caissons may be either rectangular or circular in layout, depending on operational and constructional considerations.

Because of the inherent funnel shaped excavations, open caissons with a rectangular layout will not assume a position on their cutting edges which is as uniform as can be expected if circular caissons are used. This results in an increased risk of deviation from the design position. If a rectangular shape is necessary, it should therefore be a compact one. Since the excavation and sinking procedures are difficult to check, and because an open caisson can be only lightly ballasted, the caisson should have thick walls, so as to make the dead load of the caisson definitely greater than the expected wall friction force, taking uplift into account. A rigid steel cutting edge is installed at the base of the outer walls. Alternatively, a cutting edge of high-strength concrete (at least B95) or steel fibre concrete is conceivable. At the bottom of the cutting edge rim, jetting pipes discharging toward the caisson interior can be of help in loosening non-cohesive excavation soil (fig. R 147-1, cross section C-D, showing jets on the water side).

The bottom edge of the partition walls must terminate at least 0.5 m above the bottom surface of the caisson cutting edges, in order to avoid loads being transmitted into the subsoil.

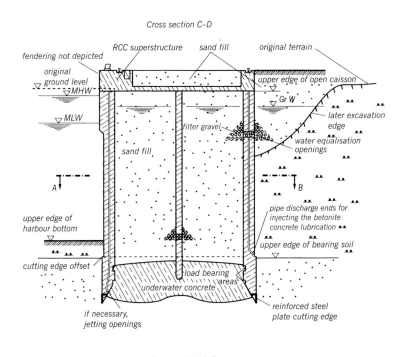

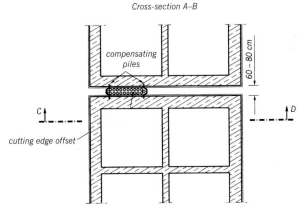

Fig. R 147-1. Design of a quay wall of open caissons with subsequent harbour dredging

The outer walls and partition walls should have reliable contact surfaces which can be easily cleaned after sinking for the reliable transmission of loads into the underwater-concrete bottom.

When constructing foundations for open caissons, the unavoidable loosening of the soil in the foundation bed and adjacent to the caisson surface encourages noticeable settlement and an incline of the completed structure. This must be allowed for in the dimensioning and structural design, as well as during the course of construction.

The second and third paragraph of R 79, section 10.5.4 apply here in their entirety. The solution of the joints shown in R 87, section 10.6, fig. R 87-1, is recommended. However, filter gravel is preferable to inflexible material for sealing the space between adjacent caissons because it can accept the movement caused by settlement without being damaged.

A space of 40 to 50 cm between adjacent caissons, as given for pneumatic caissons in R 87, section 10.6.4, is adequate considering the excavation method for open caissons only when the actual embedment is not deep and hindrances are not expected – including strata of embedded, solid, cohesive soil. If sinking is difficult, intervals should be increased to 60–80 cm. Compensating piles of adequate width, or loop-like arrangements of more deformable pile-chains may be used as closures.

10.9.4 Remarks on Construction

If construction takes place in the dry, the load-bearing capacity of the soil at the erection area must be carefully checked and observed, because it may be harmful if the soil under the cutting edges yields excessively or too unevenly. The latter condition may cause the cutting edge to break. The soil within the caisson is removed by grab buckets or pumps, while the water level inside the caisson must always be kept no lower than that of the water on the outside.

When a row of caissons is being sunk, a sequence of 1, 3, 5 ... 2, 4, 6 may be advisable, because then the same active earth pressure will act on both faces of each caisson.

The sinking of a caisson can be greatly facilitated by lubricating its outer surface, above the offset near the base, with a thixotropic liquid, such as a bentonite suspension. In order to make sure that the lubricant actually covers the entire surface, it should not be poured in from above, but should be grouted in by means of pipes placed in the caisson walls. These discharge directly above the offset near the base. A deflecting steel plate may be added to make sure that the liquid is distributed properly. (A pipe of this kind is shown on the land-side in fig. R 147-1). The injection of liquid must proceed with such care that it will not break through into the excavation space and flow away. The vertical distance up to the offset on the outer wall should be generously designed. Special care is called for if the surface of the excavation floor shows extensive subsidence after underlying loosened sand has been compacted.

After the designed embedment depth has been reached, the bottom must first be carefully cleaned. Only then may the floor slab of underwater or colcrete concrete be installed.

10.9.5 Frictional Resistance During Sinking

The suggestions regarding pneumatic caissons stated in R 87, section 10.6.6 also apply to open caissons. However, since open caissons cannot be ballasted to the same extent as the pneumatic caissons, special importance is attached to thixotropic lubrication of the outer surface at greater embedment depths. Experience shows that it reduces the mean skin friction to less than 10 kN/m^2.

10.10 Construction and Design of Quay Walls with Open Caissons in Earthquake Regions (R 148)

Section 8.2.18.1 of R 125 is applicable here. However, it must be borne in mind that flow-sensitive, non-cohesive subsoil must always be compacted outside of, as well as in the actual foundation area, before the caissons are placed. Because the soil has been loosened in the process of excavating for the foundation, a second compaction of the soil below the excavation bottom must be made in the area inside the caisson. This can be done by the closely spaced use of concrete immersion vibrators.

Active and passive earth pressures are assumed according to R 124, section 2.13.3.

Section 8.2.18.3 of R 125 must be taken into account for the anchoring of an open caisson.

In other respects, R 147, section 10.9 applies here.

10.11 Application and Design of Bored Pile Walls (R 86)

10.11.1 General

With proper construction, structural design and dimensioning, bored pile walls can also be used for waterfront structures. Besides economic and technical reasons for their use, they also fulfil the demand for safe, extensively vibration-free and/or quieter construction methods.

10.11.2 Construction

Straight or curved walls, which can be easily adapted to the particular shape required, can be erected in the layout by joining bored piles together.

The following bored pile walls can be constructed, depending on pile spacing:

(1) Overlapping bored pile wall (fig. R 86-1)

The centre-to-centre spacing of the bored piles is smaller than the pile diameter. At first, the primary piles (1, 3, 5, ...) of plain concrete are installed. They are overlapped during the construction of the intermediate reinforced secondary piles (2, 4, 6, ...). In special cases, 3 plain piles can also be arranged adjacently. The overlap is as a rule 10–15 % of the diameter, but at least 10 cm. The wall in general is then as good as watertight. Statical interaction of the piles can be presumed for loading normal to the pile wall and in the wall plane at least occasionally, but is generally not considered in the design of the bearing secondary piles. Besides being transmitted by load-distributing coping beams, vertical loads can also be distributed between the adjacent piles given adequate roughness and cleanliness of the overlapping surface, and to a certain extent also by shear forces. However, in walls with a short layout, deviation of the pile points from the wall plane outward must be prevented by fixing the pile points firmly in an especially resistant subsoil.

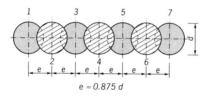

Fig. R 86-1. Overlapping bored pile wall

(2) Tangential bored pile wall (fig. R 86-2)

For practical construction requirements, the centre-to-centre spacing of the bored pile is at least 5 cm longer than the pile diameter. As a rule, each pile is reinforced. Watertightness of this wall can only be achieved by additional measures, such as overlapping columns using the jetting method (HDI) or other injection methods. The wall cannot be expected to absorb longitudinal loads by acting as a sheet.

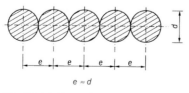

Fig. R 86-2. Tangential bored pile wall

(3) Open construction bored pile wall (fig. R 86-3)
The centre-to-centre spacing of the bored piles can amount to several times the pile diameter. The intervals are closed with structures of wood, air-placed concrete or steel sheeting.

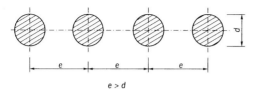

e > d

Fig. R 86-3. Open construction of bored pile wall

10.11.3 Constructing the Bored Pile Walls

The construction of closed bored pile walls as in figs. R 86-1 and R 86-2 presumes high drilling accuracy, which generally requires good guidance of the casing.

If possible, construction operations for bored pile walls should take place on dry ground, either on a natural ground or on a filled island, or on a jack-up platform.

Soil excavation is accomplished by means of cable grabs, Kelly grabs, by rotary drills, or by suction or airlift methods, inside of and from behind the advancing end of a casing. In rigid soil, the use of a casing may be omitted. Instead, water pressure or support slurries are used. Obstacles are cut away with chisels or loosened by blasting. Attention must be paid to an adequate overpressure from the water filling or support slurries in the bore hole or in the bore casing, which as a rule should be kept at least 1.5 m above the groundwater level.

With appropriate guidance, cased holes can also be drilled at an angle to the vertical.

Overlapped bored pile walls are mostly built with the aid of a machine which forces the bore casing into the ground or with a brace and pit, with a rotary and/or compressive motion. The cutting edge of the casing is used as a drill bit. The primary piles of plain concrete are suitably concreted with HOZ L or a mixture of Portland cement and fly ash (at least 25–40 %), and the sequence of operations is defined so that depending on the capacity of the drilling machine, when the secondary piles are placed, the concrete strength will normally not exceed 3 to 10 MN/m^2. The strength difference of the primary piles to be placed must be kept as small as possible in order to prevent directional deviations of the secondary piles. In temporary structures, the use of concrete of quality B10 with high filler and low cement portion can be considered for the unreinforced primary piles.

When building this type of wall in open water, sacrificed casing pipes are required above the river bottom, unless precast piles are placed in the borings and held in place by grouting or in-situ concrete. For larger free drop heights, delivery pipes are to be used down to the concrete bottom. In addition, an underwater compound should be added to the concrete to prevent segregation. DIN 4014, DIN 1045 and E DIN EN 1536 are applicable with respect to cleaning the bottom joint, placing concrete, concrete cover and design of reinforcement.

When support slurries are used, stability reducing influences from the soil and/or groundwater (e.g. elevated salinity or organic soil components) etc. must be counteracted by selection of suitable clays and/or additives.

10.11.4 Structural Remarks

In cased drilling, inadvertent turning of the reinforcing cage is unavoidable. For this reason, a radially symmetric arrangement of the reinforcement should always be used unless workmanship and inspection are very careful. Accidental moving of the reinforcing cage can be prevented by placing some freshly mixed concrete on a slab which has been installed for this purpose at the bottom of the cage.

The pile reinforcement must be stiffened to an adequate extent in order to maintain the required concrete cover and to exclude deformation of the reinforcing cage.

Welded-in stiffeners (so-called "Rhön wheels") according to ZTV-K 96 [80] are of proven quality. The indicated minimum measures are adequate only for piles of slight diameter to about 1 m. For large-diameter piles (approximately $d = 1.30$ m), for example at 1.60 m spacing, stiffener rings 2 $\varnothing$ 28 mm BSt 420 S with 8 spacers $\varnothing$ 22 mm, $l = 400$ mm are recommended, which are welded to each other and to the longitudinal pile reinforcement.

The piles are designed on the basis of DIN 4014 and DIN V ENV 1997-1. A limitation of the crack widths is only to be checked under "severe" chemical attack according to DIN 4030, insofar as no protection measures have been undertaken.

As a rule, unless the piles are held by an adequately stiff superstructure with only a small distance to the anchoring level, the wall requires wales to distribute the anchor load. If anchored overlapping or tangent bored pile walls have been placed in granular soil of medium density or in semi-rigid cohesive soil, wales may be omitted, provided that at least every other pile or other spandrel between piles is anchored. At the same time, an adequate length of the pile wall at both ends should have a wale to hold the anchor rods.

Connections to adjacent structural members should, if possible, be made only by means of the reinforcement at the pile head and in the rest of the wall, only in special cases and then only by means of recesses or special built-in connections.

10.12 Application and Design of Diaphragm Walls (R 144)

10.12.1 General

The remarks in R 86, section 10.11.1 also apply to diaphragm walls.

In-situ concrete walls, which are built in sections by the slurry trench method, are called diaphragm walls. A special grab excavator or cutter first digs a trench between guide walls, the trench is filled continuously with supporting slurry. After the slurry has been cleaned of impurities and homogenised, the reinforcement is installed and concrete is poured using the tremie method, forcing the slurry to the surface, where it is pumped away.

The diaphragm walls are described in detail in DIN 4126-100 and E DIN EN 1538. Detailed data is given there on:

- supplying the slurry ingredients; preparation, mixing, swelling, storage, placing, homogenising and reprocessing of the supporting slurry,
- reinforcement, concreting, and
- stability of the slurry-supported trench.

The remarks in DIN 4126-100 and E DIN EN 1538 are also to be carefully observed for all diaphragm walls at waterfront structures. In other respects, additional reference is made to the following literature [82] to [88].

Diaphragm walls are generally erected continuous and practically watertight, with thicknesses of 60, 80 and 100 cm, or thicknesses of 120 and 150 cm in the case of quay walls with large differences in the ground surface elevation. Where high loads prevail, a diaphragm wall of T-shaped elements standing side by side can also be used instead of a simple one. Since the corners between the wall and the bending T-web tend to break, particularly in the upper zone (low hydrostatic pressure and also low flow pressure), this must be taken into account when designing the guide walls (adequate depth) and in rating the soil masses and concrete masses. Execution with such T-shaped elements in very loosely stratified or soft soils is only recommended with additional measures, such as previous soil improvement.

In the layout, long diaphragm walls are shown as straight line sections. A curved course of a wall is shown in a chord line. The possible lengths of an individual element (segment) is limited by the stability of the slurry-supported trench. Where high groundwater level, a lack of cohesion in the soil, neighbouring heavily loaded foundations, vulnerable utility lines or the like are encountered, the normal maximum length of 10 m is reduced to a minimum of about 2.80 or 3.40 m, corresponding to the usual opening widths of a diaphragm wall grab.

In suitable soil and with sound construction, diaphragm walls can transmit large horizontal and vertical loads into the subsoil. Junctions to other vertical or horizontal structural members are possible with special junction elements which are concreted in or subsequently doweled in position, if need be by recesses. Well exposed concrete surfaces can be achieved with inserted precast parts, whose use is however limited to a depth of 12–15 m on account of the high dead weight involved.

10.12.2 Verification of Stability of the Open Trench

In order to assess the stability of the open trench, the equilibrium is investigated at a slip wedge. The dead weight of the soil and any surcharges from neighbouring structures, construction vehicles or other live loads and the external water pressure act as loads. The pressure of the slurry, the full friction on the slip surface which leads to active earth pressure, and the friction on the side surfaces of the slip wedge as well as any cohesion, act as resisting forces. In addition, the resisting force of the stiffened guide wall may be taken into account. This force is especially significant for high-lying failure planes, because the shear stress is hardly effective here in non-cohesive soil. In deep failure planes, the influence of the guide wall is so small as to be negligible.

As regards the stability of the open trench and safeguarding the excavated walls from failure, reference is also made to DIN V 4126-100 and to [82] and [83]. This verification is not required to E DIN EN 1538.

Verification of failure of a sliding body must be provided for all depths at least as per DIN V 1054-100, LS 1C, load case 2, insofar as loads from structures exist. The highest groundwater levels expected during the construction work must be taken into consideration.

In tidal areas, the critical external water level must be stipulated or determined on the basis of the intended supporting slurry level. If it is anticipated that the allowable external water level will be exceeded, e.g. following storm flooding, an open trench must be filled in good time.

10.12.3 Composition of the Slurry

A thixotropic clay or a bentonite suspension is used as slurry for supporting the trench walls. With respect to its composition, suitability test, processing with mixing and swelling times, desanding etc., reference is made to DIN 4126-100 and to DIN 4127.

It must be kept in mind particularly that the ionic equilibrium of the clay suspension changes unfavourably with the influx of salts, which may occur at structures in sea water or in strongly saline groundwater. Flocculation occurs, which can cause a decrease in the bracing capacity of the slurry. Salt-water resistant bentonite suspension must therefore be used when building diaphragm walls in such areas. The following mix formulas have proven effective in actual practice:

(1) The slurry is mixed with fresh water (tap water), 30 to 50 kg/m^3 Na-bentonite and 5 kg/m^3 CMC (carboxy-methyl-cellulose), a protective colloid.
(2) The slurry is mixed with sea water, 3 to 5 kg/m^3 of a biopolymer and an additive of 5 kg/m^3 clay of salt-resistant minerals, such as attapulgit or sepiolith. A suitable synthetic polymer may be used in place of the biopolymer.

There is a large variety of mix formulas. In any case, suitability tests must be carried out before using any mix for construction purposes. These tests must take into account the salinity of the water, the soil conditions and any other special features (e.g. where the water passes through coral). The contamination of a slurry under salt water conditions is best shown by the rise in filtrate water elution (see DIN 4127). Special care is also necessary with soil contamination, soil particles of peat or brown coal and the like. Unfavourable effects however can be at least partially neutralised by adequate additives. Suitability tests of the supporting slurries are urgently recommended.

10.12.4 Details for Construction of a Diaphragm Wall

In general, excavation of a diaphragm wall segment starts from the ground surface between the guide-walls, which as a rule are 1.0 to 1.5 m high and consist of lightly reinforced concrete. Depending on the soil conditions and loads from the excavation and pulling out equipment for the shuttering tube, they are designed as continuous wall sections, reciprocally supported outside the excavation area, or as angular retaining walls. Existing structural members are suitable as guide walls, if they reach to an adequate depth and can absorb the pressure of the supporting slurry and other occurring loads, and allow the operation of excavation and pulling out equipment.

The slurry is enriched with more and more ultra-fine particles as the excavation work progresses and must therefore be constantly checked for suitability and, if the values are no longer reached, be replaced. Generally, the support slurry can be used several times. Routine checks are to be made of the density, filtrate water elution, sand level and flow limit of the supporting slurry at the construction site.

According to DIN V 4126-100, placing of the reinforcement and the concreting work must proceed immediately after uninterrupted excavation of the soil, particularly to avoid disturbing the ground by relaxation of tension. If reinforcement and concreting work is not possible immediately, the trench base must be cleared of any subsidence of soil particles before reinforcement work takes place.

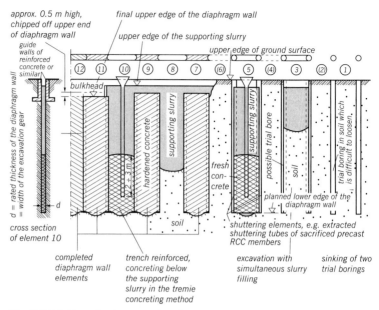

Fig. R 144-1. Example of construction of a diaphragm wall

Details of the construction of diaphragm walls may be seen in fig. R 144-1.

In some cases however stepwise production of the segments in the sequence 1, 2, 3 etc., is preferable. The shuttering elements should be as narrow as possible to keep the reinforcement-free zone small.

10.12.5 Concrete and Reinforcement

Reference is made here above all to the detailed remarks in DIN V 4126-100. For the concreting unfavourable reinforcement concentrations or recesses in the steel must be avoided when designing the reinforcement. Profiled reinforcing steels are preferable because of their better bonding properties. In order to obtain the required concrete cover of at least 5 to 10 cm, depending on the use as trench sheeting or as a permanent structure, an ample number of large-surface bar spacers should be used.

The following are recommended as a rule for the design of the minimum reinforcement:

- on the tension side in vertical direction:
  5 ⌀ 20/m for ribbed reinforcing bars BSt 500 S,

- on the compression side in vertical and horizontal direction in general:
  5 ⌀ 20/m for ribbed reinforcing bars BSt 500 S.

Attention must be paid to adequate installation stiffness of the cages, particularly in the case of minimum reinforcement.

10.12.6 Suggestions for the Calculation and Design of Diaphragm Walls

Due to their stiffness against bending and slight deformations, diaphragm walls as a rule must be designed for increased active earth pressure. The assumption of the active earth pressure is only to be justified when the required displacement for full activation of the shear stresses in the failure planes exists as a result of adequate elasticity at the foot of the wall and the supports or of adequate yielding of the anchoring and of the horizontal bending.

For high differences in elevation with head displacements in the cm-range, e.g. for quay walls for sea-going vessels with a difference in elevation ≥ 20 m, active earth pressure redistribution as per R 77, section 8.2.2, is possible. The possible deformation behaviour is to be taken into account in every single case.

Full fixity of the foot of the wall in the ground is generally not attainable with upper anchoring or support, because of the stiffness of the wall. It is therefore advisable in a wall calculation according to BLUM to take only partial fixity into account, or to figure with elastic foot fixity according to the coefficient of subgrade reaction method or the coefficient of compressibility method. If on the other hand, the finite element method is used, correct material characteristics are of paramount importance in the input data. The angle of wall friction in the active and passive zone is essentially dependent on the type of soil, progress of work and standing time of the open trench. Coarse grained soils result in extreme roughness of the excavation wall, whereas fine grained soils lead to comparatively smooth trench wall surfaces. Slower progress of work and longer standing times encourage deposits from the slurry (formation of filter cake). The angle of wall friction, contingent essentially on type of soil, working speed, trench standing time and measures to remove filter cake, can therefore be assumed to be between

$$\delta_{aD} = 0 \quad \text{and} \quad \delta_{aD} = \pm \tfrac{1}{2} \cdot \varphi'_D .$$

Wales for supports or anchorings can be formed by placing additional transverse reinforcement in the elements. If the wall elements are not too wide, one centred support or anchoring is sufficient; for wider elements, two or more are required, which are equally spaced symmetrically to the centre line.

Determination of reinforcement is made according to DIN 1045, section 17.6 and DIN 19 702 with observation of the following supplements regarding limitation of the crack width under working load. When applying table 15, linear interpolation may be used between the indicated diameter limiting lines.

The structural members are to be classified according to DIN 1045, table 10, as per table R 144-1 (see also R 72, section 10.2).

Components	Expected crack width	Verification as per crack formula
Temporary excavation walls of reinforced concrete	Normal	Not required
Permanent reinforced concrete structures, constantly under water	Slight	Required
Permanent reinforced concrete structures in the tidal range, above the water level or in weakly aggressive waters or soils	Very slight	Required

Table R 144-1. Expected crack widths and verification requirements

For individual members composed of watertight concrete (WU concrete) as per DIN 1045, additional verification must be provided that the concrete compressive zone determined according to stage II is at least 15 cm thick when the member is in service.

For smaller thicknesses, the comparison stress is to be verified in accordance with DIN 1045, section 17.6.3, whereby

$$\text{tol.}\ \sigma_v = \sqrt[3]{1.0 \cdot 0.1 \cdot \beta_{wN}^2} \quad [\beta_{wN} \text{ and } \sigma_v \text{ in MN/m}^2]$$

may be assumed.

In view of the unfavourable effect of a probable residual film of the slurry on steel, or of fine sand deposits, the bond stresses for horizontal reinforcing bars must meet the requirements of table 19, line 2 (bonding zone II) of DIN 1045.

Vertical bars can as a rule be assumed according to bonding zone I, but it is recommended to increase the anchoring lengths in the point and butt sections of the diaphragm wall.

10.13 Application and Construction of Impermeable Diaphragm Walls and Impermeable Thin Walls (R 156)

10.13.1 General

Impermeable diaphragm and impermeable thin walls consist of a material of slight permeability which is frequently fabricated from a mixture of clay, cement, water, fillers and additives. This material is placed in the required thickness in a flowable consistency, according to different methods in natural soil or in fills of higher permeability and hardens to a body of low permeability.

With the use of these impermeable diaphragm walls, the entry of groundwater or other (often more harmful) liquids is limited to a very low extent. Impermeable thin walls cannot be used if larger pressure gradients are to be expected during the construction and setting period.

The following have proven themselves as fields of application in connection with port facilities and waterfront structures of most diverse types:

- Enclosures for dumpings of environmentally harmful spoil from maintenance dredging in harbours, harbour approaches, stretches of rivers in industrial areas and the like.
- Sealing of waterfront embankments on the shore and along inland waterways against high outer water or dammed river water levels.
- Protection of water-loaded embankments against erosion, suffosion, groundwater entry and discharge etc.
- Separation of industrial facilities, tank farm areas etc., from surrounding harbour and groundwater as protection against spreading of harmful liquids.
- Restraining or reducing of land-sided water pressure difference on waterfront structures.
- Closing of larger excavations in port areas, so that the water table can be lowered without endangering adjacent waterfront structures and other facilities.

In the selection of the most expedient construction method, besides economic considerations attention is to be paid to the limit of application of the available method. The essential points hereby are:

- Depth of the impermeable diaphragm wall.
- Thickness of the impermeable diaphragm wall, contingent on its resistance to erosion. The hydraulic gradient i, the coefficients of permeability k for the cured impermeable wall compound and the bordering ground, the unconfined compressive strength and, last but not least, the duration of loading are of decisive importance here.
- Suitability of the intended construction method for erection of a completely watertight wall (with overlapping of the individual wall elements) under the prevailing soil conditions and obstacles to be expected.
- Safe embedment in slightly permeable subsoil.

10.13.2 Method of Construction

(1) Impermeable diaphragm wall [89], [90], [155], [190] to [196], [7]
The impermeable wall is built according to the diaphragm wall method (DIN V 4126-100 and DIN 4127 and R 144, section 10.12.1 to 10.12.4), but the thicknesses are as a rule 60 cm or more. Excavation depths of more than 100 m are known. Furthermore, the thickness of the impermeable diaphragm wall is contingent on the different movements of the wall, which are to be expected under loads from water pressure difference in the area of changing strata.

The method may also be employed for inclusion of blocks of rock or similar, as these obstacles can be broken up in the trench and excavated. By checking the excavation spoil, the reliable connection to the impermeable stratum can be checked. In rock, the embedding of the wall is attained by chiselling or cutting.

The wall is generally constructed by sections without shuttering elements between adjacent sections, whereby the work proceeds with an overlap of 25 to 50 cm in the area of the section edges (fig. R 156-1).

Two methods have proven effective for the construction of impermeable diaphragm walls:

(2) Single-phase method
The trench is sunk under protection of a supporting slurry admixed with cement, which serves at the same time as sealing compound, and which remains in the trench after the excavation (phase 1) and hardens there. The fabrication of each section in one phase favours the necessary rapid progress of construction, so that there is no impairment to setting of the sealing compound which starts after about six hours. If the construction work is slowed down, work can progress in the trench up to 12 hours if special compounds are used. Higher slurry consumption rates occur when passing through highly permeable soil layers, or if the grab is operated too quickly, thus preventing the formation of a filter cake.

For long excavation times and when passing through fine-graded soils, the suspension can become so enriched with solids that the grab "floats" and cannot get deep enough. This also occurs with high filtrate water elution in deep trenches.

A special form of the single-phase method is the dry diaphragm wall, used for earth dams [90], also called trench wall. Here no supporting slurry is used, as the wall section being always only 1–2 meters deep sticks fast and can be filled with clay concrete.

(3) Two-phase method
The trench is prepared in the customary manner with use of a supporting slurry. After reaching the final depth, the sealing compound is applied using the tremie method in a second step (phase 2) and the supporting

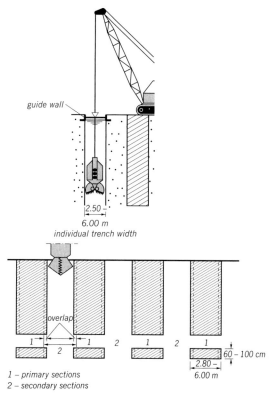

Fig. R 156-1. Construction of an impermeable diaphragm wall

slurry withdrawn. The difference in density between the impermeable wall compound and the support slurry containing soil should be 0.5–0.6 t/m^3 as otherwise the compound replacement cannot be assured.

This method requires more expenditure than the single-phase method. The impermeable wall compound however is generally more homogeneous than in the single-phase method. The two-phase method is therefore used, for instance, when work is progressing slowly because of obstacles or for large depths (> 30 m).

(4) Impermeable thin wall
To build impermeable thin walls, a steel section of 500 to 1000 mm web height (thin wall pile) is introduced into the soil strata to be sealed, and compacted, preferably with vibration hammers. The trench thus created is grouted with impermeable-wall material on withdrawal of the steel section. By continued, overlapping repetition of this procedure, a con-

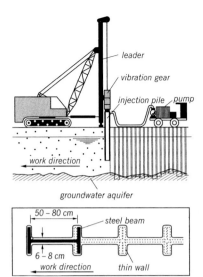

Fig R 156-2. Construction of a thin wall

tinuous thin wall is created. Instead of I-section steel profiles, other shapes can be used (e.g. depth vibrators with extra straps).

A mixture with the greatest possible density, of cement, clay or bentonite, rock flour or fly ash and water is used as the impermeable-wall material.

Impermeable thin walls are preferably used for temporary works.

The impermeable thin-wall compound is placed under pressure and in so doing penetrates into the voids of loosely deposited soils.

The minimum thickness of the diaphragm wall is vital for evaluating its effectiveness; this is determined by the section thickness of the pile point or other precast parts for making the trench. The individual work steps are executed with overlapping [91]. Adequate overlapping of neighbouring sections must be maintained (fig. R 156-2).

The advantage of the impermeable thin walls as compared to the impermeable diaphragm walls lies in the lower costs and in lesser time needed for the construction.

Its use is limited, however

- when encountering soil with obstacles which is unsuitable for driving and vibrating,
- when the depth of the wall is more than 15 to 30 m, depending on the nature of the subsoil, if customary gear is used,
- when comparatively high water pressure difference acts on a diaphragm wall of slight thickness,

- when a soil stratum that tends to flow lies above a coarse-grained permeable layer of soil, no sealing compound may flow downward to compensate for the suspension loss of the lower layer. This effect can be increased through driving and vibration influences. Then there is the risk that no impermeable-wall compound penetrates into the transition zone,
- when the natural soil tends to strong flow and/or settlement,
- when the transition to the groundwater aquifer is recognised only with difficulty, especially at varying granulometric composition in the transition area. In addition to observing the soil brought to the surface by the driving or vibration gear, suitable measurings and further observations are then required.

10.13.3 Basic Materials of the Impermeable-wall Compounds

(1) Clay

In addition to suitable natural clays and clay powder, bentonite is preferably used.

Commercial bentonite has varying properties. The flow properties and the water-binding capacity of bentonite suspensions are furthermore evidently changed by cement additive. Attention must be paid to this when preparing the impermeable-wall compound.

(2) Cement

Commercial cements have proven effective. Blast furnace cements with a high foundry sand component are particularly advantageous, or Portland cement/fly ash mixtures with high ash level.

(3) Fillers

Basically, all neutral sands, dusts, powders and granulates may be used whose maximum aggregate size remains suspended in the clay-cement suspension. In respect of the settling tendency, there are advantages in fillers with low-density. On the other hand however, the self-compacting flow is favoured by high density. When impermeable diaphragm walls are erected in the single-phase method, the prevailing soil determines the enrichment of fines. When the two-phase method is used, the grain size of the filler to the impermeable-wall material (contingent on the diaphragm wall thickness and the compound processing plant) may be in the magnitude of up to 30 mm. No special attention need be paid to the granulometric composition in case of fine fillers. It must be regular in case of coarse grain mixtures.

(4) Impermeable-wall compounds resistant to harmful substances

Silicate binders are used for impermeable-wall compounds which are resistant to harmful substances. These impermeable-wall compounds have a high density and must be installed using the two-phase method.

(5) Water
The mixing water must be neutral. Acid water may lead to flocculation of the bentonite and, just like acid-reacting filler, reduce the liquid limit of the impermeable-wall material. Light alkalising of the mixing water with slight soda or caustic soda additive has proven effective.

(6) Additives
Where difficult groundwater conditions prevail, protective colloids are recommended to stabilise the clay suspension. These additives reduce the filtering-out of the water and at the same time increase the liquid limit of the supporting slurry.

10.13.4 Demands on the Impermeable-wall Compound

The material properties and behaviour of the impermeable-wall compound in the placing and final stage are to be co-ordinated with the type of the impermeable-wall and its purpose and to be verified by tests. Impermeable-wall ready-mixtures have proven successful for sometime now, as supplied by various cement manufacturers.

Co-operation with an institute experienced in material tests of impermeable-wall compound is recommended.

(1) Placing stage
The flow behaviour of the impermeable-wall compound is determined by the liquid limit τ_F. It must be so high that the granular components contained in the impermeable-wall compound remain securely in suspension at least till the start of setting.

In the single-phase method, the upper limit of the liquid limit τ_F of the impermeable-wall compound is to be defined so that the excavation is not impeded, and not too much impermeable-wall compound is removed.

For the two-phase method, the upper limit of τ_F results from the demand for self-compacting flow in the supporting slurry. The complete displacement of the supporting suspension also presupposes that the impermeable-wall compound has an essentially greater density than the supporting slurry. It is therefore necessary to replace the supporting slurry with a cleaned or fresh suspension before placing the impermeable-wall compound.

The impermeable-wall compound must be stable. It must absorb or discharge as little water as possible and may settle to only a limited extent during the setting phase. Partial sedimentation of the cement and increased filtration of water of the suspension is only desired under load from hydrostatic water pressure, for partial increase of the cement portion in the heavier loaded lower area of the impermeable diaphragm wall.

The setting behaviour of the impermeable-wall compound must ensure that the setting process is not disturbed by excavation work in the single-phase method.

(2) State after bonding of the wall
A permeability coefficient k of the impermeable-wall compound should be verified by suitability tests after 28 days up to a hydraulic gradient of $i = 20$. The permeability requirements for the impermeable-wall compound and those for the set wall differ. The permeability of the wall compound produced in the laboratory from the raw materials (see section 10.13.5) should be smaller by one to the power of ten than the permeabilities ascertained from the wall samples. Since the volume of percolating water infiltrating per unit of time is also of significance, the desired ratio of k to wall thickness d (conductivity coefficient in s^{-1}) should be indicated in the design for the maximum occurring hydraulic gradient $i = \Delta h/d$ (Δh = water level difference between interior and exterior).

Usually, an unconfined compressive strength in accordance with DIN 18 136 of 0.2 to 0.3 MN/m^2 after 28 days is striven for in impermeable diaphragm walls in samples taken from the mixer; the values for thin walls are 0.4 to 0.7 MN/m^2. A substantial increase of the strength is not desirable, as this would decrease the deformability of the impermeable-wall compound, and deformations in the soil could lead to local damage with greater erodibility. Where possible, the deformability of the impermeable-wall compound should be of the same magnitude as that of the soil.

In view of the erosion resistance of the impermeable-wall compound, according to latest findings, the hydraulic gradient must remain limited to $i = 20$ for continuous loads of a single-phase impermeable diaphragm wall with a strength of 0.2–0.3 MN/m^2.

In impermeable thin walls, whose sealing impermeable-wall compound as a rule contains more solid matter than for impermeable diaphragm walls, a hydraulic gradient of $i = 30$ should not be exceeded for continuous loads with the confirmed minimum thickness.

Influences which could effect the behaviour of the walls in working state, for instance consolidation, shrinkage, etc., are to be taken into account properly in the material composition of the walls.

10.13.5 Testing Methods for the Impermeable-wall Compound
DIN 4126-100 and DIN 4127 are decisive for testing.

The flow behaviour of the impermeable-wall compound can be ascertained by measuring the liquid limit with the ball harp apparatus or the outlet times according to the MARSH funnel test.

The stability of the impermeable-wall compound in the placing stage may be seen in the degree of settlement (percentage settlement) which as a rule is not to be more than 3 %.

The workability limit up to which the impermeable-wall compound may be moved without damage due to excavation work, is determined through the strength of test cylinders made from impermeable-wall compound, moved for appropriate periods.

The permeability is determined in the laboratory in accordance with DIN 18 130. If need be, permeability to chemicals is also to be determined in respect of durability. As customary, 28- or 56-day old samples are investigated.

In practical application, the higher permeability in the completed structure is to be taken into account.

The unconfined compressive strength is determined according to DIN 18 136, whereby the dimensions of the samples are to be co-ordinated with the material being investigated. It is a reliable value for evaluating the bearing capacity, where this is of significance.

As a rule, the age of the samples is 28 or 56 days.

Samples from the completely set wall should be taken for investigation only in special cases, and then in co-operation with an experienced institute which fixes the extent and type of sampling.

10.13.6 Site Tests of the Impermeable-wall Compound

The properties of the impermeable-wall compound before being placed are compared with the nominal values determined from the suitability test.

For impermeable diaphragm walls according to the single-phase method, samples taken from various depths in the trench with grab or special sampling tins are to be investigated.

The site check will be carried out as in a field laboratory, by measuring the following properties:

1. Weight density
2. Support characteristics   (ball harp apparatus or pendulum apparatus)
3. Flow characteristics   (Marsh funnel)
4. Stability   (settling test, filter press)
5. Sand components at various depths
6. pH value

The water permeability of the finished wall is expediently determined as a function of time, starting after approx. 28 days, on reserved samples in the laboratory in accordance with DIN 18 130, and/or by sinking tests in boreholes [144] which should be the basis of acceptance.

10.14 Inventory Before Repairing Concrete Components in Hydraulic Engineering (R 194)

10.14.1 General

The aspects dealt with below also basically apply to checking concrete components as part of the enquiries in accordance with R 193, section

15.1. In some cases, the individual investigations listed go beyond the scope of normal testing and are therefore given particular mention here. Construction operations for repair purposes are likely to be successful only if they take accurate account of the causes of the fault or damage. Since several causes are usually involved, a systematic investigation of the actual situation should first be carried out by a qualified engineer.

Since a correct assessment of the causes of faults or damages is an important prerequisite for a lasting repair, some recommendations for determining the actual situation and for fault finding are also given (see also subsection no. 1.2 of DIN 31 051).

(1) Description of item
- Year of construction,
- Stresses arising from use, operation, environment,
- Existing stability verifications,
- Soil investigations,
- Construction drawings,
- Special features of the erection of the structure.

(2) Inventory of components affected by the damage
- Nature, position and dimensions of components,
- Building materials used (type and quality grade),
- Description of damage (nature and extent of damage with dimensions of damaged areas),
- Documentation (photographs and sketches).

(3) Necessary investigations
The nature and extent of the investigations necessary for trouble shooting are laid down on the basis of the conclusions arising from sections 10.14.1 (1) and 10.14.1 (2).

10.14.2 Investigations on the Structure

More detailed information on the actual condition of the structure can be obtained from the following investigations on the structure:

(1) Concrete
- Discolouring, moisture penetration, organic growth, efflorescence/ honeycombing, concrete flaking, flaws,
- Surface roughness,
- Adhesive strength,
- Watertightness, nests of aggregates,
- Depth of carbonisation,
- Chloride content (quantitative),
- Crack propagation, widths, depths, lengths,
- Crack movements,
- Condition of joints.

(2) Reinforcement
- Concrete covering,
- Corrosion attack, degree of rusting,
- Reduction in cross-section.

(3) Prestressing elements
- Concrete covering,
- Condition of grouting (ultrasound, radiography, endoscopy if necessary),
- Condition of tensile steel,
- Existing degree of pretension,
- Grouting mortar (SO_3 content).

(4) Components
- Deformation,
- Forces,
- Vibration behaviour.

(5) Sampling on structure
- Efflorescence/honeycombing material,
- Concrete components,
- Drilling cores,
- Drilling dust,
- Reinforcing components.

10.14.3 Investigations in the Laboratory

(1) Concrete
- Apparent density,
- Porosity/capillary action (as per DIN 52 103),
- Water absorption (as per DIN 52 617),
- Water penetration depth (WU),
- Wear resistance (as per DIN 52 108),
- Micro air pore content,
- Chloride content (quantitative in different depth zones),
- Sulphate content,
- Compressive strength (as per DIN 1048),
- Modulus of elasticity (as per DIN 1048),
- Mixture proportions (as per DIN 52 170),
- Granular composition,
- Disk strength (as per DIN 1048),
- Depth of carborisation,
- Surface tensile strength (as per DIN 1048) in various depth horizons.

(2) Steel
- Tensile test (as per DIN 488, part 3),
- Fatigue test.

10.14.4 Theoretical Investigations

Structural calculations of the bearing stability and deformation behaviour of the structure or individual components before and after repair. Estimation of carborisation progress and/or preliminary chloride enrichment with or without repair.

10.15 Repair of Concrete Components in Hydraulic Engineering (R 195)

10.15.1 General

Hydraulic structures are subject to particularly unfavourable environmental stresses arising from physical, chemical and biological factors. In addition, in harbour facilities and other areas which must be kept accessible for operational reasons, for example, the effects of anti-icing salt and other harmful impurities must be anticipated, along with occasional cargo items which are harmful to concrete.

Apart from live loads and impact and friction forces from vessels, the physical effects result primarily from the repeated drying out and moistening of the concrete, the constant temperature fluctuations with severe frost effects on water-saturated concrete and the action forms of ice. Apart from the effects of anti-icing salt and cargo goods resulting from requirements for use in some cases, chemical stresses are caused primarily by the salinity of the sea water. Chlorides which have penetrated into the concrete can destroy the passive layer of the reinforcement. In areas with a prevailing adequate presence of oxygen and moisture in the concrete, for example above the water exchange zone, corrosion can thus be caused in the reinforcement. Biological stresses occur primarily through the growth of algae and resulting metabolic products.

Such factors can lead to cracks and surface damage to the concrete and corrosion damage to the reinforcement. Particularly at risk are components in the spray and water exchange areas, especially components in contact with sea water and those in the direct vicinity of coasts in strongly salt-laden air. Fig. R 195-1 shows a diagram of the attack by sea water on reinforced concrete.

If the damage discovered indicates that repair work is necessary on concrete components, an engineering expert according to the instructions for protection and repair of concrete components [134] must always be consulted to determine the actual condition, assess the damage and plan the repair measures. The lasting success of the work depends significantly on the expertise of the personnel involved, quality and suitability of materials used, and the care exercised in both execution and monitoring.

Protection and repair measures on components which can be protected from contact with water at least while the work is being carried out (work above water) should be carried out according to ZTV-W LB 219 [138]. ZTV-RISS [159] should be used as standard work when filling cracks and hollows in such parts.

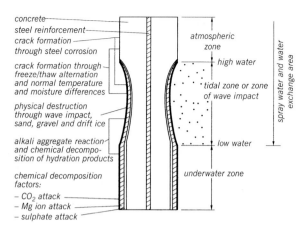

Fig. 195-1. Diagram to show the attack of sea water on reinforced concrete as per [133]

For protection and repair measures which have to be carried out under water (work under water), trial repairs should verify that the intended work can lead to the desired result under the marginal conditions of the particular case concerned. Verification of the quality of the construction materials should be provided in the form of suitability and quality tests, brought into line with the prevailing component conditions.

Measures in the context of cathodic corrosion protection should be planned on the basis of [179].

Repair work should only be carried out by companies with adequate expertise and experience in this field, complying with the requirements regarding material and personnel according to [134], [138] respectively [159].

10.15.2 Assessment of the Actual Situation

The effects of the faults and damage on the stability, serviceability and durability of the structure must be assessed on the basis of a careful inventory in accordance with R 194, section 10.14.

The actions and loads affecting the structure requiring repairs should be researched as accurately as possible, because this results in the requirements to be fulfilled by the materials or construction method to be used.

10.15.3 Planning the Repair Work

10.15.3.1 General

The repair requirements result from a comparison of the actual situation as per section 10.15.2 with the intended required state at the end of the repair work. The repair targets for the repair work should be defined as exactly as possible. According to [134] and [138], a differentiation should be made between measures for the protection and repair of the concrete and measures for the restoration or preservation of the corrosion protection in the reinforcement. When it comes to filling cracks, it should be clarified whether the cracks should be simply closed or sealed, or whether the edges of the crack should be bonded together in a positive or expanding manner.

The possible effects of protection and repair work on the durability and bearing capacity of the component or the whole structure must be investigated. Special attention here should be paid to any unfavourable changes in the building physics and changes in the bearing behaviour (increase in the dead weight, redistribution of loads etc.).

When elaborating the repair concept, special consideration should be given to the basically different marginal conditions for reinforcement corrosion in structures above and below water, particularly the different exposure to the oxygen required for the corrosion process.

Cracks must be investigated to see which causes are behind the crack formation and which loads or deformations have to be anticipated here in future.

10.15.3.2 Repair Plan

An engineering expert should draw up a repair plan for all repair work. The plan shall describe all relevant details for execution of the repairs, from preparation of the subbase via type and quality of materials being used, selected construction method and subsequent treatment through to quality assurance.

The repair plan should be drawn up where possible on the basis of [138] respectively [134].

The repair plan should contain details referring to the following points, among others:

Repair principles/fundamental solutions as per [134] and [138].

Requirements for contractors/staff, for example
- Suitability certificate for the nozzle operator for injection concreting work,
- Suitability certificate for handling plastics in concrete construction work.

Subbase preparation
- Aim of the subbase pre-treatment and type of pre-treatment method,
- Extent to which concrete is removed respectively the reinforcement exposed,
- Degree to which rust has to be removed from the reinforcement.

Concrete replacement
- Type and quality of materials and methods to be used, for example
  – Concrete
  – Injected concrete
  – Injected mortar/injected concrete with plastic additive (SPCC)
  – Cement mortar/concrete with plastic additive (PCC)
- Formwork,
- Layer thicknesses,
- Additional reinforcement,
- Construction joints.

Cracks
- Filling material,
- Filling methods.

Joints
- Preliminary work,
- Type of joint sealing material,
- Execution.

Surface protection systems
- Type of system,
- Layer thicknesses.

Subsequent treatment
- Type,
- Duration.

Quality assurance
- Basic tests,
- Suitability tests,
- Quality monitoring.

10.15.4 Performance of Repair Work

10.15.4.1 General

The performance of repair work for measures to be carried out above water is described in detail in ZTV-W LB 219 or ZTV-RISS.

Before applying cement-bonded concrete replacement material (concrete, injected concrete, SPCC, PCC), the concrete subbase should be adequately pre-wetted (for the first time 24 hours in advance). However, before placing the concrete replacement, the contact surfaces should have dried to such an extent that they appear matt-damp.

Adequate subsequent treatment is of decisive importance for the success of repair work. Cement-bonded concrete replacement should be treated in the first few days after placing by watering measures. This applies particularly to lesser layer thicknesses of concrete replacement using PCC.

In view of the differing basic materials involved, differences in colour between the old concrete and concrete replacement must always be expected for local repairs using cement-bonded concrete replacement.
The following details generally do not apply to measures with cathodic corrosion protection.

10.15.4.2 Subbase Preparation

(1) General
The plan should not state the type of subbase pre-treatment but the aim to be fulfilled with the subbase pre-treatment.
At the end of the subbase pre-treatment work, it should be checked whether the concrete subbase possesses the friction strengths stipulated for the intended repair work.

(2) Work above water level
To create a good bond, the concrete subbase must be uniformly sound and free of separative, inherent or foreign substances. Loose and brittle concrete together with all foreign substances, such as growth of algae, barnacles, oil or paint residues are to be removed. The extent to which concrete then has to be removed and the reinforcement exposed in order to achieve the repair aim depends on the basic solution selected according to [134] or [138] and should be stated in the repair plan.
Before placing a cement-bonded concrete replacement, on conclusion of subbase preparation work, as a rule cones of aggregate grains should be exposed near to the surface measuring ≥ 4 mm in diameter.
On conclusion of subbase preparation work, loose corrosion products on exposed reinforcement and exposed installed structures should be removed. For corrosion protection by restoring the alkali milieu as per [134] or [138], the degree to which rust is removed must comply at least with the normal purity degree Sa 2, of Sa 2 ½ for corrosion protection by coating the reinforcement, as per DIN 55 928. Rust removal from reinforcement in the case of chloride-induced reinforcement corrosion may only be carried out by high-pressure water jetting (≥ 600 bar).

The following methods can be used for subbase preparation, depending on the purpose:

- mortising,
- cutting,
- grinding,
- blasting with   – solid blasting agents,
                    – water/sand mixture,
                    – high-pressure water.

Material removed during subbase pre-treatment and mixtures resulting from the process are to be disposed of properly according to the waste management regulations.

(3) Work under water
The information according to section 10.15.4.2 (2) applies accordingly. Depending on the purpose, subbase pre-treatment can be carried out using the following methods:

- hydraulically driven cleaning equipment,
- underwater blasting with – solid blasting agents,
  – high-pressure water.

10.15.4.3 Repairs with Concrete

(1) General
Repairs with concrete are to be given preference particularly for large-surface repairs with larger layer thicknesses, from both a technical and economic point of view.

(2) Work above water level
Repairs with concrete should be carried out on the basis of ZTV-W LB 219 [138]. This stipulates specific requirements for the composition and properties of the concrete, contingent on the loads affecting the component.

(3) Work under water
This work is to be carried out according to section 10.15.4.3 (1). Proper placement and compaction of the concrete without segregation is to be safeguarded by adding a suitable stabiliser certified by the DIBt (German Institute for Construction Technology, Berlin) or according to the rules for underwater concrete as per DIN 1045 section 6.5.7.8.

10.15.4.4 Repairs with Injected Concrete

(1) General
The injected concrete method has proven effective for the repair of concrete structures in hydraulic engineering and is probably the most frequently used repair method.

(2) Work above water level
See section 10.15.4.3 (2). [138] stipulates specific requirements for the composition of the concrete mixture and the properties of the finished injected concrete as a function of the actions to the component.
ZTV-W B 219 makes a basic differentiation between injected concrete in layers of up to approx. 5 cm thick, which can be placed without reinforcement, and injected concrete in layers of more than 5 cm thick, which

must also be reinforced and connected to the structure by means of anchoring elements.

The surface of the injected concrete is to be left in its rough injected state. If a smooth or specially structured surface is required, mortar or injected mortar is to be applied after the injected concrete has hardened, and processed accordingly.

(3) Work under water
Not applicable

10.15.4.5 Repairs with Plastic-modified Injected Concrete (SPCC)

(1) General
The use of SPCC can be advantageous particularly for thinner layers, because the plastic additives improve certain concrete properties, such as water retention value, adhesion strength or water tightness. In addition, the plastic additives also make it possible to achieve a deformation behaviour which can be compared with the old concrete. Only plastic additives which are insensitive to moisture may be used.

(2) Work above water level
See section 10.15.4.3 (2). The SPCC layer thickness should be between 1 and 5 cm for wire-spread application. Additional reinforcement is not required for these thicknesses.

The surface of the SPCC should be left in its rough injected state. If a smooth or specially structured surface is required,

– in the case of a single layer, hardening of the SPCC is to be followed by application of a mortar which is compatible with SPCC and which can then be processed accordingly,
– in the case of several layers, the last injected layer is processed accordingly.

(3) Work under water
Not applicable.

10.15.4.6 Repairs with Cement Mortar/Concrete with Plastic Additives (PCC)

(1) General
PCC is suitable particularly for the repair of small eruption areas. PCC is applied to the contact surface by hand or machine. However, in contrast to injected concrete or SPCC, compaction is by hand in both cases. Only plastic additives which are insensitive to moisture may be used.

(2) Work above water level
See section 10.15.4.3 (2). The PCC layer can be up to approx. 10 cm thick for local repairs.

(3) Work under water
Special products are available for this application.

10.15.4.7 Repairs with Reaction Resin Mortar/Reaction Resin Concrete (PC)

(1) General
PC are practically impervious to water vapour. This is just one of the main reasons why the use of PC is restricted to local repairs or work under water.

(2) Work above water level
PC should only be used in exceptional cases and only for local repairs [136].

(3) Work under water
Special products are available for this application.

10.15.4.8 Sheathing of Components

(1) General
The damaged component is sheathed with a watertight covering which is sufficiently resistant to anticipated mechanical, chemical and biological attacks. The protective covering may be placed around the component to be protected, either with or without an adhesion bond. The aim of the procedure is to prevent the ingress of water, oxygen or other substances between the covering and the component. The procedure can be used both above and below the water.

(2) Cleaning and preparing the subbase
The work is performed in accordance with section 10.15.4.2 (2) or 10.15.4.2 (3).

(3) Sheathing of concrete with a pre-produced concrete shell attachment
Requirements relating to the pre-produced part:
- Watertight, capillary-pore-free concrete (water/cement ≤ 0.4) with high frost-resistance,
- Layered reinforcement.

The space between the shell attachment and the pre-treated concrete is filled by squeezing in a low-shrinkage cement-lime mortar with high frost resistance.

(4) Sheathing of concrete with a pre-produced fibre-reinforced concrete shell attachment
Suitable fibres:
- Steel fibres,
- Alkali-resistant glass fibres.

Requirements and execution in accordance with section 10.15.4.8 (3).

(5) Sheathing of concrete with in-situ fibre-reinforced concrete
As for section 10.15.4.8 (4).

(6) Sheathing of concrete with a plastic shell, can be used for supports
Requirements relating to the plastic shell:

- Resistant to UV radiation (above water level only),
- Resistant to surrounding water,
- Watertight and adequately diffusion-tight,
- If necessary, adequate mechanical resistance to anticipated influences, e.g. ice load, bed load and contact with vessels.

The space between the shell and the existing concrete is filled as in section 10.15.4.8 (3).

(7) Winding flexible foil around the component, can be used for supports
Clean and prepare the subbase in accordance with section 10.15.4.2 (2) respectively 10.15.4.2 (3).
Treat the reinforcement with corrosion protection and fill damaged areas above water as per section 10.15.4.3 to 10.15.4.7. Wind flexible foil around the supports.
Requirements relating to the system:

- Resistance to UV radiation,
- Resistance to the surrounding water,
- Imperviousness to water and gas,
- Adequate mechanical resistance to anticipated external effects, such as ice load,
- Watertight interlocks between foil wheels and tight upper and lower connections with the support, so that neither fluid nor gaseous substances can penetrate between the foil and subbase.

10.15.4.9 Coating the Component

(1) General
As an additional measure against penetration of harmful substances into the concrete, particularly chlorides and carbon dioxide in the case of steel and pre-stressed concrete components (when no sheathing is provided in accordance with 10.15.4.8), a coating on the cleaned, prepared component which eventually has been repaired with concrete may prove effective.

(2) Work above water level
See section 10.15.4.3 (2).
Coating may only be used when there is no risk of moisture penetrating from the back.

(3) Work under water
Special products are available for this application.

10.15.4.10 Filling Cracks
The work should be carried out on the basis of ZTV-RISS [159] as far as possible. Cracks in hydraulic engineering components with their frequently high water saturation levels can best be filled with cement lime/cement suspensions for positive connections and polyurethane for stretching connections.

10.15.4.11 Formation and Sealing of Joints
- Clean joints and extend existing joint gap if necessary,
- Repair damaged edges with epoxy resin mortar,
- Install joint sealing in accordance with the applicable specifications and directives.

The closure and filling of joints can be carried out accordingly as per DIN 18 540. [137] is to be observed for joints in traffic areas.

11 Pile-founded Structures

11.1 General

The pile-founded structures dealt with in the following sections are always to be calculated with the horizontal butt deformations, and to be checked for serviceability. In the case of large differences in surface elevation, it is worth considering whether battered anchoring as per chapter 9 should be used as well as the pile bents, to limit the deformation. Piles can be exposed to changing loads (tension/pressure) from hawsers pulling on bollards, lateral crane impact and tidal influences. Suitability of the chosen pile system must therefore be checked for alternating loads. The recommendations for statical calculations featured in this section are based on the safety concept of verification of theoretical bearing capacity in limit states, using partial safety factors (see section 0.1).
For the design of quay walls, superstructures and pile-founded structures, please also refer to R 17, section 10.1.

11.2 Determining the Active Earth Pressure Shielding on a Wall Below a Relieving Platform Under Average Ground Surcharges (R 172)

A wall can be more or less shielded from the active earth pressure by means of a relieving platform, depending above all on the location and width of the platform, as well as on the shear strength and compressibility of the soil behind the wall and below the bottom of the structure, to have a favourable influence on earth pressure distribution which is applicable to ascertaining the internal forces.

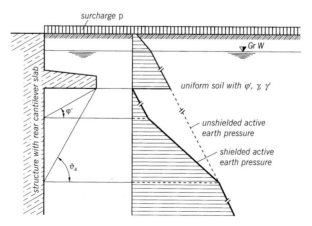

Fig. R 172-1. Solution according to LOHMEYER with uniform soil

With uniform, non-cohesive soil and average ground surcharges, the active earth pressure shielding can be determined according to LOHMEYER [93], fig. R 172-1. As can be easily confirmed by CULMANN investigations, the use of the LOHMEYER method is well applicable under the foregoing prerequisites.

With stratified, non-cohesive soil, the assumptions according to fig. R 172-2 respectively R 172-3 offer approximative solutions, whereby

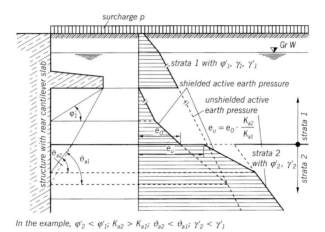

In the example, $\varphi'_2 < \varphi'_1$; $K_{a2} > K_{a1}$; $\vartheta_{a2} < \vartheta_{a1}$; $\gamma'_2 < \gamma'_1$

Fig. R 172-2. Solution according to LOHMEYER, expanded for stratified soil (solution possibility 1)

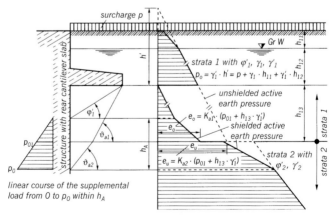

Fig. R 172-3. Solution according to LOHMEYER, expanded for stratified soil (solution possibility 2)

the calculation as per R 172-3 can be carried out electronically even for multiple changes of strata.

In cases of doubt, with multiple changes of strata, the active earth pressure can be ascertained with the help of the expanded CULMANN's method as per R 171, section 2.4.

If the soil also has a cohesion c', the shielded-off active earth pressure can be taken approximately in such a manner that at first the shielded-off active earth pressure distribution is determined without taking c' into account, and is subsequently superposed with the cohesion influence:

$$\Delta e_{ac} = c' \cdot K_{ac}$$

(K_{ac}: coefficient for active earth pressure when taking account of cohesion, see DIN V 4085-100). This procedure however, is only permissible when the percentage of cohesion is slight in comparison with the total active earth pressure. A more accurate determination is also possible here with use of the expanded CULMANN's method as per R 171, section 2.4.

The same applies to calculating the effect of earthquakes, taking into account R 124, section 2.13.

The calculations according to figs. R 172-1 to R 172-3 may not be used in cases in which several relieving platforms have been installed one above the other. Furthermore, independent of the shielding, the total stability of the structure is to be verified in the corresponding limit states to DIN V 1054-100, whereby full active earth pressure is to be applied in the relevant reference planes.

11.3 Active Earth Pressure on Sheet Piling in Front of Pile-founded Structures (R 45)

11.3.1 General

With increasing frequency, pile-founded structures with rear sheet piling must often be reinforced by driving an additional front sheet piling wall in order to allow for greater water depths. This new front sheet piling will be stressed by the earth support pressure of the existing front sheet piling, and often by soil stresses from the bearing piles, which may begin to take effect at an elevation immediately below the new harbour bottom.

Also in cases of new constructed pile-founded structures, it can occur that a front sheet piling lies in the area influenced by bearing pile forces. The loads acting on the sliding wedges and on the new sheet piling can only be approximated. The following statements apply initially to non-cohesive soil for ascertaining the internal forces.

It is presumed that verification of overall stability has been provided according to DIN V 4084-100 for limit state 1 C, and that the embedment depth of the new front quay wall is thus defined.

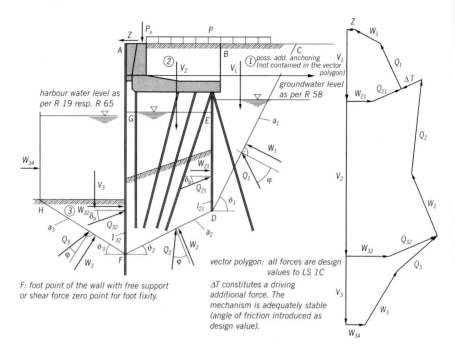

Fig. R 45-1a. Overall stability of an embankment with superstructure. Example for a failure mechanism.

In this context, fig. R 45-1a shows an example of a possible failure mechanism. Further failure mechanisms in which primarily the incline of the outer failure lines of the sliding wedges 1 and 3 are varied, must show whether the vector polygon always can be closed without supporting additional force ΔT.

In fig. R 45-1a, the symbols have the following meaning:

V_1, V_2, V_3 — Dead weight of the sliding wedges 1, 2, 3, including their share of the changing and permanent live loads and vertical water loads

W_1, W_2, W_3 — Water pressure forces on the outer failure planes a_i of sliding wedges 1, 2, 3, normal to the failure planes a_i

Q_1, Q_2, Q_3 — Failure plane surfaces of the outer failure planes a_i of the sliding wedges 1, 2, 3 under the angle of friction φ inclined to the vertical on the sliding plane

Q_{21}, Q_{32} — Failure plane forces of the inner failure planes i_{21}, i_{32}, under δ_p for sheet piling, under φ in the soil inclined

504

W_{21}, W_{32}	Water pressure forces, acting on the vertical inner failure plane i_{21} (DEB), i_{32} (FG) from left-hand side to right-hand side
W_{34}	Water pressure force on the perpendicular to H, acting to the right
P_v	Vertical load
Z	Horizontal load
ΔT	Additional force for fulfilling the equilibrium
a_1, a_2, a_3	Outer failure planes of the sliding wedges 1, 2, 3
i_{21}, i_{32}	Inner failure planes between sliding wedges 2 and 1, 3 and 2
φ	Angle of friction
δ_p	Passive angle of wall friction
$\vartheta_1, \vartheta_2, \vartheta_3$	Angle of the failure planes

11.3.2 Load Influences

The active earth pressure acting on the new front sheet piling is influenced by:

(1) The active earth pressure from the soil behind the quay wall. It is generally referred to the plane of a possibly existing rear sheet piling or to a vertical plane through the rear edge of the superstructure (rear reference plane). It is calculated with plane sliding surfaces for the existing level of the ground and the existing surcharge. The presumed angle of wall friction has to comply with DIN V 4085-100.

(2) The foot support reaction of the existing rear sheet piling, if any (Q_1 in figs. R 45-1b and d, inclined under δ_a to the horizontal)

(3) The flow pressure caused by the difference between groundwater and harbour water level, which imposes load on the sections of earth behind the existing front sheet piling.

(4) The dead load of the earth mass lying between the existing front sheet piling and the rear plane of reference, acting together with the active earth pressure according to (1). In the case of existing rear sheet piling, this is the lower support pressure required for equilibrium of the rear sheet piling, the same as that which is introduced into the soil between the two sheet piling walls.

(5) The pile forces which result from the vertical and horizontal superstructure load. In calculating the pile forces, the reaction at the upper support of the front sheet piling must be taken into consideration when an additional anchoring system independent of the pile-founded structures is not used (fig. R 45-1a).

(6) The resistance Q of the soil between the front sheet piling and the rear plane of reference defined in (1), for forward movement of the structure (fig. R 45-1b to d).

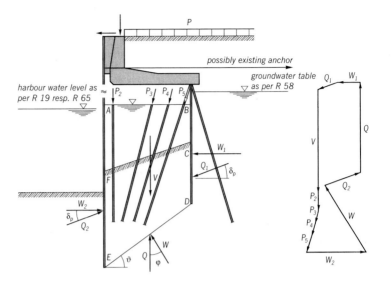

Fig. R 45-1b. Investigation of various failure mechanisms to ascertain the actions on a new front sheet piling wall for an embankment with superstructure, 1st example

In fig. R 45-1b, the symbols have the following meaning:

V	Dead weight of the soil *FCDE* including water load *ABDE*
W_1, W_2	Water pressure forces from right, left
W	Water pressure force on the failure plane *ED*
Q	Failure plane force of the failure plane *ED*
Q_1	Characteristic value of the required foot support force of the rear sheet piling
Q_2	Supporting force required for equilibrium
$P_2 - P_5$	Axial forces absorbed in the sliding wedge *CDEF*
φ	Angle of friction
δ_p	Passive angle of wall friction
ϑ	Failure plane angle

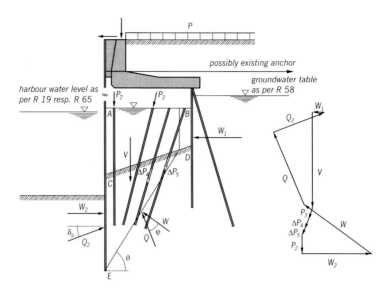

Fig. R 45-1c. Investigation of various failure mechanisms to ascertain the actions on a new front sheet piling wall for an embankment with superstructure, 2nd example

In fig. R 45-1c, the symbols have the following meaning:

V	Dead weight of the soil *CDE* including water load *ABDE*
W_1, W_2	Water pressure forces from right, left
W	Water pressure force on the failure plane *ED*
Q	Failure plane force of the failure plane *ED*
Q_2	Supporting force required for equilibrium
$P_2 - P_3$	Axial forces absorbed in the sliding wedge *CDE*
$\Delta P_4, \Delta P_5$	Axial forces from skin friction, absorbed in the sliding wedge *CDE*
φ	Angle of friction
δ_p	Passive angle of wall friction
ϑ	Failure plane angle

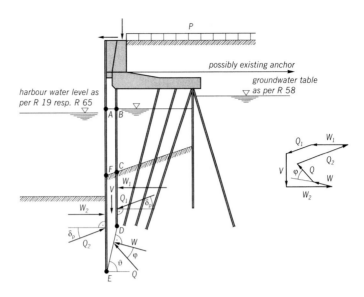

Fig. R 45-1d. Investigation of various failure mechanisms to ascertain the actions on a new front sheet piling wall for an embankment with superstructure, 3rd example

In fig. R 45-1d, the symbols have the following meaning:

V	Dead weight of the soil *FCDE* including water load *ABDE*
W_1, W_2	Water pressure forces from right, left on the section surfaces *DB, EA*
W	Water pressure force on the failure plane *ED*
Q	Failure plane force of the failure plane *ED*
Q_1	Characteristic value of the required foot support reaction of the existing front sheet piling wall
Q_2	Supporting force required for equilibrium
φ	Angle of friction
δ_p	Passive angle of wall friction
ϑ	Failure plane angle

11.3.3 Assumed Loads for Determining the Active Earth Pressure on the New Front Sheet Piling

The assumed loads for new sheet piling driven in front of an existing quay wall are shown in fig. R 45-1b to d. To simplify the calculation, in this case it is recommended to use the partial safety factors for actions on the internal forces and support reactions of the new front sheet piling. Changing actions should be enlarged by the ratio γ_Q/γ_G, so that only multiplication with the partial safety factor for constant action γ_G is necessary for calculation of the design values of the internal forces and support reactions.

The value of the foot support reaction Q_1 of the rear sheet piling when present (section 11.3.2 (2)) is taken from the characteristic values of the support reaction and internal forces for this wall. It is taken as action on the earth body between this wall and the new front sheet piling. An inner section is considered performed on the active side of the new front sheet piling which cuts the sheet piling walls P_1 and P_2 together with the piles P_3 to P_5 underneath the pile-founded structure with relieving platform.

Fig. R 45-1b shows the statical system and one of the failure mechanisms to be examined together with the corresponding vector polygon. The mechanism (see figs. R 45-1c and d) which requires the greatest resisting force Q_2 is decisive. This force is then to be taken as the characteristic active earth pressure on the new front sheet piling.

To calculate the design values of the internal forces in the front wall, Q_2 is to be multiplied with the partial safety factor for constant actions. In selecting the direction for Q_2, the boundary conditions for active earth pressure on the new front wall may not be infringed. The distribution of active earth pressure on the new front wall depends on the wall's scope of movement. The distribution may be selected according to section 8.2 or as per DIN V 4085-100. The pile forces are to be determined from a stability verification of the superstructure with the characteristic values of the actions. In the earth body between the two walls, the forces which are carried away here via skin friction and point pressure are to be introduced (ΔP-forces in the figs. R 45-1b to d).

The effect of a flow force resulting from differences in water level between groundwater and harbour water is to be taken into account by using the water pressure forces in all examined failure lines or element limits in connection with the specific weight of the saturated soil. Figs. R 45-1c and d show examples of further mechanisms for examination which are characterised by steeper failure lines through to a failure line between the foot or shear force zero point for a fixed new front wall and the foot of the existing front wall (fig. R 45-1d). The force Q_1 is now the support force of the existing front sheet piling and results in an analogue manner to Q_1 for the rear sheet piling.

11.3.4 Calculation for Cohesive Soils

A similar procedure can be used here. In consolidated soils with an effective cohesion c', in addition the cohesion force $C' = c' \cdot l$ is taken into consideration along the corresponding failure plane examined. In the case of water saturated soils exposed to loads for the first time, the value c' is replaced by the value c_u, whereby $\varphi' = 0$.

11.3.5 Loads from Water Pressure Difference

The water pressure difference acting on the front sheet piling depends, among other factors, on the soil conditions, the height of the soil behind the wall and the availability of a drainage system. In new constructions with only front sheet piling and with soil reaching up to the underside of the relieving platform, the water pressure difference is considered as acting directly on the sheet piling as described in R 19, section 4.2. If in cases with a rear sheet piling, the earth surface is below the free water table, and if the new wall has a sufficiently large quantity of water drainage openings, R 19 applies to assumed water pressure difference acting on the old sheet piling and the assumed flow pressure according to section 11.3.2 (3). In such cases, a water pressure difference of half the height of the waves to be anticipated in the harbour is assumed as a precautionary measure directly on the front sheet piling. As a rule, it is usually adequate to presume a difference in water table of 0.5 m as characteristic value.

11.3.6 Anchor Pre-stressing

In order to keep the degree of deformation low, the anchoring of subsequently driven front sheet piling can be pre-stressed allowing for a displacement distance of the anchor connection point which is in line with local conditions.

11.4 Calculation of Plane Pile-founded Structures (R 78)

There are basically two versions of plane pile-founded structures available to compensate for differences in terrain elevation (see also drawings in R 200, section 6.8.2):

- Pile-founded structure standing free over an underwater embankment with sheet piling on the land side which compensates for the remaining difference in ground elevation at the embankment head.

- Pile-founded structure with sheet piling on the water side:
  – reinforcement structure in front of and over existing pile-founded structure to enlarge the design depth, whereby an existing embankment is usually preserved. The sheet piling should be taken down to a certain depth – at least to the foot of the embankment – in an open design to prevent any build up of differential water pressure,
  – construction of a new pile-founded structure with fully backfilled sheet piling. The superstructure platform positioned on the piles effectively screens the earth pressure on the sheet piling from surcharges (fig. R 78-1).

The bearing structure cross section of the above mentioned waterfront structures, which are single dimensioned line structures in the longitudi-

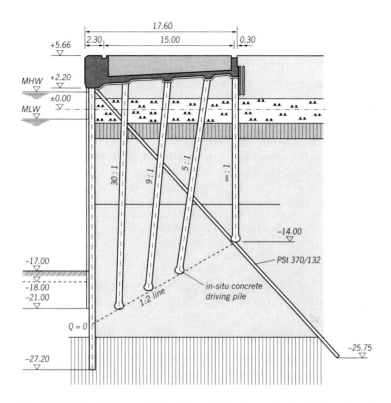

Fig. R 78-1. Executed example of a pile-founded structure with sheet piling on the water side

nal direction, is a plane pile-founded structure, whose action and resistance is stated per continuous meter or per system grid dimension.
Bollard and fender actions can also be distributed proportionally on the calculation cross section using the plate effect of the relieving platform within a structural section.
The internal forces for piles and superstructure can be ascertained in the plane case in an elementary manner:
The statical problem of the platform plate with the piles arranged in its plane can be illustrated correctly by means of an elastically supported continuous beam. The piles are illustrated as springs in the direction of the pile axis, the corresponding spring stiffness values result from the pile characteristics.
If these are piles with great expanding stiffness, for example in-situ concrete driven piles, it is possible to calculate with a rigid support.

A plane frame bearing structure consisting of several pile sticks and a bar representing the superstructure is equivalent to the elastic supported continuous beam.

In other words, only relatively simply software programs are required for computer-assisted determination of the internal forces for piles and superstructure in the plane case. Even these are capable of carrying out the various required computations with variable input parameters to optimise individual structures, as well as various limit considerations, all at reasonable expense.

Fixed or hinged mounting of the piles in the soil and on the superstructure platform, lateral bedding (and thus pile bending) as well as axial bedding can be covered appropriately by 2D standard software.

The idealisation of the "rigid" pile head platform (NÖKKENTVED [94] or SCHIEL [95]) on which the classical computation methods for relieving platforms are based can be approximately simulated in the computer by taking a very large numerical value for the bending stiffness of the superstructure, which is represented by the value "infinite" in the numerical computation. However, because the presumption of the "rigid" pile head platform is no longer applicable for elastic superstructure deformations, the classical methods basically produce incorrect results here.

The calculation of waterfront structures with "rigid" superstructure will accordingly only be applicable in special cases, e.g. when existing old quay structures with solid pile head blocks are to be re-computed (fig. R 157-1).

Widely differing pile dimensions and materials can be illustrated simply in the data record of a rod structure program by varying the spring stiffnesses of elastic bearings.

The main advantage of a program-controlled calculation is however the fact that the components can be optimised so that they have about the same load level as each other:

for example, the field widths of the superstructure platform can be changed so that the pile loads are approximately the same in all rows – a highly economical approach to dimensioning the piles.

11.5 Design and Calculation of General Pile-founded Structures (R 157)

A general pile-founded structure can be seen as a three-dimensional rod structure on which a relieving platform is positioned as superstructure. The acting loads are distributed on all piles through the combined bearing action of the superstructure as platform and plate, for effective carrying off of the loads. Construction elements and loads must be described in a spatial Cartesian coordinate system.

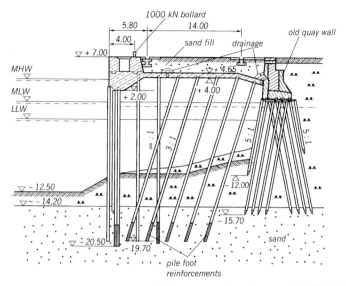

Fig. R 157-1. Executed example of a forward extension of a quay wall with elastic RCC relieving platform on steel piles

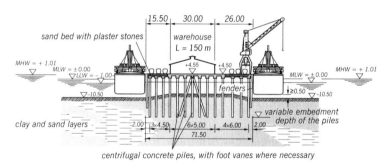

Fig. R 157-2. Executed example of a free harbour pier with elastic RCC relieving platform on centrifugal concrete piles

11.5.1 Special Structures as Part of Line Structures

If the pile-founded structure walls described in R 78, section 11.4 have non-constant elements in partial areas, such as buckling formations and corners, moles and peer heads, or if there are offset sections and differences in elevation at the front edge of the quay, for example in the case of Ro-Ro ramps, special structures are required: the foundation piles required here must be arranged so as to allow for the geometry of these special structures, providing the available space under the relieving platform.

A plane arrangement of the piles is then as a rule no longer possible, on the contrary highly complicated areas are produced here with a large number of intersection points and spatial, mutually penetrating piles whose connection points to the superstructure can be positioned on several different elevations.

11.5.2 Free-standing Pile-founded Structures

Free-standing, high pile-founded structures with superstructure platform are mainly used in the following cases:

- Soil with adequate bearing capacity is only present at a greater depth,
- Essentially free passage for wave action is to be created,
- Reduction of wave energy by the arrangement of embankments or pile-supported structures instead of quay walls or other harbour structures with vertical walls,
- Economic aspects take priority.

The relieving platforms of quay structures are mainly constructed from reinforced concrete, whereas steel profiles are usually used for foundation piles.

However, in warm regions both reinforced concrete or pre-stressed concrete piles are frequently used, because here the specific requirements resulting from the effects of ice action are not necessary.

The way in which the superstructure platform lies on the foundation piles can be selected according to functional requirements or on the basis of the specific dimensions of the superstructure: here the whole range of possible bearing is conceivable, from statically determined through to bearing with very few degrees of freedom, whereby extensive entrapment of the possible degrees of freedom can be advantageous with regard to minimum system displacements.

Held degrees of freedom of the bearings are expressed in calculation of the system only in the larger number of equations to be solved simultaneously. However, this it not relevant when suitable computing programs are used, because this is only a computer problem and not a principle structural problem.

Dimensioning of the piles and determination of the driving depths must take account of the fact that no lasting pile settlements occur even when assuming the least favourable combination of actions.

Part of the foundation piles must be arranged as pile bents so that the relieving platform cannot shift noticeably in a horizontal direction. At the same time, this thus reduces the bending moment load in all piles, if no other horizontal actions with a great influence occur, such as e.g.:

- lateral pressure on piles from flowing soil masses,
- strong currents, ice pressure, ice impact and the like.

If the above-mentioned assumptions are applicable, the piles can be calculated with sufficient accuracy with hinged bearings on the relieving platform and at the pile foot point, even if they are not designed as such structurally.

Corresponding structural measures at the pile head are to be arranged to cover any occurring, unwanted fixity. This can possibly be necessary in the case of large changes in length of the superstructure platform caused by differences in temperature.

Shifts in the pile heads caused by shrinkage during the construction period can be catered for by making up partial sections of the long superstructure platform with wide concreting joints arranged in between. These joints are then sealed with construction concrete once the majority of the shrinkage has ceased. These so-called shrinkage seals can usually keep the influence of shrinkage on the construction within harmless limits.

Relieving platforms are usually designed to be only 50 to 75 cm thick, so that they are to be seen as susceptible to bending, referred to length changes in the piles. This can help to prevent the transfer of larger bending moments when at the same time the pile arrangement is sensibly adjusted to possible load positions. Actions occurring at each particular load site are thus carried off essentially by the immediately adjacent piles, so that only a minimum action is noticed by piles standing further away.

The platform thickness is selected on the basis of the bending rating and also with regard to safeguarding a statically and structurally flawless transfer of the shear forces from the concrete platform as pile load into the pile heads; for head fixity, this also applies to the subsequent pile bending moments.

There should be no direct traffic on the relieving platforms, because a sand fill on the platform offers further operational and constructional advantages:

Utility lines, ducts etc. can be accommodated within the sand layer, and this also makes it possible to reduce the loads on the relieving platform and the piles from locally acting influences, thanks to this compensating layer.

In particular, sand fills from a thickness of 1.0 m on are generally extremely favourable, because this thickness makes it possible to accommodate all necessary lines within the sand fill, and no vibration actions from vehicle operations need to be taken into account for the substructure.

If the actions from road and rail traffic are assumed to be evenly distributed surface loads, the layer thickness must be selected according to R 5, section 5.5.

11.5.3 Statical System and Calculation

An appropriate illustration of the relieving platforms with spatial pile structures for a statical calculation results in a model consisting of the elastic platform on elastic supports, i.e. from a statical point of view and when it comes to carrying off the load, the pile-founded relieving platform acts as a flat slab which is supported on the piles without or with support head formations. As in R 78, section 11.4, the piles are idealised as elastic springs with axial direction of action, and the pile length between the hinged points is taken to be the elastic length.

If it is not possible to restrict the procedure to approximate solutions, such as elastic continuous beam calculations with load carried off in two orthogonal directions as per R 78, more demanding computing methods have to be used which can only be carried out using computers because of the numerical workload involved.

These computing methods are available in the form of software packages for platforms and folding structures, and work, for example, using the deformation method or finite element method.

These tools can be used to solve tasks with any degree of complexity for an elastic relieving platform on elastic piles, with adequate accuracy without any fundamental difficulties.

In addition, programs for bearing platforms on elastic support provide a possibility for calculating the current bearing structure with good approximation. Relieving platforms with reinforcing beams over the pile heads are taken into account realistically as bearing platforms with beams of different stiffness.

An adequate illustration of the spatial pile structure serving as support for the platform or bearing platform, must be elaborated individually with the elements and tools provided in the software.

The loads acting on the whole relieving platform and pile-founded structure are then the most favourable when the piles are arranged in such a way that the negative bending moments on all supports in a basic direction as well as the maximum pile loads are approximately of the same magnitude.

However, this ideal optimum cannot always be achieved because of the frequent presence of marginal disturbances, such as influences from crane operations, hawser pull, vessel impact and the like, particularly in view of the fact that the pile positioning is also affected by structural and execution irregularities.

11.5.4 Structural Advice (see also [96])

In order to arrive at the most economical possible total solution, the following are to be observed, among others:

- The berthing forces of larger vessels are absorbed by fendering in front of the platform with heavy berthing contact panels (fig. R 157-2). If

this is not completely possible, the fender must convey part of the impact action into the pile-founded structure.
- A heavy mooring bollard may also be installed at the wall coping in the fender area.
- Locally limited horizontal forces, for example from line pull and vessel impact, will be spread over all piles in a block through the superstructure platform which is very stiff in its level.
- In the case of small ships, fender piles are to be installed as protection for the structure piles and the ships themselves.
- Craneway beams are integrated in the reinforced concrete relieving platform as a structural component.
- Vertical loads from crane operations are taken up by additional piles in the craneway axis
- In order to disturb the bending moment progression as little as possible, folds in the relieving platform may only be located over a row of piles (fig. R 157-1).
- In tidal areas, it is practical to plan the lower edge of the relieving platform above MHW to remain independent of the normal tide when producing the relieving platform (figs. R 157-1 and R 157-2).
- Wave action is to be taken into account when designing the formwork for the relieving platform.
- The rows of vertical and battered piles are staggered against each other (fig. R 157-1).
- Horizontal keying is to be arranged as a rule between the blocks of relieving platforms.
- In the case of a long block of a free-standing peer platform, the horizontal longitudinal forces are absorbed by means of pile bents arranged in the middle of the block with the piles having the flattest possible batter.
- In the case of a long and also wide block, the horizontal forces in the transverse direction are absorbed by additional pile bents at both block ends (approx. in the longitudinal axis of the structure), again with the piles having the flattest possible batter.
- As a result of the absorption of the horizontal forces in the above described manner, the stresses in the relieving platform and piles are minimised.
- The relieving platform of a large pier bridge can be concreted on adjustable or portable formwork which is supported chiefly on vertical piles.

Components which deviate from the standard cross section and thus disturb construction progress, such as the nodes of the battered pile bents named above, require additional measures:

As far as possible, the battered piles are driven in from the relieving platform after concreting, through driving recesses left in the superstructure, and the pile bent is connected by means of local reinforced concrete seals in secondary concrete.

The driving recesses can consist of the shrinkage seals named in section 11.5.2, which may have to be widened locally for this purpose.

It must be checked to what extent the stability of the blocks separated by joints is guaranteed, and whether temporary stiffening over these recesses may be necessary.

11.6 Wave Pressure on Pile Structures (R 159)

Dealt with in section 5.10.

11.7 Verification of Overall Stability of Structures on Elevated Pile-founded Structures (R 170)

Dealt with in section 3.4.

11.8 Design and Dimensioning of Pile-founded Structures in Earthquake Areas (R 127)

11.8.1 General

Reference is made to R 124, section 2.13 regarding the general effects of earthquakes on pile-founded structures, tolerable stresses and the required safety factors. However, the risk of resonance phenomena should be checked for high and slender structures.

When designing pile-founded structures in earthquake areas, it must be taken into account that as a result of the seismic effects, the superstructure with its full, live loads and other masses result in additional horizontal mass forces on the structure and its foundation. The cross section must therefore be designed to reach an optimum solution between the advantage of shielding the active earth pressure through the relieving platform and the disadvantage of the added forces to be absorbed from seismic acceleration

11.8.2 Active and Passive Earth Pressure, Water Pressure Difference, Live Loads

The remarks in sections 2.13.3, 2.13.4 and 2.13.5 in R 124 apply accordingly. It must be observed that in the case of an earthquake, the influence of the live load prevailing behind the relieving platform including the additional soil dead load – resulting from the additional horizontal seismic actions – acts at a flatter angle than normal so that the shielding is less effective.

11.8.3 Absorption of the Horizontal Mass Forces of the Superstructure

The horizontal mass forces arising from an earthquake can act in any direction. At right angles to the quay wall, they can generally be absorbed without any difficulties by means of battered piles.

There are possibly problems involved in accommodating pile bents in the longitudinal direction of the structure. If the soil backfill of a front sheet piling reaches directly under the relieving platform, the longitudinal horizontal loads can also be carried off by embedding the piles in the ground by pile bending. However, it must be verified that the occurring displacement is not too great. Limit values of about 3 cm apply here.

A superstructure built over an embankment results in a considerable diminution of the total active earth pressure loads acting on the structure.

In this case, the relieving platform should be designed as light as possible, to achieve a minimum of horizontal mass forces.

11.9 Bracing of the Tops of Steel Pipe Driving Piles (R 192)

When driving steel piping, there is a risk of the tops bulging in the upper section of the pipe, particularly in the case of pipes with relatively small wall thicknesses. This can mean that the pipes cannot be brought down to the planned depth. In order to prevent bulging, the pile head must be braced in such cases. Various methods are available for this; the following have proven effective:

(1) Welding several steel brackets of approx. 0.80 m in length vertically on to the external wall of the pipe (fig. R 192-1). This method is relatively simple and economical, as welding takes place on the outside only.

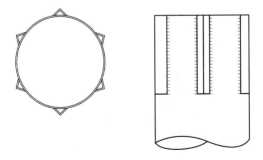

Fig. R 192-1. Bracing with brackets welded on externally

(2) Welding steel plates of approx. 0.80 m in length in the head of the pipe to form a cross (fig. R 192-2). However, this method is more labour-intensive than the method described under (1).

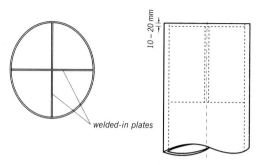

Fig. R 192-2. Bracing with welded-in plates

12 Embankments

12.1 Embankments in Seaports and Inland Harbours with Tide (R 107)

12.1.1 General

In port areas where bulk cargo is handled, at off-shore berths and in the vicinity of harbour entrances and turning basins, the shores can be sloped and permanently stabilised even if the tidal range and other water level fluctuations are quite large, provided that no heavy or long-lasting mud deposits can be expected. However, certain construction fundamentals must be observed if extensive and continuous maintenance work is to be avoided.

In selecting the slope of the embankment, any technical advantages of a flat slope must be weighed against the disadvantages of the increase in the area of the embankment which needs to be stabilised, as well as the inefficient land use of this increased area. In other words, a slope should be chosen which results in a proper balance between construction and maintenance costs on one hand, and the value and possible productive use of the land on the other.

As large seagoing vessels may not, as a rule, proceed under their own power in harbours, it is principally the large tugboats and inland vessels, as well as occasionally the small seagoing vessels and motor coasters, which are capable of eroding the shore by the action of the screw-race, the bow and stern waves, and this damage may extend to about 4 m below the water level (see R 83, section 7.6).

Water movements induced by navigation may be subdivided into screw-race, draw down on both sides of the ship (limited by the bow and the stern waves), return current, follow-up current and secondary wave systems. These components form individual loads on the waterfront and its revetment. The return current acts predominantly along the embankment below the still water table while, for example, the transversal stern wave and the secondary wave system act on and in the vicinity of the still water table.

The bank revetments must be designed to be able to resist the shear forces and seepage forces which result from more frequently occurring loads.

The question as to which loading component is a determining factor for the design always depends on the expected extent of navigation as well as on ship capacity and the cross-section of the shipping channel. Dumped material is frequently used as revetment. Very good, usable formulas have been developed in experiments and basic research in recent years for the design of these and other waterfront protection materials (for example see [100, 100a]).

The groundwater flow which occurs due to passing ships and the tides also results in loads on the embankment. In this case, a distinction must be made between the upward water pressure below the revetment of the bank, which is the stronger, the less permeable the revetment is, and the hydraulic gradient in the filter layer and in the subsoil (fig. R 107-2).

Bank protection with hand-set pitching for example, must be designed in a way that both sliding and lifting of the stones due to upward water pressure is avoided. In particular, unsteady pore water pressures caused by dynamic hydraulic loads (waves, draw down) must be taken into consideration.

Furthermore, any material transport from the subsoil or from the filter layer must be avoided. In order to achieve this, the structure of the filter layers is chosen in such a manner that the seepage force is sufficiently lowered, or a sand-tight revetment with sufficient weight is used. Besides granular filters, geotextile filters are also suitable (see R 189, section 12.4). In both cases, permeability and filter reliability (filter function and separation function) are characteristic design requirements.

Finally, special attention require all transition constructions and all interfaces to the revetment material or to the subsoil, where changes of the thickness of the construction or of the loads must be taken into consideration. In practice, many cases of damage can be attributed to design and/or execution errors in the transition construction.

The filter layer joints to the structures must also be planned and executed with special care.

The effects of chemicals must be considered in some cases when designing a revetment.

Either a permeable or an impermeable revetment must be chosen, depending on the design. Up to now, the former has been preferred in German North Sea ports. The latter, in connection with asphalt grouted rubble, was developed in the Netherlands and is occasionally used.

The type of design may be selected on the basis of anticipated construction costs. If the wave action is stronger than that used as a basis for the design, the impermeable revetment presents the advantage of lower maintenance costs.

12.1.2 Design with Permeable Revetment

Fig. R 107-1 shows a typical construction in Bremen.

The transition from the unrevetted to the revetted section consists of a horizontal berm, 3.00 m wide, covered with rip-rap. Above this berm, the stone revetment is placed on a 1 : 3 slope. A concrete beam 0.50 m by 0.60 m of B 25 lies flush with the port ground surface, forming the upper boundary of the revetment. The lower part of the revetment consists of heavy rip-rap which is dumped in a layer about 0.70 m thick, to an elevation just above MLW.

Extending upward from this point is a cover layer, about 0.50 m thick, consisting of rip-rap which has been carefully placed so as to form an

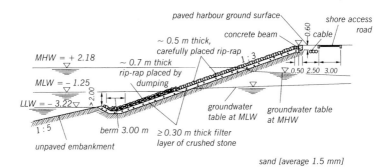

Fig. R 107-1. Standard construction of a harbour embankment in Bremen, permeable revetment, example

interlocking protective rock cover which will not be damaged by wave energy. Stone used in revetment must be solid, hard, dense and resistant to the effects of light, frost and weather.

Under the stone revetment, a continuous filter layer will be installed, which forms the desired rough transition from revetment to sand subsoil and prevents the sand from being eroded. The design of its layer structure and of the grain sizes depends on the subsoil, on the cross-section and on the expected wave load. This must therefore be determined from individually case to case.

Geotextiles are also suitable filter layers (see R 189, section 12.4).

A 3.00 m wide shore access road, constructed for heavy vehicles (fig. R 107-1) is to be laid out 2.50 m inboard of the aforementioned concrete beam, for maintenance work on the stone revetment and as a walkway to the ship berths. Where required, the cables for power supply to the port facilities and the shore beacons, as well as for the telephone lines, etc., may be buried in the strip between the concrete beam and the shore access road.

Fig. R 107-2 shows a typical embankment cross-section in the port of Hamburg. In this solution, an abutment consisting of crushed bricks approx. $3.50 \text{ m}^3/\text{m}$ is placed at the revetment toe on a plastic grid. This is covered by the revetment in two layers for most of the embankment height.

The embankment revetment is taken generally down to only 0.70 m under MLW taking account of prevailing soil conditions. Any scouring below can be repaired by simply adding more crushed bricks. Although ice action can tear pieces of crushed brick away, the costs involved in repairing such damage is comparatively negligible in Hamburg.

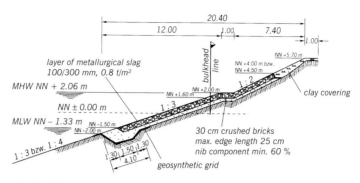

Fig. R 107-2. Construction of a harbour embankment in Hamburg with permeable revetment, example

In order to take account of the idea of a natural design of waterfront embankments with this kind of crushed stone revetments, in Hamburg an embankment has been developed with a vegetation pocket. The normal cross section as per fig. R 107-2 is expanded in the area of NN + 0.40 m by approx. 8.00 to 12.00 m, depending on space available. The resulting pocket is filled with approx. 0.40 to 0.50 m clay; here the substructure of crushed bricks is reinforced to a thickness of 0.50 m. Between clay and crushed bricks, a 0.15 m thick layer of Elbe sand is inserted as aerating zone.

The vegetation pocket is planted depending on the habitat horizons with bulrushes (*Schoeno-Plectus Tabernaemontani*), sedge (*Carex Gracilis*) and reeds (*Phragmitis Australis*). Above the berm at NN +2.0m, this also refers to the normal stone revetment, willow cuttings are inserted in the revetment. Such planting activities should be performed in April/May because of the better growth conditions.

The special design with vegetation pocket, however, is only possible in areas with sufficient space or low swell from ship traffic and wave action. Fig. R 107-2.1 shows a corresponding cross section.

Fig. R 107-3 shows a construction characteristic of the Port of Rotterdam with permeable revetment. Apart from the revetment itself, it is generally identical to the Rotterdam solution with impermeable revetment, so that the remarks made in section 12.1.3 are also applicable here with regard to further details. Recently, concrete columns frequently have been used instead of natural stone.

Fig. R 107-4 shows a new solution for a permeable harbour embankment. Permanent, trouble-free functioning of the filter layer or the filter blanket is the principal prerequisite for durability of all shore protection constructed with permeable revetments.

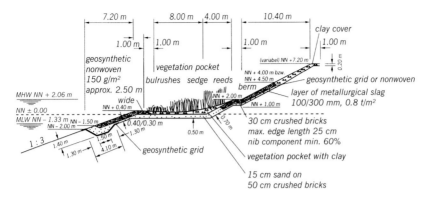

Fig. R 107-2.1. Construction of a harbour embankment with permeable revetment and vegetation pocket in Hamburg, example

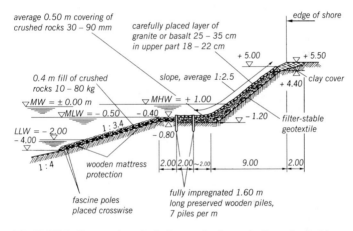

Fig. R 107-3. Construction of a harbour embankment in Rotterdam, with permeable revetment

12.1.3 Design with Impermeable Revetment

12.1.3.1 Fundamentals and Determination of Stability

Above all, the following conditions are applicable for the design:

(1) The slope of the protection zone should be as steep as possible without endangering the stability of the embankment.
(2) As far as possible, machinery should be used for the construction work.

525

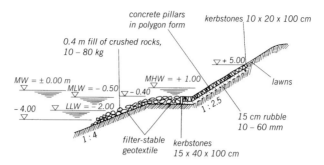

Fig. R 107-4. Construction of a harbour embankment near Rotterdam, with permeable revetment

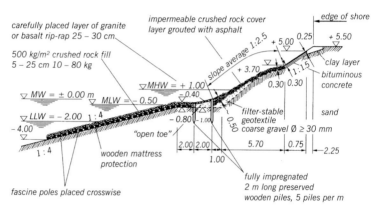

Fig. R 107-5. Construction of a harbour embankment in Rotterdam, with impermeable revetment

Rubble fill grouted with asphalt makes a suitable revetment, together with a bituminous concrete cover if necessary. Fig. R 107-5 shows a typical impermeable revetment as developed and tested in Rotterdam.

With a watertight revetment, in contrast to a fully permeable one, excess water pressure of a certain extent must be recognised and appropriately considered in design and construction. The magnitude of the excess water pressure depends on the magnitude and rapidity of the changes of the harbour water level at the embankment and on the according groundwater level behind the revetment, which is strongly influenced by the permeability of the soil and the filter layers underlying the revetment. This excess pressure reduces the friction between the revetment and the material underlying it.

When the component parallel to the surface of the embankment of the self weight of the revetment exceeds the friction resistance at this surface, additional stresses occur in the revetment, which can lead to deformations (tear and compression) of the revetment. These viscous deformations of the revetment are undesirable, especially since the revetment will creep downward as a result.

As the viscous behaviour of the revetment in itself under prevailing conditions has not been adequately studied up to the present, safety from creeping requires that the frictional resistance may in no case be less than the component of the self weight of the revetment parallel to the surface of the embankment (creep criterion).

As a criterion in unusual load cases, it is required that the component of the self weight of the revetment normal to the embankment must always be larger than the maximum hydraulic uplift, so that the cover layer will never be lifted (lift criterion). These extraordinary load cases however, are of such brief duration and so rare that viscous deformations need not be feared and are therefore not taken into account. The development of the water pressure below the cover layer can be determined in the individual cases with a steady or an assumed approximately steady flow, by the use of flow nets (see also R 113, section 4.7).

In this context, the following should be taken into account:

(1) The course of the outer water levels at mean tide and at storm tide.
(2) Water level fluctuations resulting from undertow, surge and other waves.
(3) The groundwater levels in the adjacent foreshore, as they depend on the outer water levels.
(4) The permeability of soil and fill.
(5) Cross section and composition of the bank protection.

12.1.3.2 Remarks on the Construction

The construction shown in fig. R 107-5 has a so-called "open toe" for reducing the excess water pressure. This "toe" consists of coarse gravel fill ≥ 30 mm $\varnothing$, confined between two rows of tightly spaced wooden piles, which have been fully impregnated with an environment-friendly preservative and are 2.00 m long and approx. 0.2 m thick. Subsequently, the coarse gravel layer in the "open toe" at the lower end of the rubble fill grouted with asphalt is covered with a permeable layer of paving stones 25 to 35 cm of granite or basalt.

Under the coarse gravel layer, which also extends under a substantial part of the watertight cover layer, there is a geotextile filter.

The submerged part of the revetment, which in sand has a slope of 1 : 4, begins with the 2.00 m wide berm adjacent to the "open toe". The embankment is covered with a wooden mattress to a point about 3.50 m below MLW. On top of this, tightly wrapped fascines are placed horizontally, parallel to the shore. The top layer consists of crushed stone to a thickness between 0.30 and 0.50 m. The surcharge of the cover layer shall be about 3 to 5 kN/m^2, depending on the strength of the wave action.

The asphalt grouted crushed stone revetment extends from the land-side row of wooden piles, to about 3.7 m above MLW, and has a mean slope of 1 : 2.5. In individual cases, its thickness must be adapted to the magnitude of the water pressures acting on its bottom. In the normal case, the thickness tapers from 0.5 m at the bottom to 0.3 m at the top.

The stones have an individual weight of 10 to 80 kg.

The upper edge of the revetment, for a vertical height of about 1.3 m, has a slope of 1 : 1.5. The lower portion of this zone is paved with bituminous concrete to a thickness of 0.30 to 0.25 m, while the remainder, up to the elevation of the shore, is in the same slope covered only with clay to a height of 0.50 m. This unpaved strip will facilitate subsequent installation of pipes and other service lines.

The following asphalt mixes by weight have proven efficient in practice:

For poured asphalt:
  graded sand   72 %
  filler   13 %
  asphalt-bitumen 80/100   15 %

For bituminous concrete:
  graded travel 8 to 16 mm   47 %
  graded sand   39.5 %
  filler   7 %
  asphalt-bitumen 80/100   6.5 %

Reliable drainage, which will also prevent washout of the subgrade, is likewise the most important prerequisite for preserving the stability of an embankment with impermeable revetment.

12.1.4 Revetments with Cement Grout

It must be taken into consideration that the revetment is rigid after the grout has hardened. Therefore, certain differences result in design, construction and behaviour compared to the revetments with asphalt grout or of bituminous concrete dealt with in section 12.1.3.

12.2 Embankments Under Quay Wall Superstructures Behind Tight Sheet Piling (R 68)

12.2.1 Loading on the Embankments

Besides the static soil loads, water currents in the longitudinal direction of the quay wall and groundwater flowing transverse to the structure may affect the embankments. The latter is particularly detrimental when the groundwater table in the embankment is higher than the surface water level, so that the seepage flow exits from a slope (R 65, section 4.3). The inclination of the embankment and its protection must therefore be adapted to the position of the relevant water table, the magnitude and frequency of the water level fluctuations, the lateral groundwater flow, the subsoil and the structure as a whole, so that the stability and the erosion-resistance of the embankment are ensured.

12.2.2 Risk of Silting up Behind the Sheet Piling

There is a risk of silting up behind sheet piling in tidal areas if the ingress of outside water is not reliably prevented by appropriate measures. This can cause great additional costs and static consequences. Generally, the ingress of outside water is accepted and the aggradation of silt is prevented by escape openings in the sheet piling just above the toe of the embankment. The spacing and the cross sections of the escape openings must be selected in accordance with the prevailing conditions. The area of hydraulic influence of the openings must be planned with particular care for the inward and outward tidal flow conditions. In some cases, facilities for siphoning off the silt must be provided if not a better solution is selected, according to which the soil extends till below the relieving platform.

12.3 Partially Sloped Bank Construction in Inland Harbours with Extreme Water Level Fluctuations (R 119)

Dealt with in section 6.5.

12.4 Use of Geotextile Filters in Embankment and Bottom Protection (R 189)

12.4.1 General

Geotextiles in the form of woven, non-woven and composite fabrics are used for embankment and bottom protection.

Hitherto, plastics such as polyacryl, polyamide, polyester, polyethylene and polypropylene have proven suitable as rot-resistant materials for geotextile filters. Information on their properties is given in [129].

The advantage of their use lies in the mechanical prefabrication, as a result of which very uniform material properties can be achieved. Geotextiles are also suitable for installation underwater, provided certain installation regulations and product requirements are observed.

Geotextile filters for embankment and bottom protection are subject to dynamic hydraulic loads. The geotextile filters are therefore to be designed with particular care regarding the soil, the cover layer and the loads.

Geotextile filters can certainly meet the filtering task of mineral layers but, because of their low weight, do not have the same effect as mineral filters in respect of stability of a bank protection. In contrast to the mineral grain filter therefore, only the weight of the cover layer and any mineral intermediate layer can be used for the verification of stability.

Sufficient experience has now been gained in the use of geotextiles in hydraulic engineering, and this has led to the formulation of regulations in respect of material requirements, testing of material properties and structural design, and to certain requirements relating to the construction method.

12.4.2 Design Principles

With regard to mechanical and hydraulic filter efficiency, installation stresses such as tensile and penetration forces, and durability to friction stresses with unbonded cover layers, geotextile filters for bank and bottom protection can be designed in accordance with the regulations stated in [128] and [129]. [129] contains design rules for dynamic loads based on experience with static hydraulic loads. [128] contains design rules based on turbulent flow tests ("soil type method"), aimed at dynamic hydraulic loads. Both recommendations are based essentially on German experience. International experience and design principles are to be found in [145], [146] and [147], for example.

When selecting a suitable geotextile filter, on the one hand the mechanical and hydraulic filter efficiency must be guaranteed, on the other hand there must be adequate resistance to the installation loads, particularly for installation under water. For these purposes, relatively thick ($d \geq 4.5$ mm) or heavy ($g \geq 650$ g/m^2) geotextiles have proven efficient; these can be designed according to [128].

Where conditions are more favourable than the boundary conditions stated above, lighter and thinner geotextiles can also be used. These are also designed according to [129].

The following regulations can be used for preliminary design or in simple cases. Specific designs can be based on [128] or [129].

12.4.3 Tensile Strength

In waterways of category IV with placement during vessel traffic, the tensile strength at the failure limit in accordance with DIN 53 857 must be:

$\sigma \geq 1200$ N/10 cm in the longitudinal and transverse directions.

12.4.4 Penetration Resistance

With cover layers of stone fill materials of over 30 kg stone weight, the penetration resistance must be verified [130].

12.4.5 Abrasion Resistance

If abrasive movements of the cover layer stones can occur under wave or flow loads, the abrasion resistance of the geotextile must be verified [130].

12.4.6 Use of Supplementary Layers

In non-cohesive, fine-grained soils, the risk of soil liquefaction and erosion below the geotextile exists under the effect of waves in the water exchange zone and below. This may lead to deformation of the revetment as a result of soil transference. In order to prevent this, the weight of the cover layer must be sufficiently large and the mechanical filter efficiency must be guaranteed [128]. Coarsely grained intermediate layers (cushion layers) between cover layer and geotextile (fig. R 189-1a) contribute to a more even load on the subsoil. Recently, two-ply geotextiles filled with sand (so-called sand-mats) have proven efficient installation aids for installation under water (fig. R 189-1b). An example is described in [169].

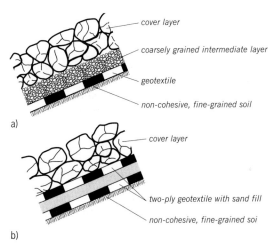

Fig. R 189-1. Alternative design of filter layers

12.4.7 General Design Information

Before installation, delivery in accordance with the contract must in all cases be checked following the appropriate delivery conditions, e.g. listed in [130] and [131]. The geotextiles supplied must be carefully stored and protected from UV radiation, the effects of weather and other damaging influences.

In order to rule out operational defects, correct positioning of the top and bottom side must be ensured when placing multilayer geotextile filters (composite fabrics) with filter layers graded by pores. Additionally, in order to avoid soil displacement, the fabric must not fall in folds. Nailing to the subsoil at the top edge of the embankment is allowed only if the geotextile will not be subject to strain during the further course of construction.

Immediately after placement, particularly with wet placement, geotextiles must be fixed by a cover layer or an intermediate filter layer. Apart from the planned cover layer, the placement of additional material on the filter is not allowed. Geotextiles should not be placed at temperatures below +5°C. Particularly important regarding soil retaining capacity of the geotextile filter is the careful joining of the individual strips, which is possible by sewing or by overlapping. In the case of sewing, the strength of the seam must comply with the required minimum strength of the geotextiles. If placing in the dry with an embankment inclination of 1 : 3 or smaller, the overlaps must be at least 0.50 m or, if placing in the wet and with all steeper embankments, 1.00 m.

In the case of soft subsoil, a check must be made as to whether in certain cases greater overlaps should be used. Construction site seams and overlaps should always run in the downward direction of the embankment. If in exceptional cases this is not possible, the strip which is lower on the bank must be on top of the other strip. If geotextile filters are equipped with a bottom supplementary layer, the supplementary layer must not be present in the overlapping area on the strip which is on top. Transversal shortening of the geotextile strips when placing stone layers must not cause uncovered areas.

When installing geotextile filters underwater, even with operating traffic, the following aspects should also be observed to attain a crease-free geotextile filter placement, a level placement without distortion and a complete cover with sufficient overlapping:

- The construction site must be marked in such a way that it may be passed by all vessels at reduced speed.
- The subgrade must be carefully prepared and must be free of stones.
- The placing equipment must be positioned in such a way that currents and draw down from passing ships cannot impair the placement procedure (possibly on stilts).
- The risk of the geotextile strips floating must be prevented by suitable placement techniques. If necessary, the geotextile must be pressed on the soil during placement. The interval in time and space between placing the geotextile and placing the stone layer must be kept to a minimum.

- Fixing of the geotextile strips on the placing equipment must be released when placing the stone layer. Stone layers are to be placed on embankments with geotextiles from bottom to top.

Underwater placement should be allowed only if the contractor has proved that he can fulfil the requirements, and if this is possible under constant supervision by divers.
Otherwise, reference is made to the provisions for use in accordance with [132].

13 Dolphins

13.1 Design of Resilient Multi-pile and Single-pile Dolphins (R 69)

13.1.1 Design Principles and Methods

The design calculation for resilient dolphins for use as berthing dolphins is such that a permissible maximum impact force, operationally practicable deflection, the required embedment depth and the necessary cross-section dimensions are ascertained with a given subsoil for the required working capacity. For use as mooring dolphins, the applicable hawser pull must also be absorbed. The problem is thus indeterminate, and it is important to solve it in such a way that the best results are obtained as regards both technical, operational and economical aspects.

Resilient dolphins can be calculated by taking account of the dolphin width b at right angles to the direction of force. The following assumptions are recommended for determination of the passive earth pressures:

Densities
The effective densities for both impact and hawser pull load are taken to be the density γ'_k of the particular submerged soil layer.

Angle of wall friction
The angle of wall friction δ_p for the passive earth pressure may be used up to $\delta_p = -2/3\, \varphi'_d$ in calculating all dolphin stresses when utilising plane failure surfaces, if the condition $\Sigma V = 0$ is fulfilled (fig. R 69-1). Otherwise, the angle of wall friction of the passive earth pressure should be reduced. For mooring dolphins, where applicable an unfavourably upward acting vertical component of hawser pull is to be taken into consideration.

The vertical load V acting from top to bottom, taking uplift into consideration, can be taken into account as consisting of the weight of the dolphin, the mass of earth within the perimeter of the dolphin, and also the vertical characteristic skin friction on the side surfaces $a \cdot t$ parallel to the direction of deflection of the dolphin and the vertical component of the equivalent force C according to the calculation of the embedment depth.

Verification of bearing capacity
The bearing capacity is to be verified with the following partial safety factors for predominantly static action:

(1) Berthing dolphin
    impact force P_{impact} $\gamma_{Q\,sup} = 1.0$
    passive earth pressure
        angle of internal friction φ'_k $\gamma_\varphi = 1.1$
        cohesion c'_k resp. $c_{u,k}$ $\gamma_c = 1.3$
        yield point of the steel $f_{y,k}$ $\gamma_M = 1.0$

(2) Mooring dolphin
hawser pull, wind load and current pressure P_{stat} $\gamma_{Q\,sup}$ = 1.0
passive earth pressure
    angle of internal friction φ'_k γ_φ = 1.1
    cohesion c'_k resp. $c_{u,k}$ γ_c = 1.3
    yield point of the steel $f_{y,k}$ γ_M = 1.5

For not predominantly static action, attention must be paid to the decrease in the fatigue strength compared to the static strength.

The statements made in R 20, section 8.2.6.1 (2) apply to the capacity of the parent material and of the circumferential welds to withstand stresses. If the utilisation of the capacity to withstand stresses due to not predominantly static action is higher than laid down in DS 804 for steel grades S 355 J2G3 as per DIN EN 10 025 (formerly St 52-3), then verification is required of the permissible stress amplitudes of the endurance limit in the parent material and at the welds.

The fatigue strength depends greatly on the quality of the steel surface. The fatigue strength can decrease by up to 50 % when subject to corrosion, which must be taken into account particularly for facilities being built in tropical marine conditions.

As the fatigue strength of welded connections is practically independent of the steel grade, the use of heat treated fine-grained structural steels should be avoided if possible in areas where not predominantly static actions occur, if there are welds transverse to the direction of the principle stresses.

13.1.2 Dolphins in Non-cohesive and Cohesive Soils

The passive earth pressure is calculated as spatial passive earth pressure according to DIN V 4085-100, section 7.7. The passive earth pressure loads ΔP_{pi} for the individual layers i result from the mean value of the two passive earth pressure ordinates multiplied by the pressed dolphin partial surface $b \cdot \Delta h$ (fig. R 69-1). In cohesive soil layers, only the shear parameter c_u (with $\varphi_u = 0$, undrained test) must be used because of the "rapid" loading by impact force P.

The equivalent force C_h can be calculated according to fig. R 69-1 while neglecting the active earth pressure influences from the condition $\Sigma H = 0$, as is usual when calculating dolphins, using the equation:

$$C_h = P_{ph} - P_h$$

For the condition $\Sigma V = 0$, it can be assumed to be inclined up to $\delta'_p = +^2/_3\, \varphi'$ to the normal to the dolphin axis.

The additional embedment depth Δt (fig. R 69-1) required to absorb the equivalent force C_h can be calculated with reference to R 56, section 8.2.9 and the symbols used there.

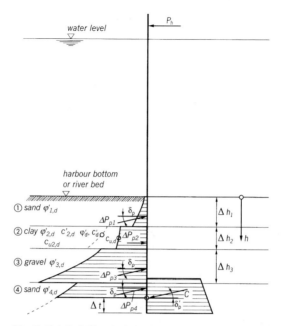

Fig. R 69-1. Dolphin calculation in stratified soil. Spatial passive earth pressure and passive earth pressure ordinates spa e_{ph} according to DIN V 4085-100

For a sloping harbour bottom or river bed, the angle of ground slope (+ or $-\beta$) is to be taken into account in the passive earth pressure coefficient K_p (fig. R 69-2). This should be between $\beta = +\,^1/_3\,\varphi_d$ and $-^2/_3\,\varphi_d$, whereby the angle of wall friction should be $\delta = -^1/_3\,\varphi_d$.

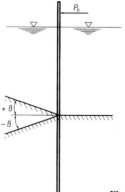

Fig. R 69-2. Dolphin at a sloping harbour bottom or river bed

13.2 Spring Constant for the Calculation and Dimensioning of Heavy Fendering and Heavy Berthing Dolphins (R 111)

13.2.1 General

The spring constant c [kN/m] is defined as the ratio between the impact load P [kN] and the elastic deformation f [m] occurring in its line of action:

$$c = P/f$$

It is of special significance for the calculation and dimensioning of heavy fenderings and elastic berthing dolphins at berths for large ships. It determines the maximum impact force and deflection for the working capacity required for the energy absorption of the approaching ship according to the demands of practice. The maximum allowable impact force for the type of ship concerned may not be exceeded.

Information concerning the allowable impact force which a particular ship can absorb by its hull or inner structural framing should be obtained from the shipping line concerned, or from German Lloyd. In general, concentrated loads should be avoided. Structural elements to distribute the pressure (fender "aprons") are therefore required for heavy impact forces.

Not only the basic static and impact values are to be considered in selecting the spring constant c, but other aspects as well, especially with regard to navigational and structural requirements. Even under clearly defined conditions, only a narrow choice of values exists for the spring constant, but this limited latitude should be exploited in the selection of the correct spring constant. In high grade fender systems and shock absorbers, provisions are made for accidental excessive impact by incorporating failure elements, break-off bolts and the like. By these means, both the ship and the structure are protected against damage. For the allowable effects of actions in the various loading cases, as well as the suitable steel grades and qualities, see R 112, section 13.4.

Since the spring constant corresponds to the stiffness of the fenders or the dolphin, a high spring constant produces hard berthing of a ship, a low spring constant a soft one and thus a less risky one.

13.2.2 Determining Factors for the Selection of the Spring Constant

13.2.2.1 The magnitude of the required energy absorption capacity A [kNm] is determined by the vessel sizes encountered and their berthing speeds (see R 128, section 13.3. in conjunction with R 40, section 5.3).

Therefore, the smallest actual value of the spring constant min c will be determined by the formula

$$\min c = \frac{2A}{\max f^2}$$

insofar as the horizontal deflection f on reaching the yield point is stipulated or limited with max f for nautical, port-operational or structural reasons.

13.2.2.2 A further criterion for the minimum stiffness of a berthing dolphin with or without simultaneous mooring tasks results from the static loadability of the structure by P_{stat} when classified in loading case 2. If a safety factor of 1.5 is chosen because of the inability to make an accurate determination of the maximum acting static load, the spring constant is:

$$\min c = \frac{1{,}5 \cdot P_{stat}}{\max f}$$

13.2.2.3 The upper limiting value of the spring constant c is determined by the maximum allowable impact force P_{impact} between ship's hull and fender or dolphin during the berthing manoeuvre, provided max f has not already been determined for the maximum allowable impact force P_{impact}:

$$\max c = \frac{P_{impact}^2}{2A}$$

Permanent dynamic influences, above all from strong waves, may also be of importance for selecting the spring constant, under certain circumstances.

13.2.3 Special Conditions

The equations in section 13.2.2 fix only the limits within which the spring constant is to be selected. The final determination must take the following aspects into account:

13.2.3.1 Except for special circumstances, such as a desired larger energy absorption capacity at berths for large vessels, or the remarks under section 13.2.3.2, max f in general should not exceed about 1.5 m, because the impact between ship and dolphin during the berthing manoeuvre will otherwise be so gentle that the ship's master will not be able to judge the movement and position of the ship properly in relation to the dolphin.

13.2.3.2 When calculating with the static load P_{stat} of the dolphin, the mutual dependency in the system comprising fender-ship-hawser must be considered. This is especially important at berths which are exposed to high winds and/or long swells. In such cases, just as for berths in the open sea, model tests should always be carried out.

As shown by past experience, the following should be observed:

(1) Stiff hawsers, i.e. short lines or steel hawsers, require stiff fenders.
(2) Soft hawsers, i.e. long lines or manila, nylon, propylene and polyamide lines require soft fenders.

In case (2), smaller stresses always result for both hawsers and dolphins.

13.2.3.3 The maximum allowable impact force P_{impact} between ship and berthing dolphin is determined on the one hand by the type of ship being berthed, and on the other by the design and construction of the dolphin, especially if it is equipped with berthing aprons and the like. Another requirement at berths for large ships is that the berthing pressure between ship and dolphin shall not exceed 200 kN/m² or in some cases even 100 kN/m². A higher berthing pressure can be allowed if it is proven that the hull and the inner structural framing of the berthing ships are able to absorb this. Reference is made to section 13.2.1, third paragraph, in respect of stressing of the hull.

13.2.3.4 If a berth is such that a ship must moor simultaneously at rigid structures and at elastic berthing dolphins, the greatest possible value of the spring constant of the dolphin is to be selected. If the fender structure thereby becomes too stiff for the maximum allowable impact force at the berthing of the ship, a full separation between fender and mooring structure should be made. In any case, thorough investigations are necessary. The same holds good for the degree of softness of exposed berthing dolphins and protective dolphins in front of piers, at jetties, guide walls and lock entrances.

13.2.3.5 If a berth is to be equipped with dolphins differing in their capacity to absorb energy, the dolphins should be designed, if at all feasible, so that the yield point of the material in every dolphin will be reached at the same horizontal deflection. Uniform stressing of all dolphins is ensured for forces acting centrally on the ship, especially by wind and waves. Furthermore, a uniform type of dolphin pile can the be used as a rule for the entire berth.

When differing water levels occur due to tide, wind effects etc., the dolphins should be equipped with a berthing apron, to ensure a fairly constant elevation where the ship's berthing pressure is to be absorbed. Uniform aprons for heavy dolphins of different capacities at a berth should be used only when no appreciable dynamic stresses occur due to wind or swell, which could damage the lighter dolphins.

13.2.3.6 Proceeding from the dimensioning of the heavy dolphin with the energy absorption capacity A_h and the spring constant c_h, the following is then applicable for the lighter one with the energy absorption capacity A_l and the spring constant c_l:

$$c_l = c_h \cdot \frac{A_l}{A_h}$$

If the stiffness of the lighter dolphins determined by this formula is too small, the dolphins should be designed to meet the existing requirements. Valid for heavy dolphins is then:

$$c_h = c_l \cdot \frac{A_h}{A_l}$$

13.2.3.7 Fig. R 111-1 shows the magnitude of the spring constant c, as well as that of the deflection f contingent on energy absorption capacity A and on impact force P_{impact} for berthing dolphins. In the normal case, a spring constant c should be selected that lies between the curves for $c = 500$ and 2000 kN/m, as close as possible to the curve for $c = 1000$ kN/m.

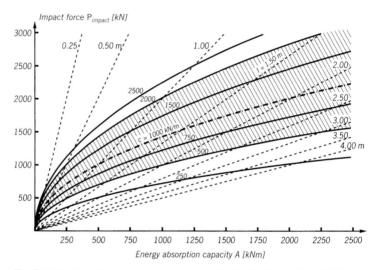

Fig. R 111-1. Magnitude of spring constant c and deflection f at berthing dolphins, contingent on energy absorption capacity A and impact force P_{impact}.

13.3 Impact Forces and Required Energy Absorption Capacity of Fenders and Dolphins in Seaports (R 128)

13.3.1 Determination of Impact Forces

According to R 111, section 13.2.1, the maximum allowable impact force $P_{impact} = c \cdot f$ [kN] is equal to the product of the spring constant and the maximum allowable deflection of the elastic berthing dolphins, fenders, shock absorbers or the like at the point of contact with the ship. The deflection f is generally limited to max. 1.50 m at berths of large ships for nautical reasons (see also R 111, section 13.2.3.1).

13.3.2 Determination of the Required Energy Absorption Capacity

13.3.2.1 General

When berthing, the movement of a ship generally consists of transverse and/or longitudinal motion and rotation around its centre of gravity, so that at first in general, only one dolphin or fender is struck (fig. R 128-1). Governing the approach energy thereby is the impact velocity of the ship v_r at fender, whose magnitude and direction results from the vector addition of the velocity components v and $\omega \cdot r$. At a full friction grip between ship and fender, the impact velocity of the ship, which is then identical with the deformation velocity of the fender, is reduced to $v_r = 0$ in the course of the impact. The centre of gravity of the ship however will generally remain in motion, even if partly in a changed magnitude and rotational direction.

The ship therefore retains a portion of its original energy of motion, even at the time of maximum fender deformation. Under certain conditions, this can lead to the situation that after contact with the first fender, the ship turns to the second one, which can produce a still stronger berthing impact.

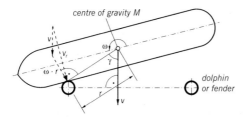

Fig. R 128-1. Depiction of a berthing manoeuvre

13.3.2.2 Numerical Determination of the Required Energy Absorption Capacity [102]

The portion of the energy of motion of the ship (at fully elastic impact or fender) to be stored temporarily by one fender in the course of the berthing impact, or to be fully absorbed (at fully inelastic impact or plastic fender) represents the energy absorption capacity the fender must possess, in order to avoid damage to ship and/or fender. In the general case depicted in fig. R 128-1, this energy absorption capacity comes to:

$$A = \frac{G \cdot C_M \cdot C_S}{2 \cdot (k^2 + r^2)} \cdot [v^2 \cdot (k^2 + r^2 \cdot \cos^2 \gamma) + 2 \cdot v \cdot \omega \cdot r \cdot k^2 \cdot \sin \gamma + \omega^2 \cdot k^2 \cdot r^2]$$

For case $\gamma = 90°$, this assumption is simplified to:

$$A = \frac{1}{2} \cdot G \cdot C_M \cdot C_S \cdot \frac{k^2}{k^2 + r^2} \cdot (v + \omega \cdot r^2)$$

$$= \frac{1}{2} \cdot G \cdot C_M \cdot C_S \cdot \frac{k^2}{k^2 + r^2} \cdot v_r^2.$$

The symbols used in the preceding formulas mean:

A = energy absorption capacity [kNm],
G = water displacement of the berthing ship as per R 39, section 5.1 [t],
k = mass radius of gyration of the ship [m],
It can generally be taken to be $0.25 \cdot l$ for large ships with high block coefficient,
l = length of ship between perpendiculars [m],
r = distance of ship's centre of gravity from point of contact with fender [m],
v = translatory velocity of motion of the ship's centre of gravity at time of first contact with the fender [m/s],
ω = turning speed of ship at time of first contact with the fender [angle in radian measure per sec = 1/s],
γ = angle between the velocity vector v and the distance r [degree],
v_r = resulting impact velocity of the ship on the fender [m/s],
C_M = mass factor according to the following comment [1],
C_S = stiffness factor according to the following comment [1].

The mass factor C_M includes the influence of hydrodynamic pressure, suction and water friction, which the turbulent water (hydrodynamic mass) exerts on the ship as it begins to stop.
According to COSTA [102], the following is of sufficient accuracy:

$$C_M = 1 + 2 \cdot \frac{t}{b}$$

The symbols therein mean:

t = draft of the ship [m]
b = ship's beam [m]

The stiffness factor C_S takes into account a reduction of the impact energy through deformations in ship's hull, depending on the condition of ship and fendering in reciprocal interaction. It may be assumed to be in the magnitude $C_S = 0.9$ to 0.95. The upper limit value applies to soft fenders and smaller ships with stiffer sides, the lower one for hard fenders and larger ships with relatively soft ship's side. Reference is also made to R 111, section 13.2.

13.3.2.3 Pointers

If a ship is towed to a berth with tug help, it may be presumed that it is hardly still running under own power in direction of its longitudinal axis, and that the ship's side lies practically parallel to the line of the fenders during berthing. For the inner fender points of a row of fenders installed at a quay wall, a greater distance occurs inevitably between ship's centre of gravity and contacted fenders. In this case, the velocity vector v can therefore be assumed at right angles to the distance r ($\gamma = 90°$) and the simplified formula for determining A may be used. Reference is made in particular to table R 40-1, section 5.3. In the calculation of the outer fender points of a row of fenders on the contrary, no use may be made of the simplified calculation assumption, because ship's centre of gravity here in direction of the fender line can also approach close to the fender point. Besides, it must be considered in all cases that the ships cannot always be brought to their berths centrally. A distance between the ships' centre of gravity and the centre of the berth of $e = 0.1 \cdot l \leq 15$ m (parallel to fender line) should always be used as a basis for the calculations of berthing dolphins, for example at a tanker berth.

13.3.3 Practical Application

If the theoretically required energy absorption capacity has been determined according to this method, there must be an adjustment between the quantity A, the allowable impact force P_{impact} and the resulting desired smallest spring constant min c. This can be achieved with the data given in R 111, sections 13.2.2.2. and 13.2.2.3 as well as with consideration of practical aspects. R 111, section 13.2.3.3 gives the maximum allowable impact force P_{impact} expressed in berthing force per m² hull plating for larger ships. Special attention is called to the interrelationship of the three factors A, P_{impact} and f as shown in fig. R 111-1.

As has been pointed out in R 111, section 13.2.3.5, second paragraph, the most efficient utilisation of the calculated capacity for energy absorption requires that the ship's berthing force be transmitted to the dolphin or other fendering system at a point whose elevation remains constant. In the case of elastically deflecting dolphins, the centre of gravity of the berthing force can, if necessary, be prevented from drifting downward by fitting the dolphin with a suitable fender apron which provides a fixed point at which the impact force will be transmitted.

13.4 Use of Weldable Fine-grained Structural Steels for Elastic Berthing and Mooring Dolphins in Marine Construction (R 112)

13.4.1 General

If dolphins of high energy absorption capacity are required, it is practical to make them of higher-strength, weldable fine-grained structural steels. Material details are given in R 67, section 8.1.6.

13.4.2 Stresses to be Considered

In the design and calculation of dolphins, a distinction must be made as to whether they are stressed:

(1) predominantly by static loads (vessel impact, line pull) or
(2) by not predominantly static loads.

The following criterion can be used in determining the loads in question:

Dolphins are stressed predominantly static when the proportion of the alternating stresses from waves or swells is low in comparison to the stresses from vessel impact and line pull, and when in examining the occurrence of stress reversal, the following stresses are not exceeded when using the design wave height H_{des} (= the characteristic wave height $H_{1/3}$) over a design period of 25 years:

- 30 % of the respective minimum yield point $f_{y,k}$ of the parent metal, insofar as no butt welds run transverse to the principal direction of stress,
- 100 MN/m^2 for the parent metal, if butt welds run transverse and continuous groove fillet welds run longitudinally to the principal direction of stress,
- 50 MN/m^2 for the parent metal, where groove fillet welds end or fillet welds run transverse to the principal direction of stress.

In areas where heavy swells occur, the loads on the dolphins are not predominantly static, unless they are protected from the effects of swells by other means, such as prestressing against another structure, for example.

In the case of not predominantly static loads, this loading case is generally applicable for dimensioning.

Attention is drawn to the general effect of the stress and stability problems on piles with larger cross-sectional shape and thin walls, and especially in loading cases where stresses up to yield point are allowed. A static check should be made for this when using large-sized pipes (see DIN 18 800-4).

13.4.3 Structural Design

13.4.3.1 The type of action determines the basic requirements for:

(1) choice of steel grade, allowable effects of actions and cross-sectional shape of the individual piles,
(2) manufacturing and material thickness, and
(3) structural design and welding techniques.

13.4.3.2 For practical reasons, the upper part of a dolphin, e.g. the uppermost pipe section, is composed of weldable, fine-grained structural steel of lower strength. This simplifies welded connections of tie beams and other structural parts.

A wall thickness is to be chosen which will permit all necessary welding work to be carried out at the site without any preheating if possible. This is especially important in tidal harbours and where there is heavy wave action.

13.4.3.3 Welds between the individual parts (e.g. pipe sections) shall be made in the shop and as far as possible, in parts of the dolphin with low stresses.

13.4.3.4 In the case of not predominantly static actions, special significance is attached to welded joints transverse to the flexural tensile stress. Therefore, the following is to be observed for circumferential welds and butt joints transverse to the direction of the applied force:

(1) In case of differing wall thicknesses at the welded joint, the transition from the thicker plate to the thinner one is to be milled flat at a taper of 4 : 1, or flatter if possible. At least rough analytical verification is to be provided for large-sized pipes for the force transition.
(2) Surface layers are to be formed without notches. The overheight of the weld shall not exceed that of the material thickness by more than 5 %, if possible.
(3) At piles without access, a full root weld is required. The transition between seam and plate is to be kept flat, without damaging splatter notches. If backing rings are used, they must be of ceramic material.
(4) At piles with access, the joints are to be welded on both sides. Root layers are to be veed out.
(5) At dolphin piles of high-strength, weldable fine-grained structural steels, a filler metal is to be used whose quality corresponds to that of the parent metal, as recommended by the manufacturer.
(6) Welding material and workmanship must be such that the values for heat penetration specified by the manufacturer will be observed.
(7) Directives of the Steel-Iron Material paper 088 [103] are to be followed.

(8) Welding has been completed, limited local stress relieving with appropriate temperature control is possible. Otherwise, section 13.4.3.3 is to be observed.

13.4.3.5 R 99, section 8.1.18 is also valid for all welding work. Circumferential welds and butt welds must be subjected to non-destructive testing and X-rayed if possible.

13.4.3.6 Dolphin piles of fine-grained structural steels generally have extended delivery dates and should therefore be ordered well in advance. It is recommended that a certain number of spare piles be purchased to allow for any possible damage during construction or in service.

14 Experience with Waterfront Structures

14.1 Average Service Life of Waterfront Structures (R 46)

As a result of changes in port operations or harbour traffic, waterfront structures must frequently be reinforced, deepened or replaced, long before they are worn out or obsolete. Therefore their service life is often much shorter than their structural life, so that sometimes only 25 years can be expected for such installations, compared to the usual average service life of waterfront structures of about 50 years.

The best possible estimation of the respective average service life serves well, for example, in preparing benefit/cost investigations, in the design as well as in the internal operations calculation and fiscal assessment. However, the data are not intended as a basis for fixing the valuation in damage cases, since in this case it is not a matter of a statistically calculated value, but rather of the actually applicable value at the occurrence of the damage.

In adjustment to the anticipated average service life, construction types are to be given preference which can later be reinforced with acceptable costs and a minimum of operational disturbances, and in such cases can be conformed to a deepened harbour bottom.

Waterfront structures for which an especially short service life is expected should be constructed in a manner that they may easily be demolished and rebuilt.

The a.m. comments on the medium service life do not apply to damage or failure due to average, and generally do not apply to all types of dolphins.

In the case of average, the remaining service life of the damaged dolphin apart from the structural design depends particularly on the type and the intensity of use, so that general recommendations for determining the remaining service life cannot be given.

14.2 Operational Damage to Steel Sheet Piling (R 155)

14.2.1 Causes of Damage

Besides the predominantly static loads used as a basis for design assumptions, steel sheet pilings in the waterfront area of ports are mostly also subject to strong dynamic stresses deriving from shipping and cargo handling operations. This danger grows with increasing opening width of the sheet piling troughs.

The box-like construction of modern push lighters, motor cargo boats and motor tankers easily leads to damage to the sheet piling as a result of impact on corners and edges of the ship's hull during berthing manoeuvres. Another cause of damage can be the catching of pointed objects, such as ship's anchors in the sheet piling troughs or crane hooks under the capping beams.

Especially extreme impairments of the waterfront structure occur when there is a head-on impact of the vessel in averages, which can take place when the vessel deviates from its course or when the coupling lines of a pushing unit rupture.

When handling aggressive materials such as salts, these may come into contact with the sheet piling and incite corrosion. This also applies to aggressive groundwater.

14.2.2 Extent of Damages

By contact with the ship's hull, bulges with compressive and upset zones can occur. Cracks, ruptures, holes and interlock failure, the latter mainly in Z-shaped sheet piles, can also be found especially in older sheet piling bulkheads, which do not yet possess today's steel grades. Furthermore, capping beam plates can be torn off. Overloading of the waterfront structure or corrosion can lead to torn off anchor connection constructions and subsequent displacement of the sheet piling. Eventually, sheet-piling bulkheads may also tear in the area of the anchor connections by increased anchor forces.

The influence of aggressive materials leads to a reduction of the thickness of steel plates, especially when they have acted in the tidal zone for a long period.

14.2.3 Elimination of Damages

If damage to a sheet piling bulkhead has been proved, its extent should at first be ascertained accurately to verify if individual structural members or the stability of the entire waterfront structure have become endangered.

The damage can be repaired by welding on plates and angles or by placing steel plate elements, which bridge a single sheet piling trough or more if necessary, provided that the sheet piling steel is weldable. If a sheet piling section has been weakened in the trough area, a solution may be installing reinforcing sections, which do not protrude beyond the alignment and connect them to the existing sheet piling. In case of major damage, it may prove necessary to extract individual sheet piles or entire sections of the wall, and to replace them with new material. Where the damage is minor, it often suffices to cut out sheet piles and to increase the height of the piles in question with new material. Special care must be taken if an anchored sheet pile is concerned and the anchor must be loosened. It must then be checked whether auxiliary anchoring must be installed. In case of very heavy damage, especially if the sheet piling has deviated from its alignment to a great extent, it may be necessary to secure a section of the waterfront by driving a new sheet piling bulkhead in front of the old one. In any case, it must be determined if the sheet piling damage has resulted in the outflow of soil and whether hollow areas could have developed behind the sheeting.

14.2.4 Preventative Measures for Avoiding or Reducing Damage

When dimensioning a sheet piling structure, it must be investigated to what extent additions must be made beyond the statical/structural requirements, under consideration of port operations. These may be, for instance, that the next stronger pile section is used for either all sheet piles or only for the water-sided ones. Reinforcing plates can be welded onto the sheet pile backs. A suitable example is the armoured sheet piling which is dealt with in R 176, section 8.4.16.

Early recognition of irregularities in the sheet piling structure can prevent or reduce major damage. It is therefore recommended to carry out control measurements and underwater investigations by divers, possibly with underwater television cameras (R 193, section 15.1).

14.3 Steel Sheet Piling Waterfront Structures Under Fire Loads (R 181)

14.3.1 General

Fire loads can be initiated by burning substances floating on the water, e.g. in the case of burning products escaping as a result of an accident, or by a fire on land. A burning ship lying up at the waterfront can also be the cause of temperature loads on the waterfront structure.

The increase in temperature can be derived from the unit temperature curve in accordance with DIN 4102, which applies to an enclosed burning area. Accordingly, the maximum temperature of approx. 1100 °C is reached after 180 minutes. However, in the case of waterfront structures, much more favourable local circumstances are to be found. This is a fire in the open air, in which the heat generated can escape without restriction. In the case of a fire on the water surface, the heat is additionally absorbed by the water.

In the case of fires in the open air, temperatures of 800 °C are not exceeded in the flame zone [119]. It can generally be assumed that the reduced maximum value of 800 °C will definitely not be reached in waterfront structures located in the open air, provided the fire is extinguished within three hours.

Therefore, fire fighting measures should aim at reducing the fire temperature as quickly as possible and preventing contact with further combustible substances. In addition, the steel sheet piling should be relieved, if possible, e.g. by removing live loads and stored items.

Otherwise, fire protection measures must comply with the regulations under public law. The fire protection requirements should be determined and stipulated on the basis of local circumstances. For this reason, reference is made in particular to the "Directives for Requirements relating to Facilities for Handling Hazardous Fluids in the Area of Waterways" [120].

14.3.2 Effects of Fire Loads on a Waterfront Structure of Steel Sheet Piling

If the steel is subject to loads by high temperatures, its mechanical properties will change. For example, the yield strength will be reduced and elongation at fracture will be increased, if the temperatures in the steel exceed 100 °C. In this case the critical steel temperature (crit T) is of particular importance. That is the temperature at which the yield strength of the steel reduces to the steel stress existing in the sheet piling. In sheet piling steels in accordance with R 67, section 8.1.6, crit T is 500 °C.

If in the stressed area, the permissible stress is not exploited, crit T will increase up to 650 °C if the existing stress is only $1/3$ of the permissible stress.

If the temperatures remain below 500 °C, there is no hazard to the stability of the structure in any circumstances with steel sheet piling. After cooling, the initial values assumed in the static calculation can again be used as the basis for the mechanical properties.

In fire loads, only the air-side surface of the sheet piling is affected, provided it is above the water level. All remaining surfaces are not subject to loads.

If it is surrounded by air and/or water, the sheet piling surface opposite to the fire-affected side contributes to cooling the surface subject to load. However, this also applies to soil material, particularly with groundwater in permeable soil.

Since the zone with the greatest heat increase generally does not lie in the area in which the steel sheet piling is subject to full load, the critical temperature increases. With singly-anchored sheet piling with a small over-anchor section, the area of greatest stress is generally located below the water level and is thus not subject to the fire load.

The circumstances are similarly favourable in the case of an unanchored sheet piling fixed in the soil and standing in water. Conversely, it should be noted in the case of sheet piling fixed at the top that the restraining moment may lie in the area of the fire loading above water level.

With anchored sheet piling restrained in the soil or at the top, flow articulations may form in the area subject to fire loads. On the one hand, these may increase deflection, but on the other hand they may increase the safety against the failure state. Sheet piling of this type with statically undetermined bearing, should therefore be assessed more favourably than statically determined sheet piling with free earth support and single anchorage.

The greater the wall thickness for the same surface, the more slowly the component will heat up and the greater the duration of fire resistance will be, though this has little effect on the relatively thin sheet piles. Owing to the plane and therefore smaller surface with an increased steel mass and owing to the insulating effect of the gap between it and the load-bearing sheet piling, armoured sheet piling (R 176, section 8.4.16) is more resistant to the effects of heat than non-armoured sheet piling, provided the armouring is not used to contribute to load bearing. The connection construction of anchored sheet piling is distinguished by the fact that it has a relatively small surface with a great piling wall thickness, so that the construction heats up more slowly. The anchoring element located in the ground behind the sheet piling is not subject to fire load and can divert a limited proportion of the heat. However, in particularly endangered areas of waterfront structures, it is recommended that a special protection for the anchor connections should always be provided.

14.3.3 Investigations of Fire Loads on an Existing Sheet Piling Structure
A burning oil slick floating on the surface of the water set fire to a 2.70 m high free sheet piling surface equipped with a reinforced concrete capping beam with nosing and subsequently with a paved embankment in accordance with fig. R 129-1, section 8.4.5, for up to an hour. After the fire, the following was discovered:
In the area of the fire, the steel sheet piling exhibited pronounced colour changes. However, the mechanical and technological properties of investigated steel samples from the section of the sheet piling subject to the heaviest loading still complied with the Technical Conditions of Supply. Accordingly, the steel has suffered no loss of quality as a result of the fire.
Measurements up to 8 m below the upper edge of the sheet piling revealed that no unusual deformation had occurred under the fire load. According to the static calculation, the sheet piling was stressed at the elevation of the water level with $^1/_3$ yield limit.
There was visible damage on the surface of the reinforced concrete capping beam in the form of local 1 to 2 cm thick chips. The steel nosing had come apart from the concrete capping beam almost all the way round and was cracked at the welded butt points. In the area of the ladder niche, the reinforced concrete capping beam was subject to such intense loading that the reinforcement was exposed in places and cracks could be observed. Some of the access ladders were distorted, apparently as a result of scarfing of the entire surface.
Damage had also occurred in the paved embankment and in the revetments. The sandstone paving with mortared joints had bulged in places, so that the mortared joints were cracked. Repair work needed to be performed.
On a dolphin standing detached in the water and designed as an enclosed steel pile, the all-round fire loading had led to a bulge approx. 4 m above

the water level or 6 m below the upper edge of the dolphin. After the fire, the mechanical and technological properties of the steel grade corresponded to sheet piling special steel StSpS (new designation S 355 GP). The individual samples taken showed the same results, although they were taken from different elevations which had also been subject to different fire loads.

The damage described leads to the conclusion that the duration and intensity of the fire loads had not been minor. Nonetheless, the steel sheet piling exhibited high fire resistance, apparently since the boundary conditions had greatly diminished the effects of the fire load.

14.3.4 Conclusions

Classification of steel sheet piling structures backfilled with soil in accordance with DIN 4102, part 2, is not necessary [126].

15 Supervision of Structures

15.1 Observation and Inspection of Waterfront Structures in Seaports (R 193)

15.1.1 General

Periodic observation and inspections of waterfront structures are necessarily taking account of individual cases, in order to check the stability, ability of operation and structural condition so that any damage which could impair the safety of the structure for normal traffic may be detected in good time. Careful and regular inspection of the structures facilitates timely maintenance work so that major repair costs or premature renewal of the structure can be avoided.

The observation of structures consists in monitoring the structure without large-scale aids. The inspection of the structure is carried out by an expert engineer using adequate equipment; he must also be able to judge the static and the constructional and hydro-mechanical conditions of the structure.

Pertinent regulations given to certain administrations are to be taken into account here, such as:

– DIN 1076
– DIN 19702
– VV-WSV 2101 [139]
– VV-WSV 2301 [163]
– ETAB [45]

The faults detected during inspection of the structure are to be assessed regarding presumed or determined causes. The further procedure is to be stated in the records.

15.1.2 Records and Reports

The results of observation and inspection together with implemented measures and their costs are to be recorded in a suitable form as an important aid for evaluating necessary maintenance measures.

The basis for observation and inspection are the construction plans "as built" and, where applicable, the structure's log book, which contains the major data such as cross sections, presumed loads, water levels and soil characteristic values, static calculations, completed modifications and maintenance work, together with the data and results of observations and inspections.

On completion of the structure, it should be surveyed. This should be repeated after about a year to ascertain any initial shifts and settlements. Further subsequent surveys are to be carried out as part of inspection of the structure.

15.1.3 Performing Observations and Inspections

15.1.3.1 Observations

The observation of the structure includes regular monitoring and checking. While monitoring the following items are to be noted:

– surface damage or changes,
– settlement, sagging, shifts,
– changes in joints and structural connections,
– absence of or damage to equipment components,
– improper use.

A check should also include actual visual inspection of normally inaccessible members of the structures such as gangways, paths and shafts, with a detailed tour round the structure by boat. Attention is to be paid to leaks, proper functioning of any drainage systems and signs of scouring or deposits in front of the wall. Reference is made to recommendations R 80, section 7.1 and R 83, section 7.6.

15.1.3.2 Inspection of the Structure

The first inspection of the structure should take place during acceptance of the structure, and a second before the end of the period(s) of guarantee.

Inspection of the structure should in any case be carried out by an engineer who is well acquainted with the structure, who can assess the static, constructional and hydro-mechanical conditions of the structure and also provide guidance for any necessary diving activities.

A visual inspection should provide an initial overview of the state of the structure. Joint widths, relative movements, deformation's and settlement's movements are to be measured and recorded together with any detected damage, such as cracks, concrete chips with exposed reinforcement.

In addition, the steel sheet piling bulkheads and piles are to be checked for corrosion and concrete cover in the case of reinforced concrete components.

The following measurements are required to check possible large-scale movements of the structure:

– alignments,
– levelings (settlement, uplift),
– measurements of deflection and toe displacement of the main wall, for example using inclinometer tubes.

The required depth in front of waterfront structures should be checked by depth soundings, if necessary also by harbour bottom surveying. In addition, when irregularities have been found, the sheet piling bulkhead must, where necessary, be examined by divers for possible leaks.

15.1.4 **Intervals of Inspection**

The frequency of inspections of waterfront structures depends on the age of the structure, its general condition, the construction materials used, the subsoil conditions, environmental influences and operational requirements and loads. As guidance to safeguard proper observation and inspection based on the listed features, it is recommended to lay down the intervals of inspections in the structure documentation. The inspection periods stipulated in DIN 1076 can be used for general guidance and should be adjusted to the individual needs in each particular case.

In the case of extraordinary loads which may have caused damage to the structure, for example ship damage, extreme loads, for example by high water or fire, the possible effects on the bearing capacity of the structure have to be approved by special investigations.

Annex I Bibliography

I.1 Annual Technical Reports

The basis of the collected publication is the Annual Technical Report of the Committee for Waterfront Structures, published in the journals "DIE BAUTECHNIK" (since 1984 "BAUTECHNIK") and "HANSA", in issues:

HANSA	87 (1950) No. 46/47, p. 1524
DIE BAUTECHNIK	28 (1951), No. 11, p. 279 – 29 (1952), No. 12, p. 345
	30 (1953), No. 12, p. 369 – 31 (1954), No. 12, p. 406
	32 (1955), No. 12, p. 416 – 33 (1956), No. 12, p. 429
	34 (1957), No. 12, p. 471 – 35 (1958), No. 12, p. 482
	36 (1959), No. 12, p. 468 – 37 (1960), No. 12, p. 472
	38 (1961), No. 12, p. 416 – 39 (1962), No. 12, p. 426
	40 (1963), No. 12, p. 431 – 41 (1964), No. 12, p. 426
	42 (1965), No. 12, p. 431 – 43 (1966), No. 12, p. 425
	44 (1967), No. 12, p. 429 – 45 (1968), No. 12, p. 416
	46 (1969), No. 12, p. 418 – 47 (1970), No. 12, p. 403
	48 (1971), No. 12, p. 409 – 49 (1972), No. 12, p. 405
	50 (1973), No. 12, p. 397 – 51 (1974), No. 12, p. 420
	52 (1975), No. 12, p. 410 – 53 (1976), No. 12, p. 397
	54 (1977), No. 12, p. 397 – 55 (1978), No. 12, p. 406
	56 (1979), No. 12, p. 397 – 57 (1980), No. 12, p. 397
	58 (1981), No. 12, p. 397 – 59 (1982), No. 12, p. 397
	60 (1983), No. 12, p. 405 – 61 (1984), No. 12, p. 402
	62 (1985), No. 12, p. 397 – 63 (1986), No. 12, p. 397
	64 (1987), No. 12, p. 397 – 65 (1988), No. 12, p. 397
	66 (1989), No. 12, p. 401 – 67 (1990), No. 12, p. 397
	68 (1991), No. 12, p. 398 – 69 (1992), No. 12, p. 710
	70 (1993), No. 12, p. 755 – 71 (1994), No. 12, p. 763
	72 (1995), No. 12, p. 817 – 73 (1996), No. 12, p. 844

I.2 Books and Papers

[1] Report of the Sub-Committee on the Penetration Test for Use in Europe, 1977. (Copies of this report are available from: The Secretary General, ISSMFE, Department of Civil Engineering, King's College London Strand, WC 2 R2LS, UK).

[2] SANGLERAT: The penetrometer and soil exploration. Amsterdam/London/New York: Elsevier Publishing Company, 1972.

[3] LANGEJAN, A.: Some aspects of the safety factor in soil mechanics, considered as a problem of probability. Proc. 6. Int. Conf. Soil Mech. Found. Eng. Montreal 1965, Vol. 2, p. 500.

[4] ZLATAREW, K.: Determination of the necessary minimum number of soil samples. Proc. 6. Int. Conf. Soil Mech. Found. Eng. Montreal 1965, Vol. 2, p. 130.

[5] ROLLBERG, D.: Bestimmung des Verhaltens von Pfählen aus Sondier- und Rammergebnissen, Forschungsberichte aus Bodenmechanik und Grundbau FBG 4, Technische Hochschule Aachen, 1976.

[6] ROLLBERG, D.: Bestimmung der Tragfähigkeit und des Rammwiderstands von Pfählen und Sondierungen, Veröffentlichungen des Instituts für Grundbau, Bodenmechanik, Felsmechanik und Verkehrswasserbau der Techn. Hochschule Aachen, 1977, No. 3, pp. 43–224.

[7] GRUNDBAU-TASCHENBUCH, 4th edition, Part 1-1990, Part 2-1991, Part 3-1992 and 5th edition, part 1-1996, Ernst & Sohn Verlag für Architektur und technische Wissenschaften, Berlin.

[8] KAST, K.: Ermittlung von Erddrucklasten geschichteter Böden mit ebener Gleitfläche nach CULMANN, Bautechnik 62 (1985), No. 9, p. 282.

[9] MINNICH, H. and STÖHR, G.: Erddruck auf eine Stützwand mit Böschung und unterschiedlichen Bodenschichten. Die Bautechnik 60 (1983), No. 9, p. 314.

[10] KREY, H.: Erddruck, Erdwiderstand und Tragfähigkeit des Baugrundes, 5th edition, Berlin: Ernst & Sohn 1936 (out-of-print); see in [7].

JUMIKIS: Active and passive earth pressure coefficient tables. Rutgers, The State University, New Brunswick/New Jersey: Engineering Research Publication (1962) No. 43.

CAQUOT, A., KÉRISEL, J. and ABSI, E.: Tables de butée et de poussée. Paris: Gauthier-Villars, 1973.

[11] BRINCH HANSEN, J. and LUNDGREN, H.: Hauptprobleme der Bodenmechanik. Berlin: Springer, 1960.

[12] BRINCH HANSEN, J. and HESSNER, J.: Geotekniske Beregninger, Copenhagen, Teknisk Forlag, 1959, p. 56.

[13] HORN, A.: Sohlreibung und räumlicher Erdwiderstand bei massiven Gründungen in nichtbindigen Böden. Straßenbau u. Straßenverkehrstechnik 1970, No. 110, Bundesminister für Verkehr, Bonn.

[14] HORN, A.: Resistance and movement of laterally loaded abutments. Proc. 5. Europ. Conf. Soil Mech. Found. Eng. Madrid, Vol. 1 (1972), p. 143.

[15] WEISSENBACH, A.: Der Erdwiderstand vor schmalen Druckflächen. Mitt. Franzius-Institut TH Hannover 1961, No. 19, p. 220.

[16] VORLÄUFIGE RICHTLINIEN für das Bauen in Erdbebengebieten des Landes Baden-Württemberg (Nov. 1972), Bekanntmachung des Innenministeriums No. V 7115/107 dated 30.11.1972.

[17] TERZAGHI, K. VON and PECK, R. B.: Die Bodenmechanik in der Baupraxis. Berlin/Göttingen/Heidelberg: Springer, 1961.

[18] DAVIDENHOFF, R.: Zur Berechnung des hydraulischen Grundbruches. Die Wasserwirtschaft 46 (1956), No. 9, p. 230.

[19] KASTNER, H.: Über die Standsicherheit von Spundwähnden in strömenden Grundwasser. Die Bautechnik 21 (1943), No. 8 and 9, p. 66.

[20] SAINFLOU, M.: Essai sur les digues maritimes verticales. Annales des Ponts et Chaussées, tome 98 II (1928), translated: Treatise on vertical breakwaters, US Corps of Engineers (1928).

[21] CERC (Us Army Coastal Engineering Research Centre). Shore Protection Manual, Washington 1984.

[22] MINIKIN, R.: Wind, Waves and Maritime Structures. London: Charles Griffin & Co. Ltd., 1963.

[23] WALDEN, H. and SCHÄFER, p. J.: Die windverzeugten Meereswellen, Part II, Flachwasserwellen. No. 1 and 2. Einzelveröffentlichungen des Deutschen Wetterdienstes, Seewetteramt Hamburg, 1969.

[24] Schüttrumpf, R.: Über die Bestimmung von Bemessungswellen für den Seebau am Beispiel der südlichen Nordsee. Mitteilungen des Franzius-Instituts für Wasserbau und Küsteningenieurwesen der Technischen Universität Hannover, 1973, No. 39.
[25] Partenscky, H.-W.: Auswirkungen der Naturvorgänge im Meer auf die Küsten – Seebauprobleme und Seebautechniken. Interocean 1970, Vol. I.
[26] Longuet-Higgins, M. S.: On the Statistical Distribution of the Heights of Sea Waves. Journal of Marine Research, Vol. XI, No. 3 (1952).
[27] Kohlhase, S.: Ozeanographisch-seebauliche Grundlagen der Hafenplanung, Mitteilungen des Franzius-Instituts für Wasserbau und Küsteningenieurwesen der Universität Hannover (1983), No. 57.
[28] Wiegel, R. L.: Oceanographical Engineering. Prentice Hall Series in Fluid Mechanics, 1964.
[29] Silvester, R.: Coastal Engineering. Amsterdam/London/New York: Elsevier Scientific Publishing Company, 1974.
[30] Hager, M.: Untersuchungen über Mach-Reflexion an senkrechter Wand. Mitteilungen des Franzius-Instituts für Wasserbau und Küsteningenieurwesen der Technischen Universität Hannover, (1975), No. 42.
[31] Berger, U.: Mach-Reflexion als Diffraktionsproblem. Mitteilungen des Franzius-Instituts für Wasserbau und Küsteningenieurwesen der Technischen Universität Hannover, (1976), No. 44.
[32] Büsching, F.: Über Orbitalgeschwindigkeiten irregulärer Brandungswellen. Mitteilungen des Leichtweiß-Instituts für Wasserbau der Technischen Universität Braunschweig, (1974), No. 41.
[33] Siefert, W.: Über den Seegang in Flachwassergebieten. Mitteilungen des Leichtweiß-Instituts für Wasserbau der Technischen Universität Braunschweig, (1974), No. 40.
[34] Battjes, J. A.: Surf Similarity. Proc. of the 14th International Conference on Coastal Engineering. Copenhagen 1974, Vol. I, 1975.
[35] Galvin, C. H. Jr.: Wave Breaking in Shallow Water, in Waves on Beaches. New York Ed. R. E. Meyer, Academic Press. 1972.
[36] Führböter, A.: Einige Ergebnisse aus Naturuntersuchungen in Brandungszonen. Mitteilungen des Leichtweiß-Instituts für Wasserbau der Technischen Universität Braunschweig, (1974), No. 40.
[37] Führböter, A.: Äußere Belastungen von Seedeichen und Deckwerken. Hamburg, Vereinigung der Naßbaggerunternehmungen e. V., 1976.
[38] Morison, J. R., O'Brien, M. P., Johnson, J. W. and Schaaf, S. A.: The Force Exerted by Surface Waves on Piles. Petroleum Transaction, Amer. Inst. Mining Eng. 189, (1950).
[39] MacCamy, R. C. and Fuchs, R. A.: Wave Forces on Piles: A Diffraction Theory. Techn. Memorandum 69, U. S. Army, Corps of Engineers, Beach Erosion Board, Washington D. C., Dec. 1954.
[40] Reports of the International Waves Commission, PIANC-Bulletin No. 15 (1973) and No. 25 (1976), Brussels.
[41] Hafner, E.: Bemessungsdiagramme zur Bestimmung von Wellenkräften auf vertikale Kreiszylinder. Wasserwirtschaft 68 (11978), No. 7/8, p. 227.
[42] Hafner, E.: Kraftwirkung der Wellen auf Pfähle. Wasserwirtschaft 67 (1977), No. 12, p. 385.
[43] Streeter, H. L.: Handbook of Fluid Dynamics. New York, 1961.

[44] KOKKINOWRACHOS, K., in: "Handbuch der Werften", Vol. 15, Hamburg, 1980.
[45] Bundesverband öffentlicher Binnenhäfen, Empfehlungen des Technischen Ausschusses Binnenhäfen, Neuss.
[46] EAK 1993 Empfehlungen für Küstenschutzwerke: Ausschuß für Küstenschutzwerke der DGEG und der HTG., "Die Küste" No. 55-1993 with correction page in "Die Küste" No. 57-1994, Westholsteinische Verlagsanstalt Boyens & Co, Heide i. Holst.
[47] BURKHARDT, O.: Über den Wellendruck auf senkrechte Kreiszylinder. Mitt. Franzius Institut Hannover, No. 29, 1967.
[48] DET NORSKE VERITAS: Rules for Design, Construction and Inspection of Fixed Offshore Structures, 1977.
[49] DIETZE, W.: Seegangskräfte nichtbrechender Wellen auf senkrechte Pfähle. Bauingenieur 39 (1964), No. 9, p. 354.
[50] DANTZIG, D. VON: Economic Decision Problems for Flood Prevention. "Econometrica" Vol. 24, No. 3, p. 276, New Haven 1956.
[51] REPORT of the DELTA COMMITTEE. Vol. 3 Contribution II. 2, p. 57. The Economic Decision Problems Concerning the Security of the Netherlands against Storm Surges (Dutch Language, Summary in English). Den Haag 1960, Staatsdrukkerij en uitgeversbedrijf.
[52] Richtlinien für Regelquerschnitte von Schiffahrtskanälen, Bundesverkehrsministerium, Abt. Binnenschiffahrt und Wasserstraßen, 1994.
[53] Beziehung zwischen Kranbahn und Kransystem, Ausschuß für Hafenumschlagtechnik der Hafenbautechnischen Gesellschaft e. V. Hansa 122 (1985) No. 21, p. 2215, and 22, p. 2319.
[54] KRANZ, E.: Die Verwendung von Kunststoffmörtel bei der Lagerung von Kranschienen auf Beton. Der Bauingenieur 46 (1971), No. 7, p. 251.
[55] DE KONING, J.: Boundary Conditions for the Use of Dredging Equipment. Paper of the Course: Dredging Operation in Coastal Waters and Estuaries, Delft/the Hague, 1968, May.
[56] KOPPEJAN, A. W.: A Formula combining the TERZAGHI load-compression relationship and the BUISMAN secular time effect. Proceedings 2nd Int. Conf. on Soil Mech. and Found. Eng. 1948.
[57] HELLWEG, V.: Ein Vorschlag zur Abschätzung des Setzungs und Sackungsverhaltens nichtbindiger Böden bei Durchnässung. Mitt. Institut für Grundbau und Bodenmechanik, Universität Hannover, 1981, No. 17.
[58] KWALITEITSEISEN vor Hout (K. V. K. 1980).
[59] WIRSBITZKI, B.: Kathodischer Korrosionsschutz im Wasserbau. Hafenbautechnische Gesellschaft, e. V., Hamburg 1981.
[60] WOLLIN, G.: Korrosion im Grund- und Wasserbau. Die Bautechnik 40 (1963), No. 2, p. 37.
[61] BLUM, H.: Einspannungsverhältnisse bei Bohlwerken. Berlin: Ernst & Sohn, 1931.
[62] ROWE, O. W.: Anchored Sheet-Pile Walls. Proc. Inst. Civ. Eng. London 1952, Paper 5788.
[63] ROWE, p. W.: Sheet-Pile Walls at Failure. Proc. Inst. Civ. Eng. London 1956, Paper 6107 and relevant discussion 1957.
[64] ZWECK, H. and DIETRICH, Th.: Die Berechnung verankerter Spundwände in nichtbindigen Böden nach ROWE [62], Mitteilungsblatt der Bundesanstalt für Wasserbau, Karlsruhe 1959, No. 13.

[65] BRISKE, R.: Anwendung von Erddruckumlagerungen bei Spundwandbauwerken. Die Bautechnik 34 (1957), No. 7, p. 264 and No. 10, p. 376.
[66] BRINCH HANSEN, J.: Spundwandberechnungen nach dem Traglastverfahren. Internationaler Baugrundkursus 1961. Mitteilungen aus dem Institut für Verkehrswasserbau, Grundbau und Bodenmechanik der Technischen Hochschule Aachen, Aachen (1962), No. 25, p. 171.
[67] LAUMANS, W.: Verhalten einer ebenen, in Sand eingespannten Wand bei nichtlinearen Stoffeigenschaften des Bodens. Baugrundinstitut Stuttgart, Mitteilung 7 (1977).
[68] OS, P. J. VAN: Damwandberekening: Computer model of Blum. Polytechnisch Tijdschrift, Editie B, 31 (1976), No. 6, pp. 367–378.
[69] FAGES, R. and BOUYAT, C.: Calcul de rideaux de parois moulées et de palplanches (Modèle mathématique intégrant le comportement irréversible du sol en état élastoplastique. Exemple d'application, Étude de l'influence des paramètres). Traveaux (1971), No. 439, pp. 49–51 and (1971), No. 441, pp. 38–46.
[70] FAGES, R,. and GALLET, M.: Calculations for Sheet Piled or Cast in Situ Diaphragm Walls (Determination of Equilibrium Assuming the Ground to be in an Irreversible Elasto-Plastic State). Civil Engineering and Public Works Review (1973), Dec.
[71] SHERIF, G.: Elastisch eingespannte Bauwerke, Tafeln zur Berechnung nach dem Bettungsmodulverfahren mit variablen Bettungsmoduli. Berlin/München/Düsseldorf: Ernst & Sohn, 1974.
[72] RANKE, A. and OSTERMAYER, H.: Beitrag zur Stabilitätsuntersuchung mehrfach verankerter Baugrubenumschließungen. Die Bautechnik 45 (1968), No. 10, pp. 341–350.
[73] LACKNER, E.: Berechnung mehrfach gestützter Spundwähnde, 3rd edition, Berlin: Ernst & Sohn, 1950. See also in [7].
[74] KRANZ, E.: Über die Verankerung von Spundwänden. Berlin: Ernst & Sohn, 1953, 2nd edition.
[75] WIEGMANN, D.: Messungen an fertigen Spundandbauwerken. Vortr. Baugrundtag. Dt. Ges. für Erd- und Grundbau, May 1953, Hamburg 1953, pp. 39–52.
[76] BRISKE, R.: Erddruckverlagerung bei Spundwandbauwerken. 2nd edition, Berlin, Ernst & Sohn, 1957.
[77] BEGEMANN, H. K. S. PH.: The Dutch Static Penetration Test with the Adhesion Jacket Cone (Tension Piles, Positive and Negative Friction, the Electrical Adhesion Jacket Cone), LGM-Mededelingen (1969), Vol. 13, No. 1, 4 and 13.
[78] SCHENK, W.: Verfahren beim Rammen besonders langer, flachgeneigter Schrägpfähle. Der Bauingenieur 43 (1968), No. 5.
[79] LEONHARDT, F.: Vorlesungen über Massivbau, Part 4, 2nd edition, Berlin/Heidelberg/New York: Springer, 1978.
[80] ZTV-K 96 Zusätzliche Technische Vertragsbedingungen für Kunstbauten. Der Bundesminister für Verkehr (Abt. Straßenbau, Abt. Binnenschiffahrt und Wasserstraßen). Verkehrsblatt Verlag Dortmund.
[81] PRIEBE, H.: Bemessungstafeln für Großbohrpfähle. Die Bautechnik, 59 (1982), No. 8, p. 276.
[82] WEISS, F.: Die Standfestigkeit flüssigkeitsgestützter Erdwände. Bauingenieur-Praxis. Berlin/München/Düsseldorf: Ernst & Sohn 1967, No. 70.
[83] MÜLLER-KIRCHENBAUER, H., WALZ, B. and KILCHERT, M.: Vergleichende Untersuchung der Berechnungsverfahren zum Nachweis der Sicherheit gegen Gleitflächenbildung bei suspensionsgestützten Erdwänden. Veröffentlichungen des Grundbauinstituts der TU Berlin, No. 5, 1979.

[84] FEILE, W.: Konstruktion und Bau der Schleuse Regensburg mit Hilfe von Schlitzwänden. Der Bauingenieur 50 (1975), No. 5, p. 168.
[85] LOERS, G. and PAUSE, H.: Die Schlitzwandbauweise – große und tiefe Baugruben in Städten. Der Bauingenieur 51 (1976), No. 2, p. 41.
[86] VEDER, CH.: Beispiele neuzeitlicher Tiefgründungen. Der Bauingeneiur 51 (1976), No. 3, p. 89.
[87] VEDER, CH.: Die Schlitzwandbauweise – Entwicklung, Gegenwart und Zukunft. Österreichische Ing. Z. 18 (1975), No. 8, p. 247.
[88] VEDER, CH.: Einige Ursachen von Mißerfolgen bei der Herstellung von Schlitzwänden und Vorschläge zu ihrer Vermeidung. Bauingenieur 56 (1981), No. 8, p. 299.
[89] CARL, L. and STROBL, TH.: Dichtungswände aus einer Zement-Bentonit-Suspension. Wasserwirtschaft 66 (1976), No. 9, p. 246.
[90] LORENZ, W.: Plastische Dichtungswände bei Staudämmen. Vorträge Baugrundtagung 1976 in Nürnberg, Deutsche Gesellschaft für Erd- und Grundbau e. V., p. 389.
[91] KIRSCH, K. and RÜGER, M.: Die Rüttelschmalwand – Ein Verfahren zur Untergrundabdichtung. Vorträge Baugrundtagung 1976 in Nürnberg, Deutsche Gesellschaft für Erd- und Grundbau e. V., p. 439.
[92] KAESBOHRER, H.-O.: Fortschritte an der Donau im Dichtungsverfahren für Stauräume. Die Bautechnik 49 (1972), No. 10, p. 329.
[93] BRENNECKE/LOHMEYER: Der Grundbau, 4th edition, Vol. II, Berlin: Wilhelm Ernst & Sohn 1930.
[94] NÖVVENTVED, C.: Berechnung von Pfahlrosten. Berlin: Ernst & Sohn, 1928.
[95] SCHIEL, F.: Statik der Pfahlgründungen. Berlin: Springer, 1960.
[96] AGATZ, A. and LACKNER, E.: Erfahrungen mit Grundbauwerken. Berlin: Springer, 1977.
[97] TECHNISCHE LIEFERBEDINGUNGEN für Wasserbausteine – Ausgabe 1997 (TLW) – des Bundesministers für Verkehr, Verkehrsblatt (1997), No. 19, p. 552 cont.
[98] UFERSCHUTZWERKE aus Beton, Schriftenreihe der Zementindustrie. Verein deutscher Zementwerke e. V., Düsseldorf, (1971), No. 38.
[99] FINKE, G.: Geböschte Ufer in Binnenhäfen, Zeitschrift für Binnenschiffahrt und Wasserstraßen (1978), No. 1, p. 3.
[100] Report of the PIANC Working Group I-4, Guidelines for the design and construction of flexible revetments incorporating geotextiles in marine environment, Supplement to PIANC-Bulletin No. 57, Brussels 1987.
[100a] Report of the PIANC Working Group II-21 "Guidelines for the design and construction of flexible revetments incorporating geotextiles in marine environment", Supplement to PIANC-Bulletin No. 78/79, Brussels 1992.
[101] BLUM, H.: Wirtschaftliche Dalbenformen und deren Berechnung. Die Bautechnik 9 (1932), No. 5, p. 50.
[102] COSTA, F. V.: The Berthing Ship. The Dock and Harbour Authority. Vol. XLV (1964), Nos. 523 to 525.
[103] STAHL-EISEN-WERKSTOFFBLATT 088, Schweißbare Feinkornbaustähle, Richtlinien für die Verarbeitung. Düsseldorf: Verlag Stahleisen.
[104] ZULASSUNGEBESCHEID für hochfeste, schweißgeeignete Feinkornbaustähle StE 460 und StE 690, Institut für Bautechnik, Reichpietschufer 1, D-1000 Berlin 30.
[105] TECHNISCHE LIEFERBEDINGUNGEN für geotextile Filter (TLG) – 1993 Edition – des Bundesministers für Verkehr, Verkehrsblatt 1993, No. 9, p. 372 cont.
[106] TECHNISCHE LIERFERBINDUNGEN für Stahlspundbohlen (TLS) – 1985 Edition – des Bundesministers für Verkehr, Verkehrsblatt 1985, No. 24.

[107] BYDIN, F. I.: Development of certain questions in area of river's winter regime, III. Hydrologic Congress, Leningrad, 1959.
[108] SCHWARZ, J., HIRAYAMA, K., WU, H. C.: Effect of Ice Thickness on Ice Forces, Proceedings Sixth Annual Offshore Technology Conference, Houston, Texas, USA 1974.
[109] KORZHAVIN, K. N.: Action of ice on engineering structures, English translation, U. S. Cold Region Research and Engineering Laboratory, Trans. T. L. 260.
[110] GERMANISCHER LLOYD: Vorschriften für Konstruktion und Prüfung von Meerestechnischen Einrichtungen, Vol. 1 – Meerestechnische Einheiten – (Seebauwerke). Hamburg: Eigenverlag des Germanischen Lloyd, July 1976.
[111] Ice Engineering Guide for Design and Construction of Small Craft Harbors. University of Wisconsin, Advisory Report SG-78-417.
[112] HORN, A.: Bodenmechanische und grundbauliche Einflüsse bei der Planung, Konstruktion und Bauausführung von Kaianlagen. Mitt. d. Inst. f. Bodenmechanik und Grundbau, HSBw München, No. 4, und Mitt. des Franzius-instituts für Wasserbau (1981), No. 54, p. 110.
[113] HORN, A.: Determination of properties for weak soils by test embankments. International Symposium "Soil and Rock Investigations by in-situ Testing"; Paris (1983), Vol. 2, p. 61.
[114] HORN, A.: Vorbelastung als Mittel zur schnelleren Konsolidierung weicher Böden. Geotechnik (1984), No. 3, p. 152.
[115] SCHMIEDEL, U.: Seitendruck auf Prähle. Bauingenieur 59 (1984), p. 61.
[116] FRANKE, E. und SCHUPPENER, B.: Horizontalbelastung von Pfählen infolge seitlicher Erdauflasten. Geotechnik (1982), p. 189.
[117] DBV-MERKBLATT. Begrenzung der Rißbildung im Stahlbeton- und Spannbetonbau. Deutscher Beton-Verein e. V.
[118] ZUSÄTZLICHE TECHNISCHE VERTRAGSBEDINGUNGEN – WASSERBAU (ZTV-W) für Wasserbauwerke aus Beton und Stahlbeton (Leistungsbereich 215), edition 1998, VkBl. 17, p. 895.
[119] RÜPING, F.: Beitrag und neue Erkenntnisse über die Errichtung und Sicherung von großen Mineralöl-Lagertanks für brennbare Flüssigkeiten der Gefahrenklasse AI. Dissertation, Hannover 1965.
[120] RICHTLINIEN FÜR ANFORDERUNGEN AN ANLAGEN ZUM UMSCHLAG GEFÄHRDENDER FLÜSSIGER STOFFE IM BEREICH DER WASSERSTRASSEN. Erlaß des Bundesministers für Verkehr vom 24.7.1975, Verkehrsblatt 1975, p. 485.
[121] MAYER, B. K., KREUTZ, B., SCHULZ, H.: Setting sheet piles with driving aids, Proc. 11th Int. Conf. Soil Mech. Found. Eng. San Francisco, 1985.
[122] ARBED, S. A., Luxembourg: Europäisches Patent 09.04.86, Patenterteilung am 9.4.86, Patentblatt 86/15.
[123] PARTENSCKY, H.-W. Binnenverkehrswasserbau, Schleusenanlagen. Berlin, Heidelberg, New York, Tokyo: Springer-Verlag, 1986.
[124] HAGER, M.: Vorlesungen Verkehrswasserbau I der RWTH Aachen.
[125] KANAL- UND SCHIFFAHRTSVERSUCHE 1967. Schiff und Hafen, 20 (1968), No. 4–9. See also 27th Mitteilungsblatt der Bundesanstalt für Wasserbau, Karlsruhe, Sept. 1968.
[126] TUNNEL-SONDERAUSGABE April 1987. Internationale Fachzeitschrift für unterirdisches Bauen. Gütersloh: Bertelsmann.
[127] DEUTSCH, V., und VOGT, M.: Die zerstörungsfreie Prüfung von Schweißverbindungen – Verfahren und Anwendungsmöglichkeiten. Schweißen und Schneiden 39 (1987), No. 3.

[128] MERKBLATT "ANWENDUNG VON GEOTEXTILEN FILTERN AN WASSERSTRASSEN (MAG)" 1993, edited by Bundesanstalt für Wasserbau, Karlsruhe. English: Code of practice: Use of Geotextile Filters on Waterways.
[129] "ANWENDUNG UND PRÜFUNG VON KUNSTSTOFFEN IM ERDBAU UND WASSERBAU", DVWK-Schriften Heft 76, 1986, Verlag Paul Parey, Hamburg, see (165).
[130] RICHTLINIEN FÜR DIE PRÜFUNG VON GEOTEXTILEN FILTERN IM VERKEHRSWASSERBAU (RPG). Karlsruhe: Bundesanstalt für Wasserbau, 1994.
[131] TECHNISCHE LIEFERBEDINGUNGEN FÜR GEOTEXTILE FILTER (TLG) – Edition 1993 – english: Technical supply conditions for geotextile filters. EG notification no.93/72/D dated 9.03.1993.
[132] ZUSÄTZLICHE TECHNISCHE VERTRAGSBEDINGUNGEN – Wasserbau (ZTV-W) für Böschungs- und Sohlensicherungen (Leistungsbereich 210), edition 1991, sections 2.2.1, 3.2.1, 4.4.1, draft 1999.
[133] CONCRETE INTERNATIONAL. Detriot 1982, p. 45–51.
[134] Richtlinie für Schutz und Instandsetzung von Betonbauteilen, Teile 1 bis 4; Deutscher Ausschuß für Stahlbeton DafStB 1990, 1991, 1992.
[135] DIERSSEN, G., GUDEHUS, G.: Vibrationsrammungen in trockenem Sand. Geotechnik 3/1992, p. 131.
[136] ZUSÄTZLICHE TECHNISCHE VERTRAGSBEDINGUNGEN UND RICHTLINIEN FÜR SCHUTZ UND INSTANDSETZUNG VON BETONBAUTEILEN (ZTV-SIB 90), Bundesverkehrsministerium, Abt. Straßenbau, Bonn.
[137] FGSV-820, Merkblatt für die Fugenfüllung in Verkehrsflächen aus Beton. Forschungsgesellschaft für Straßen- und Verkehrswesen e. V., edition 1982.
[138] ZUSÄTZLICHE TECHNISCHE VERTRAGSBEDINGUNGEN – WASSERBAU (ZTV-W) für Schutz und Instandsetzung der Betonbauteile von Wasserbauwerken (Leistungsbereich 219), Bundesminister für Verkehr, Abt. Binnenschiffahrt und Wasserstraßen, Bonn, 1997.
[139] VV-WSV 210 BAUWERKSINSPEKTION, published by Bundesminister für Verkehr, Bonn, 1984, distributed by Drucksachenstelle der Wasser- und Schiffahrtsdirektion Mitte, Hannover.
[140] REPORT OF PIANC-WORKING GROUP II-9 "Development of Modern Marine Terminals", Supplement to the PIANC-Bulletin No. 56, Brussels, 1987.
[141] BJERRUM, L.: General Report, 8, ICSMFE (1973), Moscow, Vol. 3, p. 124.
[142] WROTH, C. P.: "The interpretation of in situ soil tests", 1984, Géotechnique 34, No. 4, p. 449–489.
[143] SELIG, E. T. and MCKEE, K. E.: "Static and dynamic behaviour of small footings", Am. Soc. Civ. Eng., Journ. Soil Mech. Found. Div., Vol. 87 (1961), No. SM 6, Part I, p. 29–47 (cf. Horn, A.- Bauing. (1963) 38, No. 10, p. 404).
[144] HORN, A.: "Insitu-Prüfung der Wasserdurchlässigkeit von Dichtwänden" (1986), Geotechnik 1, p. 37.
[145] RANKILOR, p. R.: "Membranes in ground engineering" (1981), Wiley & Son.
[146] VELDHUYZEN VAON ZANTEN, R.: "Geotextiles and Geomembranes in Civil Engineering" revised edition(1994), Balkema, Rotterdam/Boston.
[147] KOERNER, R. M.: "Design with Geosynthetics", Prentice-Hall 4th edition(1998), Englewood Cliffs, N. Y.
[148] HAGER, M.: Eisdruck, Chapt. 1.14 Grundbautaschenbuch, 5th edition, part 1, Ernst & Sohn, Verlag für Architektur und technische Wissenschaften, 1996.
[149] MERKBLATT ANWENDUNG VON KORNFILTERN AN WASSERSTRASSEN (MAK), Bundesanstalt für Wasserbau (BAW), edition 1989.

[150] KUNSTSTOFFMODIFIZIERTER SPRITZBETON. Merkblatt des Deutschen Betonvereins e. V., Wiesbaden.
[151] MITTEILUNGSBLATT DER BUNDESANSTALT FÜR WASSERBAU (BAW) No. 67, Karlsruhe 1990.
[152] HEIN, W.: Korrosion von Stahlspundwänden im Wasser, Hansa, 126, annual set 1989, No. ¾, Schiffahrtsverlag "Hansa", C. Schroedter & Co., Hamburg.
[153] RICHTLINIE FÜR DIE PRÜFUNG VON BESCHICHTUNGSSTOFFEN FÜR DEN KORROSIONSSCHUTZ IM STAHLWASSERBAU (RPB), edition 1995, Bundesverkehrsministerium, Abt. Binnenschiffahrt und Wasserstraßen, Bonn.
[154] FÉDÉRATION EUROPÉENNE DE LA MANUTENTION, Section I, Rules for the design of hoisting appliances, Booklet 2: Classification and loading on structures and mechanisms F. E. M. 1.001. 3rd edition, 1987. Deutsches National-Komitee Frankfurt/Main.
[155] DEUTSCHER VERBAND FÜR WASSERWIRTSCHAFT UND KULTURBAU e. V. (DVWK): Dichtungselemente im Wasserbau DK 626/627 Wasserbau; DK 69.034.93 Abdichtung Hamburg, Berlin, Verl. Paul Parey 1990.
[156] HENNE, J.: Versuchsgerät zur Ermittlung der Biegezugfestigkeit von bindigen Böden, Geotechnik 1989, No. 2, p. 96 cont.
[157] SCHULZ, H.: Mineralische Dichtungen für Wasserstraßen, Fachseminar "Dichtungswände und Dichtsohlen", June 1987 in Braunschweig, Mitteilungen des Instituts für Grundbau und Bodenmechanik, Techn. Universität Braunschweig, No. 23, 1987.
[158] SCHULZ, H.: Conditions for day sealings at joints, Proc. of the IX. Europ. Conf. on Soil Mech. and Found. Eng., Dublin, 1987.
[159] ZUSÄTZLICHE TECHNISCHE VERTRAGSBEDINGUNGEN UND RICHTLINIEN FÜR DAS FÜLLEN VON RISSEN IN BETONBAUTEILEN (ZTV-RISS 93), Bundesverkehrsministerium, Abt. Straßenbau, Bonn
[160] HTG Kaimauer-Workshop SMM '92 Conference: Kaimauerbau, Erfahrungen und Entwicklungen, Beiträge, HANSA 129, annual set 1992, No. 7, p. 693 cont. and No. 8, p. 792 cont.
[161] DÜCKER, H. P. und OESER, F. W., Der Bau von Umschlag und Werft-Kaimauern, erläutert am Beispiel von 4 Neubauprojekten, der Bauingenieur 59 (1984), p. 15 cont.
[162] SPARBOOM, U.: Über die Seegangsbelastung lotrechter zylindrischer Pfähle im Flachwasserbereich; Mitteilungen des Leichtweiß-Instituts der TU Braunschweig, No. 93, Braunschweig 1986.
[163] VV WSV 2301 Damminspektion, published by Bundesminister für Verkehr, Bonn, distributed by Drucksachenstelle of the WSD Mitte, Hannover.
[164] ZUSÄTZLICHE TECHNISCHE VERTRAGSBEDINGUNGEN (ZTV-W) für Technische Bearbeitung (LB 202) Bundesverkehrsministerium, Abt. Binnenschiffahrt und Wasserstraßen, Bonn, 1993, VkBl. 21, p. 727.
[165] Deutscher Verband für Wasserwirtschaft und Kulturbau e. V. (DYWK): Anwendung von Geotextilien im Wasserbau DVWK Leaflet 221/1992.
[166] DAVIDENHOFF, R. and FRANKE, O. L.: Untersuchungen der räumlichen Sickerströmung in einer umspundeten Baugrube im Grundwasser, Bautechnik 42, No. 9, Berlin, 1965.
[167] DAVIDENHOFF, R.: Deiche und Erddämme, Sickerwasser-Standsicherheit, Werner-Verlag Düsseldorf, 1964.
[168] DAVIDENHOFF, R.: Unterläufigkeit von Bauwerken, Werner-Verlag Düsseldorf, 1970.
[169] ALEXY, M., FÜHRER, M., KÜHNE. E., Verbesserung der Schiffahrtsverhältnisse auf der Elbe bei Torgau: Vorbereitung, Ausführung und Erfolgskontrolle, Jahrbuch der Hafenbautechnischen Gesellschaft e. V., Hamburg, 50th Volume, 1995, p. 71 cont.

[170] RÖMISCH, K.: Propellerstrahlinduzierte Erosionserscheinungen in Häfen, HANSA, 130th annual set 1993, No. 8 and – Spezielle Probleme – HANSA; 131st annual set 1994, No. 9.
[171] PIANC Report of the 3rd International Wave Commission, Supplement of the Bulletin No. 36, Brussels 1980.
[172] Report of the PIANC Working Group II-12: analysis of Rubble Mound Breakwaters, Supplement to PIANC-Bulletin No. 78/79, Brussels 1992.
[173] CIRIA/CUR: Manual on the use of rock in coastal and shoreline engineering. CIRIS Special Publication 83, CUR Report 154, Rotterdam, A. A. Balkema 1991.
[174] BRUNN, P.: Port Engineering, London 1980.
[175] STÜCKRATH, T.: Über die Probleme des Unternehmers beim Hafenbau. Mitteilungen d. Franziusinstituts für Wasserbau und Küsteningenieurwesen der TU Hannover, No. 54, Hannover 1983.
[176] ALBERTS, D. and HEELING, A.: Wanddickenmessungen an korrodierten Stahlspundwänden; statistische Datenauswertung, Mitteilungsblatt d. BAW, No. 75, Karlsruhe 1996.
[177] ZUSÄTZLICHE TECHNISCHE VERTRAGSBEDINGUNGEN – WASSERBAU (ZTV-W) für den Korrosionsschutz im Stahlwasserbau (Leistungsbereich 218), Bundesminister für Verkehr, Abt. Binnenschiffahrt und Wasserstraßen, Bonn, 1995, VkBl. 6, p. 175.
[178] ZUSÄTZLICHE TECHNISCHE VERTRAGSBEDINGUNGEN – WASSERBAU (ZTV-W) für den Korrosionsschutz im Stahlwasserbau (Leistungsbereich 220), Bundesminister für Verkehr, Abt. Binnenschiffahrt und Wasserstraßen, Bonn, 1999, VkBl., p. 707.
[179] Kathodischer Korrosionsschutz für Stahlbeton, Hafenbautechnische Gesellschaft e.V (HTG), Hamburg 1994.
[180] Allgemeine Verwaltungsvorschrift zum Schutz gegen Baulärm – Geräuschemissionen sowie – Emissionsmeßverfahren, Carl Haymanns Verlag KG; Köln 1971.
[181] Richtlinie 79/113/EWG vom 19.12.1978 zur Angleichung der Rechtsvorschriften der Mitgliedsstaaten betreffend die Ermittlung des Geräuschemissionspegels von Baumaschinen und Baugeräten (Amtsbl. EG 1979 No. L 33 p. 15).
[182] 15th Verordnung zur Durchführung des BImSchG dated 10.11.1986 (Baumaschinen-LärmVO).
[183] VDI-Richtlinie 2714 (01/88) Schallausbreitung im Freien – Berechnungsverfahren.
[184] VDI-Richtlinie 3576: Schienen für Krananlagen, Schienenverbindungen, Schienenbefestigungen, Toleranzen.
[185] Empfehlungen und Berichte des Ausschusses für Hafenumschlagtechnik (AHU) der Hafenbautechnischen Gesellschaft e.V, Hamburg.
[186] EAB-100: Empfehlungen des Arbeitskreises "Baugruben" (EAB) auf der Grundlage des Teilsicherheitskonzeptes EAB-100; published by Deutsche Gesellschaft für Geotechnik (DGGT), Essen, Ernst & Sohn Verlag für Architektur und technische Wissenschaften, Berlin. 1996.
[187] RADOMSKI, H.: Untersuchungen über den Einfluß der Querschnittsform wellenförmiger Spundwände auf die statischen und rammtechnischen Eigenschaften, Mitt. Institut für Wasserwirtschaft, Grundbau und Wasserbau der Universität Stuttgart, No. 10, Stuttgart 1968.
[188] CLASMEIER, H.-D.: Ein Beitrag zur erdstatischen Berechnung von Kreiszellenfangedämmen, Mitt. Institut für Grundbau und Bodenmechanik, Universität Hannover, No. 44/1996.

[189] FEDDERSEN, I.: Das Hyperbelverfahren zur Ermittlung der Bruchlasten von Pfählen, eine kritische Betrachtung. Bautechnik 1982, No. 1, p. 27 cont.
[190] BAUMANN, V.: Das Soilcrete-Verfahren in der Baupraxis, Vorträge der Baugrundtagung 1984 der DGEG in Düsseldorf, p. 43 cont.
[191] BAUER, K.: Einsatz der Bauer-Schlitzwandfräse beim Bau der Dichtwand am Brombachspeicher und an der Sperre Kleine Roth, Tiefbau-BG 10/1985, p. 630 cont.
[192] STROBL, Th. and WEBER, R.: Neuartige Abdichtungsverfahren im Sandsteingebirge, Vorträge der Baugrundtagung 1986 der DGEG.
[193] STROBL, Th.: Ein Beitrag zur Erosionssicherheit von Einphasen-Dichtungswänden, Wasserwirtschaft 7/8 (1892), p. 269 cont. and Erfahrungen über Untergrundabdichtungen von Talsperren. Wasserwirtschaft 79 (1989), No. 7/8.
[194] GEIL, M.: Entwicklung und Eigenschaften von Dichtwandmassen und deren Überwachung in der Praxis, s+t 35, 9/1981, p. 6 cont.
[195] KARSTEDT, J. and RUPPERT, F.-R.: Standsicherheitsprobleme bei der Schlitzwandbauweise, Baumaschinen und Bautechnik 5/1980, p. 327 cont.
[196] MESECK, H., RUPPERT, F.-R., SIMONS, H.: Herstellung von Dichtungsschlitzwänden im Einphasenverfahren, Tiefbau, Ingenieurbau, Straßenbau 8/79, p. 601 cont.
[197] ROM Recomendaciones para Obras Maritimas (English version) Maritime Works Recommendations (MWR): Actions in the design of maritime and Harbor Works (ROM 0.2-90), Ministerio de Obras Publicas y Transportes, Madrid 1990.
[198] JAPAN SOCIETY OF CIVIL ENGINEERING: The 1995 Hyogoken-Nanbu Earthquake – Investigation into Damage to Civil Engineering Structures – Committee of Earthquake Engineering, Tokyo. 1996
[199] Report of the joint Working Group PIANC and IAPH, in cooperation with IMPA and IALA, PTC II-30: Approach Channels – A Guide for Design; supplement to PIANC-Bulletin No. 95, Brussels 1997.
[200] Report of the PIANC Working Group PTC I-16: Standardisation of Ships and Inland Waterways for River/Sea Navigation; Supplement to PIANC-Bulletin No. 90, Brussels 1996.
[201] Verordnung über die Schiffs- und Schiffsbeháltervermessung (Schiffsvermessungsverordnung SchVmV) dated July 5 1982 (BGBL. I p. 916), changed by the Erste Verordnung zur Änderung der Schiffsvermessungsverordnung dated September 3 1990 (BGBL. I. p. 1993).

I.3 Technical Provisions

The standards (EN and DIN, including prestandards), the Regulations of the German Federal Railways (DS), the Regulations of the German Committee for Steel Constructions (DASt-Ri) and the Steel Iron Material table of the "Verein Deutscher Eisenhüttenleute" (SEW) are authoritative in their respective latest version.
P = part, S = Supplement.

I.3.1 Standards

DIN EN 196-1	Test methods for cement – Part 1: Determination of Strength; German Version EN 196-1: 1994
DIN EN 196-1	–; –; Part 2: Chemical Analysis of Cement; German Version EN 196-2: 1994
DIN EN 196-3	–; –; Part 3: Determination of Setting Time and Volume Stability; German Version EN 196-3: 1994
DIN V ENV 206	Concrete; Performance, Production, Placing and Compliance Criteria
DIN EN 440	Welding Consumables: Wire Electrodes and Deposits for Gas-Shielded Metal Arc Welding of Non-Alloy and Fine Grain Steels, Classification
DIN EN 450	Fly Ash for Concrete – Definitions, Requirements and Quality Control; German Version EN 450: 1994
DIN 488 P 1	Reinforcing Steels; Grades, Properties, Marking
DIN 488 P 3	–; Reinforcing bar steel; tests
DIN EN 499	Welding Consumables: Covered Electrodes for Manual Metal Arc Welding on Non-Alloy and Fine Grain Steels, Classification
DIN EN 756	Welding Consumables: Wire and Wire-Flux Combinations for Submerged Arc Welding of Non-Alloy and Fine Grain Steels – Classification; German Version EN 756: 1995
DIN EN 996	Pile Driving Equipment – Safety Requirements; German Version EN 996: 1995
DIN 1045	Concrete and Reinforced Concrete; Design and Construction
DIN 1048-1	Testing Concrete; Testing of Fresh Concrete
DIN 1048-5	–; –; Hardened Concrete, (Specimens Prepared in Mould)
DIN 1052 P 1	Timber Structures; Design and Construction
DIN V 1054-100	Soil: Verification of the Safety of Earthworks and Foundations – Part 100: Analysis in accordance with the partial safety factor concept
DIN 1055 P 1	Design Loads for Buildings; Stored Materials, Building Materials and Structural Members, Dead Load and Angle of Friction
DIN 1055 P2	–; Soil Characteristics, Weight Density, Angle of Friction, Cohesion, Angle of Wall Friction
DIN 1055 P3	–; Live Loads
DIN 1055 P4	–; Live Loads, Wind Loads of Structures Unsusceptible to Vibration
DIN 1072	Road and Foot Bridges; Design Loads
S 1	–; –; Explanations
DIN 1075	Concrete Bridges, Dimensioning and Construction
DIN 1076	Engineering Structures Relating to Highways and Paths; Monitoring and Testing
DIN 1080 P1	Terms, Symbols and Units Used in Civil Engineering; Principles
DIN 1080 P 6	–; Soil Mechanics and Foundation Engineering
DIN 1084 P1	Control (Quality Control) of Concrete Structures and Reinforced Concrete Structures; Concrete B II on Building Sites
DIN 1164-1	Cement – Part 1: Components, Requirements
DIN 1301 P 1	Units, Names, Symbols

E DIN EN 1536	Execution of Special Geotechnical Work – Bored Piles; German Version pr EN 1536:1994
E DIN EN 1538	Execution of Special Geotechnical Work – Diaphragm Walls; German Version pr EN 1538:1994
DIN 1681	Cast Steels for General Engineering Purposes; Technical Delivery Conditions
DIN 1913 P 1	Covered Electrodes for Joint Welding Unalloyed and Low-Alloyed Steel; Classification, Designation, Technical Delivery Conditions
DIN V ENV 1991-1	Basis of Design and Actions on Structures Part 1: Basis of design
DIN V ENV 1992-1	Design of Concrete Structures Part 1: General Rules
DIN V ENV 1993-1	Design of Steel Structures, Part 1: General Rules
DIN V ENV 1993-5	Design of Steel Structures, Part 5: Steel Piles
DIN V ENV 1994-1	Design of Composite Steel and Concrete Structures, Part 1: General Rules
DIN V ENV 1995-1	Design of Timber Structures, Part 1: General Rules
DIN V ENV 1996-1	Design of Masonry Structures, Part 1: General Rules
DIN V ENV 1997-1	Geotechnical Design, Part 1: General Rules
DIN V ENV 1998-1	Earthquake Resistant Design of Structures, Part 1: General Rules
DIN V ENV 1999-1	Design of Aluminium Structures, Part 1: General Rules
DIN 4014	Bored Piles; Manufacture, Design and permissible Loading
DIN V 4017-100	Soil: Calculation of Design Bearing Capacity of Soil beneath Shallow Foundations – Part 100: Analysis in accordance with the Partial Safety Factor Concept
DIN 4018	Soil: Calculation of the Bearing Pressure Distribution under Spread Foundations
DIN V 4019-100	Soil: Analysis of Settlements – Part 100: Analysis in accordance with the Partial Safety Factor Concept
DIN 4020	Geotechnical Investigations for Construction Purposes
S 1	–; –; Application Aids, Explanations
DIN 4021	Subsoil; Exploration by Diggings and Borings as well as by Sampling
DIN 4022 P 1	Subsoil and Groundwater; Designation and Description of Soil Types and Rock; List of Soil Courses for Boring without Continuous Gaining of Core Trials in Soil and Rock
DIN 4022 P 2	–; –; Designation and Description of Soil Types and Rock, List of Soil Courses for Boring in Rock
DIN 4022 P3	–; –; List of Soil Courses for Boring in the Soil (Loose Rock)
DIN 4023	Subsoil and Water Drilling; Drawing of the Results
DIN 4026	Driven Piles; Manufacture, Dimensioning and Permissible Loading
S	–; –; Explanations
DIN 4030-1	Evaluation of Liquids, Soils and Gases Aggressive to Concrete; Principles and Limit Values
DIN 4030-2	–;–; Sampling and Analysis of Water and Soil Samples
DIN 4049-1	Hydrology; Concepts
DIN 4049-2	–; –; Concepts for Water Quality
DIN 4049-3	–; –; Concepts for Quantitative Hydrology
DIN 4054	Corrections of Waterways; Terms

DIN V 4084-100	Soil; Calculation of Slope and Embankment Failure and Overall Stability of Retaining Structures – Part 100: Analysis in accordance with the Partial Safety Factor Concept
DIN V 4085-100	Soil; Calculation of Earth Pressure – Part 100: Analysis in accordance with the Partial Safety Factor Concept
DIN 4093	Subsoil; Ground Treatment by Grouting; Planning, Grouting Procedure and Testing
DIN 4094	Subsoil; Exploration by Penetration Tests
S 1	–; –; –; Application Aids, Explanations
DIN 4096	Subsoil; Vane Testing, Dimensions of Apparatus, Mode of Operation, Evaluation
DIN 4099-P 1	Welding of Reinforcing Steel; Execution and Tests
DIN 4102 P 1–9	Fire Behaviour of Construction Materials and Components
DIN 4102 P 11–18	
DIN 4125	Grounted Anchors – Temporary Soil Anchors and Permanent Soil Anchors – Analysis, Structural Design and Testing
DIN V 4126-100	Diaphragm walls – part 100: Calculation according to the Partial Safety Factor Concept
DIN 4127	Foundation Engineering: Diaphragm Wall Clay for Supporting Liquids; Requirements, Tests, Delivery, Quality Control
DIN 4128	Small Diameter Injection Piles (In-situ Concrete Piles and Composite Piles); Construction Procedure, Design and Permissible Loading
DIN 4132	Craneways; Steel Structures, Principles for Calculation, Design and Construction
S	–; –; Explanations
DIN 4149 P 1	Buildings in German Earthquake Zones; Design Loads, Dimensioning, Design and Construction of Conventional Buildings
S	–; –; Allocation of Administrative Areas to Earthquake Zones
DIN 4150-2	Vibrations in Buildings; Effects on Persons in Buildings
DIN 4150-2/A1	–; –; Amendment A1
DIN 4150-3	–; –; Effects on Structures
DIN 4226 P 1–4	Aggregates for Concrete
DIN 4227 P 1	Prestressed Concrete; Structural Components made of Ordinary Concrete with Partial or Total Prestressing
DIN 4227 P 5	–; Injection of Cement Mortar into Prestressing Concrete Ducts
DIN 4301	Ferrous and Non-ferrous Metallurgical Slag for Civil Engineering and Building Construction Use
DIN 5901	Flat Bottom Rails – Dimensions, Sectional Properties, Steel Grades
DIN 8559 P 1	Welding Additives for Shielded Arc Welding; Wire Electrodes, Welding Rods and Solid Rods for Shielded Arc Welding of Unalloyed and Alloyed Steels
DIN 8563 P 1	Quality Assurance of Welding Operations, General Principles
DIN 8563 P 2	–; Requirements Regarding the Firm
DIN 8563-10	Quality Assurance of Welding Operations; Stud Welded Joints on Structural Steels; Stud Welding with Drawn-Arc Ignition and Ring Ignition
DIN EN 10 002-1	Tensile Testing of Metallic Materials; Method of Test at Ambient Temperature; German Version EN 10 002-1:1990 and AC 1:1990

DIN EN 10 025	Hot Rolled Products of Non-Alloyed Structural Steels; Technical Delivery Conditions, German Version EN 10 025: 1990/A1:1993
DIN EN 10 028-1	Flat Products Made from Steel for Pressure Purposes; Part 1: General Requirements, German Version EN 10 028-1: 1992. Other parts: -2: 1992; -3: 1992; -4: 1995; -5: 1996; -6: 1996
DIN EN 10 045-1	Metallic Materials; Charpy impact test; Part 1: Test Method, German Version EN 10 045-1: 1990
DIN EN 10 113-1	Hot-Rolled Products in Weldable Fine Grain Structural Steels, Part 1: General Delivery Conditions; German Version EN 10113-1: 1993
DIN EN 10 113-2	–; –; Part 2: Delivery Conditions for Normalised/Normalised Rolled Steels; German version EN 10113-2: 1993
DIN EN 10 113-3	–; –; Part 3: Delivery Conditions for Thermomechanical Rolled Steels; German Version EN 10113-3: 1993
DIN EN 10 204	Metallic Products: Types of Inspection Documents (1991 + A1 1995)
DIN EN 10 248-1	Hot Rolled Sheet Piling of Non Alloy Steels, Technical Delivery Conditions
DIN EN 10 248-2	–; –; Tolerances on Shape and Dimensions
DIN EN 10 249-1	Cold Formed Sheet Piling of Non Alloy Steels, Technical Delivery Conditions
DIN EN 10 249-2	–; –; Tolerances on Shape and Dimensions
DIN EN 12 063	Execution of Special Geotechnical Work, Sheet Piling Construction 1/1997
DIN 15 018 P 1	Cranes; Principles for Steel Structures; Stress Analysis
DIN 15 018 P 2	–; Structures of Steel, Design Principles
DIN 15 018 P 3	–; –; Stress Analysis of Mobile Cranes
DIN 15 019 P 1	Cranes; Stability for All Cranes except Non-Rail Mounted Mobile Cranes and except Floating Cranes
DIN 15 019 P 2	–; –; Stability for Non-Rail Mounted Mobile Cranes, Test Loading and Calculation
DIN 16 776 P 1	Plastic Moulding Materials; Polyethylene (PE) Moulding Materials, Classification and Designation
DIN 16 925	Extruded Panels of High-Density Polyethylene (PE-HD); Technical Delivery Conditions
DIN 17 120	Welded Circular Tubes made of General Structural Steels for Steel Constructions; Technical Delivery Conditions
DIN 18 121 P 1	Subsoil; Testing Procedure and Testing Equipment; Water Content; Determination by Drying in Oven
DIN 18 121 P 2	–; –; Field Tests
DIN 18 126	Subsoil; Investigation and Testing – Determination of Density of Non-Cohesive Soils for Maximum and Minimum Compactness
DIN 18 127	Subsoil; Testing Procedures and Testing Equipment
DIN 18 128	Subsoil; Testing Procedures and Testing Equipment; Determination of Annealing Loss
DIN 18 130 P 1	Subsoil; Testing Procedures and Testing Equipment; Determination of Water Permeability Coefficient; Laboratory Tests
DIN 18 134	Subsoil; Testing Procedures and Testing Equipment; Plate Load Test
DIN 18 136	Subsoil; Test Procedures and Testing Equipment; Unconfined Compression Test

DIN 18 137 P 1	Subsoil; Investigation of Soil Samples, Determination of Shear Resistance; Definitions and General Testing Conditions
DIN 18 137 P 2	–; –; –; Triaxial Test
DIN 18 195 P 1–10	Seals for Structures
DIN 18 196	Earthworks and Foundation Engineering; Soil Classification for Civil Engineering Purposes
DIN 18 300	VOB Order for Placing of Contracts for Construction Work, Part C: General Technical Contractual Conditions for Construction Work, Earth Works
DIN 18 311	VOB Order for Placing of Contracts for Construction Work, Part C: General Technical Contractual Conditions for Construction Work, Dredging Works
DIN 18 540	Sealing of Outside Wall Joints with Joint Sealing Agents in Building Construction
DIN 18 551	Shotcrete; Production and Quality Control
DIN 18 800-1	Steel Structures; Design and Construction
DIN 18 800-2	–; –; Analysis of Safety against Buckling of Linear Members and Frames (11/1990)
DIN 18 800-4	–; –; Analysis of Safety against Buckling of Shells (11(1990)
DIN 18 801	Steel Construction in Building, Dimensioning, Design, Construction
DIN 18 807-1	Trapezoidal Sheeting in Building; Trapezoidal Steel Sheeting; General Requirements and Determination of Bearing Capacity by Calculation
DIN 19 702	Stability of Solid Structures in Hydraulic Engineering
DIN 19 703	Locks for Waterways for Inland Navigation – Principles for Dimensioning and Equipment
DIN 19 704-1	Hydraulic Steel Structures; Part 1: Criteria for Design and Calculation
DIN 19 704-2	–; –; Part 2: Design and Manufacturing
DIN 19 704-3	–; –; Part 3: Electrical Equipment
DIN EN 29 073-1	Textiles; Test Methods for Nonwovens; Part 1: Determination of Mass per Unit Area (ISO 9703-1: 1989); German Version: EN 29 073-1: 1992
DIN EN 29 073-3	–; –; Part 3: Determination of Tensile Strength and Elongation (ISO 9073-3: 1989), German Version EN 29 073-3: 1992
DIN EN 29 073-4	–; –; Part 4: Determination of Wear Resistance (ISO 9073-4: 1989); German Version EN 29 073-4: 1992
DIN 31 051	Maintenance; Terms and Measures
DIN 50 049	Documents of Materials Testing; in future: DIN EN 10 204
S 1	–; –; Suggestions for the Configuration of Documents
DIN 50 929-1	Corrosion of Metals; Probability of Corrosion of Metallic Materials when Subject to Corrosion from the Outside; General
DIN 50 929-3	–; –; –; Buried and Underwater Pipelines and Structural Components
DIN 51 043	Trass; Requirements, Tests
DIN 52 103	Testing of Natural Stone and Rock Granulation. Determination of Water Absorption and Saturation Value
DIN 52 108	Testing Inorganic, Non-Metallic Materials; Wear Test with Grinding Wheel according to Böhme, Grinding Wheel Method

DIN 52 170 P 1–4	Determination of Composition with Hardened Concrete
DIN 52 617	Determination of Water Absorption Coefficients of Materials
DIN 53 375	Testing of Plastic Films; Determination of Friction Behaviour
DIN 53 435	Testing of Plastics; Bending Test and Impact Bending Test on Dynstat Test Samples
DIN 53 452 (EN ISO 178)	Testing of Plastics; Bending Test
DIN 53 453 (EN ISO 179)	Testing of Plastics; Impact Bending Test
DIN 53 455 (EN ISO 527-1)	Testing of Plastics; Tensile Test
DIN 53 456 (EN ISO 2039-1)	Plastics, Determination of Hardnesses; Ball Penetration Test
DIN 53 479	Testing of Plastics and Elastomers; Determination of Density
DIN 53 504	Testing of Rubber and Elastomers; Determination of Ultimate Tensile Strength, Elongation at Break and Stress Values by a Tensile Test
DIN 53 505	Testing of Rubber, Elastomers and Plastics; Shore A and D Hardness Testing
DIN 53 507	Testing of Rubber and Elastomers; Determination of the Tear Strength of Elastomers; Strip Test
DIN 53 508	Testing of Rubber and Elastomers; Accelerated Ageing
DIN 53 509-1	Testing of Rubber and Elastomers; Determination of Resistance to Ozone Cracking, Static Conditions
DIN 53 509-2	–; –; Part 2: Reference Method for the Determination of Ozone Concentration in Test Chambers
DIN 53 516	Testing of Rubber and Elastomers; Determination of Abrasion Resistance
DIN 53 857 P 1	Testing of Textiles; Single Strip Tensile Test on Textile Materials, Fabric and Woven Ribbon
DIN 53 894 P 1	Testing of Textiles; Determination of Dimensional Change of Textile Materials, Ironing with a Damp Cloth on Ironing Presses
DIN 55 928 P 4	Corrosion Protection of Steel Structures by Organic and Metal Coatings; Preparation and Testing of Surfaces
DIN 55 928 P 5	–; Coating Materials and Protective Systems
DIN 86 076	Elastomer Sealing Sheets, Seawater Resistant, Oil resistant; Dimensions, Requirements, Testing

I.3.2 Regulations of the German Federal Railways (DS)

DS 804	Specifications for Railway Bridges and Other Engineering Constructions (VEI)
DS 836	Specifications for Earthwork Structures (VE)

I.3.3 Regulations of the German Committee for Steel Constructions (DASt-Ri)

DASt-Ri 006	Welding over Finished Coatings in Steel Construction
DASt-Ri 007	Delivery, Processing and Use of Weather-Resistant Structural Steels
DASt-Ri 012	Check of Safety against Bulging for Plates; only in combination with DIN 18 800-1 (3.81)
DASt-Ri 013	Check of Safety against Bulging for Shells; only in combination with DIN 18 800-1 (3.81)
DASt-Ri 014	Recommendations for Prevention of Terracing Failure in Welded Constructions made of Structural Steel
DASt-Ri 015	Steel Beams with Slender Webs
DASt-Ri 016	Dimensioning and Design of Bearing Structures made of Thin-Walled Cold-Formed Components
DASt-Ri 017	Safety against Bulging for Shells – Special Cases
DASt-Ri 103	National Application Document (NAD) for DIN V ENV 1993, part 1-1 (11/93)
DASt-Ri 104	National Application Document (NAD) for DIN V ENV 1994, part 1-1 (2/94

I.3.4 Steel-Iron Material Table of the "Verein Deutscher Eisenhüttenleute" (SEW)

SEW 088	Fine-Grained Structural Steels Suited for Welding; Regulations for Processing, especially for Smelt Welding Supplement: Determination of Cooling Time $t_{8/5}$ for Identification of the Welding Temperature Cycles

Annex II List of Conventional Symbols

II.1 Symbols

In the following, most of the symbols and abbreviations used in the text, formulas and figures are specified. They conform as far as possible to DIN 1080. The units conform to DIN 1301. The water level designations correspond to DIN 4049 and DIN 4054.
Other formula symbols are summarised in the relevant geotechnical DIN standards.

Symbol	Definition	Dimension
A	Anchor force	MN/m or MN etc.
A	Energy absorption capacity (working capacity)	kNm
A	Cross-sectional area	m²
C	Chemical symbol for carbon	
C	Equivalent force	kN/m
C, c	Wave celerity (or velocity)	m/s
C'	Cohesion force along the failure plane in drained soil (consolidated state)	kN/m or MN/m
C_D	Drag coefficient taking into account the resistance against flow pressure	1
C_M	Mass factor	1
C_M	Inertia coefficient taking into account the resistance against acceleration of water particles	1
C_s	Stiffness factor	1
C_u	Cohesion force along the failure plan in undrained soil (unconsolidated state)	kN/m or MN/m
D	Degree of density of substances	1
D	Pile diameter	m
D_{pr}	Degree of compaction according to Proctor	1
E	Modulus of elasticity (Young's modulus)	MN/m²
F	Cross-sectional area	m²
G	Water displacement of ship as mass	t
GRT	Gross register tonnage	2.83 m³
GT	Gross tonnage	1
GS	Cast steel acc. to DIN 1681	
H	Greatest freebord height of ship	m
H	Wave height	m
H_b	Wave height at moment of breaking	m
I	Moment of inertia	m⁴
I_p	Plasticity index	%
K_a	Active earth pressure coefficient	1
K_{ah}	Horizontal active earth pressure coefficient	1
K_o	Coefficient for determining active earth pressure at rest	1

Symbol	Definition	Dimension
K_p	Passive earth pressure coefficient	1
K_{ph}	Horizontal passive earth pressure coefficient	1
L	Wave length	m
L_o	Length of deep water wave	m
$L_ü$	Overall length of ship	m
M	Moment	MNm
Mn	Chemnical symbol for manganese	
N	Normal force	MN etc.
N	Newton: Unit of force	N
kN	Kilonewton = $10^3 \cdot$ N	kN
MN	Meganewton = $10^6 \cdot$ N	MN
NN	Mean sea level	m
P	Surcharge, wave load, ice load, force	MN/m or MN
P	Chemical symbol for phosphorus	
P_1	Active earth pressure on the anchor wall, or on the assumed anchor wall	MN/m
$P_{1...n}$	Pile force influences	MN/m
P_a	Active earth pressure	MN/m
P_p	Passive earth pressure	MN/m
imp P	Impact force	MN, kN
Q	Shear force	MN/m
Q'	Limit tensile load	MN/m
$R_{1,2}$	Soil reactions in the failure plane for earth wedges 1 or 2	MN/m
R_a	Soil reaction at the active earth wedge failure plane	MN/m
R_d	Design value of resistances	kN/m²
Re	REYNOLDS number	1
R_k	Characteristic values of resistances from the soil and from structural components	kN/m²
S	Chemical symbol for sulphur	
S	Statical moment (areal moment first grade)	m³
S_d	Design value of actions	kN/m²
Si	Chemical symbol for silicon	
S_k	Characteristic values of actions	kN/m²
T	Resultant force	MN/m
T	Wave period	s
V	Vertical load	MN/m
W	Probability	%
W	Section modulus (of resistance)	m³
W	Free water surcharge, water pressure	MN/m
W	Effective dead load of the entire earth wedge above the lower failure plane	MN/m
W	Effective dead load of earth wedge	MN/m
$W_{1...n}$	Equivalent loads of passive earth pressure	MN/m

Symbol	Definition	Dimension
W_a	Effective dead load of the earth edge above the active earth pressure failure plane	MN/m
W_i	Wind load component	kN
a	Acceleration	m/s²
a	Half mean tidal range	m
a	Welding seam thickness	mm
b	Width	m
c	Spring constant	kN/m
c	Wave celerity (or velocity)	m/s
c'	Cohesion in drained (dewatered) state of soil	kN/m²
c'_f	Cohesion in state of failure	kN/m²
c_c	Apparent cohesion due to capillary stress	kN/m²
c'_k	Characteristic value of effective cohesion	kN/m²
c_f	Shape coefficient	1
c'_r	Cohesion in state of sliding	kN/m²
c_u	Cohesion in undrained (non dewatered) state of soil	kN/m²
d	Thickness	m
d	Pile thickness	cm
d	Water depth	m
dB	Decibel	dB
d_f	Water depth at structure	m
d_s	Depth of groundwater or rearward harbour water level	m
d_w	Water depth at a full wave length from structure	m
dwt or DWT	Deadweight tonnage in English tons (1 ton = 1016 kg)	ton
e_A	Initial void ratio	1
e_{ah}	Horizontal component of active earth pressure	MN/m²
e_{ph}	Horizontal component of passive earth pressure	MN/m²
e_u	Ordinate of the unshielded earth pressure	MN/m²
f	Deflection	m
$f_{y,k}$	Yield point	N/mm²
g	Acceleration due to gravity = 9.81	m/s²
h	Height, rise of water level	m
h'	Depth of soil flowed through on land side to the bottom of the water course	m
$h_{wü}$	Hydrostatic difference in level	m
i	Hydraulic gradient	1
k	Coefficient of permeability	m/s
k	Radius of gyration of the ship	m
k	Wave number 2 π/L	1/m
k_e	Coefficient of eccentricity	1
k_h	Seismic coefficient	1
k_l	Load coefficient for wind load component W_l	kN s²/m⁴
k_t	Load coefficient for wind load component W_t	kN s²/m⁴

577

Symbol	Definition	Dimension
l	Length	m
l_a	Length of the anchor pile	m
l_k	Upper pile length, statically ineffective	m
l_r	Minimum anchoring length	m
l_s	Length of the anchor pile point	m
n	Degree of porosity, pore number	1
n_{pr}	Pore number with optimum water content in Proctor test	1
$p_{failure}$	Mean base pressure at soil failure	MN/m^2
p_d	Dynamic wave pressure ordinate	kN/m^2
p_D	Pressure due to the water particle velocity caused by the flow resistance per unit length of pile	kN/m^2
pH-value	Negative decadically logarithm of hydrogen-ion concentration	grammes-ion/litre
p_M	Inertial pressure	kN/m
q	Discharge	$m^3/s\ m$
q	Load per linear meter	MN/m
q_u	Unconfined compressive strength of undrained soil	MN/m^2
r	Radius	m
s	Subsidence, settlement	cm
s	Displacement path	m
t	Embedment depth	m
t	Draft of the ship	m
t	Time coordinate	s, d, a
Δt	Additional embedment required for absorption of equivalent force	m
t_0	Theoretical embedment up to the operation line of the equivalent force C	m
u	Horizontal component of orbital velocity of water particles	m/s
u	Pore pressure	kN/m^2
u	Depth of zero pressure point N below the river bed	m
v	Velocity	m/s
v_r	Resulting velocity	m/s
w	Ordinate of water pressure	kN/m^2
w	Water content	1
w_e	Course coefficient	1
$w_ü$	Water pressure difference	kN/m^2
x	Depth of theoretical point F of sheet piling below N	m
Δx	Extra length for assumed equivalent force C	m
α	Angle of slope of the bottom	degree
α	Reduction value for moment in span	1
α	Inclination of the wall	degree
α	Angle of wind direction	degree
α_t	Coefficient of thermal expansion	$°C^{-1}$

Symbol	Definition	Dimension
β	Angle of slope	degree
β, β_s	Yield point	MN/m²
β_{wN}	Concrete nominal strength	N/mm²
γ	Partial safety factor	1
γ	Weight density of soil above water	kN/m³
γ'	Submerged weight density of soil	kN/m³
γ_R	Partial safety factor for resisting variable	1
γ_S	Partial safety factor for active variable	1
γ_W	Weight density of water	kN/m³
δ_a	Angle of wall friction of active earth pressure	degree
δ_p	Angle of wall friction of passive earth pressure	degree
η	Safety coefficient	1
ϑ	Phase angle	degree
ϑ_a	Angle of inclination of the active earth pressure failure plane	degree
ϑ_p	Angle of inclination of the passive earth pressure failure plane	degree
ν	Kinematic viscosity	m²/s
ξ	Breaker coefficient	1
ρ_d	Dry (mass) density	t/m³
ρ_{pr}	Mass density with optimum water content according to Proctor test	t/m³
ρ_w	Mass density of water	t/m³
σ	Normal stress	kN/m²
σ'	Effective normal stress	kN/m²
σ_v	Comparison stress	kN/m²
τ	Shear stress	kN/m²
τ'_f	Ultimate shear strength	kN/m²
τ'_r	Sliding shear strength (Residual shear strength)	kN/m²
φ	Angle of internal friction of soil	degree
φ'	Effective angle of internal friction	degree
φ'_f	Effective angle of internal friction in state of failure	degree
φ'_k	Characteristic value of angle of internal friction	degree
φ'_r	Effective angle of internal friction in state of sliding	degree
φ_u	Angle of internal friction of undrained soil	degree
ω	Angular frequency, turning speed of ship	1/s
ω	Wave angular frequency	1/s

II.2 Abbreviations

Symbol	Definition
abs	Absolute
allow	Allowable
approx	Approximate
cal	Calculation value (calculated)
crit	Critical
des	Designated
ef	Effective
exist	Existing (actual)
imp	Impact
max	Maximum
min	Minimum
poss	Possible (potential)
red	Reduced
req	Required
stat	Statical

II.3 Symbols for Water Levels

Symbol	Definition
	Water levels without tide
GrW or GW	Groundwater level (water table)
HaW	Normal harbour water level
LHaW	Lowest harbour water level
HHW	Highest water level
HNW	Highest navigable water level
HW	High water level
MHW	Mean high water level
MW	Mean water level
MLW	Mean low water level
LW	Low water level
LLW	Lowest water level
	Water levels in tidal areas
HHW	Highest tidal high water level
MHWS	Mean spring tide high water level
MHW	Mean tidal high water level
MW	Mean tidal water level
T ½ W	Half-tide water level
MLW	Mean tidal low water level
MLWS	Mean spring tide low water level
LLW	Lowest tidal low water level
SKN	Chart zero (corresponds approximately to MLWS)

Annex III List of Key Words

A	Section
Acceptance conditions for fender elastomers	6.14.2
Acceptance conditions for steel piles	8.1.7
Acceptance conditions for steel sheet piles	8.1.7
Access ladders	6.11
Actions	0.2, 5
Active and passive earth pressures	2
Active earth pressure in a steep, paved embankment of a partially sloping bank construction	2.5
Active earth pressure in saturated, non- or partially consolidated, soft cohesive soils	2.6
Active earth pressure in waterfront structures with replaced soil	2.8.2
Active earth pressure lines	2.4.1
Active earth pressure on sheet piling in front of pile-founded structures	11.3
Active earth pressure redistribution with high prestressing	8.4.13.2
Active earth pressure redistribution with stability verification for the lower failure plane	8.4.9
Active earth pressure redistribution, consideration in the calculation of sheet piling	8.2.2, 8.2.3, 8.2.6.1
Active earth pressure shielding on a wall below a relieving platform	11.2
Active earth pressure using the CULMANN method	2.4
Adhesion	8.2.4
Amount of displacement for mobilisation of Pp	2.10
Amount of displacement required for the mobilisation of partial passive earth pressures in non-cohesive soils	2.10
Anchor cells in double-wall cofferdams	8.3.2
Anchor connection height	8.4.10
Anchor force, increasing the	8.2.2
Anchor piles	9
Anchor piles, connection to sheet piling structures	8.4.6
Anchor piles, driven grouted	9.4
Anchor piles, driven piles with grouted skin (VM piles)	9.6
Anchor piles, driven steel piles	9.5
Anchor piles, flat battered, driven	8.4.3, 9.3
Anchor piles, limit tension load	9.4
Anchor piles, safety factors	9.3
Anchor piles, transmission of horizontal loads via pile bents, slotted wall plates, frames and large bored piles	9.10
Anchor piles, tubular grouted piles	9.7
Anchor plate, anchor wall	8.4.7
Anchor vane piles	9.5.1
Anchor wall fixed in the earth	8.2.13, 8.4.9.5
Anchor wall, bearing stability	8.2.6.2
Anchor wall, floating	8.4.10.3

	Section
Anchor wall, stability verification at lower failure plane	8.4.9
Anchor wall, staggered design	8.2.13, 8.2.14
Anchor walls for quay wall corners	8.4.11
Anchor, prestressed	8.4.13
Anchoring elements	9.2
Anchoring of sheet piling	8.4
Anchoring of sheet piling in unconsolidated, soft cohesive soils	8.4.10
Anchoring safety of piles and anchors	9.3
Anchoring with piles of small diameter	9.8
Anchoring, stability verification, anchoring bodies	8.4.9
Angel of wall friction in sheet piling	8.2.4
Angel of wall friction, negative, for passive earth pressure	8.2.4, 8.2.5
Angle of wall friction of passive earth pressure for an inclined embankment	8.2.5
Apparent cohesion	2.2
Area loads, distribution width	5.5.3
Armoured steel sheet piling	8.4.16
Armouring of steel sheet piling	8.4.16
Artesian pressure, relieving	4.7
Artesian water	2.7
Artesian water under the harbour bottom or river bed, influence on excess water pressure	2.7.4
Artesian water under the harbour bottom or river bed, influence on the active earth pressure	2.7.3
Artesian water under the harbour bottom or river bed, influence on the passive earth pressure	2.7.1
Assessment of the subsoil for the driving of sheet piles and steel piles	1.9
Auxiliary anchoring	8.4.7
Axial compressive resistance of the sheet piling	8.2.11.4
Axial tensile resistance of the sheet piling	8.2.11.5

B

Backfill	1.6
Backfill of non-cohesive soil, degree of density	1.6
Backfilling of waterfront structures	7.3, 7.4
Backfilling of waterfront structures in the dry	7.4.2
Backfilling of waterfront structures under water	7.4.3
Backfilling, light, for sheet piling	7.11
Base plates, bearing safety	8.2.6.2, 8.4.2.5
Batter pile anchoring at quay wall corners	8.4.12
Bearing capacity of container cranes	5.14.2, 5.14.3
Bearing capacity of general cargo harbour cranes	5.14.1, 5.14.3
Bearing capacity of the vessels	5.1
Bearing capacity, ultimate limit state	0.2
Bearing piles in combined steel sheet piling	8.1.4, 8.1.12, 8.2.10.4
Bearing stability verification of sheet piling	8.2.6
Bedrock, driving steel sheet piles in	8.2.15

	Section
Bentonit	10.13.2, 10.13.3
Berthing pressure of vessels	5.2
Berthing velocities of vessels	5.3
Blasting of rocky soil as preparation for driving	8.1.10, 8.2.15.3
Blockwork construction for quay walls	10.7, 10.8
Blockwork construction in earthquake areas	10.8
Bollards	5.12, 5.13, 6.1
Bollards for seagoing vessels	5.12
Bollards in inland harbours	5.13
Bollards, edge bollards	6.1.2
Bored pile walls	10.11
Bored piles, bored pile walls	10.11
Borers, marine	8.1.1.5
Borings and boring depths, layout	1.2
Borings, distance between	1.2
Borings, intermediate	1.2.3
Borings, principal	1.2.2
Bottom deepening, later	6.8
Bottom protection, geotextile filters	12.4
Bottom seals under water, mineral	7.14
Box caissons as waterfront structures	10.5
Bracing of the tops of steel pipe driving piles	11.9
Breakwaters, rubble mound	7.10
Brushwood, suspended fender	6.15
Burning off the tops of driven steel sections for load-bearing welded connections	8.1.19
Butt coverage	8.1.18
Butt joints	8.1.18

C

Caissons, open, for quay walls	10.9
Caissons, pneumatic, as waterfront structures	10.6
Calculation and design of sheet piling	8.2
Calculation method according to MORISON	5.10.2
Calculations of waterfront structures	0.3
Calculations, statical	0.1, 0.2, 0.3
Capping beam wale	8.4.4.2
Capping beams, verification of bearing stability	8.2.6.2
Car transport ships	5.1.1.6
Cargo handling operations, quay design	6.6
Categories, geotechnical	0.2.4
Cellular cofferdams as excavation enclosures	8.3.1
Cellular cofferdams as permanent waterfront structure	8.3.1
Characteristic value	0.2, 0.3
Coatings of steel sheet piling	8.1.8.4
Cofferdams	8.3.1, 8.3.2, 8.3.3
Cohesion in cohesive soils	1.1, 2.1

	Section
Cohesion, apparent, in sand	2.2
Cohesion, stability verification in the lower failure plane	8.4.9
Cohesive, non- or partially consolidated soils, determining active earth pressure	2.7
Cohesive, unconsolidated soils, sheet piling anchoring	8.4.10
Cohesive, unconsolidated soils, sheet piling design	8.2.16
Combined pipe sheet piling	8.1.4
Combined steel sheet piling	8.1.4, 8.1.12, 8.1.13, 8.2.10
Compression wale	8.3.2, 8.4.1.1, 8.4.11.3
Concrete composition for quay walls in seawater	10.1.1
Concrete composition for reinforced concrete sheet piling	8.1.2.2
Concrete cover of bearing reinforcement steels	8.1.2.3, 10.2.3, 10.11.3, 10.11.4, 10.12.5, 10.14.2
Condition $\Sigma V = 0$	8.2.4.2, 8.2.9.1, 8.2.11.6, 13.1.2
Conduits in the area of flood protection walls	4.9.8
Connection of expansion joint seal	6.18
Connection of steel anchor piles	8.4.15
Connection of steel sheet piling to a concrete structure	6.19
Consideration of axial loads in sheet piling	8.2.7
Consolidation of soft, cohesive soils	7.7, 7.13
Construction joints	10.2.4
Container cranes	5.14.2
Corner pile for timber sheeting	8.1.1.3
Corrosion of steel sheet piling	8.1.8
Corrugated steel sheet formwork	10.3.3
Counter-measures for corrosion	8.1.8
Covering for embankments in seaports	12.2
Crane loads of container cranes	5.14.2
Crane loads of general cargo harbour cranes	5.14.1
Crane loads of vehicular cranes	5.5.5
Crane rails, installation	6.17
Crane rails, outboard	6.1
Craneways, foundations of	6.16
Craneways, transversible	7.17.4
Critical shear strength	1.5
Cu value, determination	1.4
Cu values	1.4.2
CULMANN method for calculating active earth pressure	2.4
Cutter suction dredge, cutter wheel-suction dredge	7.2, 7.5

D

Damage to steel sheet piling, operational	14.2
Danger of sand abrasion on sheet piling	8.1.9
DARCY's law	4.7
Deadweight tons	5.1
Deepening of the harbour bottom	6.8
Deflection of sheet piling without anchorage	8.2.1.2(8)

	Section
Degree of compaction D_{pr} according to Proctor	1.6.1
Degree of density of dumped, non-cohesive soils	1.8
Degree of density of hydraulically filled, non-cohesive soils	1.7
Degree of density of non-cohesive backfill for waterfront structures	1.6
Degree of density, checking and investigation	1.6
Delivery conditions for steel sheet piles, technical	8.1.6, 8.1.7
Density of water	2.7, 2.9
Design and driving of timber sheeting	8.1.1
Design depth of the harbour bottom	6.7
Design of quays and superstructures, structures on pile foundations	10.1
Design value, nominal value	0.2.2
Design wave	5.6.5
Designs, soil properties	1.1
Determination of active earth pressure for saturated, non- or partially consolidated, soft cohesive soils	2.6
Diaphragm wall	10.12, 10.13
Diaphragm wall as impermeable wall	10.13
Diaphragm walls, impermeable	10.13
Diffraction from waves	5.6
Dimension deviations in interlocks	8.1.6.6
Dimensions for container cranes	5.14.3
Dimensions for general cargo harbour cranes	5.14.1
Dispensing with timber fenders	6.15
Disposal facilities in waterfront structures	6.13.5
Distribution width of live loads on quay walls	5.5.5
Dolphins	13.1, 13.2, 13.3, 13.4
Dolphins, design of resilient multi-pile and single-pile dolphins	13.1
Double-anchored sheet piling	8.2.3
Double-wall cofferdams as excavation enclosures	8.3.2
Double-wall cofferdams as waterfront structures	8.3.2
Drainage in sheet piling	4.2, 4.4, 4.5, 4.6
Drainage in waterfront structures in seaports	6.13.5
Drains, vertical	7.7
Dredging equipment	7.1, 7.2
Dredging in front of quay walls	7.1
Dredging in seaports	7.1
Dredging of underwater slopes	7.5
Dredging tolerances	6.7, 7.2
Driven battered anchor piles	9.3
Driven grouted anchor piles	9.6
Driven steel piles, welded joints	8.1.18
Driving as embedment method for sheet piles and steel piles	1.9.3.1
Driving assistance for steel sheet piling by means of loosening blasting	8.1.10
Driving assistance through loosening blasting	8.1.10
Driving damage to sheet piling	8.1.16
Driving equipment	1.9.4, 8.1.11.3, 9.5.3
Driving in bedrock or rock-like soils	8.2.15

	Section
Driving observations	8.1.13
Driving of combined steel sheet piling	8.1.12
Driving of corrugated steel sheet piles	8.1.11
Driving of reinforced concrete sheet piling	8.1.2.5
Driving of sheet piles	1.9, 8.1.1, 8.1.2, 8.1.4, 8.1.11, 8.1.12, 8.1.13, 8.1.15
Driving of sheet piles and steel piles, assessment of the subsoil	1.9
Driving of steel piles at low temperatures	8.1.15
Driving of steel sheet piles at low temperatures	8.1.15
Driving of timber sheet piles	8.1
Driving piles, steel pipe	11.9
Driving sequence, driving procedure	8.1.12.5
Driving units of the sheet piling, driving elements	8.1.14
Driving, low-noise	8.1.14
Dynamic effects	2.13.1.6, 5.8.3, 7.5.1
Dynamic hydraulic loads	12.1.1, 12.4.1
Dynamic influences	5.11.4, 13.2.2.3
Dynamic loads	8.4.8.3, 12.4.2
Dynamic penetration resistance	1.7.4
Dynamic penetration tests	1.2.1, 1.6.1
Dynamic penetration, heavy	1.7.4
Dynamic penetration, light	1.7.4
Dynamic penetrometer tests	1.2.1, 1.6.1
Dynamic penetrometers	1.9.2, 7.8.3
Dynamic pressure	5.7.2, 5.7.3, 5.7.4
Dynamic soundings	1.7.4
Dynamic stresses	1.6.2, 13.2.3.5, 14.2.1
Dynamic water pressure	5.7.3
Dynamically stressed	8.2.6.1

E

Earthquakes, effects of	2.13, 10.6, 10.8, 10.10, 11.2, 11.8
Earthwork and dredging	7
Edge bollards	6.1
Effect of artesian water under the harbour bottom or river bed, on active and passive earth pressures	2.7
Effects of earthquakes on the design and dimensioning of waterfront structures	2.13, 8.2.17
Elastomers, fenders and buffers of	6.14.2
Electric power supply facilities in seaports	6.17.3
Embankment protection, geotextile filters	12.4
Embankments in seaports and inland harbours	12.1
Embankments under quay wall superstructures behind tight sheet piling	12.2
Embedment depth for anchor walls	8.2.14
Embedment depth of steel sheet piling	8.2.10
Embedment depth with partial or full fixity of the point of the sheet piling	8.2.9
Embedment depth, determination of	8.2.9, 13.1.2
Embedment depth, selection of	8.2.8

	Section
Embedment gear for sheet piles and steel piles	1.9.4
Embedment methods for steel piles and sheet piles	1.9.3
Energy absorption capacity of dolphins and fenders	13.2, 13.3
Equipment of waterfront structures in seaports with supply and disposal facilities	6.13
Equipotential lines	4.7
Equipping of berths for large vessels with quick release hooks	6.10
Equippment of waterfront structures	6.1
Equivalent force	8.2.9
Erosion, foundation failure due to	3.3
Eurocode	0.1
Excavation enclosures, cellular cofferdams as	8.3.1
Excavation enclosures, double-wall cofferdams as	8.3.2
Excess pressure, artesian	2.7
Expansion joint, connection of seal	6.18
Expansion joints	8.4.3.5, 8.4.5.4, 10.1.5
Experience with waterfront structures	14
Expert opinions on subsoil examinations	1.3

F

	Section
Facing of concrete structures	10.1.4
Failure by heave	3.2
Failure displacement	2.10.2
Failure of the anchoring soil, safety against	8.4.9
Failure plane, lower	8.4.9
Fall of the canal water level	6.4
Fender designs	6.14
Fender facilities, wind load influence	5.11
Fendering, calculation of	13.2
Fenders for berths for large vessels (brushwood suspended fender)	6.14
Fenders in inland harbours	6.15
Field tests	1.4.2
Fill material for moles and breakwaters	7.10
Filter weephole for sheet piling structures	4.4
Filter, geotextile	12.4
Filters	4.4, 4.5, 4.6, 4.9
Fine-grained structural steels	13.4
Fire loads for steel sheet piling	14.3
Fixed sheet piling	8.2.1, 8.2.2, 8.2.10
Flap valves for waterfront structures in tidal areas	4.5
Flat cells	8.3.1.2
Flexible construction for anchor connection	8.4.15
Floating wharves in seaports	6.20
Flood protection walls in seaports	4.9
Flow net, flow lines	2.9, 3.2, 4.7
Flow pressure	4.3, 5.10, 11.3.2
Fluctuating stress	8.2.6.1, 8.2.6.2, 13.4

	Section
Formwork in marine environment	10.3
Foundation failure due to erosion; its occurrence and its prevention	3.3
Foundation of craneways	6.16
Full portal cranes	5.14.1.2

G

General cargo harbour cranes, dimensions and loads	5.14
Geotechnical categories	0.2.4
Geotextile filter	12.4
Grain deformation	7.8.1
Grain failure	7.8.1
Grain readjustment	7.8.1
Grain splintering	7.8.1
Gravel filter	4.4, 4.5, 4.6
Gross space number	5.1
Gross tonnage	5.1
Groundwater flow	4.7
Groundwater flow net	2.9, 3.2, 4.7
Groundwater flow, assessment	4.7
Groundwater level, mean, in tidal areas	4.1
Groundwater lowering for temporary stabilisation of waterfront structures	4.8
Groundwater, artesian	2.7
Grouted driven anchor piles, grouted skin piles	9.6
Grouting compound for anchor piles	9.6.3
Grouting compound for sheet piling interlocks	8.1.20.3
Guide piles	6.5

H

Half portal cranes	5.14.1.3
Harbour bottom, design depth	6.7
Harbour bottom, nominal depth	6.7
Harbour cranes	5.14
Hardwoods, tropical	8.1.1
Hawse forces on bollards in inland harbours	5.14
Hawser forces on bollards for seagoing vessels	5.12
Head equipment of steel anchor piles	9.9
Heavy dynamic penetration	1.7.4
High-strength weldable structural steels for dolphins	13.4
Hinged connections	8.4.15
Hinged plate	8.4.15
Hinged supports	8.4.14
Hook coupling of steel sheet piling	8.1.6.5
Hydraulic dredge	7.1
Hydraulic fill tolerances	7.2
Hydraulic filling of port areas behind waterfront structures	7.3
Hydrodynamic approach	6.20.3
Hydrodynamic influences	0.3

	Section
Hydrodynamic pressure	13.3.2.2
Hyperbola method	9.4.5

I

Ice compression strength	5.15
Ice impact	5.15, 5.16
Ice loads on piles	5.15.3
Ice loads on waterfront structures	5.15.2
Ice pressure	5.15, 5.16
Ice thickness	5.15, 5.16
Impact coefficient	5.5
Impact factor	5.5
Impact forces on dolphins and fendering	13.2, 13.3
Impermeable diaphragm walls	10.13
Impermeable thin wall	10.13.2(4)
Impermeable thin walls	10.13
Impermeable wall compounds	10.13
Inclined anchors	8.4.1.4
Inclined embankment, angel of wall friction of passive earth pressure	8.2.5
Increasing the passive earth pressure in front of waterfront structures	2.11
Indentation of structural sections	10.1.5
Initial state	1.1
Inland harbours, design and loading of bollards	5.13
Inland harbours, design and standard dimensions of waterfront structures in	6.3
Inland harbours, design of waterfront areas in	6.6
Inland harbours, embankments in	12.1
Inland harbours, fenders in	6.15
Inland harbours, partially sloped waterfront construction in	6.5
Inland vessels, dimensions of	5.1.3
Inspection of steel sheet piles	8.1.7, 15.1
Inspection of waterfront structures in seaports	15.1
Installation of crane rails	6.17
Interlock damage as a result of driving load, repair	7.4.4, 8.1.16
Interlock dimensions, deviations in	8.1.6.6
Interlock joining in steel sheet piling	8.1.5
Interlock sealing	8.1.20, 8.1.21.3
Intermediate borings	1.2.3
Intermediate piles between mixed steel sheet piling	8.1.4, 8.1.2, 8.2.10.4
Introduction of forces from steel anchor piles into a reinforced concrete superstructure	9.9
Inventory before repairing concrete components in hydraulic engineering	10.14
Inventory of concrete components in hydraulic engineering	10.14

J

Jacking in sheet piles	1.9.3.3
Jetting when driving steel sheet piles	8.1.23
Joint design for sheet piling connection to a concrete structure	6.19

	Section
Joints in quay walls of concrete and reinforced concrete	10.1.5

K

Keying of the structural blockwork	10.1.5, 10.7
Killed steels for steel sheet piling	8.1.6.4, 8.1.15, 8.1.18.2, 8.2.6

L

Laboratory test	1.4.1
Ladders	6.15
Landings for stairs in seaports	6.12.3
Layout and depth of borings and penetrometer tests	1.2
Leaks in sheet piling	8.1.18
Length of a wall section	5.12, 10.1.5
Light backfilling for sheet piling	7.11
Light dynamic penetration	1.7.4
Limit skin friction for driven, grouted anchor piles and sheet piling	9.4
Limit state of bearing capacity	0.2
Limit tension load of anchor piles	9.4
Live loads, vertical	5.5
Load case 1, 2 and 3	5.4.1, 5.4.2, 5.4.3
Load cases for waterfront structures	5.4, 8.2.6
Loading of waterfront structures	5
Loads for container cranes	5.14.2
Loads for general cargo cranes	5.14.1
Loads on mooring and fender facilities	5.11.4
Loads on pile driving trestles	8.1.17
LOHMEYER method	11.2
Longitudinal forces in the sheet piling, in the steel wales	8.2.12
Loosening blasting	8.1.10, 8.1.11.4
Loosening blasting as driving assistance	8.1.10
Low-noise driving of sheet piles and steel piles	8.1.14

M

Main anchoring	8.4.7
Mean characteristic soil properties	1.1
Mean groundwater level	4.1
Mineral bottom seals under water	7.14
Mixed gravel filter	4.5
Moles, narrow partition, in sheet piling construction	8.3.3
Moles, rubble mound	7.10
Moment influences in stability examinations for the lower failure plane	8.4.9
Mooring dolphins of weldable structural steels	13.4
Mooring dolphins, spring constants for	13.2
Mooring facilities	6.3.5
Mooring facilities, wind load influence	5.11
Mooring hooks	5.4.2, 5.13
Mooring ring	5.13

	Section
Movements of ships, effects of waves from	5.9
Multi-pile dolphins, calculation of resilient	13.1

N

Narrow partition moles in sheet piling construction	8.3.3
National application document (NAD)	0.1
Negative wall friction angle	8.2.4
Noise protection	8.1.14
Nominal depth of the harbour bottom	6.7
Nominal value	0.2.2
Non-cohesive soils, soil values	1.1
Nosing	8.4.4, 8.4.6, 10.1.3

O

Observation of waterfront structures in seaports	15.1
Observations during the installation of steel sheet piles, tolerances	8.1.13
Operational damage to steel sheet piling	14.2
Overall stability of structures on elevated pile-founded structures	3.4
Overlapping plank sheeting in timber sheet piles	8.1.1

P

Partial safety, partial safety factors	0.1, 0.2.1
Partially sloped bank construction in inland harbours with extreme water level fluctuations	12.3
Partially sloped waterfront construction in inland harbours	6.5
Partially sloping bank construction, active earth pressure in a paved embankment	2.5
Passive earth pressure in front of differences in elevation in soft cohesive soils, with rapid loading on the land side	2.12
Passive earth pressure, measures to increase	2.11
Passive earth pressure, mobilisation	2.10
Passive earth pressure, resulting course of	8.2.9
Patented steel cable anchors	8.2.6.4
Paving of embankments	12.1
Penetrometer tests, layout and depth	1.2
Percolating groundwater	2.9
Pile driving trestles	8.1.17
Pile founded structures, live loads	5.5
Pile founded structures, sheet piling loads at	11.3
Pile groups, limit tension load for	9.4.4
Pile lengths, required, for driven, grouted anchor piles	9.6.2
Pile spacing for driven, grouted anchor piles	9.6.3
Pile structures, safety factors	5.10.8
Pile structures, wave pressure on	5.10
Pile-founded structures	11
Pile-founded structures in earthquake areas	11.8
Pile-founded structures, design and calculation	11.5

	Section
Pile-founded structures, elevated, overall stability	3.4
Pile-founded structures, general	11.1
Pile-founded structures, plane	11.4
Piling with connected interlocks	8.1.4.2
Pipe sheet piling, combined	8.1.4
Plane pile founded structures	11.4
Plate load bearing test	1.4.2
Plug formation	8.2.11.4
Plunging breakers	5.6
Pneumatic caissons as waterfront structures	10.6
Point resistance	8.2.9.1, 8.2.11
Polyethylene (PE) for sliding battens and panels	6.14.2.2
Porosity n	1.6
Preloading, consolidation by	7.13
Preparation of reports and expert opinions on subsoil examinations under difficult conditions	1.3
Pressure difference	4.2, 4.3
Pressure load of sheet piling	8.2.11
Prestressed steels for anchor piles	9.2, 9.4.1
Prestressing of anchors of high-strength steels	8.4.13
Principal borings	1.2.2
Profile selection for steel sheet piling	8.1.3.2
Protruding corner structures with round steel anchoring	8.4.11
Protruding quay wall corners with batter pile anchoring	8.4.12

Q

	Section
Quality requirements for steels in steel sheet piles	8.1.6
Quay loads from cranes and other transhipment equipment	5.14
Quay wall corners	8.4.11, 8.4.12
Quay wall corners, batter pile anchoring	8.4.12
Quay wall corners, round steel anchoring	8.4.11
Quay wall surcharges, quay wall live loads	5.5
Quay wall, dynamic stress	8.2.6.1
Quay wall, predominantly alternating stress	8.2.6.1
Quay wall, predominantly static stress	8.2.6.1
Quay walls in blockwork construction	10.7, 10.8
Quay walls in inland harbours, standard dimensions	6.3
Quay walls in open caisson construction	10.9, 10.10
Quay walls in seaports, cross section dimensions	6.1
Quay walls in seaports, observation and inspection	15.1
Quay walls with open caissons in earthquake regions	10.10
Quay walls, design of	8.4.16
Quick release hooks	6.10

R

	Section
Railings for stairs in seaports	6.12.4
Receding waves, loads from	5.8

	Section
Recess bollard	5.13
Rectangular sheeting for timber sheet piles	8.1.1
Re-design of waterfront structures in inland harbours	6.9
Redistribution of active earth pressure	8.2.2, 8.2.3
Refraction in waves	5.6
Regions subject to mining subsidence, waterfront structures in	8.1.21
Reinforced concrete capping beams for waterfront structures	8.4.5
Reinforced concrete roadway slabs on piers	10.4
Reinforced concrete sheet piling, concrete	8.1.2.2
Reinforced concrete sheet piling, dimensions	8.1.2.4
Reinforced concrete sheet piling, driving	8.1.2.5
Reinforced concrete sheet piling, range of application	8.1.2.1
Reinforced concrete sheet piling, reinforcement	8.1.2.3
Reinforced concrete sheet piling, watertightness to prevent loss of soil	8.1.2.6
Reinforced concrete wales	8.4.3
Reinforced concrete waterfront structures	10.2
Reinforcement for reinforced concrete sheet piling	8.1.2
Reinforcement of waterfront structures	6.8
Reinforcing plates	8.1.5.7, 14.2.4
Relief well	4.6
Relieving artesian pressure	4.6
Relieving platform, determining of the active earth pressure shielding	11.2
Repair of concrete structures in hydraulic engineering	10.14, 10.15
Repair of interlock damage	8.1.16
Repairing damage to a burst interlock	8.1.16
Reports on subsoil examinations	1.3
Reynold's number	5.10.4.1
Rolling tolerances for interlocks in steel sheet piles	8.1.6
Rotation	8.2.1.2
Round steel anchoring at quay wall corners	8.4.11
Round steel anchors	8.2.6.3, 8.4.7.3
Rubble mound breakwaters	7.10
Rubble mound moles	7.10

S

Safety against earthquake influence	2.13.6, 10.8.3
Safety against failure by heave	3.2
Safety against failure of anchoring soil	8.4.9.7
Safety clearance for dredging	6.3, 6.7
Safety concept	0.2
Safety factors	5.6.5, 5.10.8
Safety factors for anchoring	9.3
Safety factors for pile structures	5.10.8
Safety from foundation failure with cofferdams	8.3.1.3, 8.3.2.2
Safety of anchor piles	9.1
Safety of anchoring	8.4.9
Sand drains, bored, jetted, driven	7.7.4

	Section
Sand, apparent cohesion	2.2
Scour and scour protection at waterfront structures	7.6
Seagoing vessels, dimensions	5.1
Sealing of piling interlock joints	8.1.20, 8.1.21.3
Sealing of reinforced concrete sheet piling	8.1.2.6
Sealing of steel sheet piling	8.1.20, 8.1.21.3
Sealing of timber sheeting	8.1.1.5
Seaports, bollards in	5.12
Seaports, box caissons as waterfront structures	10.5
Seaports, dolphins and fenders in	13.2, 13.3, 13.4
Seaports, electric power supply facilities	6.17.3
Seaports, embankments in	12.1, 12.2
Seaports, open caissons in	10.9
Seaports, pneumatic caissons	10.6
Seaports, protruding quay wall corners	8.4.11
Seaports, stairs in, landings	6.12.3
Seaports, stairs in, railings	6.12.4
Seaports, top elevation of the waterfront structures in	6.2
Seaports, water supply facilities in	6.13.2
Section length	5.12, 10.7, 10.1.5
Service life of waterfront structures	14.1
Serviceability	0.2
Settlement and subsidence in moles and breakwaters	7.10
Shear parameters	1.4, 1.5, 2.0, 2.6, 2.12
Shear parameters of the drained soil in the state of failure or sliding	1.5
Shear strength cu	1.4
Shear stresses	8.1.5.2
Shear test, direct and triaxial	1.4.1
Shear-resistant interlock joining	8.1.5
Sheet piling, light backfilling	7.11
Sheet piling anchoring in unconsolidated, soft cohesive soils	8.4.10
Sheet piling anchors, thread of	8.4.8
Sheet piling drainage	4.2, 4.4, 4.5
Sheet piling in unconsolidated, soft cohesive soils, especially in connection with undisplaceable structures	8.2.16
Sheet piling load under water pressure difference	4.2, 11.3.4
Sheet piling point, theoretical	8.2.9
Sheet piling steels	8.1.6
Sheet piling structures	8
Sheet piling structures, single anchored, in earthquake areas	8.2.18
Sheet piling under influence of earthquakes	8.2.18.2
Sheet piling under vertical loads	8.2.6, 8.2.11
Sheet piling waterfront on canals for inland vessels	6.4
Sheet piling without anchorage	8.2.1
Sheet piling without anchorage, fully fixed	8.2.1
Sheet piling, anchoring	8.4
Sheet piling, angel of wall friction	8.2.4

	Section
Sheet piling, axial compressive resistance	8.2.11.4
Sheet piling, axial tensile resistance	8.2.11.5
Sheet piling, bearing stability verification	8.2.6
Sheet piling, calculation and design	8.2
Sheet piling, calculation of	8.2.1, 8.2.2, 8.2.3
Sheet piling, danger of sand abrasion	8.1.9
Sheet piling, double anchored	8.2.3
Sheet piling, dynamically stressed	8.2.6
Sheet piling, fixed in the ground	8.2.1, 8.2.2, 8.2.10
Sheet piling, fully fixed in the ground without anchorage	8.2.1
Sheet piling, leaks in	8.1.18
Sheet piling, longitudinal forces in	8.2.12
Sheet piling, material and construction	8.1
Sheet piling, predominantly alternating stress	8.2.6
Sheet piling, predominantly fluctuating stresses	8.2.6
Sheet piling, predominantly static stress	8.2.6
Sheet piling, single-anchored	8.2.2
Sheet piling, stiffeners	8.4
Sheet piling, threads of sheet piling anchors	8.4.8
Sheet piling, vertical load bearing capacity	8.2.11
Ship dimensions	5.1
Shoaling effect of waves	5.6.4
Single-pile dolphins, calculation of resilient	13.1
Skin friction, limit load for driven, grouted anchor piles	9.6
Sliding battens and panels of polyethylene (PE)	6.14.2.2
Sliding safety with pneumatic caissons, open caissons and box caissons	10.5, 10.6, 10.9
Soil compaction to increase the passive earth pressure	2.11.3
Soil compaction using drop weights	7.12
Soil liquefaction	8.2.18.1
Soil properties, mean characteristic	1.1
Soil replacement for waterfront structures	2.8, 2.11.2, 7.9
Soil replacement procedure for waterfront structures	7.9
Soil replacement to increase the passive earth pressure	2.11.2
Soil stabilisation to increase the passive earth pressure	2.11.5
Soil surcharge to increase the passive earth pressure	2.11.4
Spilling breakers	5.6
Spring constant for dolphins and fenders	13.2
Stabilisation of waterfront structures by groundwater lowering	4.8
Stability for fixed earth support of sheet piling	8.4.9.4
Stability for the lower failure plane with driven, grouted anchor piles	9.6
Stability in cohesive soils	8.4.9.2
Stability in differing layers	8.4.9.3
Stability in non-cohesive soils	8.4.9.1
Stability of a fixed earth support anchor wall	8.4.9.5
Stability of sheet piling without anchorage	8.2.1
Stability verification for sheet piling under vertical load	8.2.7

	Section
Stability with redistribution of active earth pressure	8.4.9
Stability, increasing by groundwater lowering	4.8
Staggered anchor walls	8.2.13, 8.2.14
Staggered embedment depth of steel sheet piling	8.2.10
Stairs in inland harbours	6.5
Stairs in seaports, landings	6.12.3
Stairs in seaports, railings	6.12.4
Standard cross-section of waterfront structures in seaports	6.1, 6.2
Standard cross-sections of waterfront structures in inland harbours	6.3
Statical calculations	0.1, 0.2, 0.3
Steel anchor piles, flexible connection	8.4.15
Steel anchor piles, head equipment	9.9
Steel cable anchors, design, bearing safety	8.4.10.2
Steel capping beams	8.4.4
Steel grades of sheet piles	8.1.6
Steel nosing	8.4.4, 8.4.6
Steel piles, acceptance conditions	8.1.7
Steel piles, burning off the tops	8.1.19
Steel piles, driving at low temperatures	8.1.15
Steel piles, works acceptance	8.1.7
Steel pipe driving piles, bracing	11.9
Steel sheet pile connection to a concrete structures	6.19
Steel sheet piles, acceptance conditions	8.1.7
Steel sheet piles, burning off the tops	8.1.19
Steel sheet piles, delivery conditions, works acceptance	8.1.6, 8.1.7
Steel sheet piles, driving at low temperatures	8.1.15
Steel sheet piles, driving corrugated	8.1.11
Steel sheet piles, Driving in bedrock	8.2.15
Steel sheet piles, hook coupling	8.1.6.5
Steel sheet piles, jetting when driving	8.1.23
Steel sheet piles, observations during the installation of	8.1.13
Steel sheet piles, quality requirements for steels	8.1.6
Steel sheet piles, types of interlock	8.1.6.5
Steel sheet piles, vibration of U- and Z-shaped	8.1.22
Steel sheet piles, welded joints	8.1.18
Steel sheet piles, works acceptance	8.1.7
Steel sheet piling wales of reinforced concrete	8.4.3
Steel sheet piling wales, calculation and design	8.4.2, 8.4.3
Steel sheet piling wales, design	8.4.1
Steel sheet piling, armoured	8.4.16
Steel sheet piling, capacity to absorb horizontal longitudinal forces acting parallel to the shore	8.2.12
Steel sheet piling, coating	8.1.8.4
Steel sheet piling, combined	8.1.4, 8.1.12, 8.2.10
Steel sheet piling, damage to, operational	14.2
Steel sheet piling, driven into bedrock or rock-like soils	8.2.15
Steel sheet piling, driving assistance by means of loosening blasting	8.1.10

	Section
Steel sheet piling, driving combined	8.1.12
Steel sheet piling, embedment depth	8.2.10
Steel sheet piling, fire load	14.3
Steel sheet piling, interlock joining	8.1.5
Steel sheet piling, profile selection	8.1.3.2
Steel sheet piling, sealing	8.1.20, 8.1.21.3
Steel sheet piling, shear-resistant interlock joining	8.1.5
Steel sheet piling, staggered embedment depths	8.2.10
Steel sheet piling, watertightness	8.1.20
Steel wales for sheet piling, design	8.4.1, 8.4.2
Steel wales for sheet piling, verification of bearing capacity	8.4.2
Steel wales, longitudinal forces in	8.2.12
Stiffeners of sheet piling	8.4
Structural materials of the steel sheet piling	8.1.3, 8.1.6
Subsidence of non-cohesive soils	7.8
Subsoil	1.1, 1.2, 1.3, 1.4, 1.5, 1.6, 1.7, 1.8, 1.9
Subsoil examinations, reports and expert opinions	1.3
Subsoil investigations	1.2
Subsoil, assessment of and embedment methods for sheet piles	1.9.3
Subsoil, characteristics for driving of sheet piles and steel piles	1.9.2
Supervisions of structures	15
Supply facilities in waterfront structures	6.13
Supporting slurry for diaphragm walls	10.12.3
Supporting slurry for plastic impermeable walls	10.13.2
Surcharges on sheet piling	8.2.11
Surging waves, loads from	5.8
Surging/collapsing breakers	5.6
Suspended brushwood fenders	6.14.3

T

Taper piles	8.1.11.4
Taper sheeting for timber sheet piles	8.1.1
Tension load, limit, of anchor piles	9.4
Tension piles	9.2
Tension wale	8.4.1.1, 8.4.11.3
Thin walls, impermeable	10.13
Threads of sheet piling anchors	8.4.8
Timber piles, fender piles	6.12, 6.14, 6.15
Timber sheet piles, Design of point	8.1.1
Timber sheet piles, overlapping plank sheeting	8.1.1
Timber sheet piles, rectangular sheeting	8.1.1
Timber sheet piles, Taper sheeting	8.1.1
Timber sheet piles, V-jointing	8.1.1
Timber Sheet piles, watertightness	8.1.1.4
Timber sheeting	8.1.1
Timber sheeting, sealing	8.1.1.5
Timber, technical quality coefficients	8.1.1

	Section
Top elevation of waterfront structures in seaports	6.2
Top steel nosing for reinforced concrete walls and capping beams at waterfront structures	8.4.6
Translation	8.2.1.2
Transversible craneways	6.17.4
Triaxial tests	1.4.1
Types of interlocks of the steel sheet piles	8.1.6.5

U

Ultimate shear strength	1.5
Unconsolidated cohesive soils	8.2.16
Unconsolidated soft cohesive soils, sheet piling anchoring	8.4.10
Underwater slopes, dredging	7.5
Undrained shear strength cu	1.4
Use of geotextile filters in embankment and bottom protection	12.4

V

Vane piles	9.5
Vane shear tests	1.4.2
Verification of bearing capacity of steel wales	8.4.2
Verification of overall stability of structures on elevated pile-founded structures	3.4
Verification of stability of anchoring at lower failure plane	8.4.9
Vertical drains	7.7
Vertical live loads on waterfront structures	5.5
Vertical load bearing capacity of sheet piling	8.2.11
Vessel berthing velocity	5.3
Vessel pressure, berthing pressure	5.2
Vibration as embedment method for sheet piles and steel piles	1.9.3.2
Vibration of U- and Z-shaped steel sheet piles	8.1.22
V-jointing in timber sheet piles	8.1.1

W

Wale bolts	8.2.6.3, 8.4.1.3, 8.4.2.5
Wale joints	8.4.1.2
Wales for protruding quay wall corners	8.4.11
Wales of reinforced concrete for sheet piling with driven steel anchor piles	8.4.3
Wales, bearing safety	8.2.6.2
Wales, sheet piling- of reinforced concrete	8.4.3
Walkways	6.1
Wall friction angle at an anchor wall	8.4.9.7
Wall friction angle in dolphins	13.1.2
Water displacement by ships	5.1
Water level fluctuations	6.5
Water levels for flood protection walls in seaports	4.9.2
Water levels, water pressure, drainage	4
Water pressure difference	4.2, 4.3, 4.8, 5.4

	Section
Water pressure difference at flood protection walls	4.9.3
Water pressure difference at waterfront structures with soil replacement	2.8.3
Water pressure difference in the water-side direction	4.2
Water pressure difference on sheet piling in front of built-over embankments	4.3
Water pressure difference, sheet piling load from	4.2, 11.3.4
Water supply facilities in seaports	6.13.2
Water tightness in reinforced concrete sheet piling	8.1.2.6
Water-cement ratio	8.1.2.2, 10.2.3
Waterfront areas in inland harbours, design	6.6
Waterfront construction, partially sloped, in inland harbours	6.5
Waterfront structures in inland harbours, re-design of	6.9
Waterfront structures in regions subject to mining subsidence	8.1.21
Waterfront structures with soil replacement and fouled or disturbed dredge pit bottom	2.8
Waterfront structures, cellular cofferdams as	8.3.1
Waterfront structures, configuration of cross-section	6
Waterfront structures, double-wall cofferdams as	8.3.2
Waterfront structures, equipment	6
Waterfront structures, quays and superstructures of concrete and reinforced concrete	10
Waterfront structures, safety through groundwater lowering	4.8
Waterfront structures, supply facilities in	6.13
Waterfront structures, vertical live loads	5.5
Waterfront structures, wave pressure on vertical	5.7
Watertightness in timber sheet piles	8.1.1.4
Watertightness of steel sheet piling	8.1.20
Wave load on an individual pile	5.10.3
Wave movement, effects	5.6, 5.8, 5.9
Wave pressure on pile structures	5.10, 11.6
Wave pressure, wave height	5.6, 5.7
Wave theory, designations	5.6, 5.10.1
Waves, breaking	5.6, 5.7
Waves, reflecting	5.7
Wear coat on quay walls	10.1.4
Weight density of soils	1.1
Weldability of the sheet piling steels	8.1.6.4, 8.1.18
Welded joints in driven steel piles	8.1.18
Welded joints in steel sheet piles	8.1.18
Welded joints, butt joints	8.1.18
Wells, open	10.9
Wheel loads of container cranes	5.14
Wheel loads of general cargo harbour cranes	5.14
Wind loads on moored ships	5.11
Works acceptance of steel piles	8.1.7
Works acceptance of steel sheet piles	8.1.7